LS-33211

TRUTHLIKENESS

SYNTHESE LIBRARY

STUDIES IN EPISTEMOLOGY,

LOGIC, METHODOLOGY, AND PHILOSOPHY OF SCIENCE

Managing Editor:

JAAKKO HINTIKKA, *Florida State University, Tallahassee*

Editors:

DONALD DAVIDSON, *University of California, Berkeley*
GABRIËL NUCHELMANS, *University of Leyden*
WESLEY C. SALMON, *University of Pittsburgh*

VOLUME 185

ILKKA NIINILUOTO

Department of Philosophy, University of Helsinki

TRUTHLIKENESS

D. REIDEL PUBLISHING COMPANY

A MEMBER OF THE KLUWER  ACADEMIC PUBLISHERS GROUP

DORDRECHT / BOSTON / LANCASTER / TOKYO

Library of Congress Cataloging-in-Publication Data

Niiniluoto, Ilkka.
  Truthlikeness.

  (Synthese library; v. 185)
  Bibliography: p.
  Includes indexes.
  1. Truth. 2. Logic. I. Title. II. Title: Truth likeness.
BC171.N54   1987   121   87–4311
ISBN 90–277–2354–0

Published by D. Reidel Publishing Company
P.O. Box 17, 3300 AA Dordrecht, Holland

Sold and distributed in the U.S.A. and Canada
by Kluwer Academic Publishers,
101 Philip Drive, Norwell, MA 02061, U.S.A.

In all other countries, sold and distributed
by Kluwer Academic Publishers Group,
P.O. Box 322, 3300 AH Dordrecht, Holland

All Rights Reserved
© 1987 by D. Reidel Publishing Company, Dordrecht, Holland
No part of the material protected by this copyright notice may be reproduced or
utilized in any form or by any means, electronic or mechanical
including photocopying, recording or by any information storage and
retrieval system, without written permission from the copyright owner

Printed in The Netherlands

*To Petro,
Riikka-Maria,
and Atro*

CONTENTS

PREFACE   xi

CHAPTER 1. DISTANCE AND SIMILARITY   1
            1.1. Metric Spaces and Distances   1
            1.2. Topological Spaces and Uniformities   18
            1.3. Degrees of Similarity   22
            1.4. The Pragmatic Relativity of Similarity
                  Relations   35

CHAPTER 2. LOGICAL TOOLS   39
            2.1. Monadic Languages L_N^k 39
            2.2. Q-Predicates   43
            2.3. State Descriptions   47
            2.4. Structure Descriptions   50
            2.5. Monadic Constituents   51
            2.6. Monadic Languages with Identity   58
            2.7. Polyadic Constituents   61
            2.8. Distributive Normal Forms   72
            2.9. First-Order Theories   77
            2.10. Inductive Logic   80
            2.11. Nomic Constituents   91

CHAPTER 3. QUANTITIES, STATE SPACES, AND LAWS   103
            3.1. Quantities and Metrization   103
            3.2. From Conceptual Systems to State
                  Spaces   106
            3.3. Laws of Coexistence   109
            3.4. Laws of Succession   114
            3.5. Probabilistic Laws   118

CHAPTER 4.	COGNITIVE PROBLEMS, TRUTH, AND INFORMATION	122
	4.1. Open and Closed Questions	122
	4.2. Cognitive Problems	126
	4.3. Truth	134
	4.4. Vagueness	143
	4.5. Semantic Information	147
CHAPTER 5.	THE CONCEPT OF TRUTHLIKENESS	156
	5.1. Truth, Error, and Fallibilism	156
	5.2. Probability and Verisimilitude	160
	5.3. Approach to the Truth	164
	5.4. Truth: Parts and Degrees	172
	5.5. Degrees of Truth: Attempted Definitions	179
	5.6. Popper's Qualitative Theory of Truthlikeness	183
	5.7. Quantitative Measures of Verisimilitude	192
CHAPTER 6.	THE SIMILARITY APPROACH TO TRUTHLIKENESS	198
	6.1. Spheres of Similarity	199
	6.2. Targets	204
	6.3. Distance on Cognitive Problems	209
	6.4. Closeness to the Truth	217
	6.5. Degrees of Truthlikeness	222
	6.6. Comparison with the Tichý—Oddie Approach	232
	6.7. Distance between Statements	242
	6.8. Distance from Indefinite Truth	256
	6.9. Cognitive Problems with False Presuppositions	259
CHAPTER 7.	ESTIMATION OF TRUTHLIKENESS	263
	7.1. The Epistemic Problem of Truthlikeness	263
	7.2. Estimated Degrees of Truthlikeness	268
	7.3. Probable Verisimilitude	278
	7.4. Errors of Observation	280
	7.5. Counterfactual Presuppositions and Approximate Validity	284

CHAPTER 8.	SINGULAR STATEMENTS	290
	8.1. Simple Qualitative Singular Statements	290
	8.2. Distance between State Descriptions	297
	8.3. Distance between Structure Descriptions	302
	8.4. Quantitative Singular Statements	303
CHAPTER 9.	MONADIC GENERALIZATIONS	310
	9.1. Distance between Monadic Constituents	310
	9.2. Monadic Constituents with Identity	321
	9.3. Tichý—Oddie Distances	323
	9.4. Existential and Universal Generalizations	335
	9.5. Estimation Problem for Generalizations	341
CHAPTER 10.	POLYADIC THEORIES	346
	10.1. Distance between Polyadic Constituents	346
	10.2. Complete Theories	362
	10.3. Distance between Possible Worlds	365
	10.4. First-Order Theories	368
CHAPTER 11.	LEGISIMILITUDE	372
	11.1. Verisimilitude vs Legisimilitude	372
	11.2. Distance between Nomic Constituents	374
	11.3. Distance between Quantitative Laws	382
	11.4. Approximation and Idealization	394
	11.5. Probabilistic Laws	403
CHAPTER 12.	VERISIMILITUDE AS AN EPISTEMIC UTILITY	406
	12.1. Cognitive Decision Theory	406
	12.2. Epistemic Utilities: Truth, Information, and Truthlikeness	410
	12.3. Comparison with Levi's Theory	416
	12.4. Theoretical and Pragmatic Preference	420
	12.5. Bayesian Estimation	426

CHAPTER 13.	OBJECTIONS ANSWERED	442
	13.1. Verisimilitude as a Programme	442
	13.2. The Problem of Linguistic Variance	446
	13.3. Progress and Incommensurability	460
	13.4. Truthlikeness and Logical Pragmatics	469
NOTES		474
BIBLIOGRAPHY		490
INDEX OF NAMES		508
INDEX OF SUBJECTS		514

PREFACE

The modern discussion on the concept of truthlikeness was started in 1960. In his influential *Word and Object*, W. V. O. Quine argued that Charles Peirce's definition of truth as the limit of inquiry is faulty for the reason that the notion 'nearer than' is only "defined for numbers and not for theories". In his contribution to the 1960 International Congress for Logic, Methodology, and Philosophy of Science at Stanford, Karl Popper defended the opposite view by defining a comparative notion of verisimilitude for theories.

The concept of verisimilitude was originally introduced by the Ancient sceptics to moderate their radical thesis of the inaccessibility of truth. But soon *verisimilitudo*, indicating likeness to the truth, was confused with *probabilitas*, which expresses an opiniotative attitude weaker than full certainty. The idea of truthlikeness fell in disrepute also as a result of the careless, often confused and metaphysically loaded way in which many philosophers used — and still use — such concepts as 'degree of truth', 'approximate truth', 'partial truth', and 'approach to the truth'.

Popper's great achievement was his insight that the criticism against truthlikeness — by those who urge that it is meaningless to speak about 'closeness to truth' — is more based on prejudice than argument. Indeed, no one had seriously tried to give a precise definition for this concept. In Popper's view, the realist correspondence conception of truth had already been 'saved' by Alfred Tarski in the 1930s. So he made in 1960 his own attempt to make the notion of truthlikeness respectable — in spite of his general misgivings about the 'scholasticism' of the programme of formal explication within analytical philosophy of science. And he also cherished the hope that the proposal to build a fallibilist theory of science upon the concept of verisimilitude would finally help us to get rid of formal systems of inductive logic, as developed by Rudolf Carnap.

Popper's attempt at explication failed, as papers published in 1974 by David Miller and Pavel Tichý showed. Nevertheless, Sir Karl had succeeded in making an intuitively convincing distinction between truthlikeness (as some kind of combination of truth and information content) and probability. Further, he had clearly separated the *logical* or semantical problem of truthlikeness (i.e., what does it mean to say that a theory is closer to the truth than another?) from the *epistemic* problem (i.e., on what evidential grounds can one rationally and conjecturally claim that one theory is closer to the truth than another?).

But something essential was missing from Popper's qualitative and quantitative definitions of truthlikeness: perhaps slightly paradoxically, this was the idea of *likeness* or *similarity*. Already in 1974, a new programme for explicating verisimilitude was initiated by Pavel Tichý, Risto Hilpinen, and myself, and was soon joined by Raimo Tuomela and Graham Oddie. The basic idea of this 'similarity approach' is that the truthlikeness of a statement h depends on the similarities between the states of affairs allowed by h and the true state of the world.

This book gives a self-contained and comprehensive exposition of the similarity theory of truthlikeness. It summarizes all the main work done in this field since 1974 — with the exception of the use of higher-order logic that Graham Oddie develops in his forthcoming book *Likeness to Truth* (D. Reidel, 1986). These two works — which have been written simultaneously and independently of each other — are the first book-length treatises on truthlikeness. In spite of some remaining disagreements between Oddie and me, I hope that our books will convince even the prejudiced readers that the concept of verisimilitude is by no means 'meaningless' or 'absurd'. On the contrary, I claim, it is a fascinating and rewarding subject of study within logical pragmatics.

In a companion volume to the present work, *Is Science Progressive?* (D. Reidel, 1984), I have argued that the concept of truthlikeness is an indispensable ingredient of critical scientific realism. It has important applications as a tool within the history and the philosophy of science — and also a great significance for our understanding of the nature of knowledge-seeking enterprises like science and their role in human culture. This book develops systematically the logical details for a theory of truthlikeness that is needed to support the wider philosophical theses of the essays in *Is Science Progressive?*.

In brief, I am confident that the concept of truthlikeness, which has a long but not entirely honourable history, will have a long and bright future. Perhaps all the readers do not share my optimism. But at least I hope that I have done a service for them too. Even if many things still

PREFACE   xiii

remain to be done, the frequently voiced complaint that no one has yet given a sufficiently explicit account of truthlikeness, which would make it possible to evaluate the arguments and theses of scientific realists, can now be forgotten. While it cannot be predicted to what extent, if at all, the theory of truthlikeness in the particular form developed here will survive the test of time, at least it is now open to evaluation both by the supporters and the critics of realism.

My own interest in truthlikeness arose from my earlier work on induction, where I attempted to apply inductive logic in an 'anti-inductivist' way to problems involving theories and conceptual change (Niiniluoto and Tuomela, 1973). I had a vague idea that Hintikka's measure of corroboration might have something to do with the epistemic problem of verisimilitude — more than Popper's own measure of corroboration. Moreover, Larry Laudan's (1973) excellent historical survey of the thesis that science "approaches to the truth" had convinced me that the concept of truthlikeness is indispensable for a fallibilist and realist theory of scientific progress.

Further stimulus for solving the logical problem of verisimilitude came in the autumn of 1973, when the news about Miller's refutation of Popper's comparative concept of truthlikeness reached Finland. Laudan took the failure of Popper's definition as one of the motives for developing a model of scientific progress which denies that science is a truth-seeking activity. Some of Popper's followers eventually withdrew back to a formulation of critical rationalism with truth and content — but without verisimilitude. But the supporters of the new similarity approach instead wanted to face the challenge of finding a workable account of degrees of truthlikeness.

The first wave of results about truthlikeness, obtained in 1974—79 by Tichý, Miller, myself, Tuomela, and Oddie, relied heavily on a specific tool-box in philosophical logic: Carnap's Q-predicates and state descriptions for monadic languages, and Hintikka's constituents for first-order logic. Already this was disappointing to some critical rationalists who had hoped for a simpler way of saving verisimilitude. What is more, I suggested that a solution to the epistemic problem of verisimilitude can be obtained by calculating expected degrees of truthlikeness relative to a system of inductive probabilities.

In this spirit, I argued in 1978 that the task of explicating the notion of truthlikeness is "important for all supporters of the 'critical' (as opposed to 'naive') scientific realism — independently of their relation to the Popperian school". In particular, "the interest in truthlikeness is not incompatible with simultaneous interest in Sir Karl's *bête noire*, inductive logic".

The systematic use of the Carnap-Hintikka tools from first-order logic had an unintended side effect. I suspect that some philosophers regarded the logical problem of truthlikeness, if not exclusively a Popperian problem, in some sense 'artificial' — not applicable to the relevant and interesting real-life scientific problems involving quantitative mathematical theories. Indeed, in the late 1970s there seemed to be a wide gulf separating those philosophers who primarily based their metascientific investigations on qualitative first-order languages and those who employed set-theoretical reconstructions of quantitative theories (among others, Suppes, Sneed, and Stegmüller). I argued in 1978 that the Sneedian structuralist programme for representing the 'empirical claims' of theories contains as a special case the problem of truthlikeness, but this remark (in spite of its correctness) was not convincing, since I was not then able to show how the treatment of truthlikeness can be translated into cases with quantitative statements.

The situation changed with the second wave of results in 1980—83: Roger Rosenkrantz defined truthlikeness for probabilistic laws; I observed that some standard results about Bayesian statistical estimation can be interpreted in terms of the estimated degrees of truthlikeness; I realized that, as Carnap's qualitative conceptual spaces are simply countable partitions of the state spaces of quantitative theories, there is a uniform method for analysing approximation relations between lawlike statements relative to such (qualitative or quantitative) spaces.

These observations led me to formulate my theory of truthlikeness in an abstract framework of cognitive problems, which may be interpreted in several alternative ways. At the same time, after ten year's work in this problem area, I somewhat unexpectedly realized that the most reasonable abstract definition of the truhlikeness measure in a sense contains as a special case Isaac Levi's definition of epistemic utility in his classical *Gambling with Truth* (1967).

Hence, the solution of the logical and the epistemic problems of truthlikeness given in this book has the nature of a synthesis which merges together two major trends within the theory of scientific inference: Bayesianism (covering the inductive logic of Carnap and

Hintikka, and the personalist statistics of Savage) and Popperianism. This synthesis turns out to be also an extension Levi's cognitive decision theory.

The structure of the book is based on the following plan. Chapters 1—4 contain preliminaries which will be used mainly in Chapters 8—13. An impatient reader may start directly from Chapter 5, which is an appetizer for the main course served in Chapters 6—7. A philosopher, who does not have much taste for technical details, may get an idea of the main message of the book by reading the following sections: 4.2, 4.3, 5.1—5.4, 5.6, 6.1—6.6, 7.1—7.5, 12.1, 12.4, 13.1—13.4.

To be more specific, Chapter 1 is a general introduction to the concepts of similarity and distance. As these notions constitute the key element of my approach to truth-likeness, it is useful to collect for further reference examples and definitions of similarity relations from various fields — and in this way prepare our intuition for the later chapters (especially Chapters 6—13).

Chapter 2 summarizes the basic tools of logic, including inductive logic and modal logic, that will be needed in Chapters 8—11. Chapter 3 introduces quantitative concepts and laws within a state space framework that will be used mainly in Chapters 8.4 and 12.

Chapter 4 gives precise formulations to the concept of cognitive problem, truth, and semantical information. Together with the idea of similarity, these are the basic elements that we need for building up our definition of truthlikeness.

Chapter 5 traces the origin of the concept verisimilitude to the sceptic Carneades and his infallibilist critic St. Augustine. A history of fallibilist dynamic epistemology from Cusanus to Hegel, Peirce, Bradley, Engels, Ewing, and Popper is outlined. This critical survey allows us to distinguish the concept of verisimilitude from other related explicanda like 'partial truth' and 'degree of truth'.

Chapter 6 formulates my general definition for the degree of truthlikeness, $Tr(g, h_*)$, of a statement g relative to the 'target' h_*. Here g is a disjunction of mutually exclusive hypotheses from set $\mathbf{B} = \{h_i | i \in \mathbf{I}\}$, and h_* is the most informative true statement in $\mathbf{B}$. Measure Tr is defined in terms of a metric or distance function Δ on $\mathbf{B}$. The proposed definition is compared with a number of rivals. It is also generalized to the cases, where the target h_* is disjunctive or only counterfactually true.

Chapter 7 continues the abstract treatment by showing how the estimated degree of truthlikeness of g on evidence e, $\text{ver}(g|e)$, can be defined as the expected value of $\text{Tr}(g, h_*)$, given an epistemic probability distribution on the set **B**.

The general theory of truthlikeness is then applied to special cases: singular sentences (Chapter 8), monadic generalizations (Chapter 9), polyadic generalizations and first-order theories (Chapter 10), qualitative and quantitative laws (Chapter 11). In each case, the distance function Δ is explicitly introduced for the relevant cognitive problem. Distances between sentences are also seen to induce distance measures between structures. This observation leads to a general treatment of approximation and idealization for lawlike quantitative statements (see Chapter 11). Chapter 12 analyses cognitive decision making by taking truthlikeness as the relevant epistemic utility. Maximization of expected verisimilitude as a principle of inference is compared with Levi's acceptance rule, Popper's solution to the problem of pragmatic preference between theories, and Bayesian statistical estimation of real-valued parameters.

The final Chapter 13 replies to some possible objections to the similarity approach, and compares my metric treatment of 'closeness to the truth' with some alternatives. I argue that Miller's demand for the invariance of comparative truthlikeness relations under all one-to-one translations between conceptual frameworks is definitely too strong. I also show how it is possible to appraise cognitive progress in theory-change with meaning variance. This problem, which depends on logical and philosophical issues about translation and incommensurability, has to be left to some extent open in this work.

Standard notation from logic, set-theory, and elementary mathematics is used, but otherwise the book does not presuppose any previous knowledge from the reader. Sections, formulas, figures, examples, and notes are numbered separately for each chapter. In the same chapter they are referred to simply by their number. But when the reference, e.g., to Formula 15 of Chapter 7 occurs in Chapter 8, it has the form 'formula (7.15)'.

I have once characterized my treatment of truthlikeness as having "Popper's voice but Carnap's hands": my ambition has been to combine Popper's deep insight about the significance of verisimilitude with Carnap's uncompromising and admirable rigour in working out the details. Evidently I owe my greatest debt to these two great masters of the philosophy of science in our century.

It is equally clear that my views on the methods and tools of philosophy have been decisively influenced by two great students of scientific inference, Jaakko Hintikka and Isaac Levi.

In the different periods of my work on truthlikeness, I have profited enormously of the chance of discussing, debating, and disagreeing — through publications, correspondence, and personal contacts — with Risto Hilpinen, David Miller, Pavel Tichý, Raimo Tuomela, Graham Oddie, Isaac Levi, Roberto Festa, and David Pearce. Other important influences and contacts will be visible in the text that follows the preface.

As always, I have profited from the stimulating intellectual atmosphere of the Department of Philosophy, University of Helsinki. For practical help in preparing the manuscript, I wish to thank especially Mrs. Auli Kaipainen (for most of the typing) and Mr. Ilpo Halonen (for drawing the figures). And once again I am grateful to my family for encouragement and patience.

I dedicate this book to my children: some day they will be closer to the truth than we ever were.

ILKKA NIINILUOTO

CHAPTER 1

DISTANCE AND SIMILARITY

Intuitively speaking, a statement is *truthlike* or *verisimilar* if it is 'like the truth' or 'similar to the truth'. In other words, a truthlike statement has to 'resemble the truth' or to be 'close to the truth', but it need not be true or even probable (i.e., likely to be true).[1] The notion of *truthlikeness* or *verisimilitude* can thus be regarded as a special case of the more general concept of *similarity* or *resemblance*. For this reason, we shall give in this chapter a survey of the work on similarity and dissimilarity that has been done in the fields of philosophy, logic, mathematics, statistics, information theory, computer science, biology, anthropology, psychology, and linguistics.

1.1. METRIC SPACES AND DISTANCES

The notion of a metric space was introduced by Maurice Fréchet in 1906 as an abstraction of the familiar properties of the Euclidean plane. A metric space is a set X with a real-valued function $d: X \times X \to \mathbb{R}$ which measures the distances between the points of X. More formally, a function $d: X \times X \to \mathbb{R}$ is called a *pseudometric* on X if it satisfies the following conditions for all x, y, and z in X:

(D1) $d(x, y) \geq 0$

(D2) $d(x, x) = 0$

(D3) $d(x, y) = d(y, x)$ (symmetry)

(D4) $d(x, y) \leq d(x, z) + d(z, y)$ (the triangle inequality).

A pseudometric d on X is a *metric* on X if condition (D2) can be strengthened to

(D5) $d(x, y) = 0$ iff $x = y$.[2]

If d is a metric on X, the pair (X, d) is called a *metric space*.

In many cases, the numerical values of a metric d on X are less interesting than the results that d gives for comparative purposes. If

(X, d) is a metric space, let us say that z is *closer* to x than to y if and only if $d(x, z) < d(y, z)$. Functions $d': X \times X \to \mathbb{R}$ which satisfy conditions (D1), (D2), and (D5) are sometimes called *semimetrics*. In many cases, symmetric semimetrics are called *distance functions* even if they do not satisfy the triangle inequality. If a distance function d' is obtained from a metric d on X by a transformation which preserves the order[3] and the zero point, then d and d' agree with each other in the sense that

$$d(x, z) < d(y, z) \quad \text{iff} \quad d'(x, z) < d'(y, z)$$

for all $x, y, z \in X$. In other words, a metric d and a semimetric d' may agree on the comparative notion of closeness.

The *diameter* diam(A) of a set $A \subseteq X$ is defined by

$$\text{(1)} \quad \text{diam}(A) = \sup_{x, y \in A} d(x, y).$$

Set A is *bounded* if diam(A) $< \infty$. The metric d on X is bounded if X is bounded. In this case, the metric d can be *normalized* by defining

$$\text{(2)} \quad d_0(x, y) = \frac{d(x, y)}{\text{diam}(X)} \quad \text{for all } x, y \in X.$$

Hence, $0 \leq d_0(x, y) \leq 1$ for all x and y in X.

The following examples illustrate the possibilities of defining metrics and distances between various kinds of elements — such as real numbers, n-tuples of real numbers, functions, probability distributions, binary sequences, sets, fuzzy sets, trees, and samples.

EXAMPLE 1. Let X be the *real line* $\mathbb{R}$, and define $d_1: \mathbb{R} \times \mathbb{R} \to \mathbb{R}$ by

$$\text{(3)} \quad d_1(x, y) = |x - y| \quad \text{for all } x, y \in \mathbb{R}.$$

Then $(\mathbb{R}, d_1)$ is a metric space. Function $d_2: \mathbb{R} \times \mathbb{R} \to \mathbb{R}$ defined by

$$\text{(4)} \quad d_2(x, y) = (x - y)^2 \quad \text{for all } x, y \in \mathbb{R}$$

is not a metric on $\mathbb{R}$, since it does not satisfy (D4). However, d_2 is a distance function on $\mathbb{R}$, since it can be obtained from d_1 by the order-preserving transformation $g(z) = z^2$ (i.e., $d_2 = g \circ d_1$).

EXAMPLE 2. Let X be the n-dimensional *real space* $\mathbb{R}^n$ ($n \geq 2$), where the elements of $\mathbb{R}^n$ are n-tuples $(x_1, \ldots, x_n)$ of real numbers. For points $x = (x_1, \ldots, x_n)$ and $y = (y_1, \ldots, y_n)$ in $\mathbb{R}^n$, define

$$(5) \quad d_1(x, y) = \sum_{i=1}^{n} |x_i - y_i|$$

$$(6) \quad d_2(x, y) = \sqrt{\sum_{i=1}^{n} (x_i - y_i)^2}$$

$$(7) \quad d_3(x, y) = \max_{i=1,\ldots,n} |x_i - y_i|.$$

Then $(\mathbb{R}^n, d_i)$ is a metric space for all $i = 1, 2, 3$. Function d_1 is called the *Manhattan* or *city-block metric*, d_2 is the *Euclidean metric*, and d_3 is the *Tchebycheff metric*. An illustration of these functions for the case $n = 2$ is given in Figure 1. Another way of visualizing the difference between these metrics is to draw unit spheres, i.e., the diagram of all points which are at a given fixed distance from a given point (see Figure 2). It is easy to see that the Euclidean metric d_2 is invariant under rotation of the coordinates.

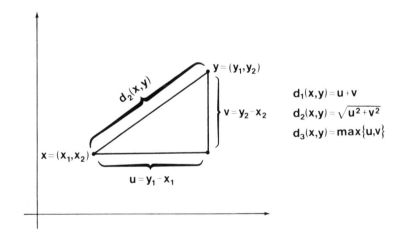

Fig. 1. Metrics on the plane $\mathbb{R}^2$.

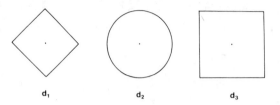

Fig. 2. Unit spheres in $\mathbb{R}^2$ for d_1, d_2, and d_3.

Functions d_1, d_2, and d_3 are special cases of the family of *Minkowski metrics* for $\mathbb{R}^n$:

$$(8) \quad \left(\sum_{i=1}^{n} |x_i - y_i|^p \right)^{1/p} \quad (p \geq 1).$$

The choice $p = 1$ gives the city-block metric d_1 and $p = 2$ the Euclidean metric d_2.[4] The Tchebycheff metric d_3 corresponds to the choice of $p = \infty$, since

$$\max_{i=1,\ldots,n} \{z_i\} = \lim_{p \to \infty} (|z_1|^p + \ldots + |z_n|^p)^{1/p}.$$

This implies that the unit spheres for the metrics (8) approach the sphere for d_3 in the limit when $p \to \infty$ (cf. Figure 2). Also the following theorem, proved by G. Polya in 1915, holds: if a_p is the element of $A \subseteq \mathbb{R}^n$ which is closest to $x \in \mathbb{R}$ according to the metric (8), and if the sequence a_p converges to $a \in A$ when $p \to \infty$, then a is the element of A which is closest to x according to d_3 (see Rice, 1964, pp. 8–10).

Metric d_3 was used for approximation problems by P. S. Laplace in 1799 and later in the 1850s by P. L. Tchebycheff. Metric d_2 was used in the least square approximation by A. M. Legendre in 1804.

Metrics d_1, d_2, and d_3 can be immediately generalized to the spaces of infinite sequences $\langle z_i | i < \infty \rangle$ of real numbers $z_i \in \mathbb{R}$ or complex numbers $z_i \in \mathbb{C}$. For $1 \leq p \leq \infty$, let

$$1^p = \left\{ \langle z_i | i < \infty \rangle | z_i \in \mathbb{C} \text{ and } \sum_{i=1}^{\infty} |z_i|^p < \infty \right\}.$$

With the definitions

$$az = \langle az_i | i < \infty \rangle$$
$$z + u = \langle z_i + u_i | i < \infty \rangle$$

for $a \in \mathbb{R}$, $z = \langle z_i | i < \infty \rangle \in 1^p$, and $u = \langle u_i | i < \infty \rangle \in 1^p$, set 1^p is a linear space. With the norm

$$\|z\|_p = \left(\sum_{i=1}^{\infty} |z_i|^p \right)^{1/p}$$

1^p is a *normed linear space*. For $1 \leq p < \infty$, 1^p is an example of a *Hilbert space*, since it is separable. On the other hand, set

$$1^\infty = \{\langle z_i | i < \infty \rangle | z_i \in \mathbb{C} \text{ and } \{z_i | i < \infty\} \text{ is bounded}\}$$

with the norm

$$\|z\|_\infty = \sup_{i < \infty} |z_i|$$

is not separable. In general, if $\|\ \|: X \to \mathbb{R}$ is a norm in a linear space X, then the function $\|z - u\|$, for $z \in X$ and $u \in X$, is a metric on X.

EXAMPLE 3. Let X be the set K of all continuous functions $f: [0, 1] \to \mathbb{R}$ defined on the closed interval $[0, 1] \subseteq \mathbb{R}$. The Minkowski metrics on K, or L_p-metrics, are defined by

$$(9) \qquad L_p(f, g) = \left(\int_0^1 |f(x) - g(x)|^p \, dx \right)^{1/p} \quad (p \geq 1)$$

for functions f and g in K. As special cases of (9), with $p = 1, 2,$ and ∞, we have

$$(10) \qquad \int_0^1 |f(x) - g(x)| \, dx$$

$$(11) \qquad \left(\int_0^1 (f(x) - g(x))^2 \, dx \right)^{\frac{1}{2}}$$

(12) $\sup\limits_{0 \leq x \leq 1} |f(x) - g(x)|.$

Metrics (10)–(12) are direct generalizations of the metrics (5)–(7).

If function f and g are integrable in the sense of Lebesgue, then (10) and (11) define only a pseudometric. However, if functions differing from each other only in a set of measure zero are identified, we obtain the L_p-spaces, i.e., normed linear spaces of functions f on $I = [0, 1]$ with the L_p-norm

$$\|f\|_p = \left(\int_I |f|^p \, d\mu \right)^{1/p},$$

where μ is the Lebesgue measure on I. Therefore, it is sometimes said that (10) corresponds to the L_1-norm, (11) to the L_2-norm, and (12) to the L_∞-norm.

The differences between the L_p-metrics can be illustrated by noting how the behavior of the function z^p on $[0, \infty]$ depends on p (see Figure 3). If $p = 1$, function z^p grows linearly — in direct proportion to the value of z. If p is very large, then z^p gives very heavy weights to large values of z. In contrast, if $p < 1$, function z^p grows slowly with increasing z. (See Rice, 1964, pp. 5–6.) This shows that, when p grows, the distance (6) between functions f and g depends more and more crucially on the maximum distance between the values $f(x)$ and $g(x)$ on $[0, 1]$; for small p, it depends primarily on the area which is left between the curves $f(x)$ and $g(x)$. (See Figure 4.)

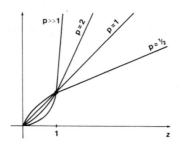

Fig. 3. Function z^p.

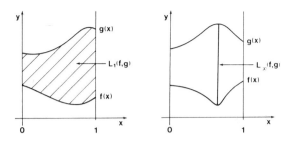

Fig. 4. Metrics L_1 and L_∞.

EXAMPLE 4. Let **x** and **y** be two discrete *random variables* on set $A = \{a_1, \ldots, a_n\}$ with the probabilities

$$P(\mathbf{x} = a_i) = p_i \geq 0 \ (i = 1, \ldots, n)$$
$$P(\mathbf{y} = a_i) = q_i \geq 0 \ (i = 1, \ldots, n),$$

where $p_1 + \ldots + p_n = 1$ and $q_1 + \ldots + q_n = 1$. Then the *directed divergence* between **x** and **y** is defined by

(13) $\quad \mathrm{div}(\mathbf{x}, \mathbf{y}) = \sum_{i=1}^{n} p_i \log \dfrac{p_i}{q_i}.$[5]

Similarly, the directed divergence between **y** and **x** is

$$\mathrm{div}(\mathbf{y}, \mathbf{x}) = \sum_{i=1}^{n} q_i \log \dfrac{q_i}{p_i},$$

which usually differs from $\mathrm{div}(\mathbf{x}, \mathbf{y})$. Definition (13) satisfies conditions (D1), (D2), and (D5). (Cf. Kullback, 1959.) To obtain a symmetric notion, satisfying condition (D3), the *divergence* between **x** and **y** may be defined by

(14) $\quad \mathrm{div}(\mathbf{x}, \mathbf{y}) + \mathrm{div}(\mathbf{y}, \mathbf{x}),$

which is equal to

$$\sum_{i=1}^{n} (p_i - q_i) \log \dfrac{p_i}{q_i}.$$

These definitions can be generalized to continuous one-dimensional random variables **x** and **y** as follows: let f_x and g_y be the density functions of **x** and **y**, respectively. Then the directed divergence between **x** and **y** is

$$(15) \quad \int_{-\infty}^{\infty} f_x(z) \log \frac{f_x(z)}{g_y(z)} \, dz$$

and the divergence between x and y is

$$(16) \quad \int_{-\infty}^{\infty} (f_x(z) - g_y(z)) \log \frac{f_x(z)}{g_y(z)} \, dz.$$

The notion of divergence, which is due to Jeffreys (1948), is a semi-metric: it does not satisfy the triangle inequality. Other distances between probability distributions include E. Hellinger's (1909) and A. Battacharya's (1946) dissimilarity coefficient

$$\cos^{-1} \int_{-\infty}^{\infty} (f_x(z) f_y(z))^{\frac{1}{2}} \, dz$$

and A. N. Kolmogorov's variational distance

$$\int_{-\infty}^{\infty} (f_x(z) - f_y(z)) \, dz$$

(cf. (10)). (See Rao, 1977.)

EXAMPLE 5. Let F and G be the *distribution functions* of two continuous random variables. Then the following definition, introduced by P. Lévy, satisfies the conditions for a metric:

$$(17) \quad d(F, G) = \inf_{h \in \mathbb{R}} \{ F(x - h) - h \leqslant G(x) \leqslant F(x + h) + h$$

$$\text{for all } x \in \mathbb{R} \}.$$

(See Gnedenko and Kolmogorov, 1954, p. 33.)

EXAMPLE 6. Let $x = (x_1, \ldots, x_n)$ and $y = (x_1, \ldots, x_n)$ be *binary sequences* of length n, i.e., sequences consisting of zeros and ones. Then the *Hamming distance* between x and y is defined as the number of digits in which x and y disagree. More formally, let $x \vee y$ be the exclusive or-operator between the elements of $\{0, 1\}$:

$$x_i \vee y_i = 1, \text{ if } x_i \neq y_i$$
$$ = 0, \text{ if } x_i = y_i.$$

Then the Hamming distance between x and y is

(18) $\quad d_H(x, y) = \sum_{i=1}^{n} x_i \vee y_i.$

This notion, which was introduced by R. W. Hamming (1950) in his study of error correcting codes in communication theory, defines a metric on the set of binary sequences of a fixed length.

EXAMPLE 7. Let $X = \mathscr{P}(Y) - \{\emptyset\}$, where $\mathscr{P}(Y)$ is the power set of some non-empty finite set Y. Thus, the elements A of X are themselves non-empty finite sets. The distance between A and B in X can be defined by

(19) $\quad \dfrac{|A \Delta B|}{|Y|}$

where $|A|$ is the cardinality of set A, and $A \Delta B$ is the *symmetric difference* of A and B:

(20) $\quad A \Delta B = (A - B) \cup (B - A).$

(Cf. Figure 5.) Thus, (19) is the normalized cardinality of the symmetric difference of A and B.

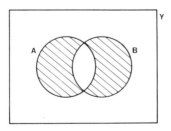

Fig. 5. The symmetric difference $A \Delta B$ of A and B.

Each subset A of Y can be correlated with its indicator function $K_A: Y \to \{0, 1\}$ defined by

$$K_A(x) = 1, \text{ if } x \in A$$
$$ = 0, \text{ if } x \notin A.$$

Assume that the members of Y have been enumerated as $Y = \{y_1, \ldots, y_n\}$. Then the function K_A can be represented as the finite binary sequence $\langle K_A(y_1), \ldots, K_A(y_n) \rangle$. In this case, $|A \triangle B|$ equals

$$(21) \quad \sum_{i=1}^{n} |K_A(y_i) - K_B(y_i)|,$$

which is the same as the Hamming distance between the sequences corresponding to sets A and B (see Example 6). Thus, definition (19) is essentially a normalized Hamming distance between sets A and B.

If A and B are *fuzzy* subsets of Y, their indicator functions K_A and K_B may take any values within the interval $[0, 1]$. Then the Hamming distance between fuzzy sets A and B can again be formally defined by (21). (See, for example, Dubois and Prade, 1980.)

A direct generalization of (19) can be obtained by replacing the partially ordered set $(X, \subseteq)$ by an arbitrary *Boolean algebra* $(\mathscr{B}, \leq)$. The pair $(\mathscr{B}, \leq)$, where $\mathscr{B} \neq \emptyset$ and $\leq$ is a binary relation in $\mathscr{B}$, is a Boolean algebra, if (i) $(\mathscr{B}, \leq)$ is a *lattice*, i.e., $\leq$ is a partial order in $\mathscr{B}$ such that all two elements a and b of $\mathscr{B}$ have a supremum (least upper bound) $a \wedge b$ and an infimum (greates lower bound) $a \vee b$ relative to $\leq$, and there is a minimum element $\mathbf{0}$ and a maximum element $\mathbf{1}$ in $\mathscr{B}$, (ii) every element a of $\mathscr{B}$ has a complement a' in $\mathscr{B}$ (satisfying $a \vee a' = a' \vee a = \mathbf{1}$ and $a \wedge a' = a' \wedge a = \mathbf{0}$), (iii) the distributive laws

$$a \wedge (b \wedge c) = (a \wedge b) \wedge c$$
$$a \vee (b \vee c) = (a \vee b) \vee c$$

hold for all $a, b, c \in \mathscr{B}$. (The last condition guarantees that the complement a' is unique.) Then the symmetric difference

$$a \triangle b = (a \wedge b') \vee (a' \wedge b)$$

is an element of $\mathscr{B}$; it defines a *Boolean valued metric*. If P is a probability measure defined on $\mathscr{B}$, then

$$(22) \quad P(a \triangle b) = P(a \wedge b') + P(a' \wedge b)$$

is a pseudo-metric on $\mathscr{B}$. It is called *Mazurkiewicz's metric* — after S. Mazurkiewicz who proposed it in 1935 as a 'stochastic distance' between the elements of the Boolean algebra of deductive systems (cf. Chapter 2.9).

EXAMPLE 8. Let $X = \mathscr{P}(Y) - \{\emptyset\}$, where (Y, d) is a finite metric space. Then the *distance* between two *sets* $A \subseteq X$ and $B \subseteq X$ is often defined by

(23) $\quad d(A, B) = \min_{\substack{x \in A \\ y \in B}} d(x, y).$

Definition (23) is not a metric, since $d(A, B) = 0$ if and only if $A \cap B \neq \emptyset$. As a special case of (23), the distance between a *point* $x \in X$ and a *set* $A \subseteq X$ is $d(\{x\}, A)$, i.e.,

$$d(x, A) = \min_{y \in A} d(x, y).$$

Hence, $d(x, A) = 0$ if and only if $x \in A$. The following metric $h: X \times X \to \mathbb{R}$ is known as the *Hausdorff distance* (also Fréchet distance) between elements A and B of X:

(24) $\quad h(A, B) = \max\{\max_{x \in A} d(x, B), \max_{y \in B} d(y, A)\}.$

Thus, $h(\{x\}, \{y\}) = d(x, y)$ for all x and y in Y, but

(25) $\quad h(\{x\}, B) = \max_{y \in B} d(x, y),$

so that $h(\{x\}, B)$ is generally different from $d(\{x\}, B)$, as defined by (23).

EXAMPLE 9. Let V be a finite non-empty set, and let V^* be the set of all finite strings of elements of V. We say that the elements of V^* are *words* over the *alphabet* V. Let w_1 and w_2 be any elements of V^*, and let a and b be any elements of V. Definite three kinds of transformations $T_S: V^* \to V^*$ (substitution), $T_D: V^* \to V^*$ (deletion), and $T_I: V^* \to V^*$ (insertion) as follows:

$$T_S(w_1 a w_2) = w_1 b w_2$$
$$T_D(w_1 a w_2) = w_1 w_2$$
$$T_I(w_1 w_2) = w_1 a w_2.$$

Then the function $d_L: V^* \times V^* \to \mathbb{R}$ defined by

(26) $\quad d_L(w_1, w_2) =$ the minimum number of transformations T_S, T_D, and T_I required to derive word w_2 from word w_1

is called the *Levenshtein metric* on V^* (cf. Fu, 1982, p. 248). It was defined by V. I. Levenshtein (1966) for the study of error-correcting parsing for formal languages and grammars. Definition (26) can be modified by giving different weights to the three types of error transformations.

A similar idea can be applied to give a metric on the set of *finite labeled trees* (see Fu, 1982, p. 292). A tree $\mathscr{A}$ is a partially ordered set $\langle A, \leqslant \rangle$ such that for all $x \in A$ the sets $B_x = \{z \in A \,|\, z \leqslant x\}$ are well-ordered by $\leqslant$ and have the same minimal element x_1. The element x_1 is called the *root* of $\mathscr{A}$; a tree is usually drawn so that its root is at the top. (See Figure 6.) The order type of B_x is called the *rank* of x. The elements of A are called *nodes*. *Terminal* nodes are those elements of A which are maximal with respect to the order $\leqslant$. Linearly ordered subsets B of A are called *branches* of $\mathscr{A}$, if they are closed upward with respect to $\leqslant$ (i.e., $B_x \subseteq B$ for all $x \in B$). A *subtree* of $\mathscr{A}$ determined by node x is the set $\{y \in A \,|\, x \leqslant y\}$. Tree $\mathscr{A}$ is finite, if A is a finite set. In this case, the rank of each node is finite, and the length of each branch of $\mathscr{A}$ is also finite. Finally, a tree $\mathscr{A} = \langle A, \leqslant \rangle$ is *labeled* with the alphabet V, if there is a mapping $f: A \to V$, which associates one element of V to each node of $\mathscr{A}$.

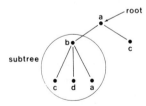

Fig. 6. A finite tree labeled with the alphabet $\{a, b, c, d\}$. The maximum rank of the nodes is 2.

Let us define five types of transformations for finite labeled trees over alphabet V:

T_1: the substitution of the label of a node by another element of V
T_2: the insertion of a new labeled node between a node and its immediate predecessor
T_3: the insertion of a new labeled node to the left of all the immediate successors of a node
T_4: the insertion of a new labeled node to the right of a node
T_5: the deletion of a node which determines a subtree with at most one branch.

The transformations T_1-T_5 are called substitution, stretch, branch, split, and deletion, respectively. They are illustrated in Figure 7. If $\mathscr{A}_1$ and $\mathscr{A}_2$ are two finite trees labeled with the alphabet V, then the function defined by

(27) $d_F(\mathscr{A}_1, \mathscr{A}_2)$ = the minimum number of transformations T_1, T_2, T_3, T_4, and T_5 required to derive $\mathscr{A}_2$ from $\mathscr{A}_1$

is called the *Fu metric*. Again, definition (27) can be modified by assigning different weights to the transformations T_1-T_5. (See also Fu, 1982, pp. 338–344.) The Fu metric can be applied to various problems in pattern recognition and cluster analysis.

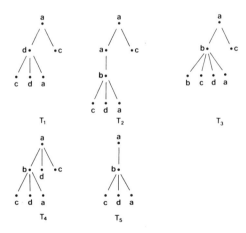

Fig. 7. Examples of trees obtainable by transformations T_1-T_5 from the tree of Figure 6.

EXAMPLE 10. Let $\mathscr{A}$ be a finite tree, and assume that the terminal nodes of $\mathscr{A}$ are correlated one-to-one with the elements of the label set $V = \{v_1, \ldots, v_n\}$. The set of labels of the terminal nodes of a subtree of $\mathscr{A}$ determined by a non-terminal node is called a *node set* of $\mathscr{A}$. Let $\{W_1, \ldots, W_{n-1}\}$ be a list of the node sets of $\mathscr{A}$ such that a set W appears in this list $b - 1$ times if b branches go through the node x determining W. (Here x is the root of the subtree corresponding to the node set W.) If $\mathscr{A}_1$ and $\mathscr{A}_2$ are two such finite trees, the terminal nodes correlated with the same label set V, and $\{W_1^1, \ldots, W_{n-1}^1\}$ and $\{W_1^2, \ldots, W_{n-1}^2\}$ are the corresponding lists of the node sets of $\mathscr{A}_1$ and $\mathscr{A}_2$, respectively, then the following function is called the *Boorman–Olivier metric*:

$$(28) \quad d_{\text{BO}}(\mathscr{A}_1, \mathscr{A}_2) = \min_{f \in \pi_{n-1}} \sum_{i=1}^{n-1} |W_i^1 \triangle W_{f(i)}^2|$$

where π_{n-1} is the set of the permutations of the set $\{1, \ldots, n-1\}$. (See Boorman and Olivier, 1973.) Any permutation f which minimizes the sum in (28) is called an *optimum pairing* of the two collections of sets $\{W_1^1, \ldots, W_{n-1}^1\}$ and $\{W_1^2, \ldots, W_{n-1}^2\}$.[6] Besides being a metric, function d_{BO} has two nice properties: if $\mathscr{A}_1$ and $\mathscr{A}_2$ differ only on a subtree, then $d_{\text{BO}}(\mathscr{A}_1, \mathscr{A}_2)$ depends only on that subtree; and if $\mathscr{A}_1$ and $\mathscr{A}_2$ differ only outside a subtree, then $d_{\text{BO}}(\mathscr{A}_1, \mathscr{A}_2)$ does not depend on the internal structure of that subtree. (See the conditions of 'additivity' and 'subtree-opacity' in Boorman and Olivier, 1973.) Definition (28) is illustrated in Figure 8.

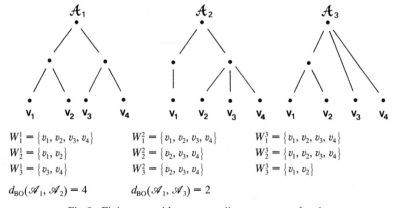

Fig. 8. Finite trees with corresponding sequences of node sets.

EXAMPLE 11. Let $\{Q_1, \ldots, Q_K\}$ be a K-fold *classification system*, defined by mutually exclusive and jointly exhaustive classificatory concepts Q_i ($i = 1, \ldots, K$). Then each individual has to belong to one and only one class in $\{Q_1, \ldots, Q_K\}$. Let s_n be a sample of n individuals such that $n_i \geq 0$ individuals belong to class Q_i ($i = 1, \ldots, K$); hence, $n_1 + \cdots + n_K = n$. Let t_n be another distribution of n individuals into the classes $\{Q_1, \ldots, Q_K\}$ such that $m_i > 0$ individuals belong to Q_i ($i = 1, \ldots, K$).

Class	Q_1	Q_2	....	Q_K	Sum
s_n	n_1	n_2	...	n_K	n
t_n	m_1	m_2	...	m_K	n

Karl Pearson (1900) suggested that the 'fit' between s_n and t_n can be tested by studying the quantity

$$(29) \quad \chi^2 = \sum_{i=1}^{K} \frac{(n_i - m_i)^2}{m_i}.$$

Here $\chi^2 = 0$ if and only if $n_i = m_i$ for all $i = 1, \ldots, K$. Usually the values m_i are determined by assuming that the probability of finding an individual in class Q_i is p_i (null hypothesis) and by identifying m_i with the expected values np_i. In this case, the quantity (29) has asymptotically (as $n \to \infty$) the χ^2-distribution with $K - 1$ degrees of freedom (cf. Cramer, 1946, pp. 416—419). On the other hand, t_n might as well be another sample of n individuals; in this case, χ^2 would be a nonsymmetric semimetric between two samples of the same size.

The χ^2-test is also widely used in the study of the independence of two variables or, equivalently, two classification systems $\{Q_1, \ldots, Q_K\}$ and $\{R_1, \ldots, R_L\}$. Let s_n be a sample of the size n such that $n_{ij} \geq 0$ individuals have been found in the class $Q_i \cap R_j$. The values constitute a two-dimensional empirical distribution or a *contingency table* (see Figure 9).

class	R_1	$\cdots$	R_j	$\cdots$	R_L	sum
Q_1	n_{11}	$\cdots$	n_{1j}	$\cdots$	n_{1L}	$n_1.$
$\vdots$	$\vdots$		$\vdots$		$\vdots$	$\vdots$
Q_i	n_{i1}	$\cdots$	n_{ij}	$\cdots$	n_{iL}	$n_i.$
$\vdots$	$\vdots$		$\vdots$		$\vdots$	$\vdots$
Q_K	n_{K1}	$\cdots$	n_{Kj}	$\cdots$	n_{KL}	$n_K.$
sum	$n._1$	$\cdots$	$n._j$	$\cdots$	$n._L$	n

$$\sum_{i=1}^{K} n_i. = \sum_{j=1}^{L} n._j = n$$

Fig. 9. Contingency table.

If the two classifications are independent, we expect to find

$$m_{ij} = \frac{n._j n_i.}{n}$$

individuals in class $Q_i \cap R_j$. The values m_{ij}, $i = 1, \ldots, K$, $j = 1, \ldots, L$, define another contingency table. Assuming independence, the quantity

$$(30) \quad \chi^2 = \sum_{i=1}^{K} \sum_{j=1}^{L} \frac{(n_{ij} - m_{ij})^2}{m_{ij}}$$

is asymptotically distributed in the χ^2-distribution with $(K-1)(L-1)$ degrees of freedom.

EXAMPLE 12. Let $(x_1, \ldots, x_n)$ be a sample of values of a one-dimensional random variable **x** with the mean μ and the variance σ^2. Let $\bar{x}$ be the sample mean

$$(31) \quad \bar{x} = \frac{1}{n} \sum_{i=1}^{n} x_i.$$

The *scatter* of sample $(x_1, \ldots, x_n)$ about the point x_0 is defined by

$$(32) \quad \text{sc}(x_0, n) = \sum_{i=1}^{n} (x_i - x_0)^2.$$

As
$$\text{sc}(x_0, n) = \text{sc}(\bar{x}, n) + n(\bar{x} - x_0)^2,$$

$\text{sc}(x_0, n)$ is minimized by choosing $x_0 = \bar{x}$. The expected value of $\text{sc}(x_0, n)$ is

$$n(\sigma^2 + (\mu - x_0)^2),$$

which is minimized by choosing $x_0 = \mu$. If $(x_1^{(1)}, \ldots, x_{n_1}^{(1)})$ and $(x_1^{(2)}, \ldots, x_{n_2}^{(2)})$ are two samples with the means $\bar{x}^{(1)}$ and $\bar{x}^{(2)}$, respectively, the scatter of the pooled sample (i.e., the two samples together) about its mean $\bar{x}$ is

$$\text{sc}(\bar{x}, n_1 + n_2) = \text{sc}(\bar{x}^{(1)}, n_1) + \text{sc}(\bar{x}^{(2)}, n_2) + \\ + n_1(\bar{x}^{(1)} - \bar{x})^2 + n_2(\bar{x}^{(2)} - \bar{x})^2.$$

These notions can be generalized to the case of k-dimensional random variables (see Wilks, 1962, pp. 540—560). Let $(z_1, \ldots, z_n)$ be a sample such that each z_i is a k-tuple $(x_{1i}, \ldots, x_{ki})$, for $i = 1, \ldots, n$. Let $z_0 = (x_{10}, \ldots, x_{k0})$. Then, for $i = 1, \ldots, k, j = 1, \ldots, k$, define

$$\bar{x}_i = \frac{1}{n} \sum_{\xi=1}^{n} x_{i\xi}$$

$$u_{ij} = \sum_{\xi=1}^{n} (x_{i\xi} - \bar{x}_i)(x_{j\xi} - \bar{x}_j).$$

Then the matrix $\|u_{ij}\|_{i,j=1,\ldots,k}$ is the internal scatter matrix of sample $(z_1, \ldots, z_n)$, and its determinant $|u_{ij}|$ is the *internal scatter* of $(z_1, \ldots, z_n)$. The *scatter* of $(z_1, \ldots, z_n)$ about z_0 is defined as the determinant

$$\text{sc}(z_0, n) = |u_{ij} + n(\bar{x}_i - x_{i0})(\bar{x}_j - x_{j0})|.$$

Let $\|u^{ij}\|$ be the inverse matrix of $\|u_{ij}\|$. Then

$$\text{sc}(z_0, n) = |u_{ij}| \cdot [1 + n \sum_{i=1}^{k} \sum_{j=1}^{k} u^{ij}(\bar{x}_i - x_{i0})(\bar{x}_j - x_{j0})]$$

which is minimized by choosing $z_0 = (\bar{x}_1, \ldots, \bar{x}_k)$.

If $(z_1, \ldots, z_n)$ is a sample from a k-dimensional normal distribution with the mean $(\mu_1, \ldots, \mu_k)$, then the following function

$$(33) \quad D^2 = (n-1) \sum_{i=1}^{k} \sum_{j=1}^{k} u^{ij}(\bar{x}_i - \mu_i)(\bar{x}_j - \mu_j),$$

defined by P. C. Mahalanobis (1936), is known as the *Mahalanobis (squared) distance* between the sample and population. Similarly, let $(z_1^{(1)}, \ldots, z_{n_1}^{(1)})$ and $(z_1^{(2)}, \ldots, z_{n_2}^{(2)})$ be two samples from the same k-dimensional normal distribution, and let $\|u^{ij}\|$ be the inverse matrix of $\|u_{ij}\|$, where $u_{ij} = u_{ij}^{(1)} + u_{ij}^{(2)}$. Then the function

$$(34) \quad D^2 = (n_1 + n_2 - 2) \sum_{i=1}^{k} \sum_{j=1}^{k} u^{ij}(\bar{x}_i^{(1)} - \bar{x}_i^{(2)})(\bar{x}_j^{(1)} - \bar{x}_j^{(2)})$$

is the Mahalanobis (squared) distance between the two samples.

1.2. TOPOLOGICAL SPACES AND UNIFORMITIES

A collection τ of subsets of set X is a *topology* of X if $\emptyset \in \tau$, $X \in \tau$, and τ is closed under unions and under finite intersections.[7] If τ is a topology of X, the pair (X, τ) is a *topological space*, and the elements A of τ are called *open sets*. The complements of open sets are *closed*.

To each metric space (X, d) one can introduce a topological structure τ_d as follows. An *open ball* of radius $r > 0$ and center $x \in X$ is defined by

$$U_r(x) = \{y \in X \mid d(x, y) < r\}.$$

Set $A \subseteq X$ is a *neighbourhood* of point $x \in X$ if there is an open ball $U_r(x) \subseteq A$. Set A is *open* if A is a neighbourhood of all its points. It follows that the open balls $U_r(x)$ form a *basis* for this topology τ_d: each non-empty open set is a union of open balls. Two metrics d_1 and d_2 on X are *topologically equivalent* if they generate the same topology, i.e., if $\tau_{d_1} = \tau_{d_2}$. For example, metric (3) generates the standard topology on $\mathbb{R}$, and metrics (5)–(7) generate the standard topology on $\mathbb{R}^n$, the so-called product topology on $\mathbb{R}^n = \mathbb{R} \times \cdots \times \mathbb{R}$. On the other hand, metrics (10) and (12) on K are not topologically equivalent.

Conversely, there are topological spaces which cannot be generated by any metric. A topological space (X, τ) is *metrizable* if there is a metric d on X such that $\tau = \tau_d$. A necessary condition for a topological space (X, τ) to be metrizable is that (X, τ) is a *Hausdorff* space, i.e., for all $x, y \in X$, $x \neq y$, there are open sets U and V in τ such that $x \in U$, $y \in V$, and $U \cap V = \emptyset$.

The notion of convergence can be defined for metric spaces in the following way. A sequence $\langle x_n | n = 1, 2, \ldots \rangle$ of elements $x_n \in X$ of a metric space (X, d) *converges* to point $x \in X$ if and only if $d(x_n, x) \to 0$, when $n \to \infty$, i.e., for all $\varepsilon > 0$ there is a n_0 such that $d(x_n, x) < \varepsilon$ when $n \geq n_0$. For example, if d_1 is defined by (3), then convergence in $(\mathbb{R}, d_1)$ reduces to the standard notion of convergence for real numbers. Convergence in the function space (K, d), where d is the L_∞-metric (12), corresponds to the notion *uniform convergence* of sequences of real functions. Further $\langle x_n | n = 1, 2, \ldots \rangle$ is a *Cauchy-sequence* if and only if for all $\varepsilon > 0$ there is a n_0 such that $d(x_n, x_m) < \varepsilon$ when $n \geq n_0$ and $m \geq n_0$. Every converging sequence is also a Cauchy-sequence, but the converse does not hold generally. A metric space (X, d) is *complete* if every Cauchy-sequence converges to a point in X. If the metric space (X, d) is not complete, one can embed it in a complete metric space $(\hat{X}, \hat{d})$ which is called the (Hausdorff) *completion* of X.

More generally, let us say that a point $x \in X$ is a *limit point* of the set $A \subseteq X$ if $A \cap U_r(x) \neq \emptyset$ for every $r > 0$. Let (X, d_1) and (Y, d_2) be metric spaces, and let $A \subseteq X$ and $f: A \to Y$. Assuming that $a \in X$ is a limit point of A, $y \in Y$ is the *limit* of function f at point a if for each neighbourhood W of y there is a neighbourhood V of a such that $f(V \cap A) \subseteq W$.

Many interesting topological notions — such as convergence — make sense even if one cannot associate a metric with the space. A. Weil defined in 1937, as an abstraction of G. Cantor's method of defining real numbers in terms of Cauchy-sequences of rationals, the notion of 'uniformity' which is stronger than topology but weaker than metric. In other words, 'uniformizable' spaces are situated between topological spaces and metrizable spaces.

A *uniformity* on a set $X \neq \emptyset$ is a collection $\mathcal{U}$ of subsets of $X \times X$ such that for all U and V in $\mathcal{P}(X \times X)$

(U1)   If $U \in \mathcal{U}$ and $U \subseteq V$, then $V \in \mathcal{U}$.

(U2)   If $U \in \mathcal{U}$ and $V \in \mathcal{U}$, then $U \cap V \in \mathcal{U}$.

(U3) If $U \in \mathcal{U}$, then $\Delta \subseteq U$.

(U4) If $U \in \mathcal{U}$, then $U^{-1} \in \mathcal{U}$.

(U5) If $U \in \mathcal{U}$, then there is $W \in \mathcal{U}$ such that $W \circ W \subseteq U$.

Here

$$\begin{aligned}\Delta &= \{(x, x) | x \in X\} \\ U^{-1} &= \{(y, x) | (x, y) \in U\} \\ W \circ W &= \{(x, z) | \exists y \in X (x, y) \in W \text{ and } (y, z) \in W)\},\end{aligned}$$

i.e., Δ is the diagonal of X, U^{-1} is the inverse of U, and $W \circ W$ is the Peircean product of W with itself. The elements U of a uniformity $\mathcal{U}$ are called its *entourages*, and the pair $(X, \mathcal{U})$ is a *uniform space*. (See Bourbaki, 1966, Ch. II.)

In a metric space (X, d), one can define a uniformity $\mathcal{U}_d$ in the following way. For each $r > 0$, let

$$V_r = \{(x, y) \in X \times X | d(x, y) < r\}.$$

(Cf. Figure 10.) Then the class $\{V_r | r > 0\}$ is a base for the uniformity $\mathcal{U}_d$, i.e., $\mathcal{U}_d$ includes the sets in $\{V_r | r > 0\}$ and all of their supersets. It follows immediately that $\mathcal{U}_d$ satisfies the conditions (U1)–(U5).

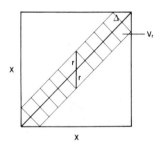

Fig. 10. Elements V_r for the base of the uniformity $\mathcal{U}_d$ on X.

Another way of defining a uniformity — which is independent of any metric notions — is illustrated in Figure 11: the base for the uniformity includes all such sets W associated with finite partitions $\{A_i\}_{i=1,\ldots,n}$ of X.

DISTANCE AND SIMILARITY

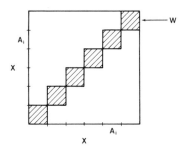

Fig. 11. Elements W for the base of a uniformity on X.

A non-empty family $\mathscr{F}$ of subsets of set X is a *filter* on X if $\emptyset \notin \mathscr{F}$ and $\mathscr{F}$ is closed under finite intersection and under supersets. If (X, τ) is a topological space (not necessarily metrizable), then a set $A \subseteq X$ is a neighbourhood of $x \in A$ if there is an open set $B \subseteq A$ such that $x \in B$. The neighbourhoods of a point $x \in X$ form a filter on X; it is called the *neighbourhood filter* of x. A filter $\mathscr{F}$ on X is said to *converge* to $x \in X$ if $\mathscr{F}$ includes the neighbourhood filter of x. Let $f: X \to X'$ be a mapping from one topological space X to another X'. Then f is said to be *continuous at point* $x_0 \in X$ if for each neighbourhood W of $f(x_0)$ there is a neighbourhood V of x_0 such that $f(V) \subseteq W$. Function f is *continuous* on X if f is continuous at each point $x \in X$. It follows that f is continuous if and only if the inverse image $f^{-1}(A)$ of every open set A in X' is open in X.

Let $(X, \mathscr{U})$ be a uniform space, and $V \in \mathscr{U}$. Elements x and y of X are said to be V-*close* if $(x, y) \in V$. Set $A \subseteq X$ is said to be V-*small* if $A \times A \subseteq V$. A filter $\mathscr{F}$ on X is a *Cauchy-filter* if $\mathscr{F}$ contains arbitrarily small elements, i.e., for each $V \in \mathscr{U}$ there is $A \in \mathscr{F}$ such that A is V-small. For example, let

$$V(x) = \{y \in X \mid (x, y) \in V\}$$

for $V \in \mathscr{U}$ and $x \in X$. Then the neighbourhood filter

$$\{V(x) \mid V \in \mathscr{U}\}$$

of x is a Cauchy-filter which is minimal with respect to set inclusion. Every convergent filter on X is a Cauchy-filter, but not conversely. The space $(X, \mathscr{U})$ is *complete* if every Cauchy-filter is convergent. Space

$(X, \mathcal{U})$ is *separating* or *Hausdorff* if

$$\bigcap_{U \in \mathcal{U}} U = \Delta.$$

Then in a separating uniform space $(X, \mathcal{U})$, each converging filter has one and only one limit. To each uniform space $(X, \mathcal{U})$, one can associate a complete separating uniform space $(\hat{X}, \hat{\mathcal{U}})$, called the *Hausdorff completion* of X, where $\hat{X}$ is the set of minimal Cauchy-filters on X.

Each uniform space $(X, \mathcal{U})$ can be given a topological structure $\tau_\mathcal{U}$ by letting $\{V(x) | V \in \mathcal{U}\}$ be a neighbourhood filter of x in the topology $\tau_\mathcal{U}$. A topological space (X, τ) is *uniformizable* if there is a uniformity $\mathcal{U}$ on X such that $\tau = \tau_\mathcal{U}$. A topological space X is uniformizable if and only if, for any point $x_0 \in X$ and any neighbourhood V of x_0, there is a continuous function $f: X \to [0, 1]$ such that

$$f(x_0) = 0$$
$$f(x) = 1 \quad \text{if} \quad x \notin V.$$

Every uniformity $\mathcal{U}$ on X can be obtained as a 'least upper bound' of the uniformities $\mathcal{U}_i$, $i \in I$, generated by a family of pseudometrics d_i, $i \in I$, where the base of $\mathcal{U}_i$ is the class of sets

$$\tilde{V}_r = \{(x, y) \in X \times X | d_i(x, y) \leq r\}$$

for $r > 0$ (cf. Figure 10). However, all uniform spaces are not metrizable. A topological space (X, τ) is a uniformizable Hausdorff space if and only if it can be embedded in a compact space. This means that there is a homeomorphism $f: X \to Y_0$ (i.e., a bijection such that f and f^{-1} are continuous) such that Y_0 is a subspace of a compact space Y.[8] Further, a uniform space $(X, \mathcal{U})$ is metrizable if and only if it is Hausdorff and $\mathcal{U}$ has a countable base.

1.3. DEGREES OF SIMILARITY

The notion of similarity has played an important role in many different fields. Many authors have suggested independently of each other — so it seems — that one can in some way or another measure 'degrees of similarity'. The similarities between these various theories of similarity have not generally been appreciated.

In traditional philosophical discussions, the concept of similarity was often associated with *analogy*. In particular, in the Thomist doctrine one can attribute properties to God only "according to analogy": God's wisdom is to God as Socrates' wisdom is to Socrates. Another standard example originates from Aristotle: "as sight is in the eye, so intellect is in the mind". The essential feature of this notion of analogy is the fact that it exhibits two relations (e.g., sight: eye, intellect: mind) which are similar to each other (cf. Bochenski, 1961, pp. 178—179). The abstract concept of *structural similarity* or *isomorphy* can be regarded as a generalization of this idea: two systems of objects A and B, structured by the relations $R \subseteq A \times A$ and $S \subseteq B \times B$, respectively, are *isomorphic* if there is a bijection $f: A \to B$ such that

$$\langle x, y \rangle \in R \quad \text{iff} \quad \langle f(x), f(y) \rangle \in S$$

for all x and y in A.

Inference by analogy (*argumentum a simile*) has traditionally been recognized as an important form of non-demonstrative inference. Many British philosophers — from Bishop Butler (1736), D. Hume (1748), and T. Reid (1785) to J. S. Mill (1843), S. Jevons (1874), and J. M. Keynes (1921) — discussed its nature by analysing, e.g., the problem whether the similarity between the planets Earth and Mars is sufficient to warrant the inference to the existence of life on Mars.[9] Its role in science — especially through the similarity of quantitative laws — was discussed by J. C. Maxwell in the 1850s (cf. Hesse, 1966). Analogical inference was also treated in several German textbooks of logic — from I. Kant (1800) and G. W. F. Hegel (1812—16) to M. W. Drobisch (1836), C. Sigwart (1873—78), W. Wundt (1880—83), B. Erdmann (1892), and J. von Kries (1916).[10] Typical formulations of arguments by analogy included

> M is P
> S is similar (ähnlich) to M
> ───────────────
> S is P

(Erdmann) and

> M has the property P
> S is like M in the properties $a, b, c, \ldots$
> ───────────────
> Therefore probably S has the property P

(Wundt).[11]

It seems that Mill (1843) was the first who tried to give a quantitative definition for the "probability derived from analogy". He noted that there will be "a competition between the known points of agreement and the known points of difference" in two objects a and b, and he essentially suggested to treat analogical inference as enumerative induction with respect to the properties of a and b:

> If, after much observation of b, we find that it agrees with a in nine out of ten of its known properties, we may conclude with a probability of nine to one, that it will possess any given derivative property of a.

(See Niiniluoto, 1981a, p. 8.) The same idea is expressed in C. Peirce's (1883) schema for 'hypothetic inference':

> M has, for example, the numerous marks P', P'', P''', etc. S has the proportion r of the marks P', P'', P''', etc. Hence, probably and approximately, S has r-likeness to M.

Thus, Peirce's notion of *r-likeness* is essentially a relative degree of similarity between two objects a and b: r is the ratio of the number of the shared properties of a and b to the number of all properties of b. A different definition for the degree of similarity (Grad der Ähnlichkeit) was given by T. Ziehen (1920): the ratio of the number of the shared properties of a and b to the number of non-shared properties of a and b (cf. Klug, 1966, p. 113). In the terms of Keynes (1921), Ziehen's definition equals the ratio between (the cardinalities of) the *positive analogy* and the *negative analogy* between a and b.

The study of the similarities between organisms or between their species is an indispensable tool in plant and animal *taxonomy*.[12] In *cladistic analysis*, inferences about the pathways of evolution are based upon 'homologies', i.e., striking similarities between species which are not merely similarities of function (e.g., the skulls of tortoise and armadillo) but are inherited from a common ancestor. It is assumed in cladistics that "the more homologies two animals share, the more recent is their common ancestor, and the more closely they are related".[13]

In *phenetics*, similarities between organisms are studied without reference to their evolution. Following the basic ideas of the French botanist Michel Adanson (1727–1806), the representatives of modern *numerical taxonomy* base their work on biological systematics upon the idea of overall phenetic similarity between organisms. The neo-Adansonian principles in taxonomy include the following: Classifications should be based on phenetic similarity; they are the better, the more characters they include. A priori, every character is of equal weight in creating natural taxa. Overall similarity between any two entities is a function of their individual similarities in each of the many characters in which they are being compared. (Sneath and Sokal, 1973, p. 5.) The earliest attempt to apply numerical measures in phenetic taxonomy was F. Heincke's method of distinguishing between races of herring in 1898. Later a great number of *coefficients of resemblance* or of *dissimilarity* have been introduced within different areas of biological systematics — from the classification of bacteria to the study of anthropology — the real break through dating from the late fifties (see Sneath and Sokal, 1973).

Let us denote taxonomic units (individuals or species) by a_j, $i = 1, \ldots, m$, let x_{ij} be the state of the character 0_i ($i = 1, \ldots, n$) for unit a_j. For two-state characters, the states may be represented by 0 and 1 (or $-$ and $+$), where

$x_{ij} = 1$ if a_j has the property 0_i
$\phantom{x_{ij}} = 0$ if a_j does not have the property 0_i.

Multistate characters may be either qualitative (e.g., x_{ij} may be the colour of a_j) or quantitative (e.g., x_{ij} may be the height of a_j). The matrix $\|x_{ij}\|_{i=1,\ldots,n; j=1,\ldots,m}$ is then the data matrix for the evaluation of the phenetic similarity or dissimilarity between the taxonomic units a_j ($j = 1, \ldots, m$). Usually it will be assumed that the values of x_{ij} are transformed by 'ranging' so that they vary between 0 and 1.[14]

Assume first that 0_i are two-state or quantitative multistate characters. Then the dissimilarity between two units a_j and a_k has been measured by

(35) $\quad \sum_{i=1}^{n} (x_{ij} - x_{ik})^2$

(F. Heincke's study of herrings in 1898); by

(36) $\quad \frac{1}{n} \sum_{i=1}^{n} |x_{ij} - x_{ik}|$

(the *mean character difference* used in J. Czekanowski's study of the

Neanderthal man in 1909 and T. Haltenorth's study of cats in 1937); by

$$(37) \quad \left(\sum_{i=1}^{n} (x_{ij} - x_{ik})^2 \right)^{\frac{1}{2}}$$

(R. Sokal's measure of *taxonomic distance* in 1961); by

$$(38) \quad \sum_{i=1}^{n} \left(\frac{|x_{ij} - x_{ik}|}{|x_{ij} + x_{ik}|} \right)$$

(the *Canberra metric* defined by G. N. Lance and W. T. Williams in 1967); and by

$$(39) \quad \left(\frac{1}{n} \sum_{i=1}^{n} \left(\frac{x_{ij} - x_{ik}}{x_{ij} + x_{ik}} \right)^2 \right)^{\frac{1}{2}}$$

(P. J. Clark's *coefficient of divergence* in 1952). It is seen immediately that these measures are variations of the city-block metric d_1 and the Euclidean metric d_2 defined by (5) and (6). More precisely, (37) is $d_2(a_j, a_k)$, (35) is $d_2(a_j, a_k)^2$, and (36) is $\frac{1}{n} d_1(a_j, a_k)$.

While the above definitions measure the dissimilarity of two objects a_j and a_k by their distance in the n-dimensional space, coefficients of similarity based on Karl Pearson's (1900) product-moment *correlation coefficient*

$$(40) \quad r_{jk} = \frac{\sum_{i=1}^{n} (x_{ij} - \bar{x}_j)(x_{ik} - \bar{x}_k)}{\sqrt{\sum_{i=1}^{n} (x_{ij} - \bar{x}_j)^2 \sum_{i=1}^{n} (x_{ik} - \bar{x}_k)^2}}$$

where

$$\bar{x}_j = \frac{1}{n} \sum_{i=1}^{n} x_{ij}; \; \bar{x}_k = \frac{1}{n} \sum_{i=1}^{n} x_{ik},$$

evaluate similarity by the (cosine of the) angle between the a_j and a_k as vectors from the origin. (40) is not a metric, and — unlike the distance measures — it is not invariant with respect to the movement of the origin.

If the taxonomic units are samples or 'clusters' of individuals, the distance between such sets J and K can be calculated, e.g., by the Mahalanobis distance D^2, defined by (34), or by applying Karl Pearson's (1926) *coefficient of racial likeness*

$$(41) \quad \left[\frac{1}{n}\sum_{i=1}^{n}\left(\frac{(\bar{x}_{iJ}-\bar{x}_{iK})^2}{(s_{iJ}^2/|J|)+(s_{iK}^2/|K|)}\right)\right]^{\frac{1}{2}} - \frac{2}{n}$$

where $\bar{x}_{iJ}$ and s_{iJ}^2 are the sample mean and variance for character i for sample J.

Assume then that 0_i $(i = 1, \ldots, n)$ are two-state qualitative characters.[15] The characters of individual a_j $(j = 1, \ldots, m)$ can be represented by a sequence

$$S(a_j) = (x_{1j}, \ldots, x_{nj})$$

or, alternatively, by the set of its positive characters

$$C(a_j) = \{0_i | x_{ij} = 1 \text{ and } i = 1, \ldots, n\}.$$

Suppose that units a_j and a_k agree on the presence of p characters (*positive matches*) and on the absence of q characters (*negative matches*), and disagree on the remaining $w = n - (p + q)$ characters. (See Figure 12.) In this case, there are $m = p + q$ *matches* and w *mismatches* between a_j and a_k. The similarity between a_j and a_k has been measured by

$$(42) \quad \frac{p}{p+w}$$

(used by P. Jaccard in 1908 for plant ecology); by

$$(43) \quad \frac{2p}{2p+w}$$

(used by L. R. Dice in 1945 as a measure for "ecologic association between species"); by

$$(44) \quad \frac{p+q}{p+q+w} = \frac{p+q}{n} = \frac{m}{m+w} = \frac{m}{n}$$

(the *simple matching coefficient* defined by R. Sokal and C. D.

Michener in 1958); and by

(45) $$\frac{m}{m+2w}$$

(the coefficient defined by D. J. Rogers and T. T. Tanimoto in 1960).

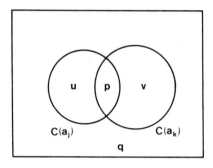

$u+v = w$ (mismatches)
$p+q = m$ (matches)
$m+w = n$

Fig. 12. Matches between units a_j and a_k.

Jaccard's and Dice's coefficients (42) and (43) differ from (44) in ignoring the number q of negative matches. If Peirce's and Ziehen's definitions are interpreted as restricted to positive characters only, then the notion of r-likeness would correspond to

$$\frac{p}{p+v}$$

and Ziehen's degree of similarity to

$$\frac{p}{w}.$$

But if they are taken to include negative matches as well, then the simple matching coefficient (44) is the same as Peirce's definition of r-likeness and Ziehen's degree of similarity corresponds to the ratio m/w. As $0 \leq m/n \leq 1$, we could measure the dissimilarity between a_j

and a_k by the complement of (44), i.e., by

$$(46) \quad 1 - \frac{m}{n} = \frac{w}{n}$$

which equals the mean character difference (36) for two-state characters. Moreover, (46) is the same as the normalized Hamming distance (19) between the sets $C(a_j)$ and $C(a_k)$. In other words, (46) is the same as $d_H(S(a_j), S(a_k))/n$, where d_H is the Hamming distance (22) between the binary sequences $S(a_j)$ and $S(a_k)$.

Similar ideas have been used in the field of *pattern recognition*. The basic problem there is to associate a pattern a (e.g., a picture or a scene) with one of the given prototypes $b_1, \ldots, b_k$; to do this, a suitable measure for the distance between a and b_i is needed.[16] In *cluster analysis*, the prototypes b_i are replaced by clusters, and a is classified into the cluster b_i such that the distance between a and b_i is minimized.[17] In K. S. Fu's (1982) syntactic approach, patterns are represented as sentences in formal tree languages, and then the recognition problem is solved by employing the metrics between trees (see Section 1 above). Alternatively, patterns may be represented by some suitably chosen features, and distances in feature spaces can be defined by the complements of measures like (42)–(44). In particular, in optical character recognition — e.g., in the automatic reading of (possibly distorted) letters or in automatic picture processing[18] — a figure is 'quantized' by dividing it into a great number of adjacent cells of equal size and by indicating its level of grey tone in each cell. This digital image can also be 'binarized' by changing the cells to white or black, depending on a given treshold value for the grey tone. The similarity between such images can then be evaluated by using, e.g., the divergence measure or the Hamming distance. The shape of the figure can be taken into account by giving special weight to certain cells or by requiring that there has to be enough matches within certain specified areas. (See Figure 13.)

The idea that essentially geometrical expressions like 'between', 'high', 'side', etc. could be used in the description of qualities was expressed already by J. F. Herbart in the 1820s, and it was later developed by physiologists and psychologists (Hermann von Helmholtz, Carl Stumpf) and by philosophers (Alexius Meinong, Edmund Husserl, Rudolf Carnap). These ideas can be illustrated by the notion of *colour space*. Most natural languages have names for six spectral colours

30   CHAPTER ONE

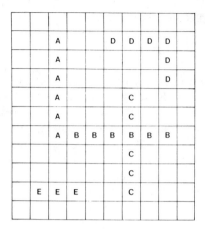

Fig. 13. A binary image is recognized as number 4 if at least five A-cells, at least five B-cells, and at least five C-cells are black, and at least four D-cells and all E-cells are white (Ullmann, 1976, p. 312).

which can be perceived in the spectrum: red, orange, yellow, green, blue, violet. A physical description shows, however, that these colours represent intervals in a continuum of light beams with different wavelengths (see Figure 14). Human eyes contain mechanisms which are sensitive to electromagnetic radiation with wavelength between 400—700 nm, but only a finite number of these colours can be visually discriminated from each other. Normally coloured light is not 'pure' or

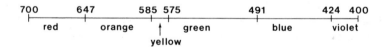

Fig. 14. The wavelengths of the spectral colours (expressed in nanometers).

'saturated', but rather a mixture of beams with different wavelengths. The surface of a physical thing is said to have a specific colour when it has the disposition to reflect certain kinds of beams when it is illuminated with white light (which is the mixture of all spectral colours). It is an established physical fact that all non-pure colours can be obtained as a superposition of three pure colours. As a consequence, all colours can be represented as a convex set in a three-dimensional space (cf. Gudder, 1977). The most well-known of such representations is the CIE system, developed by W. D. Wright and J. Guild for Commission Internationale de l'Éclairage in 1931. For a fixed level a of *brightness* (intensity), the set C_a of colours with brightness a is given in Figure 15. The extreme points in the curved arc from B to R

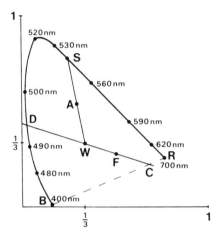

Fig. 15. Space C_a of colours with brightness a.

correspond to the pure spectral colours, and the straight line segment from B to R to different shades of purple. The point $(1/3, 1/3)$ is the white point W or W_a of C_a. The *hue* of colour A is given by the point S where the line drawn through W and A intersects the boundary of C_a. (For colour F, the hue is given for the complement colour D of C.) The *saturation* of A is defined as the ratio between the segments WA and WS. When the brightness a decreases towards zero, set C_a shrinks more and more until it becomes a single point. The white point W_a of the sets C_a, when $a \geq 0$, correspond to the achromatic colours (i.e., black, shades of grey, white). Thus, each physical, chromatic or achromatic colour with a particular brightness, hue, and saturation corresponds to a single point in the convex cone illustrate in Figure 16. Visual colours in the psychological sense correspond then to the elements of finite partitions of this cone. This suggests that the distances between colours can be defined by using metrics for the three-dimensional real space $\mathbb{R}^3$.

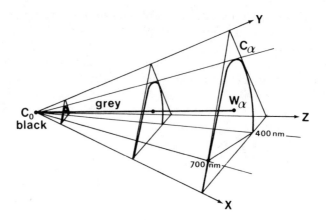

Fig. 16. Colour spaces C_a with increasing brightness $\alpha \geq 0$.

Ernst Mach argued in 1905 that "inferences from similarity and analogy are not strictly matters of logic, at least not of formal logic, but only of psychology" (Mach, 1976, p. 166). Rudolf Carnap attempted in *Der logische Aufbau der Welt* (1928) to give a phenomenalist reconstruction of the world on the basis of similarity between 'elementary experiences' (Elementarerlebnisse). This work was continued and refined by Nelson Goodman's (1951) theory of nominalistic systems and of the topological structure of qualities.[19] Psychologists, following the work of C. H. Coombs in 1952 and R. N. Shepard in 1962, have used mathematical and experimental techniques to study the 'psychological distance' of various kinds of stimuli (colour sensations, tones, words, figures, etc.). The most sophisticated work in this area has been done by Amos Tversky.[20]

Tversky's starting point is the method of modern measurement theory: ask a subject to make comparative judgments of dissimilarity and study whether these judgments can be represented by a function d which then measures degrees of dissimilarity. Function d is thus required to satisfy the condition

> x is at least as dissimilar to y as w is dissimilar to z iff $d(x, y) \geq d(w, z)$

for all stimuli x, y, w, and z. The same approach can be used in the construction of a measure sim of similarity between stimuli. One can then study the properties of functions d and sim — and also their mutual relation which is usually assumed to be of the form

(47) $\text{sim}(x, y) = 1 - d(x, y)$

when d is normalized between 0 and 1.[21]

In his earlier work, Tversky studied conditions that would guarantee the following property for the metric d between n-dimensional stimuli x and y:

(48) $d(x, y) = \varphi^{-1}\left[\sum_{i=1}^{n} \varphi(|x_i - y_i|)\right]$

where x_i and y_i are the perceived values of x and y on the ith dimension, respectively. The choice of $\varphi(z) = z^p$ ($p \geq 1$) in (48) yields the family (8) of the Minkowski metrics on $\mathbb{R}^n$. Another choice, $\varphi(z) = r^z - 1$ with $r > 1$, gives a family of *exponential metrics*

(49) $\dfrac{1}{\log r} \log\left[1 + \sum_{i=1}^{n} (r^{|x_i - y_i|} - 1)\right].$

Later Tversky (1977) has presented evidence against the metric assumptions (D3), (D4), and (D5) for the notion of dissimilarity. He points out that in many cases similarity judgments are *directional*: we say that "the son resembles the father" rather than "the father resembles the son", and "Turks fight like tigers" rather than "tigers fight like Turks". In a judgment of the form 'x is like y', usually the more 'salient stimulus' is chosen as the prototype or referent y to which the subject x is compared. Hence, the notion of similarity is *not always symmetric*.[22] He also argues that the triangle inequality should not be assumed to hold in all situations.

Tversky (1977) considers similarity as 'feature matching', i.e., each object in set X is represented by a set of features or attributes in a feature space F. This idea is familiar to us from Figure 12, so that we can use the same notation here. Thus, $C(a)$ is the set of the relevant properties of object a. Tversky makes the assumption that the ordinal relation of similarity satisfies the following conditon: the similarity between x and y depends on three factors: $C(x) \cap C(y)$, $C(x) - C(y)$, and $C(y) - C(x)$, and increases with the first of them and decreases with the second and the third. With further conditions of 'independence', 'solvability', and 'invariance', he proves the existence of a function sim: $X \to \mathbb{R}$ which preserves the ordinal similarity relations and a function $f: \mathcal{P}(X) \to \mathbb{R}$ (both unique up to a positive linear transformation) such that

$$(50) \quad \text{sim}(x, y) = \theta f(C(x) \cap C(y)) - \alpha f(C(x) - C(y)) - \beta f(C(y) - C(x))$$

for some real numbers $\theta, \alpha, \beta \geq 0$. This *contrast model* takes the similarity between x and y be a weighted difference of the common features of x and y and their distinctive features. Function sim(x, y) is symmetric if and only if $\alpha = \beta$. Function f may be taken to be additive: if $A \cap B = \emptyset$, then $f(A \cup B) = f(A) + f(B)$. As its special cases, we obtain

$$(51) \quad \text{sim}(x, y) = f(C(x) \cap C(y)), \text{ if } \theta = 1, \alpha = \beta = 0$$

$$(52) \quad \text{sim}(x, y) = -f(C(x) \triangle C(y)), \text{ if } \theta = 0, \alpha = \beta = 1,$$

both proposed by F. Restle in 1961. If f is the cardinality function (i.e., $f(A) = |A|$), then (51) is close to Jaccard's coefficient (42), and (52) is a variant of the simple matching coefficient (44).

Besides (50), Tversky mentions also the *ratio model*

$$(53) \quad \text{sim}(x, y) = \frac{f(C(x) \cap C(y))}{f(C(x) \cap C(y)) + \alpha f(C(x) - C(y)) + \beta f(C(y) - C(x))}$$

for $\alpha, \beta \geq 0$. Again sim(x, y) is symmetric if and only if $\alpha = \beta$. If f is the cardinality function, then — using the notation of Figure 12 — (53) reduces to

$$(54) \quad \frac{p}{p + \alpha u + \beta v}.$$

The choice $\alpha = \beta = 1$ gives Jaccard's measure (42) (also proposed by R. A. M. Gregson for psychometrics in 1975); the choice $\alpha = \beta = \frac{1}{2}$ gives Dice's coefficient (43) (also proposed for psychological similarity by H. Eisler and G. Ekman in 1959); the choice $\alpha = 1$ and $\beta = 0$ gives a proposal of R. R. Bush and F. Mosteller in 1951. Finally, the choice $\alpha = 0$ and $\beta = 1$ yields one interpretation of Peirce's notion of *r*-likeness.

Tversky also argues that the Equation (47) is not always valid. However, this is not very important if we are only interested in guaranteeing that the similarity and dissimilarity orders are compatible. For this purpose, it is sufficient that sim(x, y) is a strictly decreasing function of $d(x, y)$. In addition to (47), this condition is satisfied, e.g., by the function

$$(55) \quad \text{sim}(x, y) = \frac{1}{1 + d(x, y)}$$

which makes sim(x, y) to vary from 1 to 0 when $d(x, y)$ varies from 0 to ∞.

1.4. THE PRAGMATIC RELATIVITY OF SIMILARITY RELATIONS

According to *Leibniz's law* of the *identity of the indiscernibles*, two objects a and b are identical if and only if a and b share all of their properties. This suggests that a and b can be considered as 'partially identical' or 'similar' if they share some (but not all) of their properties. Most theories of similarity and dissimilarity that were surveyed in the preceding section are, indeed, based upon this idea.

What philosophical consequences does this approach to similarity have? Does it show that similarity is, perhaps in the same sense as identity, a logical notion? A good starting point for the discussion of these questions is *Goodman's seventh structure on similarity*:

Similarity cannot be equated with, or measured in terms of, possession of common characteristics. (Goodman, 1972, p. 443.)

Goodman's argument for this thesis runs in four steps. (i) If we say that two things are similar when they have *at least one property in common*, similarity would be a universal — and hence useless — relation, since every two things have some property in common. (ii) If two things are said to be similar when they have *all their properties in common*, similarity would be an empty — and hence useless — relation, since two different things cannot have all properties in common. (iii) We cannot say either that two things x and y are more alike than two others w and z if x and y have *more properties in common* than do w and z, since "any two things have exactly as many properties in common as any other two". More precisely, in a universe of n things, "each two things have in common exactly 2^{n-2} properties out of the total of 2^{n-1} properties"; in an infinite universe, each two things have an infinite number of common properties. Moreover, inclusion of higher-order properties will not change the argument. (iv) Finally, let us say that x and y are more alike than w and z if the *cumulative importance of the properties shared* by x and y is greater than that of the properties shared by w and z. But then we have to acknowledge that the "importance" of properties is "a highly volatile matter, varying with every shift of context and interest".

Goodman's argument is partly based on problematic assumptions about *properties*.[23] As a nominalist, Goodman does not accept the existence of properties — he replaces the talk about properties with talk about predicates (which are linguistic entities) or their extensions (which are classes of individuals). Then trivially two different individuals a and b share the 'property' that they both are elements of the set $\{a, b\}$ (cf. step (i) above). Similarly, a set of n elements has $2^n - 1$ nonempty subsets; 2^{n-2} of them contain both a and b; 2^{n-2} contain a but not b; 2^{n-2} contain b but not a; and $2^{n-2} - 1$ contain neither a nor b (cf. step (iii) above). It is clear that this view about properties would completely trivialize most measures of Section 1.3: for example, in Figure 12 we would always have $p = u = v$.

Goodman's step (iii) is generalized in S. Watanabe's *Theorem of the Ugly Duckling*: "An ugly duckling and a swan are just as 'similar' to each other as are two swans". More precisely, "the number of predicates shared by any arbitrary pair of objects is a constant, in so far as they are distinguishable". To see this, let λ be any set of m predicates, and let $\bar{\lambda}$ be the Boolean completion of λ, i.e., $\bar{\lambda}$ contains all predicates definable by conjunction, disjunction, and negation from the elements of λ. Then $\bar{\lambda}$ is a set of $2^{2^m} - 1$ elements with 2^m atoms[24], and any pair of distinct individuals will satisfy together $2^{2^m} - 2$ of the predicates in $\bar{\lambda}$. (Cf. Watanabe, 1969.)

An immediate objection to these arguments would be to point out that all predicates cannot be taken to define properties. This is shown, once and for all, by Russell's paradox in set theory. But this would not be a sufficient answer to the Theorem of the Ugly Duckling. The basic trouble with that theorem is the step from λ to $\bar{\lambda}$, i.e., the idea that similarities could be based on conjunctive, disjunctive, and negative properties.[25] A realist who is ready to accept some physical properties in his ontology (such as 'the property of being 5 meters long'), may with good reason suspect such entities as 'the property of being blue-or-non-round'. Even a nominalist has to admit that such complex predicates contain redundant information: if we already know that objects a and b

disagree on predicate 'M_1', then we also know that they disagree on predicate 'non-M_1' and disagree on predicate 'M_1-and-M_2'. More generally, the idea that similarity judgments should take into account *all* properties — whatever that may mean — is problematic. For example, if we are interested in the similarity between Citroen GS 2000 and Citroen CX 2200, it does not help us to know that neither of these cars is a bottle of wine — the latter piece of information, even if correct, is simply irrelevant for our purposes. In contrast, in comparing dinosaurs and crocodiles it is important to note that both are wingless: as 'having wings' is a relevant property for the similarity comparison, also the absence of this property is relevant. (Cf. the role of negative matches in formulas (44) and (45).)

It may be concluded that similarity judgments presuppose a selection of the *respects* λ which are taken to be *relevant* for the comparison — they are relative to a 'feature space' or a 'frame of reference' (Goodman). In evaluating the resemblance of two objects, only the characters in λ are relevant, not their Boolean combinations. The measures for degrees of similarity in Section 3 are in fact based upon this idea.

As there are no purely logical (i.e., syntactical or semantical) grounds for choosing the relevant features, we have to conclude that the notion of similarity is not a concept of pure logic. This conclusion is strengthened by the observation that "comparative judgments of similarity often require not merely selection of relevant properties but a weighting of their relative importance" (Goodman, 1972, p. 445). These *weights* are not needed in the rare special case, where the class of matches between a and b is set-theoretically included in the class of matches between a' and b'. But in other cases we either have to admit that some objects are incomparable in their similarity (cf. Keynes, 1921, p. 36) or to introduce weights on λ indicating the relative importance of the different respects of comparison. The message of the phenetic school in numerical taxonomy is to use equal weights on a large feature space — but even this is a choice of a system of weights. Most measures in the preceding section assume equal weights, but they can also be modified by introducing non-equal weightings. Moreover, most of these measures are liable to give different kinds of results when weightings are changed.

The metaphysical doctrine known as *essentialism* would give a way out to those who dislike the relativity of similarity judgments to weighting. The essentialists agree with the realists that there are properties, but they argue further that some properties are *intrinsically* more important than others. For example, in the Aristotelean essentialism it is assumed that the difference between man and other animals can be expressed by one essential property of man, viz. rationality. It has also been assumed that even individual things — not only species — have 'individual essences' (cf. Plantinga (1974) and Kripke (1981)). Thus, the essentialists would claim that there is a unique way of fixing the system of relevant properties and their weights. Against this view, *moderate realists* argue that the choice of such weights is a historically changing and context-dependent matter. This is the position that is taken in this book. At the same time, our account remains realistic: even though there is no framework λ which has a privileged status in the description of reality, the applicability of the predicates in λ to individual things in the world depends on the nature of these things (not conversely, as nominalism would claim).[26]

The conclusion of this section can be expressed by saying that the concept of similarity is *semantically unambiguous* but *pragmatically ambiguous*.[27] This means that degrees of similarity and comparisons of similarity are relative to two pragmatic boundary conditions: the choice of the relevant characters λ and the choice of the weights for the importance of these characters. When these conditions are fixed, each measure of similarity gives a uniquely defined concept which can be applied in an unambiguous way. Even though there are not purely logical grounds for these choices, in most contexts where the concept of similarity is needed there will be natural methods for making them.

CHAPTER 2

LOGICAL TOOLS

This chapter gives a compact summary of the basic tools from logic that will be needed later in our study. The reader is assumed to be familiar with elementary logic, but otherwise the presentation tries to be self-contained. The most central of the ideas to be introduced here include Carnap's notion of a monadic conceptual system, Hintikka's theory of distributive normal forms for first-order logic, and the Carnap–Hintikka style of approach to inductive logic.

2.1. MONADIC LANGUAGES L_N^k

Let L_N^k be a formal language with the logical vocabulary:

(i) connectives: & (conjunction)
∨ (disjunction)
~ (negation)
⊃ (implication)
≡ (equivalence)

(ii) parentheses: ((left)
) (right)

(iii) individual variables: $x_1, x_2, \ldots, x_i, \ldots$

(iv) existential quantifier: $(\exists x_i)$
universal quantifier: (x_i)

and with the non-logical vocabulary:

(v) individual constants: $a_1, \ldots, a_N$

(vi) one-place predicates: $M_1, \ldots, M_k$.

L_N^k is thus a *monadic predicate language* with N individual constants and k primitive one-place predicates. We assume that $0 \leq N \leq \omega$ and $0 < k < \omega$. In particular, language L_0^k does not have any individual constants, while L_ω^k has a denumerably infinite number of them.[1]

The *atomic formulas* of L_N^k are of the form $M_i(a_j)$ or $M_i(x_j)$. The (well-formed) *formulas* of L_N^k are finite strings of symbols defined as follows: atomic formulas are formulas; if α and β are formulas, so are $(\alpha \,\&\, \beta)$, $(\alpha \vee \beta)$, $\sim\alpha$, $(\alpha \supset \beta)$, $(\alpha \equiv \beta)$, $(\exists x_i)\alpha$ and $(x_i)\alpha$. A formula is *quantifier-free* if it does not contain any occurrences of quantifiers. A formula is *open* if it contains free variables. An arbitrary open formula with $x_1, \ldots, x_n$ as the free variables is denoted by $\alpha(x_1, \ldots, x_n)$. An open formula of the form $\alpha(x)$ is also called a *predicate expression*; it is said to be *pure* if it contains neither quantifiers nor individual constants. The result of substituting a term t (i.e., individual constant or individual variable) to all the free occurrences of variable x in formula $\alpha(x)$ is denoted by $\alpha(t/x)$. The *sentences* of L_N^k are those formulas which do not contain free occurrences of variables. Sentences without individual constants are called *generalizations*. If $\alpha(x)$ is a pure predicate expression, $\alpha(a_i/x)$ is called a *singular sentence*.

As a notational convention, the disjunction of formulas α_i, $i \in I$, is often written as

$$\bigvee_{i \in I} \alpha_i$$

and their conjunction as

$$\bigwedge_{i \in I} \alpha_i.$$

If $I = \{1, \ldots, n\}$, these notations can be written also as

$$\bigvee_{i=1}^{n} \alpha_i \quad \text{and} \quad \bigwedge_{i=1}^{n} \alpha_i.$$

We shall assume that a standard axiomatization of predicate logic is given for language L_N^k. If formula α is *deducible* from a class of formulas Σ (relative to the given axiomatization), we shall write $\Sigma \vdash \alpha$. If α is deducible from an empty set, i.e., if α is a theorem of monadic predicate logic, we write $\vdash \alpha$. Formulas α and β are *logically equivalent* if $\vdash \alpha \equiv \beta$, and α is *logically stronger* than β if $\vdash \alpha \supset \beta$. A formula α is *universal* (resp. *existential*) if it is logically equivalent to a formula of the form $(x)(y) \ldots (z)\beta$ (resp. $(\exists x)(\exists y) \ldots (\exists z)\beta$) where β is a quantifier-free formula.

Logical equivalence $\vdash \alpha \equiv \beta$ is an equivalence relation among the formulas of L_N^k. Let $\tilde{\alpha} = \{\beta \mid \vdash \alpha \equiv \beta\}$ be the equivalence class determined by formula α, and let $\mathscr{A}_L = \{\bar{\alpha} \mid \alpha \text{ is a sentence of } L_N^k\}$ be the set of the equivalence classes. Define

$$\bar{\alpha} \leq \bar{\beta} \quad \text{iff} \quad \vdash \alpha \supset \beta.$$

Then $(\mathscr{A}_L, \leq)$ is a Boolean algebra (cf. Example 1.7) with

$$\inf(\bar{\alpha}, \bar{\beta}) = \bar{\alpha} \wedge \bar{\beta} = \overline{\alpha \wedge \beta}$$
$$\sup(\bar{\alpha}, \bar{\beta}) = \bar{\alpha} \vee \bar{\beta} = \overline{\alpha \vee \beta}$$
$$\bar{\alpha}' = \overline{\sim \alpha}$$
$$1 = \overline{\alpha \vee \sim \alpha}$$
$$0 = \overline{\alpha \wedge \sim \alpha}.$$

$(\mathscr{A}_L, \leq)$ is called the *Lindenbaum algebra* of L_N^k (cf. Bell and Slomson, 1969, p. 40).

Let U be a non-empty set of objects, and let V be a function which maps the individual constants a_i to elements of U (i.e., $V(a_i) \in U$ for $i = 1, \ldots, N$) and the predicates M_j to subsets of U (i.e., $V(M_j) \subseteq U$ for $j = 1, \ldots, k$). Here V is called an *interpretation* for L_N^k, and $V(M_j)$ is the *extension* of predicate M_j in U. The $(N+k+1)$-tuple $\Omega = \langle U, V(a_1), \ldots, V(a_N), V(M_1), \ldots, V(M_k) \rangle$ is called a *structure* for L_N^k. Sometimes this structure is denoted simply by $\langle U, V \rangle$. Set U is called the *domain* of structure $\langle U, V \rangle$. Let $u = \langle u_i \mid i < \omega \rangle$ be an infinite sequence of elements u_i of U, and let $u(b/i)$ be the sequence obtained from u by replacing u_i with b. Following Tarski (1944), the *satisfaction* relation $\Omega \vDash_u \alpha$ ('sequence u satisfies formula α in structure Ω') is defined recursively as follows:

(1) (i) $\Omega \vDash_u M_j(a_i)$ iff $V(a_i) \in V(M_j)$
   (ii) $\Omega \vDash_u M_j(x_i)$ iff $u_i \in V(M_j)$
   (iii) $\Omega \vDash_u (\beta \,\&\, \gamma)$ iff $\Omega \vDash_u \beta$ and $\Omega \vDash_u \gamma$
   (iv) $\Omega \vDash_u (\beta \vee \gamma)$ iff $\Omega \vDash_u \beta$ or $\Omega \vDash_u \gamma$
   (v) $\Omega \vDash_u \sim \beta$ iff $\Omega \nvDash_u \beta$
   (vi) $\Omega \vDash_u (\beta \supset \gamma)$ iff $\Omega \nvDash_u \beta$ or $\Omega \vDash_u \gamma$
   (vii) $\Omega \vDash_u (\beta \equiv \gamma)$ iff $(\Omega \vDash_u \beta$ iff $\Omega \vDash_u \gamma)$
   (viii)$\Omega \vDash_u (\exists x_i)\beta$ iff $\Omega \vDash_{u(b/i)} \beta$ for some b in U
   (ix) $\Omega \vDash_u (x_i)\beta$ iff $\Omega \vDash_{u(b/i)} \beta$ for all b in U.

Here the notation $\Omega \nvDash_u \alpha$ means the same as not: $\Omega \vDash_u \alpha$. The notions of truth and falsity for sentences are then defined as follows: sentence α is *true* in structure Ω (write: $\Omega \vDash \alpha$) if some sequence u satisfies α in U; otherwise α is *false* in Ω (write: $\Omega \nvDash \alpha$).[2] Structure Ω is a *model* of sentence α if $\Omega \vDash \alpha$. Structure is a model of a set of sentences Σ (write: $\Omega \vDash \Sigma$) if $\Omega \vDash \alpha$ for all α in Σ.

Note that for sentences conditions (1)(i), (iii)–(vii) can be written without mentioning the sequence u. For example, if β and γ are sentences,

$$\Omega \vDash (\beta \& \gamma) \quad \text{iff} \quad \Omega \vDash \beta \text{ and } \Omega \vDash \gamma$$
$$\Omega \vDash (\beta \vee \gamma) \quad \text{iff} \quad \Omega \vDash \beta \text{ or } \Omega \vDash \gamma$$
$$\Omega \vDash {\sim}\beta \quad \text{iff} \quad \Omega \nvDash \beta.$$

In some applications it will be assumed that language L_N^k contains a name for each individual in U, i.e., $\{V(a_i) \mid i = 1, \ldots, N\} = U$. If this assumption holds, the truth conditions for generalizations can be written as follows:

$$\Omega \vDash (\exists x)\beta(x) \quad \text{iff} \quad \Omega \vDash \beta(a_i/x) \text{ for some } i = 1, \ldots, N$$
$$\Omega \vDash (x)\beta(x) \quad \text{iff} \quad \Omega \vDash \beta(a_i/x) \text{ for all } i = 1, \ldots, N.$$

Sentences α which are true in all structures Ω for L_N^k (i.e., for all domains U and interpretation functions V) are *valid* or *logically true*. An open formula $\alpha(x_1, \ldots, x_n)$ is valid if its universal closure $(x_1) \ldots (x_n)\alpha$ is valid. A sentence (or a set of sentences) is *inconsistent* or *logically false* if it has has no models, and *consistent* otherwise. It follows that α is inconsistent if and only if ${\sim}\alpha$ is logically true. Consistent sentences which are not logically true are called *factual*.

Monadic predicate logic is *complete*, i.e., a formula α is valid if and only if $\vdash \alpha$, and *decidable*, i.e., there is an effective method of determining of an arbitrary formula α whether it is valid or not.

In applying the formal language L_N^k the predicates M_i are assumed to designate certain *properties* of individuals. Such properties are also called *attributes, intensions,* or *concepts*. In the Carnap–Montague approach, the property F_M designated by predicate M is defined as a function which associates with each possible domain of objects U the extension of predicate M in U.[3] Intuitively, to know completely the 'meaning' of a predicate M is to know how M should be applied in all possible situations.

Two properties F_M and $F_{M'}$ *exclude* each other if $F_M(U) \cap F_{M'}(U) = \emptyset$ for all domains U. Property $F_{M'}$ *includes* property F_M if $F_{M'}(U) \subseteq F_M(U)$ for all U. For example, the properties designated by 'red' and 'blue' exclude each other, and the property designated by 'bachelor' includes that designated by 'man'. Sentences of L_N^k which express such conceptual connections are said to be *analytically true*.[4] For example, if F_M and $F_{M'}$ exclude each other, the sentence $(x) \sim (M(x)\ \&\ M'(x))$ is analytically true; if $F_{M'}$ includes F_M, the sentence $(x)(M'(x) \supset M(x))$ is analytically true. If some analytically true sentences are associated with language L_N^k, they are called *meaning postulates* for L_N^k (see Carnap, 1962, 1971). The class of all meaning postulates is denoted by $\mathscr{MP}$. It may be empty, and it is always assumed to be consistent. In a language L_N^k with meaning postulates $\mathscr{MP}$, the notation $\Sigma \vdash \alpha$ should be understood as a shorthand for $\Sigma \cup \mathscr{MP} \vdash \alpha$. In particular, $\vdash \alpha$ means the same as $\mathscr{MP} \vdash \alpha$. Moreover, the notion of validity should be defined as truth in all the structures for L_N^k which are models of $\mathscr{MP}$.

2.2. Q-PREDICATES

Conjunctions of the form

(2) $\quad (\pm)M_1(x)\ \&\ \cdots\ \&\ (\pm)M_k(x)$

where $(\pm)$ is either empty or the negation sign $\sim$, define the *Q-predicates* $Q_1(x), \ldots, Q_q(x)$ of L_N^k. The Q-predicates are mutually exclusive and jointly exhaustive:

(3) $\quad \vdash \sim (Q_i(x)\ \&\ Q_j(x))$, if $i \neq j$

$\qquad \vdash \bigvee_{i=1}^{q} Q_i(x)$.

In other words, the extensions $V(Q_i)$ of the predicates Q_i $(i = 1, \ldots, q)$ on a domain U constitute a complete q-fold *classification system* for the individuals of U:

(4) $\quad V(Q_i) \cap V(Q_j) = \emptyset$, if $i \neq j$

$\qquad V(Q_1) \cup \cdots \cup V(Q_q) = U$.

Here $V(Q_i)$ is called the *cell* determined by Q_i. These conditions together state that each element of U satisfies one and only one Q-predicate.

If the primitive predicates $M_1, \ldots, M_k$ are logically independent, the number of Q-predicates in L_N^k is $q = 2^k$. However, if some connections between these predicates are accepted as analytically true, then those Q-predicates are disregarded which are incompatible with the set $\mathscr{MP}$ of meaning postulates. For example, if sentence $(x)(M_1(x) \supset \sim M_2(x))$ is in $\mathscr{MP}$, then Q-predicates of the form $M_1(x)$ & $M_2(x)$ & $(\pm)M_3(x)$ & $\cdots$ & $(\pm)M_k(x)$ are excluded. Hence, in this case the number q of Q-predicates is less than 2^k.

Let $a(x)$ be a pure predicate expression, i.e., $a(x)$ is built up by connectives from atomic formulas containing the variable x. Then $a(x)$ can be expressed in a *Q-normal form*, i.e., as a finite disjunction of Q-predicates. In other words, there is a class of indices $I_a \subseteq \{1, \ldots, q\}$ such that

(5) $\quad \vdash a(x) \equiv \bigvee_{i \in I_a} Q_i(x).$

The simplest way of proving (5) is based on the following idea: if we write a truth-table to show how the truth-value of sentence $a(a/x)$ depends on the truth-values of the atomic sentences $M_1(a), \ldots, M_k(a)$, precisely the lines where $a(a/x)$ receives the value 'true' correspond to the Q-predicates in the normal form (5). (Cf. Carnap, 1962, pp. 93–94, 126.) Indeed, the Q-normal form for predicate expressions can be regarded as an analogue in a 'logic of properties' to the perfect disjunctive normal form in propositional logic (cf. von Wright, 1957, p. 33).

The cardinality $|I_a|$ of I_a is called the *width* of the predicate expression $a(x)$. Further, $|I_a|/q$ is the *relative width* of $a(x)$. Hence, in particular, the width of an inconsistent formula $a(x)$ is 0; the width of each Q-predicate is 1 (since $I_{Q_i} = \{i\}$); the relative width of each Q_i is $1/q$; and, assuming $\mathscr{MP} = \emptyset$, the width of each primitive predicate M_i is $q/2$, so that the relative width of each M_i is $1/2$.

For many applications it is useful to group the primitive predicates into mutually exclusive and jointly exhaustive *families* (see Carnap, 1971b, 1980).[5] For example, cars can be classified with respect to their colour, year of production, weight, efficiency of the motor, number of cylinders, etc. Each of these criteria corresponds to an *attribute space* A which is partitioned into a countable number of *regions* $\{X_1, X_2, \ldots\}$.

LOGICAL TOOLS 45

For example, the colour space can be partitioned e.g. into six regions:

$$\{\text{red, orange, yellow, green, blue, violet}\}.$$

(Cf. Chapter 1.3.) The attribute space corresponding to the number of cylinders is some initial segment of the natural numbers, e.g.,

$$\{1, 2, 3, \ldots, 50\}.$$

The space corresponding to weight is a subset of real numbers divided into intervals, e.g.

$$\{[0, 1), [1, 2), \ldots, [9\,999\,999, 10\,000\,000]\}$$

where the numbers express the weight measured in grams.

A finite number of attribute spaces, together with a countable partition of each space, constitutes a *monadic conceptual system*.[6] A monadic predicate language for such a system has to contain a predicate for each region in each attribute space. This language is finite if all the partitions of the attribute spaces are finite. Let

(6) $\quad \mathcal{M}_1 = \{M_1^1, \ldots, M_{n_1}^1\}$

$\quad\quad\ \mathcal{M}_2 = \{M_1^2, \ldots, M_{n_2}^2\}$

$\quad\quad\ \vdots$

$\quad\quad\ \mathcal{M}_k = \{M_1^k, \ldots, M_{n_k}^k\}$

be a finite collection of finite families of predicates M_j^i such that the following meaning postulates hold for all $i = 1, \ldots, k$:

(7) $\quad \bigvee_{j=1}^{n_i} M_j^i(x)$

$\quad\quad \sim(M_j^i(x)\ \&\ M_m^i(x)), \quad \text{if } j \neq m.$

In other words, within each family $\mathcal{M}_i$, the predicates M_j^i constitute a n_i-fold classification system. The Q-predicates of this language are defined by conjunctions of the form

(8) $\quad M_{j_1}^1(x)\ \&\ M_{j_2}^2(x)\ \&\ \cdots\ \&\ M_{j_k}^k(x)$

where $1 \leq j_1 \leq n_1, \ldots, 1 \leq j_k \leq n_k$. The Q-predicate (8) will be denoted by $Q_{j_1 j_2 \ldots j_k}(x)$. The total number q of Q-predicates is $n_1 \cdot n_2 \cdot \cdots \cdot n_k$, if (7) are the only meaning postulates in $\mathcal{MP}$.

As a special case of (6), each family $\mathcal{M}_i$ may be a dichotomy, i.e.,

(9) $\quad \mathcal{M}_1 = \{M_1, \sim M_1\}$
$\quad \mathcal{M}_2 = \{M_2, \sim M_2\}$
$\quad \vdots$
$\quad \mathcal{M}_k = \{M_k, \sim M_k\}.$

Then the definition (8) reduces to the earlier definition (2) of Q-predicates. Moreover, in the case (9) the number $n_1 \cdot n_2 \cdot \ldots \cdot n_k$ equals $2 \cdot 2 \cdot \ldots \cdot 2$ (k times) or 2^k.

If the attribute spaces $A_1, \ldots, A_k$ corresponding to the families $\mathcal{M}_1, \ldots, \mathcal{M}_k$ have a topological and metric structure, the regions X^i_j designated by predicates M^i_j may have unequal widths. When A_i is the space of values of a one-dimensional quantity, the width $w_i(X^i_j) = w^i_j$ of region X^i_j is simply the length of the interval X^i_j. If the colour space is divided into five (rather than six) regions

{red, orange-or-yellow, green, blue, violet},

the second region is wider than the others. More generally, to each attribute space A_i there corresponds a *width function* w_i defined on subsets of A_i. If $w_i(A_i)$ is finite, the values of w_i can be normalized by dividing them by the number $w_i(A_i)$. If $X^1_{j_1} \times X^2_{j_2} \times \cdots \times X^k_{j_k}$ is a region in the product space $A_1 \times A_2 \times \cdots \times A_k$, then its relative width can be defined as the product

(10) $\quad \displaystyle\prod_{i=1}^{k} (w_i(X^i_{j_i})/w_i(A_i)).$

The function w_i determines also the widths of the predicates M^i_j in the ith family $\mathcal{M}_i$, and formula (10) gives the relative width of the Q-predicate $Q_{j_1 j_2 \ldots j_k}$. If the families $\mathcal{M}_i$ are defined by (9) and if the relative width of each M_i within $\mathcal{M}_i$ is $1/2$, then the relative width of a Q-predicate is $1/2^k$ — which equals our earlier definition.

Carnap made also a further assumption which will prove to be very important later in our study. Suppose that for each family $\mathcal{M}_i$ there is a distance function $d_i : \mathcal{M}_i \to \mathbb{R}$ which measures the distance $d_i(M^i_j, M^i_m)$ between any pair of predicates M^i_j and M^i_m in $\mathcal{M}_i$. Then the distance between two Q-predicates $M^1_{j_1}(x) \& \cdots \& M^k_{j_k}(x)$ and

$M_{m_1}^1(x) \& \cdots \& M_{m_k}^1(x)$ can be defined by the Euclidean formula

$$d(M_{j_1}^1(x) \& \cdots \& M_{j_k}^k(x), M_{m_1}^1(x) \& \cdots \& M_{m_k}^k(x))$$

$$= \left(\sum_{i=1}^{k} d_i(M_{j_i}^i(x), M_{m_i}^i(x))^2 \right)^{\frac{1}{2}}.$$

It follows that $(\mathbf{Q}, d)$, where $\mathbf{Q}$ is the set of the Q-predicates, is a metric space (cf. Chapter 1.1). This idea will be discussed in more detail in Chapter 3.2.

2.3. STATE DESCRIPTIONS

It follows from the representation (5) that the Q-predicates are the strongest pure predicate expressions in language L_N^k. Therefore, the most complete description of an individual a_j that can be given in L_N^k is to tell which Q-predicate a_j satisfies. If this information is given for all individuals $a_1, \ldots, a_N$, we obtain a *state description* of L_N^k. In other words, state descriptions of L_N^k (where $N < \omega$) are sentences of the form

(11) $Q_{i_1}(a_1) \& \cdots \& Q_{i_N}(a_N)$.

Sentence (11) is thus a conjunction of kN atomic sentences or negations of atomic sentences. It will be denoted by $s(i_1, \ldots, i_N)$. For example, $s(1, 1, \ldots, 1)$ is the state description $Q_1(a_1) \& Q_1(a_2) \& \cdots \& Q_1(a_N)$. The set of all state descriptions of L_N^k is denoted by $\mathscr{Z}_N^k$ or simply by $\mathscr{Z}$. The number of different state descriptions in $\mathscr{Z}_N^k$ is q^N; if L_N^k does not have meaning postulates, q^N equals $(2^k)^N$. By the definitions (2) and (8) of Q-predicates, it is immediately clear that different state descriptions in $\mathscr{Z}$ are incompatible with each other and the disjunction of all state descriptions in $\mathscr{Z}$ is logically true:

(12) $\vdash \sim(s \& s')$ if $s \in \mathscr{Z}, s' \in \mathscr{Z}, s \neq s'$

   $\vdash \bigvee_{s \in \mathscr{Z}} s.$

Definition (11) for state descriptions presupposes that the number N is finite, since otherwise the conjunction (11) is infinitely long and cannot be expressed as a sentence of L_ω^k. Therefore, the state descriptions of language L_ω^k have to be defined as infinite *sets of sentences* of L_ω^k, i.e., as sets of the form

(13) $\{Q_{i_j}(a_j) \mid j < \omega\}$.

The number of state descriptions of L_ω^k is q^ω which equals 2^ω. If we want to express the state description (13) by a sentence, we have to extend L_ω^k to an *infinitary monadic language* $^{\text{inf}}L_\omega^k$, obtained from L_ω^k by allowing denumerably infinite conjunctions and disjunctions[7] — and, in particular, sentences of the form

(14) $\bigwedge_{j<\omega} Q_{i_j}(a_j)$.

The motivation for the notion of state description is based upon the idea that there is a fixed universe U which language L_N^k speaks about, and the state descriptions of L_N^k give alternative complete specifications of the 'state' of U. However, if there are individuals in U which cannot be referred to in L_N^k, the state description of L_N^k which is true in U may be misleading: for example, it may be the case that $M(a_i)$ holds for all $i = 1, \ldots, N$ but still there is some individual in U which satisfies the formula $\sim M(x)$. To avoid such situations it should be assumed that L_N^k *contains a name for each element of* U, i.e., the interpretation V maps the individual constants $\{a_i \mid i = 1, \ldots, N\}$ bijectively (i.e., one-to-one and onto) the domain U:

(15) $V(a_i) \neq V(a_j)$ if $i \neq j$
$\{V(a_i) \mid i = 1, \ldots, N\} = U$.

In this case, the cardinality $|U|$ of U must be N ($0 < N \leq \omega$). When (15) holds, language L_N^k is said to *fit* the domain U, and the true state description s_* of L_N^k is essentially the same as the model-theoretical *diagram* of the structure $\Omega = \langle U, V(a_1), \ldots, V(a_N), V(M_1), \ldots, V(M_k)\rangle$. It follows that the models of sentence s_* are precisely those structures Ω' which contain a substructure isomorphic with Ω (cf. Monk, 1976, p. 329). In discussing state descriptions in the sequel, we shall assume that L_N^k fits the universe U.

The *range* $R(\alpha)$ of a sentence α of L_N^k is the set of those state descriptions s in $\mathscr{Z}$ which entail α, i.e.,

(16) $R(\alpha) = \{s \in \mathscr{Z} \mid s \vdash \alpha\}$.

LOGICAL TOOLS 49

Some properties of this notion are listed in the following theorem;

(17) (a) $R(s) = \{s\}$ if $s \in \mathscr{Z}$
  (b) $R(\sim\alpha) = \mathscr{Z} - R(\alpha)$
  (c) $R(\alpha \wedge \beta) = R(\alpha) \cap R(\beta)$
  (d) $R(\alpha \vee \beta) = R(\alpha) \cup R(\alpha)$
  (e) $R(\alpha) = \mathscr{Z}$ iff $\vdash \alpha$
  (f) $R(\alpha) = \emptyset$ iff $\vdash \sim\alpha$.

Here (a) follows from (12) and from the consistency of each state description. One half of (b) follows again from the consistency of state descriptions: if $s \vdash \sim\alpha$, then $s \nvdash \alpha$, since otherwise $s \vdash \sim\alpha \ \& \ \alpha$. The claim

(18) If $s \nvdash \alpha$, then $s \vdash \sim\alpha$

can be proved by induction on the length of sentence α. If α is an atomic sentence of the form $M_j(a_i)$, then by definitions (2) and (11) s must contain as a conjunct either $M_j(a_i)$ or $\sim M_j(a_i)$. Since $s \nvdash M_j(a_i)$ the latter alternative must hold, so that $s \vdash \sim M_j(a_i)$. If α is of the form $\sim\beta$, where β satisfies (18), then $s \nvdash \sim\beta$ implies by the inductive hypothesis that $s \nvdash \beta$ does not hold, i.e., $s \vdash \beta$ holds. But this is equivalent to the fact that $s \vdash \sim\sim\beta$ or $s \vdash \sim\alpha$. If α is of the form $\beta \ \& \ \gamma$, where β and γ satisfy (18), then $s \nvdash \beta \ \& \ \gamma$ implies $s \nvdash \beta$ or $s \nvdash \gamma$. Hence $s \vdash \sim\beta$ or $s \vdash \sim\gamma$ which implies $s \vdash \sim\beta \vee \sim\gamma$ or $s \vdash \sim(\beta \ \& \ \gamma)$. If α is of the form $\exists x \beta(x)$, then $s \nvdash \exists x \beta(x)$ implies $s \nvdash \beta(a_i/x)$ for all individual constants a_i. Hence, by the inductive hypothesis, $s \vdash \sim\beta(a_i/x)$ for all a_i, so that $s \vdash (x) \sim \beta(x)$ or $s \vdash \sim(\exists x)\beta(x)$. This concludes the proof of (18). The rest of the proof for claims (c)—(f) is now easy.

As $s \vdash \alpha$ for all $s \in R(\alpha)$, sentence α is entailed also by the disjunction of all such state description s in $R(\alpha)$. On the other hand, α entails this disjunction, since by (12)

$$\vdash \sim \bigvee_{s \in R(\alpha)} s \equiv \bigvee_{s \notin R(\alpha)} s$$

and by (17b)

$$\vdash \bigvee_{s \notin R(\alpha)} s \equiv \bigvee_{s \in R(\sim\alpha)} s$$

and

$$\vdash \bigvee_{s \in R(\sim a)} s \supset \sim a.$$

We have thus shown that each sentence a of L_N^k can be represented as a disjunction of the state descriptions within its range $R(a)$:

(19) $$\vdash a \equiv \bigvee_{s \in R(a)} s.$$

The disjunction in (19) is called the $\mathscr{Z}$-*normal form* of a. This representation shows that the state descriptions of L_N^k ($N < \omega$) are the strongest factual sentences expressible in L_N^k. A similar theorem can be proved for the generalized state descriptions of the form (14) in language $^{\inf}L_\omega^k$.

2.4. STRUCTURE DESCRIPTIONS

Two state descriptions of the form (11) or (13) are *isomorphic* if they can be obtained from each other by permuting individual constants. In other words, state descriptions $s(i_1, \ldots, i_N)$ and $s(j_1, \ldots, j_N)$ are isomorphic if sequence $\langle j_1, \ldots, j_N \rangle$ contains the same numbers as $\langle i_1, \ldots, i_N \rangle$ but possibly in a different order. Isomorphism is an equivalence relation in the set $\mathscr{Z}$ of state descriptions. A disjunction of state descriptions containing precisely all the members of an equivalence class of the isomorphism relation is called a *structure description* of L_N^k. Hence, sentence a is (equivalent to) a structure description if the range $R(a)$ of a coincides with an equivalence class of the isomorphism relation.

For example, in language L_4^2 with four Q-predicates the state descriptions

$$s_1 = Q_1(a_1) \,\&\, Q_1(a_2) \,\&\, Q_2(a_3) \,\&\, Q_4(a_4)$$
$$s_2 = Q_2(a_1) \,\&\, Q_4(a_2) \,\&\, Q_1(a_3) \,\&\, Q_1(a_4)$$
$$s_3 = Q_4(a_1) \,\&\, Q_2(a_2) \,\&\, Q_1(a_3) \,\&\, Q_1(a_4)$$

are isomorphic with each other. (See Figure 1.)

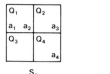

Fig. 1. Isomorphic state descriptions.

A common feature of these state descriptions is that they place two individuals in the cell (determined by) Q_1, one in Q_2, and one in Q_4.

More generally, assume that a structure description S corresponds to the equivalence class determined by a state description s of L_N^k such that s places N_i individuals in cell Q_i ($i = 1, \ldots, q$), where $N_1 + \cdots + N_q = N$. Then S is denoted by $S(N_1, \ldots, N_q)$. If language L_N^k fits the domain U of structure Ω, then

(20) $\Omega \vDash S(N_1, \ldots, N_q)$ iff $|V(Q_i)| = N_i$, $i = 1, \ldots, q$.

The total number of structure descriptions of L_N^k is

$$\binom{N + q - 1}{q - 1} = \frac{(N + q - 1)!}{N!(q - 1)!}.$$

The number of state description in the range of $S(N_1, \ldots, N_q)$ is

$$\frac{N!}{N_1! \ldots N_q!}.$$

(Cf. Carnap, 1962, pp. 138–140.)

In language L_ω^k, structure descriptions are (possibly infinite) sets of infinite sets of sentences. Again they correspond to sequences $\langle N_1, \ldots, N_q \rangle$ of numbers, where $0 \leq N_i \leq \omega$ for $i = 1, \ldots, q$, and $N_1 + \cdots + N_q = \omega$. Hence, the total number of state descriptions of L_ω^k is $\omega^{q-1} = \omega$. The width of $S(N_1, \ldots, N_q)$ is 1 if one of the N_i's is ω and the others 0; ω if one of the N_i's is ω and at least one of the others is positive and finite; 2^ω if at least two of the N_i's is ω. (Cf. Carnap, 1962, p. 140.)

2.5. MONADIC CONSTITUENTS

While the structure descriptions of L_N^k essentially tell *how many* individuals there are in the cells Q_i ($i = 1, \ldots, q$), and the state descriptions tell *which* individuals belong to the cells Q_i, the constituents indicate which cells Q_i are *empty* and which cells *non-empty*. In other words, *constituents* of L_N^k are generalizations of the form

(21) $\quad (\pm)(\exists x)Q_1(x) \ \& \ \cdots \ \& \ (\pm)(\exists x)Q_q(x).$

If $Q_{i_1}, \ldots, Q_{i_w}$ (and only these) occur in (21) without a negation sign, constituent (21) can be rewritten in the equivalent form

(22) $\quad (\exists x)Q_{i_1}(x) \ \& \ \cdots \ \& \ (\exists x)Q_{i_w}(x) \ \& \ (x)(Q_{i_1}(x) \lor \cdots \lor Q_{i_w}(x)).$

Constituent (22) is often denoted by C^w. The number w is called the *width* of C^w.

The total number t of constituents in L_N^k ($0 \leqslant N \leqslant \omega$) is $2^q - 1$; hence, it is independent of the number N. The constituent which claims that all Q-predicates have an empty extension is here excluded from consideration; otherwise the number of constituents in L_N^k would be 2^q. This means that constituents can be defined even for language L_0^k which does not have any state descriptions. The constituents of L_N^k are denoted by $C_1, \ldots, C_t$. If $\mathbf{CT}_i$ is the set of those Q-predicates that occur positively in constituent C_i, then C_i can be written in the following form, corresponding to (22) above:

(23) $\quad \bigwedge_{j \in \mathbf{CT}_i} (\exists x)Q_j(x) \ \& \ (x)\left[\bigvee_{j \in \mathbf{CT}_i} Q_j(x)\right].$

Hence, the cardinality $|\mathbf{CT}_i|$ of $\mathbf{CT}_i$ is the width of constituent C_i. If $N > 0$, a constituent C^w of width w is equivalent to a disjunction of $\binom{N-1}{w-1}$ structure descriptions and to a disjunction of

$$\sum_{i=0}^{w} (-1)^i \binom{w}{i} (w-i)^N$$

state descriptions.

The constituents of L_N^k are mutually exclusive and jointly exhaustive:

(24) $\quad \vdash \sim(C_i \ \& \ C_j), \quad \text{if } i \neq j$

$\qquad \vdash \bigvee_{i=1}^{t} C_i$

Moreover, each consistent generalization α in L_N^k is equivalent to a finite disjunction of constituents, i.e., there is a non-empty set of indices $J_\alpha \subseteq \{1, \ldots, t\}$ such that

(25) $\quad \vdash \alpha \equiv \bigvee_{i \in J_\alpha} C_i.$

The disjunction in (25) is called the *distributive normal form* of α (cf. von Wright, 1957, pp. 19, 33—42)[8], and the cardinality $|J_\alpha|$ of J_α is the *breadth* of α. The breadth of each constituent is thus 1, and the breadth of a logical truth is t.

In order to transform a generalization to its distributive normal form, one should first eliminate nested quantifiers by the following rules: if x is not free in formula α, then

(26) $\quad \vdash (\exists x)(\alpha \vee \beta) \equiv (\alpha \vee (\exists x)\beta)$
$\vdash (x)(\alpha \vee \beta) \equiv (\alpha \vee (x)\beta)$
$\vdash (\exists x)(\alpha \& \beta) \equiv (\alpha \& (\exists x)\beta)$
$\vdash (x)(\alpha \& \beta) \equiv (\alpha \& (x)\beta)$
$\vdash (\exists x)(\alpha \supset \beta) \equiv (\alpha \supset (\exists x)\beta)$
$\vdash (x)(\alpha \supset \beta) \equiv (\alpha \supset (x)\beta)$
$\vdash (\exists x)(\beta \supset \alpha) \equiv ((x)\beta \supset \alpha)$
$\vdash (x)(\beta \supset \alpha) \equiv ((\exists x)\beta \supset \alpha).$

Secondly, universal quantifiers are eliminated by the rule

(27) $\quad \vdash (x)\alpha \equiv {\sim}(\exists x){\sim}\alpha.$

Thirdly, in all subformulas of the form $(\exists x)\alpha(x)$ the predicate expression $\alpha(x)$ is replaced by its Q-normal form (5). Fourthly, existential quantifiers are distributed over disjunctions by the rule

(28) $\quad \vdash (\exists x)\left[\bigvee_{i \in I_\alpha} Q_i(x) \right] \equiv \bigvee_{i \in I_\alpha} (\exists x) Q_i(x).$

The resulting formula is built up by propositional connectives from formulas of the form $(\exists x)Q_i(x)$. Fifthly, a truth table for this formula is written showing how its truth value depends on the truth values of $(\exists x)Q_i(x)$, $i = 1, \ldots, q$. Then the rows where the value is 'truth' correspond directly to the constituents of the form (21) which belong to the distributive normal form of the given generalization.

The disjunction G^w of all constituents C^w of the same given width w ($w = 1, \ldots, q$) is called a *constituent-structure*.[9] Thus, sentence G^w tells that precisely w cells Q_i are exemplified in the universe — without telling which ones they are. The total number of constituent-structures G^w in language L_N^k is q.

EXAMPLE 1. Language L_3^1 with one primitive predicate M, three individual constants a_1, a_2, a_3, and no meaning postulates.
Q-predicates:

$$Q_1(x) = M(x)$$
$$Q_2(x) = {\sim}M(x)$$

State descriptions:

$$\begin{aligned}
s_1 &= M(a_1) \,\&\, M(a_2) \,\&\, M(a_3) & &= s(1, 1, 1) \\
s_2 &= M(a_1) \,\&\, M(a_2) \,\&\, {\sim}M(a_3) & &= s(1, 1, 2) \\
s_3 &= M(a_1) \,\&\, {\sim}M(a_2) \,\&\, M(a_3) & &= s(1, 2, 1) \\
s_4 &= M(a_1) \,\&\, {\sim}M(a_2) \,\&\, {\sim}M(a_3) & &= s(1, 2, 2) \\
s_5 &= {\sim}M(a_1) \,\&\, M(a_2) \,\&\, M(a_3) & &= s(2, 1, 1) \\
s_6 &= {\sim}M(a_1) \,\&\, M(a_2) \,\&\, {\sim}M(a_3) & &= s(2, 1, 2) \\
s_7 &= {\sim}M(a_1) \,\&\, {\sim}M(a_2) \,\&\, M(a_3) & &= s(2, 2, 1) \\
s_8 &= {\sim}M(a_1) \,\&\, {\sim}M(a_2) \,\&\, {\sim}M(a_3) & &= s(2, 2, 2)
\end{aligned}$$

Structure descriptions:

$$\begin{aligned}
S_1 &= S(3, 0) = s_1 \\
S_2 &= S(2, 1) = s_2 \lor s_3 \lor s_5 \\
S_3 &= S(1, 2) = s_4 \lor s_6 \lor s_7 \\
S_4 &= S(0, 3) = s_8
\end{aligned}$$

Constituents:

$$\begin{aligned}
C_1 &= (\exists x)M(x) \,\&\, (x)M(x) = S_1 \\
C_2 &= (\exists x){\sim}M(x) \,\&\, (x){\sim}M(x) = S_4 \\
C_3 &= (\exists x)M(x) \,\&\, (\exists x){\sim}M(x) = S_2 \lor S_3 \\
[C_4 &= {\sim}(\exists x)M(x) \,\&\, {\sim}(\exists x){\sim}M(x)]
\end{aligned}$$

Constituent-structures:

$$G^1 = C_1 \vee C_2$$
$$G^2 = C_3.$$

The relations between these sentences are illustrated in Figure 2.

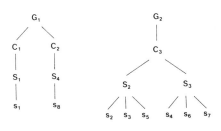

Fig. 2.

EXAMPLE 2. Language L_2^2 with two primitive predicates M_1 and M_2, two individual constants a_1 and a_2, and no meaning postulates.

Q-predicates:

$$Q_1(x) = M_1(x) \ \& \ M_2(x)$$
$$Q_2(x) = M_1(x) \ \& \ {\sim}M_2(x)$$
$$Q_3(x) = {\sim}M_1(x) \ \& \ M_2(x)$$
$$Q_4(x) = {\sim}M_1(x) \ \& \ {\sim}M_2(x)$$

State descriptions:

$$s_1 = Q_1(a_1) \ \& \ Q_1(a_2) = s(1, 1)$$
$$s_2 = Q_1(a_1) \ \& \ Q_2(a_2) = s(1, 2)$$
$$s_3 = Q_1(a_1) \ \& \ Q_3(a_2) = s(1, 3)$$
$$s_4 = Q_1(a_1) \ \& \ Q_4(a_2) = s(1, 4)$$
$$s_5 = Q_2(a_1) \ \& \ Q_1(a_2) = s(2, 1)$$
$$s_6 = Q_2(a_1) \ \& \ Q_2(a_2) = s(2, 2)$$
$$s_7 = Q_2(a_1) \ \& \ Q_3(a_2) = s(2, 3)$$
$$s_8 = Q_2(a_1) \ \& \ Q_4(a_2) = s(2, 4)$$
$$s_9 = Q_3(a_1) \ \& \ Q_1(a_2) = s(3, 1)$$

$s_{10} = Q_3(a_1)$ & $Q_2(a_2) = s(3, 2)$
$s_{11} = Q_3(a_1)$ & $Q_3(a_2) = s(3, 3)$
$s_{12} = Q_3(a_1)$ & $Q_4(a_2) = s(3, 4)$
$s_{13} = Q_4(a_1)$ & $Q_1(a_2) = s(4, 1)$
$s_{14} = Q_4(a_1)$ & $Q_2(a_2) = s(4, 2)$
$s_{15} = Q_4(a_1)$ & $Q_3(a_2) = s(4, 3)$
$s_{16} = Q_4(a_1)$ & $Q_4(a_2) = s(4, 4)$

Structure descriptions:

$S_1 = S(2, 0, 0, 0) = s_1$
$S_2 = S(1, 1, 0, 0) = s_2 \lor s_5$
$S_3 = S(1, 0, 1, 0) = s_3 \lor s_9$
$S_4 = S(1, 0, 0, 1) = s_4 \lor s_{13}$
$S_5 = S(0, 2, 0, 0) = s_6$
$S_6 = S(0, 1, 1, 0) = s_7 \lor s_{10}$
$S_7 = S(0, 1, 0, 1) = s_8 \lor s_{14}$
$S_8 = S(0, 0, 2, 0) = s_{11}$
$S_9 = S(0, 0, 1, 1) = s_{12} \lor s_{15}$
$S_{10} = S(0, 0, 0, 2) = s_{16}$

Constituents:

$C_1 = (\exists x)Q_1(x)$ & $(x)Q_1(x)$
$C_2 = (\exists x)Q_2(x)$ & $(x)Q_2(x)$
$C_3 = (\exists x)Q_3(x)$ & $(x)Q_3(x)$
$C_4 = (\exists x)Q_4(x)$ & $(x)Q_4(x)$
$C_5 = (\exists x)Q_1(x)$ & $(\exists x)Q_2(x)$ & $(x)(Q_1(x) \lor Q_2(x))$
$C_6 = (\exists x)Q_1(x)$ & $(\exists x)Q_3(x)$ & $(x)(Q_1(x) \lor Q_3(x))$
$C_7 = (\exists x)Q_1(x)$ & $(\exists x)Q_4(x)$ & $(x)(Q_1(x) \lor Q_4(x))$
$C_8 = (\exists x)Q_2(x)$ & $(\exists x)Q_3(x)$ & $(x)(Q_2(x) \lor Q_3(x))$
$C_9 = (\exists x)Q_2(x)$ & $(\exists x)Q_4(x)$ & $(x)(Q_2(x) \lor Q_4(x))$
$C_{10} = (\exists x)Q_3(x)$ & $(\exists x)Q_4(x)$ & $(x)(Q_3(x) \lor Q_4(x))$

LOGICAL TOOLS

$C_{11} = (\exists x)Q_1(x)\ \&\ (\exists x)Q_2(x)\ \&\ (\exists x)Q_3(x)\ \&$
$\qquad (x)(Q_1(x)\ \vee\ Q_2(x)\ \vee\ Q_3(x))$
$C_{12} = (\exists x)Q_1(x)\ \&\ (\exists x)Q_2(x)\ \&\ (\exists x)Q_4(x)\ \&$
$\qquad (x)(Q_1(x)\ \vee\ Q_2(x)\ \vee\ Q_4(x))$
$C_{13} = (\exists x)Q_1(x)\ \&\ (\exists x)Q_3(x)\ \&\ (\exists x)Q_4(x)\ \&$
$\qquad (x)(Q_1(x)\ \vee\ Q_3(x)\ \vee\ Q_4(x))$
$C_{14} = (\exists x)Q_2(x)\ \&\ (\exists x)Q_3(x)\ \&\ (\exists x)Q_4(x)\ \&$
$\qquad (x)(Q_2(x)\ \vee\ Q_3(x)\ \vee\ Q_4(x))$
$C_{15} = (\exists x)Q_1(x)\ \&\ (\exists x)Q_2(x)\ \&\ (\exists x)Q_3(x)\ \&\ (\exists x)Q_4(x)$
$[C_{16} = \sim(\exists x)Q_1(x)\ \&\ \sim(\exists x)Q_2(x)\ \&\ \sim(\exists x)Q_3(x)\ \&\ \sim(\exists x)Q_4(x)]$

Constituent-structures:

$$G^1 = C_1 \vee C_2 \vee C_3 \vee C_4$$
$$G^2 = C_5 \vee C_6 \vee C_7 \vee C_8 \vee C_9 \vee C_{10}$$
$$G^3 = C_{11} \vee C_{12} \vee C_{13} \vee C_{14}$$
$$G^4 = C_{15}$$

Examples of Q-normal forms:

$$\vdash M_1(x) \equiv (Q_1(x) \vee Q_2(x))$$
$$\vdash M_2(x) \equiv (Q_1(x) \vee Q_3(x))$$
$$\vdash (M_1(x) \supset M_2(x)) \equiv (Q_1(x) \vee Q_3(x) \vee Q_4(x))$$

Examples of $\mathcal{Z}$-normal forms:

$$\vdash M_1(a_1) \equiv (s_1 \vee s_2 \vee s_3 \vee s_4 \vee s_5 \vee s_6 \vee s_7 \vee s_8)$$
$$\vdash \sim(M_1(a_2) \supset M_2(a_2)) \equiv (s_2 \vee s_6 \vee s_{10} \vee s_{14})$$

Examples of distributive normal forms:

$\vdash (x) \sim M_1(x) \equiv (C_3 \vee C_4 \vee C_{10})$
$\vdash (\exists x)M_2(x) \equiv (C_1 \vee C_3 \vee C_5 \vee C_6 \vee C_7 \vee C_8 \vee C_{10} \vee$
$\qquad \vee C_{11} \vee C_{12} \vee C_{13} \vee C_{14} \vee C_{15})$
$\vdash (x)(M_1(x) \supset M_2(x)) \equiv \sim(\exists x)Q_2(x)$
$\qquad \equiv (C_1 \vee C_3 \vee C_4 \vee C_6 \vee C_7 \vee C_{10} \vee C_{13}).$

2.6. MONADIC LANGUAGES WITH IDENTITY

We have so far assumed that language L_N^k does not contain the sign $=$ for identity. Let $L_N^{k=}$ be the *monadic predicate language with identity*, obtained from L_N^k by joining $=$ to the alphabet, allowing atomic formulas of the form $t_1 = t_2$ (for any terms t_1 and t_2), adding the standard axioms for $=$, and by supplementing the definition (1) of satisfaction with the following clauses:

(29) $\quad \Omega \vDash_u a_i = a_j \quad$ iff $\quad V(a_i)$ is the same element of U as $V(a_j)$

$\quad\quad\;\; \Omega \vDash_u x_i = x_j \quad$ iff $\quad u_i$ is the same element of U as u_j.

The expressive power of $L_N^{k=}$ is greater than that of L_N^k, since the sign of identity makes it possible to define *numerical quantifiers* $\exists^{\geqslant n}$ ('there are at least n individuals such that') and $\exists^n$ ('there are precisely n individuals such that'):

(30) $\quad (\exists^{\geqslant n} x)\alpha(x) =_{df} (\exists x_1) \ldots (\exists x_n)(x_1 \neq x_2 \;\&\; x_1 \neq x_3 \;\&\; \cdots$
$\quad\quad\quad \cdots \;\&\; x_1 \neq x_n \;\&\; x_2 \neq x_3 \;\&\; \cdots$
$\quad\quad\quad \cdots \;\&\; x_{n-1} \neq x_n \;\&\; \alpha(x_1) \;\&\; \cdots$
$\quad\quad\quad \cdots \;\&\; \alpha(x_n))$

$\quad (\exists^n x)\alpha(x) =_{df} (\exists x_1) \ldots (\exists x_n)(x_1 \neq x_2 \;\&\; \cdots$
$\quad\quad\quad \cdots \;\&\; x_{n-1} \neq x_n \;\&\; \alpha(x_1) \;\&\; \cdots$
$\quad\quad\quad \cdots \;\&\; \alpha(x_n) \;\&\; (x)(\alpha(x) \supset x = x_1 \vee \cdots$
$\quad\quad\quad \cdots \vee x = x_n)),$

where $1 \leqslant n < \omega$. Further, we may define $\exists^0$ ('there are no individuals such that') by

(31) $\quad (\exists^0 x)\alpha(x) =_{df} \sim(\exists x)\alpha(x).$

Definitions (30) can be expressed in a shorter way as follows:

$$(\exists x_1) \ldots (\exists x_n) \left(\bigwedge_{\substack{i\;j \\ (i \neq j)}} x_i \neq x_j \;\&\; \bigwedge_{i=1}^n \alpha(x_i) \right)$$

$$(\exists x_1) \ldots (\exists x_n) \times$$

$$\times \left(\bigwedge_{\substack{i\;j \\ (i \neq j)}} x_i \neq x_j \;\&\; \bigwedge_{i=1}^n \alpha(x_i) \;\&\; (x)\left(\alpha(x) \supset \bigvee_{i=1}^n x = x_i \right) \right)$$

In particular, $\exists^{\geq 1}$ is the same as the ordinary existential quantifier $\exists$, and $\exists^1$ is the same as $\exists!$ ('there is one and only one individual such that'). On the other hand, the quantifiers $\exists^{\geq \omega}$ ('there are at least infinitely many individuals such that') and $\exists^\omega$ ('there are precisely ω individuals such that') cannot be expressed by sentences of $L_N^{k=}$.[10]

It is possible to express within $L_N^{k=}$ the assumption (15) that language L_N^k fits the finite domain U:

$$(32) \quad \bigwedge_{\substack{i=1 \\ (i \neq j)}}^{N} \bigwedge_{j=1}^{N} a_i \neq a_j$$

$$(x)\left(\bigvee_{i=1}^{N} x = a_i \right).$$

The sentences (32) may be taken as meaning postulates of $L_N^{k=}$ whenever $L_N^{k=}$ has been chosen so as to fit a domain U with $|U| = N < \omega$. Moreover, in language $L_0^{k=}$ (without individual constants) it is possible to express assumptions about the size of the universe, e.g. the sentence

$$(\exists^N x) x = x$$

is true in U if and only if $|U| = N$.

The notions of Q-predicate, state description, and structure description for language $L_N^{k=}$ can be defined in the same way as for L_N^k. However, in this case a constituent can tell of each Q-predicate Q_i not only whether it is empty or not but also *how many* individuals instantiate it.[11] To be more precise, note that at least d nested quantifiers are needed to express claims of the form $(\exists^{\geq n} x) Q_i(x)$, for $n = 1, \ldots, d$, or $(\exists^n x) Q_i(x)$, for $n = 0, \ldots, d-1$. Let us call the maximal number of nested quantifiers in a formula its quantificational *depth*. Then constituents of $L_N^{k=}$ *at the depth* d are sentences of the form

$$(33) \quad \bigwedge_{i=1}^{q} (D_i x) Q_i(x),$$

where $D_i x$ ($i = 1, \ldots, q$) is one of the following quantifiers

$$\exists^0 x, \exists^1 x, \ldots, \exists^{d-1}, \exists^{\geq d}.$$

Hence, constituents at the depth 1 are equivalent to ordinary monadic constituents, since they contain only quantifiers $\exists^0$ (or $\sim\exists$) and $\exists^{\geqslant 1}$ (or $\exists$) (cf. definition (21)). In the general case, to each constituent at the depth d there corresponds uniquely a sequence of numbers $\langle c_1, \ldots, c_q \rangle$, where $0 \leqslant c_i \leqslant d$ for $i = 1, \ldots, q$.

If $L_N^{k=}$ does not have any meaning postulates, then the total number of constituents at the depth d is $(d + 1)^q$. This is again independent of N, and therefore holds also for languages $L_0^{k=}$ and $L_\omega^{k=}$. On the other hand, if (32) has been accepted as a meaning postulate for $L_N^{k=}$, or (33) for $L_0^{k=}$, then all claims of the form $(\exists^{\geqslant d})Q_i(x)$ are inconsistent if $d > N$. Moreover, a constituent at the depth d is consistent if and only if it corresponds to a sequence $\langle c_1, \ldots, c_q \rangle$ with $c_1 + \cdots + c_q = N$. Hence, in these cases the constituents of $L_N^{k=}$ or $L_0^{k=}$ can be defined as the constituents at the depth N which satisfy the condition that $c_1 + \cdots + c_q = N$. But these constituents essentially are nothing else than the structure descriptions for language L_N^k. The only novelty here is that it is possible to express *without individual constants* in $L_0^{k=}$ sentences which have the same truth conditions as the structure descriptions of L_N^k.

If no prior limits are given to the size of the universe, then the constituents of language $L_0^{k=}$ remain consistent for arbitrarily great depths d — and in general these constituents are not reducible to any sentences of L_N^k. Moreover, if $C_i^{(d)}$ is a depth-d constituent, it is equivalent to a finite disjunction of depth-$(d+1)$ constituents $C_j^{(d+1)}$. Namely if $C_i^{(d)}$ claims that there are at least d individuals in cell Q_m, then the deeper constituent $C_j^{(d+1)}$ may claim either that $(\exists^d x)Q_m(x)$ or $(\exists^{\geqslant d+1} x)Q_m(x)$. Hence, if the sequence $\langle c_1, \ldots, c_q \rangle$ corresponding to $C_i^{(d)}$ contains r occurrences of the number d, then $C_i^{(d)}$ will be split into a disjunction of 2^r constituents at the depth $d+1$. Repeating this argument, we see that each depth-d constituent $C_i^{(d)}$ can be expressed as a finite disjunction of depth-$(d+e)$ constituents $C_j^{(d+e)}$, for any $e = 1, 2, \ldots$. If α is a generalization in $L_0^{k=}$ with the quantificational depth d, then α has a distributive normal form at the depth d:

$$(34) \quad \vdash \alpha \equiv \bigvee_{i \in \mathbf{J}_\alpha^{(d)}} C_i^{(d)}$$

(cf. (25)), and also a normal form at the depth $d + e$ ($e = 1, 2, \ldots$):

$$(35) \quad \vdash \alpha \equiv \bigvee_{i \in \mathbf{J}_\alpha^{(d+e)}} C_i^{(d+e)}.$$

(35) is called the *expansion* of α to the depth $d + e$.

2.7. POLYADIC CONSTITUENTS

Let us now assume that L is a first-order language whose non-logical vocabulary λ contains at least one polyadic (i.e., non-monadic) predicate. The logical vocabulary of L contains besides connectives, parentheses, individual variables, and quantifiers (cf. Section 1) also one additional symbol: , (comma). The non-logical vocabulary λ contains besides individual variables $a_1, \ldots, a_N$ ($0 \leq N \leq \omega$) and one-place predicates $M_1^1, \ldots, M_{k_1}^1$ ($0 \leq k_1 < \omega$) also n-place predicates $M_1^n, \ldots, M_{k_n}^n$ ($0 \leq k_n < \omega$), for $n = 2, 3, \ldots$, where at least one of the numbers k_n is greater than zero. Language L is then called a *polyadic first-order language of type* $\langle k_1, k_2, \ldots \rangle$.

The atomic formulas of L are of the form $M_i^n(t_1, \ldots, t_n)$, where M_i^n is a n-place predicate and $t_1, \ldots, t_n$ are terms (i.e., individual variables or constants). Other syntactical notions, including the notion of deduction, are defined as in the case of monadic languages.

An interpretation V for language L maps individual constants a_j to elements of U and n-place predicates M_i^n to n-place relations in U, i.e., $V(M_i^n) \subseteq U^n$ for all $i = 1, \ldots, k_n$. A domain U together with the interpretations of the elements of λ, i.e.,

$$\langle U, V(a_1), \ldots, V(a_N), V(M_1^1), \ldots, V(M_{k_1}^1), \ldots$$
$$\ldots, V(M_1^n), \ldots, V(M_{k_n}^n), \ldots \rangle$$

is called an *L-structure*. Again this structure may be denoted by $\langle U, V \rangle$. The value $t[u]$ of a term t for an infinite sequence $u = \langle n_i \mid i < \omega \rangle$ of elements of U is $V(a_j)$ if t is a_j and u_j if t is x_j. Then the satisfaction condition for atomic formulas can be written as follows:

(36) $\Omega \vDash_u M_i^n(t_1, \ldots, t_n)$ iff $\langle t_1[u], \ldots, t_n[u] \rangle \in V(M_i^n)$.

When conditions (2) are supplemented with (36), we obtain a definition of satisfaction for the formulas of L, and a definition of truth and falsity for the sentences of L.

Language L is *complete* in the sense that a formula α of L is a theorem if and only if it is valid. However, according to Church's Theorem, L is *not decidable*, since there is no effective method for telling whether an arbitrary formula α of L is a theorem or not.

It was shown by Jaakko Hintikka in 1953 that each generalization in a polyadic language L can be expressed as a finite disjunction of constituents, if the notion of a constituent is relativized to a finite quantificational depth. The *depth* of a formula α is defined as the maximal number of nested and connected quantifiers in α, where two quantifiers, $(\exists x)$ and $(\exists y)$ say, are *nested* if one of them occurs within the scope of the other, and *connected* if there are, within the scope of the outermost of these quantifiers, quantifiers with variables $z_1, \ldots, z_m$ such that x and z_1, z_i and z_{i+1} ($i = 1, \ldots, m$), and z_m and y occur in the same atomic formula in α.[12] For example, the depth of

$$(\exists x)(y)(M_2^1(x) \;\&\; M_1^2(y, z))$$

is one, and the depth of

$$(x)(M_2^1(x) \supset (\exists y)M_3^2(x, y))$$

is two.

To define the constituents of L, let us first generalize the definition of a Q-predicate. Let $S(\lambda, x_1, \ldots, x_n)$ be the set of all atomic formulas which can be formulated by using the predicates in λ and by the variables $x_1, \ldots, x_n$. Conjunctions of the form

$$(37) \qquad \bigwedge_{s \in S(\lambda, x_1, \ldots, x_n)} (\pm)s$$

are denoted by $Ct^{(0)}(\lambda, x_1, \ldots, x_n)$ or simply by $Ct^{(0)}(x_1, \ldots, x_n)$ with suitable subscripts. Formulas $Ct^{(0)}(\lambda, x_1, \ldots, x_n)$ will be called *attributive constituents* of the depth 0 and of the *level* n. Two different formulas of the form (37) are incompatible with each other, and the disjunction of all such formulas is logically true:

$$\vdash \sim(Ct_i^{(0)}(x_1, \ldots, x_n) \;\&\; Ct_j^{(0)}(x_1, \ldots, x_n)) \quad (i \neq j)$$

$$\vdash \bigvee_i Ct_i^{(0)}(x_1, \ldots, x_n).$$

If λ contains k monadic predicates $M_1, \ldots, M_k$ and $n = 1$, then

$$S(\{M_1, \ldots, M_k\}, x) = \{M_1(x), \ldots, M_k(x)\}$$

and (37) reduces to the definition (2) of Q-predicates. If λ contains one monadic predicate M_1^1 and one two-place predicate M_1^2, then

$$S(\{M_1^1, M_1^2\}, x_1, x_2) = \{M_1^1(x_1), M_1^1(x_2), M_1^2(x_1, x_1), M_1^2(x_1, x_2),$$
$$M_1^2(x_2, x_1), M_1^2(x_2, x_2)\}.$$

In this case, the following conjunction

$$M_1^1(x_1) \,\&\, \sim M_1^1(x_2) \,\&\, M_1^2(x_1, x_1) \,\&\, \sim M_1^2(x_1, x_2) \,\&$$
$$\&\, \sim M_1^2(x_2, x_1) \,\&\, M_1^2(x_2, x_2)$$

is an example of an attributive constituent of the form (37). As a generalization of the Q-normal form (5), it follows that each quantifier-free pure predicate expression of the form $\alpha(x_1, \ldots, x_n)$ in L can be expressed as a finite disjunction of formulas of the form $Ct^{(0)}(x_1, \ldots, x_n)$, i.e.,

$$(38) \quad \vdash \alpha(x_1, \ldots, x_n) \equiv \bigvee_{i \in I_\alpha} Ct_i^{(0)}(x_1, \ldots, x_n).$$

If language L contains N individual constants $a_1, \ldots, a_N$, then the *state descriptions* of L are sentences of the form $Ct^{(0)}(a_1, \ldots, a_N)$. They express the properties and the interrelations of all individuals $a_1, \ldots, a_N$ with the resources of the predicates in λ and without quantifiers. If L is of type $\langle k_1, k_2 \rangle$ and does not contain meaning postulates, then the total number of state descriptions in L is

$$(2^{k_1 + k_2})^N \cdot (2^{k_2})^{N(N-1)}.$$

It follows from (38) that every singular sentence of L can be expressed as a disjunction of state descriptions.

The *structure descriptions* of L can be defined in the same way as in monadic languages. If L is of the type $\langle k_1, k_2 \rangle$, a structure description of L tells essentially how many individuals belong to each predicate of the form

$$(\pm)M_1^1(x) \,\&\, \cdots \,\&\, (\pm)M_{k_1}^1(x) \,\&\, (\pm)M_1^2(x, x) \,\&\, \cdots \,\&\, (\pm)M_{k_2}^2(x, x)$$

and how many pairs of two individuals satisfy each formula of the form

$$(\pm)M_1^2(x, y) \,\&\, \cdots \,\&\, (\pm)M_{k_2}^2(x, y).$$

Constituents $C_j^{(1)}$ of the depth 1 are sentences of the form

$$(39) \quad \bigwedge_{i \in CT_j^{(0)}} (\exists x_1) Ct_i^{(0)}(x_1) \,\&\, (x_1) \left[\bigvee_{i \in CT_j^{(0)}} Ct_i^{(0)}(x_1) \right].$$

Their structure thus coincides with the definition (23) of monadic constituents, but in this case the attributive constituents $Ct_i^{(0)}(x_1)$ of level

1 contain as conjuncts not only atomic formulas of the form $M_i^1(x_1)$ or their negations but also formulas like $M_i^2(x_1, x_1)$ or $\sim M_2^3(x_1, x_1, x_1)$ which involve the polyadic predicates in λ. In other words, (39) tells what kinds of individuals there exist in the universe, when these individuals are characterized by their properties and by their relations to themselves.

Attributive constituents $Ct_i^{(1)}(x_1, \ldots, x_{n-1})$ of the depth 1 and of the level $n-1$ are sentences of the form

$$(40) \quad \bigwedge_{k \in \mathbf{CT}_i^{(0)}} (\exists x_n) Ct_k^{(0)}(x_1, \ldots, x_{n-1}, x_n) \, \&$$

$$\& \, (x_n) \left[\bigvee_{k \in \mathbf{CT}_i^{(0)}} Ct_k^{(0)}(x_1, \ldots, x_{n-1}, x_n) \right].$$

Constituents $C_j^{(2)}$ of the depth 2 are sentences telling which attributive constituents $Ct_i^{(1)}(x_1)$ of the depth 1 and of the level 1 exist in the universe:

$$(41) \quad \bigwedge_{i \in \mathbf{CT}_j^{(1)}} (\exists x_1) Ct_i^{(1)}(x_1) \, \& \, (x_1) \left[\bigvee_{i \in \mathbf{CT}_j^{(1)}} Ct_i^{(1)}(x_1) \right].$$

When definitions (40) and (41) are combined, we see that the constituent $C_j^{(2)}$ has the following structure:

$$(42) \quad \bigwedge_{i \in \mathbf{CT}_j^{(1)}} (\exists x_1) \left[\bigwedge_{k \in \mathbf{CT}_i^{(0)}} (\exists x_2) Ct_k^{(0)}(x_1, x_2) \, \& \right.$$

$$\& \, (x_2) \left(\bigvee_{k \in \mathbf{CT}_i^{(0)}} Ct_k^{(0)}(x_1, x_2) \right) \right] \, \&$$

$$\& \, (x_1) \left[\bigvee_{i \in \mathbf{CT}_j^{(1)}} \left[\bigwedge_{k \in \mathbf{CT}_i^{(0)}} (\exists x_2) Ct_k^{(0)}(x_1, x_2) \, \& \right. \right.$$

$$\& \, (x_2) \left(\bigvee_{k \in \mathbf{CT}_i^{(0)}} Ct_k^{(0)}(x_1, x_2) \right) \right] \right].$$

A constituent $C_j^{(2)}$ of the depth 2 is thus a systematic description of all

LOGICAL TOOLS 65

the different kinds of pairs of individuals that one can find in the universe.

EXAMPLE 3. Let U be the set of natural numbers 0, 1, 2, ..., and let $M(x, y)$ state that $x \leq y$. The interpretation of M in U is thus a reflexive relation. Assume that M is the only predicate in language L, i.e., $\lambda = \{M\}$. Then the depth-1 constituent of L which is true in U is simply

$$(\exists x_1)M(x_1, x_1) \ \& \ (x_1)M(x_1, x_1),$$

and the depth-2 constituent true in U is

(43) $(\exists x_1)\{(\exists x_2)(M(x_1, x_1) \ \& \ M(x_2, x_2) \ \& \ M(x_1, x_2) \ \& \ {\sim}M(x_2, x_1))$
$\& \ (\exists x_2)(M(x_1, x_1) \ \& \ M(x_2, x_2) \ \& \ M(x_1, x_2) \ \& \ M(x_2, x_1))$
$\& \ (\exists x_2)(M(x_1, x_1) \ \& \ M(x_2, x_2) \ \& \ {\sim}M(x_1, x_2) \ \& \ M(x_2, x_1))$
$\& \ (x_2)[(M(x_1, x_1) \ \& \ M(x_2, x_2) \ \& \ M(x_1, x_2) \ \& \ {\sim}M(x_2, x_1))$
$\lor \ (M(x_1, x_1) \ \& \ M(x_2, x_2) \ \& \ M(x_1, x_2) \ \& \ M(x_2, x_1))$
$\lor \ (M(x_1, x_1) \ \& \ M(x_2, x_2) \ \& \ {\sim}M(x_1, x_2) \ \& \ M(x_2, x_1))]\}$
$\& \ (\exists x_1)\{(\exists x_2)(M(x_1, x_1) \ \& \ M(x_2, x_2) \ \& \ M(x_1, x_2) \ \& \ {\sim}M(x_2, x_1))$
$\& \ (\exists x_2)(M(x_1, x_1) \ \& \ M(x_2, x_2) \ \& \ M(x_1, x_2) \ \& \ M(x_2, x_1))$
$\& \ (x_2)[(M(x_1, x_1) \ \& \ M(x_2, x_2) \ \& \ M(x_1, x_2) \ \& \ {\sim}M(x_2, x_1))$
$\lor \ (M(x_1, x_1) \ \& \ M(x_2, x_2) \ \& \ M(x_1, x_2) \ \& \ M(x_2, x_1))]\}$
$\& \ (x_1)\{\{(\exists x_2)(M(x_1, x_1) \ \& \ M(x_2, x_2) \ \& \ M(x_1, x_2) \ \& \ {\sim}M(x_2, x_1))$
$\& \ (\exists x_2)(M(x_1, x_1) \ \& \ M(x_2, x_2) \ \& \ M(x_1, x_2) \ \& \ M(x_2, x_1))$
$\& \ (\exists x_2)(M(x_1, x_1) \ \& \ M(x_2, x_2) \ \& \ {\sim}M(x_1, x_2) \ \& \ M(x_2, x_1))$
$\& \ (x_2)[(M(x_1, x_1) \ \& \ M(x_2, x_2) \ \& \ M(x_1, x_2) \ \& \ {\sim}M(x_2, x_1))$
$\lor \ (M(x_1, x_1) \ \& \ M(x_2, x_2) \ \& \ M(x_1, x_2) \ \& \ M(x_2, x_1))$
$\lor \ (M(x_1, x_1) \ \& \ M(x_2, x_2) \ \& \ {\sim}M(x_1, x_2) \ \& \ M(x_2, x_1))]\}$
$\lor \ \{(\exists x_2)(M(x_1, x_1) \ \& \ M(x_2, x_2) \ \& \ M(x_1, x_2) \ \& \ {\sim}M(x_2, x_1))$
$\& \ (\exists x_2)(M(x_1, x_1) \ \& \ M(x_2, x_2) \ \& \ M(x_1, x_2) \ \& \ M(x_2, x_1))$
$\& \ (x_2)[(M(x_1, x_1) \ \& \ M(x_2, x_2) \ \& \ M(x_1, x_2) \ \& \ {\sim}M(x_2, x_1))$
$\lor \ (M(x_1, x_1) \ \& \ M(x_2, x_2) \ \& \ M(x_1, x_2) \ \& \ M(x_2, x_1))]\}\}.$

The structure of constituent (43) can be conveniently illustrated by a tree where an arrow from one node to another indicates that the relation M holds between them. (See Figure 3.)

Fig. 3.

The generalization of these ideas to arbitrary finite depth $d > 0$ is now clear. *Constituents $C_j^{(d)}$ of the depth d are sentences of the form*

$$(44) \quad \bigwedge_{i \in \mathbf{CT}_j^{(d-1)}} (\exists x_1) Ct_i^{(d-1)}(x_1) \mathbin{\&} (x_1) \left[\bigvee_{i \in \mathbf{CT}_j^{(d-1)}} Ct_i^{(d-1)}(x_1) \right]$$

where each attributive constituent $Ct_i^{(d-1)}(x_1)$ has the structure of a tree where all maximal branches have the length d:

$$\bigwedge_{k \in \mathbf{CT}_i^{(d-2)}} (\exists x_2) Ct_k^{(d-2)}(x_1, x_2) \mathbin{\&} (x_2) \left[\bigvee_{k \in \mathbf{CT}_i^{(d-2)}} Ct_k^{(d-2)}(x_1, x_2) \right].$$

More generally, if $0 < m \leq d - 1$, then constituent $C_j^{(d)}$ contains attributive constituents $Ct_i^{(m)}(x_1, \ldots, x_{d-m})$ of the depth m and of the level $d - m$, i.e., sentences of the form

$$(45) \quad \bigwedge_{k \in \mathbf{CT}_i^{(m-1)}} (\exists x_{d-m+1}) Ct_k^{(m-1)}(x_1, \ldots, x_{d-m+1}) \mathbin{\&}$$

$$\mathbin{\&} (x_{d-m+1}) \left[\bigvee_{k \in \mathbf{CT}_i^{(m-1)}} Ct_k^{(m-1)}(x_1, \ldots, x_{d-m+1}) \right].$$

When $m = 1$, definition (45) reduces to (40) with $n = d$.

The structure of a constituent $C_j^{(d)}$ in L is thus essentially a finite set of finite trees with maximal branches of the length d. Each such branch corresponds to a sequence of d individuals that can be drawn with replacement from the universe, and each such sequence is described by a sentence of the form $C_i^{(0)}(x_1, \ldots, x_d)$ which is the most informative description in L of the properties and the interrelations between $x_1, \ldots, x_d$.

Let $x_1, \ldots, x_{d-1}, x_d$ be a sequence of variables which constitutes a branch going through a constituent $C_j^{(d)}$. Then the quantifier $(\exists x_d)$ is followed by a formula of the form $Ct_1^{(0)}(x_1, \ldots, x_{d-1}, x_d)$. Let $x_1, \ldots, x_{d-1}, x'_d$ be another branch which overlaps with the former branch up to the element x_{d-1} (see Figure 4). Then, assuming that constituent $C_j^{(d)}$ is consistent, the formula $Ct_2^{(0)}(x_1, \ldots, x_{d-1}, x'_d)$ has to agree with $Ct_1^{(0)}(x_1, \ldots, x_{d-1}, x_d)$ in all its claims about $x_1, \ldots, x_{d-1}$. This means that consistent constituents contain redundant information.

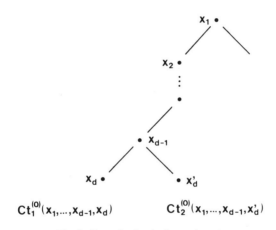

Fig. 4. Part of a depth-d constituent.

To eliminate the repetition of the same formulas in consistent constituents, we may rewrite attributive constituents $Ct_i^{(m)}(x_1, \ldots, x_{d-m})$ of the form (45) as follows:

(46) $A_i^{(m)}(x_1, \ldots, x_{d-m-1}; x_{d-m})$ & $B_i^{(m)}(x_1, \ldots, x_{d-m-1}, x_{d-m})$.

(46) is called a *reduced* attributive constituent of the depth m. Here $A_i^{(m)}$ is a conjunction of atomic formulas or negations of atomic formulas which all contain the variable x_{d-m}; it gives a complete description of the properties of x_{d-m} and the relations it bears to the

earlier individuals $x_1, \ldots, x_{d-m-1}$ in the same branch. The other conjunct $B_i^{(m)}(x_1, \ldots, x_{d-m-1}, x_{d-m})$ is then of the form (45), where now $Ct_k^{(m-1)}$ are reduced attributive constituents of the depth $m-1$.

A constituent is *reduced* if it is built up from reduced attributive constituents. The main difference between ordinary and reduced constituents is the fact that in the latter case each new element x_{d-m} is described by a sentence $A_i^{(m)}(x_1, \ldots, x_{d-m-1}; x_{d-m})$ as soon as it is introduced by an existential quantifier.

For example, the existential part of the constituent (43) looks as follows in the reduced form:

$$(\exists x_1)\{M(x_1, x_1) \& [(\exists x_2)(M(x_2, x_2) \& M(x_1, x_2) \& \sim M(x_2, x_1))$$
$$\& (\exists x_2)(M(x_2, x_2) \& M(x_1, x_2) \& M(x_2, x_1))$$
$$\& (\exists x_2)(M(x_2, x_2) \& \sim M(x_1, x_2) \& M(x_2, x_1))]$$
$$\& (x_2)[(M(x_2, x_2) \& M(x_1, x_2) \& \sim M(x_2, x_1))$$
$$\vee (M(x_2, x_2) \& M(x_1, x_2) \& M(x_2, x_1))$$
$$\vee (M(x_2, x_2) \& \sim M(x_1, x_2) \& M(x_2, x_1))]\}$$
$$\& (\exists x_1)\{M(x_1, x_1) \& [(\exists x_2)(M(x_2, x_2) \& M(x_1, x_2) \& \sim M(x_2, x_1))$$
$$\& (\exists x_2)(M(x_2, x_2) \& M(x_1, x_2) \& M(x_2, x_1))$$
$$\& (x_2)[(M(x_2, x_2) \& M(x_1, x_2) \& \sim M(x_2, x_1))$$
$$\vee (M(x_2, x_2) \& M(x_1, x_2) \& M(x_2, x_1))]\}$$

Let $L^=$ be the polyadic first-order language *with identity* which is obtained by joining the identity sign $=$ to language L (cf. Section 6). Then the constituents of $L^=$ have the same structure as in L, but now one has to assume an *exclusive interpretation of the quantifiers*.[13] Intuitively this means that constituents of $L^=$ are ramified lists of sequences of different individuals, i.e., a constituent of the depth d tells what kinds of sequences of d interrelated individual can be drawn *without replacement* from the universe. More technically, this means that quantified subformulas in constituents have to be understood as follows:

(47) $(\exists x_m)\alpha$ as $(\exists x_m)(x_1 \neq x_m \& \cdots \& x_{m-1} \neq x_m \& \alpha)$

         $(x_m)\alpha$ as $(x_m)(x_1 \neq x_m \& \cdots \& x_{m-1} \neq x_m \supset \alpha)$.

EXAMPLE 4. Let U be the set of integers $\{\ldots, -2, -1, 0, 1, 2, \ldots\}$, and let the interpretation of a two-place predicate M be the relation

< in U. The relation < in U is irreflexive, asymmetric, and transitive. Assume that M is the only predicate in language L^-. Then the true constituents of L^- at the depths 2 and 3 are illustrated in Figures 5 and 6, respectively. Examples of sequences of individuals satisfying the constituent of Figure 6 are given in Figure 7.

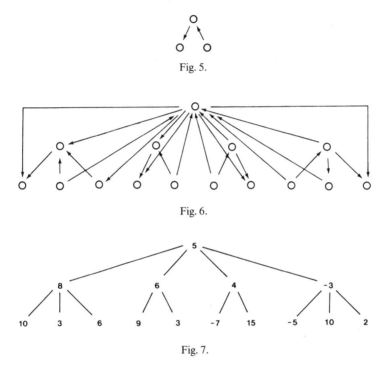

Fig. 5.

Fig. 6.

Fig. 7.

The notion of constituent can be generalized to situations where n terms $t_1, \ldots, t_n$ have been fixed and all descriptions of individuals are made as if $t_1, \ldots, t_n$ appeared higher up in the same branch. In other words, a *constituent* $C_j^{(d)}[t_1, \ldots, t_n]$ *relative to* $t_1, \ldots, t_n$ includes a complete description $A_0(t_1, \ldots, t_n)$ of the interrelations between $t_1, \ldots, t_n$ and each reduced attributive constituent $Ct_i^{(m)}(x_1, \ldots, x_{d-m-1}; x_{d-m})[t_1, \ldots, t_n]$ relative to $t_1, \ldots, t_n$ (cf. formula (46)) contains a complete account of the relations of x_{d-m} to $t_1, \ldots, t_n$. If terms $t_1, \ldots, t_n$ are individual constants $a_1, \ldots, a_n$, then $C_j^{(d)}[a_1, \ldots, a_n]$ is a sentence; if $t_1, \ldots, t_n$ are individual variables

$y_1, \ldots, y_n$, then $C_j^{(d)}[y_1, \ldots, y_n]$ is an open formula with n free variables.

Let $\Gamma_L^{(d)}$ be the set of the depth-d constituents of language L. (Similar notation is used for $\Gamma_{L^-}^{(d)}$ and $\Gamma_L^{(d)}[t_1, \ldots, t_n]$.) Then it is immediately clear that two different constituents $C_j^{(d)}$ and $C_k^{(d)}$ have to be incompatible with each other: according to definition (44), we must have $\mathbf{CT}_j^{(d-1)} \neq \mathbf{CT}_k^{(d-1)}$, and this means that, for some depth-$(d-1)$ attributive constituent $Ct_i^{(d-1)}$, one of the constituents asserts that $Ct_i^{(d-1)}$ is empty and the other that it is non-empty. From (38) it also follows that the constituents in $\Gamma_L^{(d)}$ exhaust all the logical possibilities expressible by the vocabulary of λ and with d layers of quantifiers. Hence, for all $d = 1, 2, \ldots$,

(48) $\quad \vdash \sim(C_j^{(d)} \,\&\, C_k^{(d)})$, if $j \neq k$

$\quad \vdash \bigvee_{C_j^{(d)} \in \Gamma_L^{(d)}} C_j^{(d)}.$

It may be noted that the number of constituents in $\Gamma_L^{(d)}$ grows extremely rapidly with d. Assume, for example, that λ contains only one two-place predicate M. Then there are in L^- two attributive constituents of the depth 0 and the level 1 and, hence, $2^2 - 1 = 3$ constituents of the depth 1. When $d = 2$, a new individual x_2 can be chosen in $2^3 = 8$ ways relative to the earlier individual x_1. This shows that there are

$$2 \cdot \sum_{i=1}^{8} \binom{8}{i} = 2 \cdot (2^8 - 1) = 510$$

different attributive constituents of the depth 1 and the level 1 and, hence,

$$2^{510} \approx 10^{150}$$

constituents in $\Gamma_L^{(2)}$. Further, the number of attributive constituents of the depth 2 is

$$2 \cdot \sum_{i=1}^{8} 32^i \approx 2 \cdot 34 \cdot 10^{12},$$

LOGICAL TOOLS 71

and the number of constituents in $\Gamma_{L^=}^{(3)}$ is approximately

$$2^{2 \cdot 34 \cdot 10^{12}} \approx 10^{7 \cdot 10^{11}} = 10^{700000000000}.$$

The total number of constituents of all depths $d = 1, 2, \ldots$ is infinite, i.e.,

$$\left| \bigcup_{d < \omega} \Gamma_L^{(d)} \right| = \omega.$$

Nevertheless, if λ is finite (as we have assumed), $\Gamma_L^{(d)}$ is *finite* for all $d < \omega$.

On the other hand, when $d > 1$, all constituents in $\Gamma_L^{(d)}$ are not consistent. Here polyadic languages differ essentially from monadic ones, since all depth-1 constituents in $\Gamma_L^{(1)}$ are consistent. We may give two necessary conditions for the consistency of a constituent $C_j^{(d)}$. First, the *Truncation Requirement* says that the elimination of any layer of quantifiers from a depth-d constituent should give the same depth-$(d-1)$ constituent — where the elimination of the kth layer means the following: remove all quantifiers $(\exists x_k)$ and (x_k), together with all atomic formulas containing x_k, and eliminate all unnecessary repetitions. The *Repetition Requirement* says that for each x_k occurring in a constituent $C_j^{(d)}$ (where $k < d$) there must be one element x_{k+1} covered by x_k (the 'replica' of x_k) which has precisely the same relations to earlier elements $x_1, \ldots, x_{k-1}$ and to x_k in the same branch as x_k has. Both of these requirements are motivated by the idea that the sequences of individuals described by a constituent are assumed to be drawn (with replacement) from the same domain. (Cf. Hintikka, 1973; Hintikka and Niiniluoto, 1973.) Then a constituent $C_k^{(d)}$ in $\Gamma_L^{(d)}$ is called *trivially inconsistent* if it violates either the Truncation Requirement or the Repetition Requirement. For constituents in $\Gamma_{L^=}^{(d)}$ it is enough to give only the Truncation Requirement, since with identity $=$ in the language it is not possible any more to choose the same element twice in the same sequence.

The trivial inconsistency of a constituent $C_j^{(d)}$ can be 'seen' directly from its syntactical structure. It turns out, however, that in polyadic languages there is no effective method for separating *non-trivially inconsistent* constituents from the consistent ones (cf. the next section).

Two generalizations of the notion of constituent may still be mentioned. If language L contains an infinite number of predicates, then the number of attributive constituents at all depths is likewise infinite. If we allow denumerably infinite disjunctions and conjunctions, constituents are expressible in the same form as above, but then there may be an infinite number branches going through a single node in tree structure of a constituent. Such constituents resemble closely the so-called *Scott sentences* for the infinitary language $L_{\omega_1\omega}$ (cf. Monk, 1976, p. 505).

The notion of constituent has been defined also for the infinitary language $N_{\kappa\lambda}$ of Hintikka and Rantala where — in contrast with L and languages $L_{\kappa\lambda}$ — a formula may contain infinitely long nested sequences of subformulas. Thus, the formulas of $N_{\kappa\lambda}$ are defined as trees with maximal branches of the length $\leq \lambda$, where λ is an infinite ordinal. The constituents of these languages are therefore *infinitely deep* sentences.[14]

2.8. DISTRIBUTIVE NORMAL FORMS

Two structures $\Omega = \langle U, V \rangle$ and $\Omega' = \langle U', V' \rangle$ for language L are *isomorphic*, $\Omega \approx \Omega'$, if there is a bijective mapping $f: U \to U'$ such that

$$f(V(a_i)) = V'(a_i) \quad \text{for all individual constants } a_i$$
$$\langle u_1, \ldots, u_k \rangle \in V(M_j^k) \quad \text{iff} \quad \langle f(u_1), \ldots, f(u_k) \rangle \in V'(M_j^k)$$
$$\text{for all } u_1, \ldots, u_k \in U.$$

Structures Ω and Ω' are *elementarily equivalent*, $\Omega \equiv \Omega'$, if precisely the same sentences of L are true both in Ω and in Ω'. If we define the *theory of structure* Ω in L by

(49) $\quad \text{Th}(\Omega) = \{\alpha \mid \alpha \text{ is a sentence of } L \text{ and } \Omega \vDash \alpha\}$,

then

(50) $\quad \Omega \equiv \Omega' \quad \text{iff} \quad \text{Th}(\Omega) = \text{Th}(\Omega')$.

Isomorphic structures are elementarily equivalent: if $\Omega \approx \Omega'$, then $\Omega \equiv \Omega'$. The converse holds for finite structures Ω and Ω'.

Structures $\Omega = \langle U, V \rangle$ and $\Omega' = \langle U', V' \rangle$ for L are *partially d-isomorphic*, $\Omega \approx_d \Omega'$, if there is a sequence $F_0 \subseteq \cdots \subseteq F_d$ of mappings from subsets of U into U' such that

(51) (i) $F_0 = \{f \mid f = \{\langle V(a_i), V'(a_i)\rangle \mid i = 1, 2, \ldots\}\}$

   (ii) if $f \in F_d$ and $u_1, \ldots, u_k$ belong to the domain of f,
   then $\langle u_1, \ldots, u_k \rangle \in V(M_j^k)$
   iff $\langle f(u_1), \ldots, f(u_k) \rangle \in V'(M_j^k)$

   (iii) if $f \in F_i$ $(i < d)$ and $u \in U$, there is an element $u' \in U'$ such that $f \cup \{\langle u, u' \rangle\} \in F_{i+1}$

   (iv) if $f \in F_i$ $(i < d)$ and $u' \in U'$, there is an element $u \in U$ such that $f \cup \{\langle u, u' \rangle\} \in F_{i+1}$.

Structures Ω and Ω' are *elementarily d-equivalent*, $\Omega \equiv_d \Omega'$, if the same sentences of depth $\leq d$ are true in both Ω and Ω'. Thus, if

$$\text{Th}^{(d)}(\Omega) = \{\alpha \mid \alpha \text{ is a sentence of } L \text{ of depth } \leq d \text{ and } \Omega \models \alpha\},$$

then

$$\Omega \equiv_d \Omega' \quad \text{iff} \quad \text{Th}^{(d)}(\Omega) = \text{Th}^{(d)}(\Omega').$$

The theorem of Fraisse and Ehrenfeucht shows that

(52) The following conditions are equivalent for each $d < \omega$:
  (a) $\Omega \approx_d \Omega'$
  (b) $\Omega \equiv_d \Omega'$.

(53) The following conditions are equivalent:
  (a) $\Omega \equiv \Omega'$
  (b) $\Omega \approx_d \Omega'$ for all $d < \omega$
  (c) $\Omega \equiv_d \Omega'$ for all $d < \omega$.

(See Monk, 1976, pp. 408–412; Rantala, 1987.)

By theorem (48), for each $d < \omega$ and for each structure Ω for L, there is one and only one constituent in $\Gamma_L^{(d)}$ which is true in Ω. Let us denote this unique constituent by $C_\Omega^{(d)}$. Then, of course, $\Omega \models C_\Omega^{(d)}$. What other models does $C_\Omega^{(d)}$ have? An answer to this question follows from the observation that the condition $\Omega' \models C_\Omega^{(d)}$ entails $\Omega \approx_d \Omega'$. This can be seen by choosing F_0 as in definition (51)(i) and then taking F_i as the set of mappings of the form

$$f \cup \{\langle u_1, u_1' \rangle, \ldots, \langle u_i, u_i' \rangle\},$$

where $\langle u_1, \ldots, u_i \rangle$ and $\langle u'_1, \ldots, u'_i \rangle$ are sequences of elements from U and U', respectively, which both satisfy the same attributive constituent $Ct^{(d-i)}(x_1, \ldots, x_i)$ occurring in constituent $C_\Omega^{(d)}$ (cf. Rantala, 1987).

The condition $\Omega \equiv_d \Omega'$ entails that $\Omega' \vDash C_\Omega^{(d)}$, since the depth of $C_\Omega^{(d)}$ is $\leq d$. Combining these results with (52) and (53) gives:

(54) The following conditions are equivalent for each $d < \omega$:
(a) $\Omega \approx_d \Omega'$
(b) $\Omega \equiv_d \Omega'$
(c) $\Omega' \vDash C_\Omega^{(d)}$.

(55) The following conditions are equivalent:
(a) $\Omega \equiv \Omega'$
(b) $\Omega \approx_d \Omega'$ for all $d < \omega$
(c) $\Omega \equiv_d \Omega'$ for all $d < \omega$
(d) $\Omega' \vDash C_\Omega^{(d)}$ for all $d < \omega$.

Similar results can be proved for various generalizations of the notion of constituent (cf. Section 7). For example, a countable structure Ω' is a model of the Scott sentence for a countable structure Ω in language $L_{\omega_1 \omega}$ if and only if Ω' is isomorphic to Ω. Further, the notion of partial isomorphism can be extended so that all models of a constituent of depth ω are partially isomorphic (Rantala, 1979; Karttunen, 1979).

Theorem (55) gives us a simple semantical proof of the existence of distributive normal forms in first-order logic:

(56) For each generalization α in L with the depth $\leq d$, there is a finite set of constituents $\Delta_\alpha^{(d)} \subseteq \Gamma_L^{(d)}$ such that

$$\vdash \alpha \equiv \bigvee_{C_j^{(d)} \in \Delta_\alpha^{(d)}} C_j^{(d)}.$$

The disjunction given in (56) is the *distributive normal form* of α at the depth d. To prove (56), choose

$$\Delta_\alpha^{(d)} = \{ C_\Omega^{(d)} \in \Gamma_L^{(d)} \mid \Omega \vDash \alpha \},$$

i.e., $\Delta_\alpha^{(d)}$ is the set of all constituents $C_\Omega^{(d)}$ corresponding to the models of α. Then $\Omega \vDash \alpha$ entails first that $C_\Omega^{(d)} \in \Delta_\alpha^{(d)}$ and then

(57) $$\Omega \vDash \bigvee_{C_j^{(d)} \in \Delta_\alpha^{(d)}} C_j^{(d)}.$$

LOGICAL TOOLS   75

Conversely, if (57) holds, then $\Omega \vDash C_{\Omega'}^{(d)}$, for some model Ω' of α. Hence, by (54), $\Omega' \equiv_d \Omega$, so that $\Omega \vDash \alpha$.

Theorem (56) can be generalized to constituents of language $L^=$ and to relativized constituents of the form $C_j^{(d)}[t_1, \ldots, t_n]$.

Theorem (56) shows that constituents in $\Gamma_L^{(d)}$ are the strongest generalizations of depth $\leq d$ expressible in language L. Similarly, constituents in $\Gamma_L^{(d)}[a_1, \ldots, a_n]$ are the strongest sentences of depth $\leq d$ with occurrences of the constants $a_1, \ldots, a_n$, and constituents in $\Gamma_L^{(d)}[y_1, \ldots, y_n]$ are the strongest open formulas of depth $\leq d$ with the free variables $y_1, \ldots, y_n$.

The semantical proof of the representation (56) guarantees that the distributive normal form of a sentence α contains only *consistent* constituents, since condition $\Omega \vDash C_\Omega^{(d)}$ entails the consistency of $C_\Omega^{(d)}$. There are also effective syntactical procedures for transforming an arbitrary sentence α of depth $\leq d$ to a finite disjunction of depth-d constituents, but the obtained normal forms may contain non-trivially inconsistent constituents as well (Rantala, 1987). If the cardinality $|\Delta_\alpha^{(d)}|$ of $\Delta_\alpha^{(d)}$ is called the *d-breadth* of generalization α, then there is no recursive function which would give the breadth of an arbitrary generalization α in L. The reason for this fact can be seen immediately from theorem (56). If α and β are two generalizations of depth $\leq d$, and each of them is transformed to its normal form, then

(58) $\quad \alpha \vdash \beta \quad$ iff $\quad \bigvee_{C_j^{(d)} \in \Delta_\alpha^{(d)}} C_j^{(d)} \vdash \bigvee_{C_j^{(d)} \in \Delta_\beta^{(d)}} C_j^{(d)}$

$\quad\quad\quad\quad\quad$ iff $\quad \Delta_\alpha^{(d)} \subseteq \Delta_\beta^{(d)}$.

If $\Delta_\alpha^{(d)}$ could be effectively determined for each α, then by (58) the notion of deducibility $\vdash$ would also be an effective notion in L. However, this would be in conflict with the undecidability of polyadic first-order logic.

A constituent $C_k^{(d+e)}$ of the depth $d+e$ ($e > 0$) is said to be *subordinate* to another constituent $C_j^{(d)}$ of the depth d if $C_j^{(d)}$ can be obtained from $C_k^{(d+e)}$ by omitting the last e layers of quantifiers from it. In this case, $C_k^{(d+e)}$ clearly entails $C_j^{(d)}$. Therefore, if β is the disjunction of all consistent depth-$(d+e)$ constituents subordinate to $C_j^{(d)}$, then β entails $C_j^{(d)}$. Conversely, if $C_m^{(d+e)}$ is a depth-$(d+e)$ constituent which is not subordinate to $C_j^{(d)}$, and if $C_i^{(d)}$ is the result of eliminating the last e

layers of quantifiers from $C_m^{(d+e)}$, then $C_j^{(d)}$ is incompatible with $C_i^{(d)}$ which is entailed by $C_m^{(d+e)}$. Hence, $C_j^{(d)}$ is incompatible with $C_m^{(d+e)}$ itself. This gives us the following result:

(59) Each constituent in $\Gamma_L^{(d)}$ is logically equivalent to the disjunction of all consistent constituents in $\Gamma_L^{(d+e)}$ ($e > 0$) which are subordinate to it.

According to (59), all the constituents of language L constitute a tree structure illustrated in Figure 8.

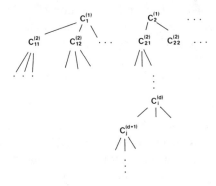

Fig. 8. Constituents of L.

If a generalization α of the depth d is first transformed into a distributive normal form at the depth d, and if each depth-d constituent in $\Delta_\alpha^{(d)}$ is expressed as a disjunction of depth-$(d+e)$ constituents, then α will be equivalent to a finite disjunction of depth-$(d+e)$ constituents:

(60) $\quad \vdash \alpha \equiv \bigvee_{C_j^{(d+e)} \in \Delta_\alpha^{(d+e)}} C_j^{(d+e)}.$

The cardinality of $\Delta_\alpha^{(d+e)}$ is then the $(d+e)$-*breadth* of α. The disjunction given in (60) is called the *expansion* of α to the depth $d+e$.

The expansion procedure can be applied also to inconsistent constituents. From the completeness of first-order logic it follows that for each inconsistent constituent $C_j^{(d)}$ there is some $e > 0$ such that all depth-$(d+e)$ constituents $C_k^{(d+e)}$ subordinate to $C_j^{(d)}$ are *trivially* inconsistent. Hence, if generalization α is represented as a disjunction of consistent or inconsistent constituents of the depth d, then there is a finite depth $d+e$ where the expansion of α contains only consistent and trivially inconsistent disjuncts. When the latter are eliminated, we have just the representation (60) of α at the depth $d+e$. However, there is no effective general method for determining how great depths $d+e$ are required before this will happen for a given α. (See Hintikka, 1973.)

2.9. FIRST-ORDER THEORIES

Let L be a polyadic first-order language without individual constants. A set Σ of sentences of L is a *first-order theory* if Σ is closed under deduction, i.e., $\Sigma \vdash \alpha$ entails $\alpha \in \Sigma$ for all sentences α of L. If we write

(61) $\quad Cn(\Sigma) = \{\alpha \mid \alpha \text{ is a sentence of } L \text{ and } \Sigma \vdash \alpha\}$,

then Σ is a theory if and only if $Cn(\Sigma) = \Sigma$. Theory Σ is *consistent* if Σ is a consistent set of sentences. Theory Σ is *axiomatizable* if there is a recursive set T of sentences of L such that $Cn(T) = \Sigma$; in this case, T is a set of *axioms* for Σ.

A theory Σ is *complete* in L if $\Sigma \vdash \alpha$ or $\Sigma \vdash {\sim}\alpha$ for all sentences α of L. It follows that theory Σ is complete if and only if all the models of Σ are elementarily equivalent (cf. Section 8). Hence, theory Σ is complete if and only if Σ is the theory $Th(\Omega)$ of some structure Ω for L, i.e., $\Sigma = Th(\Omega)$ for some Ω.

Let now $\langle C^{(d)} \mid d < \omega \rangle$ be any infinite sequence of constituents which corresponds to a branch of Figure 8. Hence,

(62) $\quad \ldots C^{(d+1)} \vdash C^{(d)} \vdash \cdots \vdash C^{(1)}$.

We shall say that such a sequence of constituents is *monotone*. Assuming (62), $\Sigma = Cn(\{C^{(d)} \mid d < \omega\})$ is a complete theory in L: if α is any sentence of the depth d, then $\Sigma \vdash \alpha$ if $C^{(d)} \in \Delta_\alpha^{(d)}$ and $\Sigma \vdash {\sim}\alpha$ if $C^{(d)} \notin \Delta_\alpha^{(d)}$. Conversely, if Σ is a complete theory in L, then $\Sigma = Th(\Omega)$ for some structure Ω for L. It follows from theorem (55) that all models of $\{C_\Omega^{(d)} \mid d < \omega\}$ are elementarily equivalent to Ω. Hence, $\Sigma = Cn(\{C_\Omega^{(d)} \mid d < \omega\})$. We have thus proved the following result:

(63) A theory Σ in L is complete iff Σ is axiomatizable by a monotone sequence $\langle C^{(d)} \mid d < \omega \rangle$ of constituents of L.

These results show that each monotone sequence $\langle C^{(d)} \mid d < \omega \rangle$ of constituents corresponds to a structure Ω for L which is unique up to elementary equivalence. The tree-type structures described by constituents $C^{(d)}$ — called *surface structures* in Hintikka (1973) — can, in this sense, be considered as *finite approximations* to the structure Ω.

Tarski (1956) showed in his Calculus of Deductive Systems that the class of first-order theories in language L has a natural lattice structure:

order: $\quad \Sigma \leq \Delta \quad$ iff $\quad \Sigma \subseteq \Delta \quad$ iff $\quad \Delta \vdash \Sigma$
intersection: $\quad \inf\{\Sigma, \Delta\} = \Sigma \cap \Delta$
sum: $\quad \sup\{\Sigma, \Delta\} = Cn(\Sigma \cup \Delta)$
complement: $\quad \Sigma' = \bigcap_{\alpha \in \Sigma} Cn(\{\sim\alpha\})$
unit: $\quad \mathbf{1} = Cn(\alpha \wedge \sim\alpha)$
null: $\quad \mathbf{0} = Cn(\alpha \vee \sim\alpha)$.

For finitely axiomatizable theories in L, this structure is a Boolean algebra, where

$$\inf\{Cn(\alpha), Cn(\beta)\} = Cn(\alpha \vee \beta)$$
$$\sup\{Cn(\alpha), Cn(\beta)\} = Cn(\alpha \mathbin{\&} \beta)$$
$$Cn(\alpha)' = Cn(\{\sim\alpha\}).$$

Let $\mathbf{M}$ be the class of all structures for a first-order language L. If Σ is a set of sentences in L, let $\text{Mod}(\Sigma)$ be its class of models:

(64) $\quad \text{Mod}(\Sigma) = \{\Omega \in \mathbf{M} \mid \Omega \vDash \Sigma\}$.

If α is a sentence of L, we write $\text{Mod}(\alpha)$ for $\text{Mod}(\{\alpha\})$. For each Σ, the class $\text{Mod}(\Sigma)$ is closed under elementary equivalence: if $\Omega \vDash \Sigma$ and $\Omega' \equiv \Omega$, then $\Omega' \vDash \Sigma$. Furthermore,

$\text{Mod}(\alpha \wedge \beta) = \text{Mod}(\alpha) \cap \text{Mod}(\beta)$
$\text{Mod}(\alpha \vee \beta) = \text{Mod}(\alpha) \cup \text{Mod}(\beta)$
$\text{Mod}(\sim\alpha) \quad = \mathbf{M} - \text{Mod}(\alpha)$
$\Sigma \subseteq \Delta \quad$ iff $\quad \text{Mod}(\Delta) \subseteq \text{Mod}(\Sigma)$
$\text{Mod}(\Sigma) \quad = \text{Mod}(Cn(\Sigma))$
$\text{Th}(\text{Mod}(\Sigma)) = Cn(\Sigma)$
$\text{Mod}(\alpha) = \text{Mod}(\beta) \quad$ iff $\quad Cn(\alpha) = Cn(\beta) \quad$ iff $\quad \vdash \alpha \equiv \beta$.

A class $\mathbf{K} \subseteq \mathbf{M}$ of structures for L is an *elementary class* if $\mathbf{K} = \text{Mod}(\alpha)$ for some sentence α of L. Further, K is Δ-*elementary* if $K = \text{Mod}(\Sigma)$ for some set of sentences Σ of L, and Σ-*elementary* if K is a union of elementary classes. Then K is elementary if and only if K is both Δ-elementary and Σ-elementary.

If the class EC of elementary classes in L is chosen as the base of the topology of $\mathbf{M}$, then $\mathbf{M}$ is a compact space where the Σ-elementary classes are open, Δ-elementary classes closed, and elementary classes clopen (i.e., open and closed). The relation $\equiv$ of elementary equivalence is an equivalence relation in $\mathbf{M}$. Let $\mathbf{M}^* = \mathbf{M}/\equiv$ be the class of the equivalence classes obtained by identifying elementarily equivalent structures. In other words, $\mathbf{M}^*$ is the quotient space of $\mathbf{M}$ modulo elementary equivalence $\equiv$. Then $\mathbf{M}^*$ will be a *Boolean space*, i.e., a compact Hausdorff space with a basis of clopen sets. Complete theories in L correspond to the points of $\mathbf{M}^*$, finitely axiomatizable theories to the clopen sets of points in $\mathbf{M}^*$, and arbitrary theories to the closed sets in $\mathbf{M}^*$. In this sense, each theory Σ in L is equivalent to the 'disjunction' of its complete extensions, i.e., all complete theories in L which entail Σ.

The class $(EC, \subseteq)$, together with ordinary set-theoretical operations, is a Boolean algebra. It is easy to see that the function $h: \mathscr{A} \to EC$ such that

$$h(\bar{\alpha}) = \text{Mod}(\alpha) \quad \text{for sentences } \alpha \text{ of } L$$

defines an isomorphism between the Lindenbaum algebra $\mathscr{A}$ of L and EC (cf. Section 1). It follows that $\mathscr{A}$ is isomorphic to the class EC^* of the clopen subsets of $\mathbf{M}^*$. This is sufficient to guarantee that $\mathbf{M}^*$ is homeomorphic to the *Stone space* $S(\mathscr{A})$ of the Lindenbaum algebra $\mathscr{A}$, where the Stone space $S(B)$ of a Boolean algebra B is defined as the class of the ultrafilters (filters maximal with respect to inclusion $\subseteq$) on B. These connections can be summarized in the following chains:

Σ is a complete theory in L
iff $\Sigma = \text{Th}(\Omega)$ for some $\Omega \in \mathbf{M}$
iff $\text{Mod}(\Sigma) = \{\Omega' \in \mathbf{M} \mid \Omega \equiv \Omega'\}$ for some $\Omega \in \mathbf{M}$
iff $\text{Mod}(\Sigma)$ corresponds to a point in $\mathbf{M}^* = \mathbf{M}/\equiv$
iff $\bar{\Sigma} = \{\bar{\alpha} \mid \alpha \in \Sigma\}$ is an ultrafilter on $\mathscr{A}_L$
iff $\bar{\Sigma}$ corresponds to a point in $S(\mathscr{A}_L)$.

Σ is a theory in L
iff $\Sigma = \text{Th}(K)$ for some $K \subseteq \mathbf{M}$
iff $\text{Mod}(\Sigma)$ corresponds to a closed subset of $\mathbf{M}^* = \mathbf{M}/\equiv$
iff $\tilde{\Sigma} = \{\tilde{a} \mid a \in \Sigma\}$ is a filter on $\mathscr{A}_L$
iff $\tilde{\Sigma}$ corresponds to a closed subset of $S(\mathscr{A}_L)$.[15]

In defining the notion of a theory, we have used the 'logistic method' which originally was developed for the purposes of metamathematics (i.e., the syntactic and semantic study of mathematical theories). Within this approach — which also has been called the *statement view* of theories (cf. Stegmüller, 1976) — a theory is defined as a deductively closed set of sentences within a formal language. The application of this idea to the empirical theories of factual scientific disciplines faces two special problems, however. First, the basic vocabulary of the language of an empirical theory may be divided into theoretical and observational terms which complicates the structure of the theory.[16] Secondly, the class of models $\text{Mod}(\Sigma)$ of theory Σ, as defined by (64), contains elements which clearly are 'unintended'. For example, if Σ is a theory in biology, $\text{Mod}(\Sigma)$ will contain structures whose domain consists entirely of numbers. For this reason, it has been suggested that the notion of a theory should include a class $\mathbf{I}$ of its *intended models*, where $\mathbf{I}$ cannot be specified by purely syntactic and semantic means. In other words, a theory is defined as a pair $\langle \Sigma, \mathbf{I} \rangle$, where Σ is a deductively closed set of sentences in language L and $\mathbf{I}$ is a class of structures for L. The *claim* of the theory $\langle \Sigma, \mathbf{I} \rangle$ is then the statement

$$\mathbf{I} \subseteq \text{Mod}(\Sigma),$$

i.e., that all intended models in $\mathbf{I}$ are models of Σ.[17] This claim reduces simply to $\Omega \vDash \Sigma$, if $\mathbf{I}$ contains only one element Ω.

2.10. INDUCTIVE LOGIC

Let $\text{Sent}(L)$ be the set of sentences of a language L, and let $\text{Sent}_0(L)$ be the set of logically consistent sentences of L. Then a function $P: \text{Sent}(L) \times \text{Sent}_0(L) \to \mathbb{R}$ is a *conditional probability measure* for language L if for all h, h_1, h_2 in $\text{Sent}(L)$ and for all e in $\text{Sent}_0(L)$:

1^0 $P(h/e) \geq 0$
2^0 $P(e/e) = 1$

LOGICAL TOOLS 81

3^0 $P(h_1 \vee h_2/e) = P(h_1/e) + P(h_2/e)$,   if $\vdash \sim(h_1 \& h_2)$
4^0 $P(h_1 \& h_2/e) = P(h_1/e)P(h_2/h_1 \& e)$,   if $h_1 \& e \in \text{Sent}_0(L)$.

P is assumed here to be an *epistemic* probability measure: $P(h/e)$ indicates a rational degree of belief in the truth of h given evidence e.[18]

While the personalist school of probability takes 1^0-4^0 to exhaust all the general principles about rational degrees of belief, the tradition of inductive logic is characterized by the attempt to impose more structure to measure P on the basis of the language L. We shall review in this section some proposals concerning 'inductive probabilities'.[19]

Let us say that *P respects logic* if the following conditions hold:

5^0 $P(h_1/e) = P(h_2/e)$,   if $\vdash h_1 \equiv h_2$
6^0 $P(h/e_1) = P(h/e_2)$,   if $\vdash e_1 \equiv e_2$.

Conditions 5^0 and 6^0 say that P 'behaves well' with respect to logical equivalence. They entail that

7^0 $P(h/e) = 1$,   if $\vdash e \supset h$.

The non-conditional probabilities of sentences h in $\text{Sent}(L)$ can be defined by

$$P(h) = P(h/t)$$

where t is an arbitrary tautology in L. Then we have by 1^0-6^0

8^0 $P(h) \geq 0$
9^0 $P(h) = 1$,   if $\vdash h$
10^0 $P(h_1 \vee h_2) = P(h_1) + P(h_2)$,   if $\vdash \sim(h_1 \& h_2)$
11^0 $P(h_1 \& h_2) = P(h_1)P(h_2/h_1)$
12^0 $P(\sim h) = 1 - P(h)$
13^0 $P(h_1) \leq P(h_2)$,   if $\vdash h_1 \supset h_2$
14^0 $P(h_1) = P(h_2)$,   if $\vdash h_1 \equiv h_2$.

Conversely, if $P: \text{Sent}(L) \to \mathbb{R}$ satisfies 8^0-11^0, 14^0, then

$$P(h/e) = \frac{P(h \& e)}{P(e)}, \text{ for } h \in \text{Sent}(L), e \in \text{Sent}_0(L),$$

defines a conditional probability measure for L satisfying 1^0-6^0.

It should be noted that conditions 5^0 and 6^0 are highly idealized if P is assumed to express the degrees of belief of some real-life agent

— such as an individual scientist or the scientific community. These conditions require that this agent is *logically omniscient*, i.e., that he is a perfect logician who is able to recognize all deductive relations between the sentences of L. For example, if the agent X does not notice that $\vdash e \supset h$, then his personal probability P_X may satisfy $P_X(h/e) = P_X(h) < 1$, even if 7^0 states that $P(h/e) = 1$. On the other hand, it is plausible to assume that $P_X(h/e) = 1$ if the agent X knows that $\vdash e \supset h$.[20]

We shall always assume conditions 5^0 and 6^0 below. For monadic languages this is perhaps not too restrictive, since the notion of deduction is decidable in such languages. For polyadic languages, Hintikka (1970b) has developed a system of *surface probabilities* $P^{(s)}$ which assign the probability 0 only to trivially inconsistent constituents and, therefore, satisfy condition 5^0 only if $h_1 \equiv h_2$ is a 'surface tautology', i.e., only if the negation of $h_1 \equiv h_2$ is trivially inconsistent at the depth of $h_1 \equiv h_2$.

A probability measure P for language L is *regular* if the converses of 7^0 and 9^0 hold:

15^0 $P(h/e) = 1$ only if $\vdash e \supset h$
16^0 $P(h) = 1$ only if $\vdash h$.

Condition 16^0 entails

17^0 $P(h) = 0$ only if $\vdash \sim h$.

It follows that in a finite monadic language L_N^k a regular probability P has to assign a non-negative value to all state descriptions: a probability measure P for L_N^k ($N < \omega$) is regular if and only if $P(s) > 0$ for all $s \in \mathcal{Z}$.

If language L contains meaning postulates $\mathcal{MP}$, then a probability measure P for L has to be relativized to $\mathcal{MP}$. This means that $P(h) = 1$ for all h in $\mathcal{MP}$. Moreover, the regularity of P means that $P(h) = 1$ only if $\mathcal{MP} \vdash h$.

Assume now that L is the finite monadic language L_N^k. By the representation (19), we have for all h in Sent(L_N^k):

(65) $P(h) = \sum_{s \in R(h)} P(s)$.

Therefore, the probability measure P for L_N^k is completely specified as soon as its values $P(s)$ for state descriptions $s \in \mathcal{Z}$ are fixed. The Bolzano–Wittgenstein measure c^+ gives the same probability to all state descriptions:

(66) $\quad c^+(s) = \dfrac{1}{q^N} \quad$ for all $s \in \mathcal{Z}$.

Carnap's c^* gives first equal probabilities to the structure descriptions and then the probability of each state description is divided equally to all state description within its range (cf. Carnap, 1950). Hence,

(67) $\quad c^*(s) = \dfrac{(q-1)!N_1! \ldots N_q!}{(N+q-1)!}$

if s belongs to the range of the structure description $S(N_1, \ldots, N_q)$.

The difference between measures c^+ and c^* is clearly visible in singular inductive inference. Assume that L_N^k fits the universe U, and e_n is a description in L_N^k of a sample of n individuals $a_1, \ldots, a_n$ from U such that $n_i \geq 0$ individuals have been found to exemplify the Q-predicate Q_i ($i = 1, \ldots, q$). Thus, $n_1 + \cdots + n_q = n$. Let h_i be the hypothesis that the next individual a_{n+1} will be found in the cell Q_i, i.e., h is $Q_i(a_{n+1})$. Then

(68) $\quad c^+(h_i/e_n) = \dfrac{1}{q}$

(69) $\quad c^*(h_i/e_n) = \dfrac{n_i + 1}{n + q}$.

As $c^+(h_i) = 1/q$, (68) shows that learning from experience becomes impossible with measure c^+. This was noted already by George Boole in 1854. The result (69) for c^* is a generalization of Laplace's *rule of succession*: if an event A has occurred m times in n trials, then the probability that A occurs in the next trial is

$$\dfrac{m+1}{n+2}$$

which is a special case of (69) with $q = 2$ (i.e., A occurs or does not occur) and $n_i = m$.

In Carnap's λ-*continuum*, probabilities of the form $P(h_i/e_n)$ are defined by

(70) $\quad c_\lambda(h_i/e_n) = \dfrac{n_i + \lambda/q}{n + \lambda}$

where λ is a real-valued parameter, $0 < \lambda < \infty$ (Carnap, 1952). (70) is a weighted average of n_i/n (i.e., observed relative frequency of individuals in cell Q_i) and of $1/q$ (i.e., the relative weight of predicate Q_i in L_N^k). Thus, λ indicates the strength of 'logical' considerations relative to 'empirical' ones within singular induction.

If $\lambda = 0$, formula (70) gives as a special case the *straight rule*

$$c_0(h_i/e_n) = \dfrac{n_i}{n}.$$

Carnap excludes this case, since c_0 is not regular. (For example, $c_0(h_i/e_1) = 1$ if $n_i = 1$.) If $\lambda \to \infty$, definition (70) reduces to c^+. If $\lambda = q$, definition (70) equals the measure c^*. In his later work, Carnap (1980) defended the view that λ should be at most q, perhaps equal to 1.

As a generalization of the λ-continuum, one may assume that the Q-predicates are defined by families of predicates, so that the relative widths γ_i of the Q-predicates Q_i ($i = 1, \ldots, q$) may differ from each other. Then (70) is replaced by

(71) $\quad \dfrac{n_i + \gamma_i \lambda}{n + \lambda}.$

(Cf. Carnap, 1971b, 1980.)

Another generalization of the λ-continuum is based upon the idea that pairs of Q-predicates may be at different distances from each other. If we have found many individuals in the cell Q_j, then the probability of $h_i = Q_i(a_{n+1})$ may be taken to depend on the distance of Q_i from Q_j. Problems concerning analogical reasoning can be treated by inductive probabilities satisfying this requirement.[21]

Carnap extends the λ-continuum to the language L_ω^k by the stipulation that

(72) $\quad c(h/e) = \lim\limits_{N \to \infty} {}_N c(h/e),$

where ${}_N c$ is the probability $c_\lambda(h/e)$ in language L_N^k. (Thus, (70) remains unchanged when $N \to \infty$.) It follows that, in Carnap's system for L_ω^k, the constituent C^q has the absolute probability 1:

(73) $\quad P(C^q) = 1$
$\quad\quad\quad P(C^w) = 0 \quad$ if $\quad w < q$.

Here C^q is the maximally wide constituent of L_ω^k which claims that all Q-predicates are instantiated in the universe U. Such a universe U is called *atomistic* (relative to the vocabulary of L_ω^k), since it is not possible to express any true universal generalizations about U in L_ω^k. This shows that probability P is not regular, since it gives the probability 1 to a factual sentence C^q. More generally, we have in Carnap's system:

(74)　　Let h be a factual universal generalization in L_ω^k, and let e be a finite conjunction of singular sentences in L_ω^k. Then $c(h/e) = 0$.

Jaakko Hintikka showed in 1964 how one can construe inductive probabilities for L_N^k ($N \leq \omega$) which avoid the undesirable features (73) and (74) of Carnap's systems. In Hintikka's (1965) *Jerusalem system*, probabilities are first divided equally to all constituents, and then the probability of a constituent is divided equally to the state descriptions within its range. In Hintikka's *combined system*, probabilities are divided equally in three steps: first to constituents, then to structure descriptions, and finally to state descriptions. In both cases we have

(75) $\quad P(C^w) = \dfrac{1}{t} \quad$ for all C^w

where t is the number of constituents. Hence, universal generalizations may receive prior probabilities greater than 0.

In Hintikka's (1966) λ-α-*continuum* the prior probabilities of constituents depend on a new parameter α which functions as an 'index of caution' with respect to inductive generalization. The greater α is, the less probable universal generalizations are — and the more probable constituent C^q is. Hintikka's suggestion is

(76) $\quad P(C^w) = \dfrac{\Gamma(\alpha + w\lambda/q)/\Gamma(w\lambda/q)}{\sum\limits_{i=1}^{q} \binom{q}{i} \Gamma(\alpha + i\lambda/q)/\Gamma(i\lambda/q)},$

where Γ is the gamma function. If $\alpha = 0$, we obtain from (76) the formula (75). If $0 < \alpha < \omega$, then

(77) $\quad P(C^w) < P(C^{w'}) \quad \text{iff} \quad w < w'$.

If $\alpha \to \infty$, then

$$P(C^q) \to 1$$
$$P(C^w) \to 0, \quad \text{if } w < q.$$

In other words, Carnap's λ-continuum is a special case of the λ-α-continuum when $\alpha = \omega$.

In the λ-α-system, the probability (70) relativized to constituent C^w is defined by

(78) $\quad P(h_i/e_n \ \& \ C^w) = \dfrac{n_i + \lambda/w}{n + \lambda}$.

Here (78) differs from (70) in that q has been replaced by the width w of constituent C^w. Probabilities of the form $P(e_n/C^w)$ can be calculated by formula (78) by applying the probability axiom 4^0. Then the posterior probabilities of constituents can be calculated by Bayes' Theorem:

$$P(C^w/e_n) = \frac{P(C^w)P(e_n/C^w)}{\sum_j P(C_j)P(e_n/C_j)}$$

where j in the sum ranges over the indices of those constituents C_j which are compatible with e_n. Through the normal form (25), i.e.,

$$\vdash h \equiv \bigvee_{i \in J_h} C_i,$$

this is sufficient to determine the posterior probability $P(h/e_n)$ for each generalization in L_N^k:

$$P(h/e) = \sum_{i \in J_h} P(C_i/e).$$

Assume now that, among the n individuals of the sample e_n, there are c *kinds* of individuals, and denote this sample by e_n^c. In other words, c is the number of indices i such that $n_i > 0$, so that it measures the *variety* of evidence e_n^c (with respect to L_N^k). Let C^c be the constituent of L_N^k which claims that precisely the same c cells Q_i are

LOGICAL TOOLS   87

instantiated in the universe U which are already exemplified in the sample e_n^c. Then the basic result of the λ-α-system tells that, as soon as $\alpha < \omega$ and $\lambda > 0$, there is one and only one constituent of L_N^k which receives asymptotically the posterior probability 1:

(79)   If $\alpha < \omega$ and $\lambda > 0$, then
$P(C^c/e_n^c) \to 1$, if $n \to \infty$ and c is fixed
$P(C^w/e_n^c) \to 0$, if $n \to \infty$, $w > c$, and c is fixed.

This result shows that a rational treatment of inductive generalizations is possible in the λ-α-system.

An important special case of the λ-α-system is obtained by choosing $\lambda = w$ in formula (78) and likewise $\lambda = q$ in formula (76). This *generalized combined system* contains the combined system as a special case with $\alpha = 0$. In this system, the formula (78) reduces to

$$\frac{n_i + 1}{n + \lambda},$$

and the probabilities $P(C^w)$ and $P(e_n^c/C^w)$ can be expressed by

(80) $$P(C^w) = \frac{(\alpha + w - 1)!}{\sum_{i=1}^{q} \binom{q}{i} [(\alpha + i - 1)!/(i - 1)!]}$$

(81) $$P(e_n^c/C^w) = \frac{(w - 1)!}{(n + w - 1)!} n_{j_1}! \ldots n_{j_c}!$$

where α is a natural number and $n_{j_1}, \ldots, n_{j_c}$ are all the non-zero numbers n_i associated with evidence e_n^c. Hence,

(82) $P(C^w/e_n^c)$
$$= \frac{(\alpha + w - 1)!/(n + w - 1)!}{\sum_{i=1}^{q-c} \binom{q-c}{i} [(\alpha + c + i - 1)!/(n + c + i - 1)!]}.$$

If h is a universal generalization in L_N^k which claims that certain b cells are empty (and is compatible with e_n^c), then

(83) $P(h/e_n^c)$

$$= \frac{\sum_{i=0}^{q-b-c} \binom{q-b-c}{i} [(a+c+i-1)!/(n+c+i-1)!]}{\sum_{i=0}^{q-c} \binom{q-c}{i} [(a+c+i-1)!/(n+c+i-1)!]}$$

(cf. Niiniluoto and Tuomela, 1973, p. 31). If a and n are sufficiently large with respect to q, then the formulas (82) and (83) can be approximated by

(84) $\quad P(C^w/e_n^c) \approx \dfrac{(a/n)^{w-c}}{(1+a/n)^{q-c}}$

(85) $\quad P(h/e_n^c) \approx \dfrac{1}{(1+a/n)^b}.$

The corresponding result for singular sentences is

(86) $P(Q_i(a_{n+1})/e_n^c) =$

$$\frac{n_i + 1}{n + c} \cdot \frac{\sum_{i=0}^{q-c} \binom{q-c}{i} [(a+c+i-1)!(n+c)!/(n+c+i)!]}{\sum_{i=0}^{q-c} \binom{q-c}{i} [(a+c+i-1)!(n+c-1)!/(n+c+i-1)!]}$$

Hence, the probability that the next individual a_{n+1} is of the same kind as the earlier ones in e_n^c is

(87) $\quad \dfrac{\sum_{i=1}^{q-c} \binom{q-c}{i} [(a+c+i-1)!(n+c)!/(n+c+i)!]}{\sum_{i=1}^{q-c} \binom{q-c}{i} [(a+c+i-1)!(n+c-1)!/(n+c+i-1)!]}$

The value of (87) approachs the limit 1, if $n \to \infty$. (Cf. Hilpinen, 1968, pp. 70–71.)

Assume that P is a probability measure for language L_ω^k which satisfies the conditions

(A1) P is regular.
(A2) P is invariant with respect to any permutation of individual constants (symmetry with respect to individuals).
(A3) $P(Q_i(a_{n+1})/e_n^c)$ depends only on the numbers q, n_i, and n.

Then it follows that $P(Q_i(a_{n+1})/e_n^c)$ is of the form (71), where γ_i is the width of Q_i ($i = 1, \ldots, q$) and $\lambda > 0$. If P satisfies the additional condition

(A4) $\gamma_i = 1/q$ for all $i = 1, \ldots, q$ (symmetry with respect to Q-predicates),

then P must belong to Carnap's original λ-continuum. In this sense, conditions (A1)–(A4) give an 'axiomatization' to the λ-continuum.[22]

If condition (A3) is weakened to the form

(A5) $P(Q_i(a_{n+1})/e_n^c)$ depends only on the numbers q, n_i, n, and c,

then (A1), (A2), (A4), and (A5) together guarantee that P belongs to a *q-dimensional system* of inductive probabilities.[23] The values of P depend on q parameters λ and δ_w ($w = 1, \ldots, q-1$), where λ corresponds to Carnap's λ and δ_w regulates the prior probability of constituents C^w of the width w. In fact, for $w = 1, \ldots, q-1$,

$$\delta_w = P(Q_j(a_{w+1})/e_w^w),$$

where Q_j is not exemplified in e_w^w. The only special case of this q-dimensional continuum where $P(C^q) = 1$ holds is obtained by choosing for all δ_w the values that these probabilities would have in the λ-continuum. In all the other cases, there will be constituents C^w such that $P(C^w) > 0$, so that the fundamental theorem (79) about posterior probabilities is satisfied as well.

These results show that Carnap's λ-continuum is the only special case of the q-dimensional system where universal generalizations cannot receive non-zero probabilities on infinite domains of individuals. As the index c for the variety of evidence e_n^c shows how many universal generalizations of L_ω^k the evidence e_n^c has already falsified, the inability of Carnap's λ-continuum to handle inductive generalization adequately results from the fact that his assumption (A3) accounts only for the *enumerative* aspects of induction and excludes the *eliminative* aspects.

The intersection of Hintikka's λ-α-system and the q-dimensional system contains, besides the λ-continuum, precisely those members of the latter where $\lambda(w)$ as a function of w is simply of the form aw for some constant $a > 0$. The choice $a = 1$ gives us Hintikka's generalized combined system as a special case of the q-dimensional system — with the exception that the q-dimensional system allows for more flexible choices of the prior probabilities of constituents than (76) and (77).

One choice of the probabilities $P(C^w)$ which violates (77) is obtained by assuming that all constituent-structures G^w ($w = 1, \ldots, q$) are equally probable and by dividing the probability of G^w equally to all constituents C^w belonging to G^w (cf. Kuipers, 1978b). A more general procedure is applicable even in the case where the number q of Q-predicates is ω (as will be the case when the number k of predicates is infinite): assume that

$$P(G^w) = \zeta_w \quad (w = 1, 2, \ldots)$$

where

$$\sum_{w=1}^{\infty} \zeta_w = 1;$$

then divide the probability ζ_w equally to all the constituents C^w of the width w.

The problem of generalizing inductive logic to polyadic first-order languages is notoriously difficult. However, Hilpinen (1966, 1971) has suggested how Hintikka's system can be extended to monadic languages $L_N^{k=}$ with identity and to languages L with relation symbols. The axiomatic treatment of the q-dimensional system also has important consequences relevant to this problem. Recall that a depth-d constituent $C_j^{(d)}$ in L is a statement which claims that certain attributive constituents $Ct_i^{(d-1)}(x)$ are exemplified (see (44)). Hence, such attributive constituents constitute a classification system; in this case, to know that an individual a_m satisfies $Ct_i^{(d-1)}(x)$ we have to consider all kinds of sequences of d individuals starting with a_m (cf. (45)). If assumptions corresponding to (A1), (A2), and (A5) hold relative to this classification system, then the results concerning the monadic q-dimensional system show qualitatively how the probabilities of depth-d constituents

$C_j^{(d)}$ behave. In particular, it will be the case that, for sufficiently large evidence consisting of ramified sequences of d individuals, there is one and only one depth-d constituent which receives asymptotically the probability 1 (cf. (79)).

Most applications of inductive logic have been restricted to probabilities of the form $P(h/e)$, where h is a singular sentence or a generalization and e is a description of a sample of individuals in the same language as h. The case, where h is a generalization in a language L' with theoretical concepts and evidence e is expressed in an observational sublanguage L of L', was studied in Niiniluoto (1976b). Probabilities of the form $P(h/e \& T)$, where e and h are in L and T is a theory in L', were analysed in Niiniluoto and Tuomela (1973).

2.11. NOMIC CONSTITUENTS

So far we have considered only extensional (first-order) languages. However, it is doubtful that such languages are sufficiently rich to express natural laws. Take, for example, a simple law of the form

(88)   All F's are G.

According to the Humean view, (88) is nothing more than a statement about the actual world and its history — it tells, for each individual x which belongs, has belonged, or will belong, to the actual world, that if x is F then x is also G. Thus, in this view, (88) can be expressed by the material general conditional:

(89) $(x)(F(x) \supset G(x))$.

This analysis fails to distinguish between *lawlike* and *accidental generalizations*. For example, it may happen that only righthanded persons will ever be elected as presidents of Finland. Still, the claim 'All presidents of Finland are right-handed' is not a law, since — as we think — it would have been *possible* to elect some left-handed candidate. That this possibility did not materialize, was only an accident in the history of the actual world. The situation is different with genuinely lawlike statements like 'Copper conducts electricity' which assert the existence of a *necessary* or *nomic* connection between two physical attributes. The difference between accidental and lawlike generalizations manifests itself in many different ways: only the latter have explanatory and predictive power, counterfactual force, and systematic connections to other nomic statements.

There are two main strategies in analysing the notion of lawlikeness.[24] In the "pragmatist" view, universal laws are generalizations of the form (89) which satisfy certain methodological or pragmatic conditions — such as confirmability, systemicity, or explanatory power. Lawlikeness is, therefore, something which is attributed to certain generalizations by the scientific community. The "realist" view takes lawlikeness to be primarily a semantic feature: nomic statements assert something more than the Humean analysis presupposes. The law 'Copper conducts electricity' supports counterfactuals of the form 'If this piece of wood were copper, it would conduct electricity', just because is speaks both of actual situations and of physically possible situations. For this reason, (89) is not an adequate formalization of the law (88): the latter requires that the generalization (89) is true, not only in the actual world, but in all physically possible worlds as well. In other words, intensional concepts are needed in order to express the law (88).

Let $\Box$ be the operator of *physical* or *nomic necessity*, and let $\Diamond$ be the corresponding operator of *physical possibility*, defined by

$$\Diamond =_{df} \sim\Box\sim.$$

If L is any first-order language, denote by $L(\Box)$ the extension of L which is obtained by adding $\Box$ (and $\Diamond$) to the vocabulary of L and by introducing the following rule: if α is a formula, then $\Box\alpha$ (and $\Diamond\alpha$) is a formula. The formula $\Box\alpha$ will be read as 'it is physically necessary that α'.

By using this notation, the law (88) can now be expressed by

(90) $\Box(x)(F(x) \supset G(x))$.

This is equivalent to the formulas

(91) $\Box\sim(\exists x)(F(x) \,\&\, \sim G(x))$

and

(92) $\sim\Diamond(\exists x)(F(x) \,\&\, \sim G(x))$,

which say that the realization of F and non-G at the same time is physically impossible.[25]

A philosopher following the pragmatist startegy may accept the use of the necessity operator, but then he thinks that claims of the form (90) do not have true values at all. Instead, they may have epistemic assertability conditions relative to the state of knowledge within the scientific community. A nomic realist, on the other hand, tries to find a semantic treatment which allows him to assign truth values to intensional statements. A systematic tool for doing this is provided by the possible worlds semantics, developed by Stig Kanger, Jaakko Hintikka, and Saul Kripke in the 1950s.[26]

Assume now that, in the class of the structures Ω for language L, a relation 'is physically possible relative to' is defined. Then the truth of intensional sentences in a structure Ω is defined by the condition

(93) $\Omega \vDash \Box\alpha$ iff $\Omega' \vDash \alpha$ for all structures Ω' which are physically possible relative to Ω.

From (93) with Tarski's definition (1) it follows that

(94) $\Omega \vDash \Diamond\alpha$ iff $\Omega' \vDash \alpha$ for some structure Ω' which is physically possible relative to Ω.

Thus, the law (90) is true in the actual world if and only if the extension of F is included in the extension of G in all physically possible worlds.

Intuitively speaking, structure Ω' is physically possible relative to structure Ω if Ω and Ω' differ at most by their singular features but satisfy the same laws. It follows that this relation is reflexive, symmetric, and transitive. This is known to imply the result that the logic of $\Box$ is the Lewis system S5. This system is axiomatized by the following principles:

$$\Box\alpha \supset \alpha$$
$$\Box(\alpha \supset \beta) \supset (\Box\alpha \supset \Box\beta)$$
$$\Diamond\alpha \supset \Box\Diamond\alpha,$$

with Modus Ponens and the following Rule of Necessitation: if $\vdash\alpha$, then $\vdash\Box\alpha$.[27] Among the important theorems of S5 we may mention:

$$\Box\Box\alpha \equiv \Box\alpha$$
$$\Diamond\Diamond\alpha \equiv \Diamond\alpha$$
$$\Box\Diamond\alpha \equiv \Diamond\alpha$$
$$\Diamond\Box\alpha \equiv \Box\alpha$$
$$\Box\alpha \equiv \sim\Diamond\sim\alpha$$
$$\alpha \supset \Diamond\alpha$$
$$\Box(\alpha \,\&\, \beta) \equiv (\Box\alpha \,\&\, \Box\beta)$$
$$\Diamond(\alpha \vee \beta) \equiv (\Diamond\alpha \vee \Diamond\beta)$$
$$(\Box\alpha \vee \Box\beta) \supset \Box(\alpha \vee \beta)$$
$$\Diamond(\alpha \,\&\, \beta) \supset \Diamond\alpha \,\&\, \Diamond\beta.$$

Assume now that L is the monadic language L_N^k with the one-place predicates $M_1, \ldots, M_k$, Q-predicates $Q_1, \ldots, Q_q$ ($q = 2^k$), and the constituents $C_1, \ldots, C_t$ ($t = 2^q - 1$). (Cf. Sections 1, 2, and 5 above.) By (5), any physically necessary generalization of the form

$$\Box(x)(M_1(x) \supset M_2(x))$$

can be expressed in terms of some Q-predicates $Q_j, j \in I$:

(95) $\quad \Box(x)\left[\bigvee_{j \in I} Q_j(x)\right].$

Further, (95) is equivalent to

$$\Box \sim (\exists x)\left[\bigvee_{j \notin I} Q_j(x)\right]$$

$$\equiv \Box \sim \bigvee_{j \notin I} (\exists x) Q_j(x)$$

$$\equiv \Box \bigwedge_{j \notin I} \sim (\exists x) Q_j(x)$$

$$\equiv \bigwedge_{j \notin I} \Box \sim (\exists x) Q_j(x),$$

where the last formula says that each Q-predicate $Q_j, j \notin I$, is necessarily empty.

Formula (95) is a "simple causal law" in the sense of Uchii (1973). The stronger sentence

(96) $\quad \bigwedge_{j \in I} \Diamond(\exists x) Q_j(x) \ \& \ \Box(x)\left[\bigvee_{j \in I} Q_j(x)\right],$

which entails (95), is called a "non-paradoxical basic causal law" by Uchii. I shall call (96) a *nomic constituent* in this work. An equivalent formulation of (96) is

(97) $\quad \bigwedge_{j \in I} \Diamond(\exists x) Q_j(x) \ \& \ \bigwedge_{j \notin I} \Box \sim (\exists x) Q_j(x)$

which says that certain kinds of individuals are physically possible and other physically impossible.

Assuming only that some Q-predicate is possibly non-empty, the number of nomic constituents is $L_N^k(\Box)$ is $2^q - 1$, i.e., the same as the number t of the ordinary constituents of L_N^k. Let us denote them by B_1, ..., B_t. It is also clear that the nomic constituents are mutually exclusive and jointly exhaustive:

$$\vdash \sim (B_i \& B_j) \quad (i \neq j)$$

$$\vdash \bigvee_{i=1}^{t} B_i.$$

The structure of nomic constituents is closely parallel to ordinary constituents. As a notational convention, let us denote by B_i that nomic constituent in $L_N^k(\Box)$ which is obtained from the constituent C_i of L_N^k (cf. (23)):

$$C_i = \bigwedge_{j \in \mathbf{CT}_i} (\exists x) Q_j(x) \& (x) \left[\bigvee_{j \in \mathbf{CT}_i} Q_j(x) \right]$$

$$B_i = \bigwedge_{j \in \mathbf{CT}_i} \Diamond (\exists x) Q_j(x) \& \Box(x) \left[\bigvee_{j \in \mathbf{CT}_i} Q_j(x) \right].$$

Then the width of B_i is $|\mathbf{CT}_i|$, i.e., the number of cells it claims to be possibly non-empty. It is interesting to note that B_i is both weaker and stronger than C_i: claims about possible existence are weaker than claims about actual existence, but claims about necessary non-existence are stronger than claims about actual non-existence.

A constituent C_m is compatible with a nomic constituent B_i if and only if $\emptyset \neq \mathbf{CT}_m \subseteq \mathbf{CT}_i$, i.e., C_m claims that some kinds of individuals possibly existing by B_i are actually existing. There are $2^{q-|\mathbf{CT}_m|}$ nomic constituents compatible with a given constituent C_m. Similarly, the number of constituents C_m compatible with a given B_i is $2^{|\mathbf{CT}_i|} - 1$. Further,

(98) $\quad \vdash B_i \equiv \bigvee \{B_i \& C_m \mid \mathbf{CT}_m \subseteq \mathbf{CT}_i\}.$

It follows that every statement in $L_N^k(\Box)$ which is expressible in a *B-normal form*, i.e., as a disjunction of nomic constituents, is also expressible in a *B &C-normal form*, i.e., as a disjunction of sentences of the form B_i & C_m. In particular, this is true of all those laws which claim that certain cells are necessarily empty. For example, the law (95) is equivalent to the disjunctions

(99) $\bigvee \{B_i \mid \emptyset \neq \mathbf{CT}_i \subseteq \mathbf{I}\}$
        $\bigvee \{B_i \,\&\, C_m \mid \emptyset \neq \mathbf{CT}_m \subseteq \mathbf{CT}_i \subseteq \mathbf{I}\}.$

The total number of mutually exclusive sentences of the form B_i & C_m is

$$\sum_{i=1}^{k} \binom{k}{i} \cdot (2^k - 1) = 3^k - 2^k.$$

As Uchii (1973) notes, nomic constituents of the form (96) are not the strongest modal statements expressible in language $L_N^k(\square)$. Consider sentences of the form

(100) $\bigwedge_{i=1}^{t} (\pm) \Diamond C_i.$

Each such sentence is equivalent to a formula

(101) $\bigwedge_{i \in \mathbf{H}} \Diamond C_i \,\&\, \square \left[\bigvee_{i \in \mathbf{H}} C_i \right]$

for some $\mathbf{H} \subseteq \{1, \ldots, t\}$. Assuming only that some constituent of L_N^k is possibly true, sentences of the form are mutually exclusive and jointly exhaustive. Their number is $2^t - 1$, which is much greater than the number t of the nomic constituents.

It is instructive to note that (101) entails the statement

(102) $\bigwedge_{\substack{i \in \bigcap \mathbf{CT}_j \\ j \in \mathbf{H}}} \square (\exists x) Q_i(x) \,\&\, \bigwedge_{\substack{i \in \bigcup \mathbf{CT}_j \\ j \in \mathbf{H}}} \Diamond (\exists x) Q_i(x) \,\&\,$

$\&\, \bigwedge_{\substack{i \notin \bigcup \mathbf{CT}_j \\ j \in \mathbf{H}}} \square {\sim} (\exists x) Q_i(x)$

In other words, if the positive conjuncts of (100), i.e., the constituents C_i, $i \in \mathbf{H}$, make jointly some existence claims and jointly some nonexistence claims, then (101) entails that each of these claims is neces-

LOGICAL TOOLS 97

sary. (See the first and third conjuncts of (102).) By dropping the first conjunct from (102), we obtain a nomic constituents of the form (97) with

$$I = \bigcup_{j \in H} CT_j.$$

The statements (100) are thus logically stronger than nomic constituents in the sense that they are able to make physically necessary existence claims. For the purposes of the philosophy of science, this extra strength is largely superfluous, since laws of nature are typically statements about physically necessary *non*-existence. This is also the reason why we shall find the concept of nomic constituent to be useful in our theory of truthlikeness (cf. chapters 3 and 11).

The laws that are expressible in language $L_N^k(\Box)$ are typical examples of what John Stuart Mill called *laws of coexistence*: for example, (90) says that, for each object x, the property F is necessarily coexistent at the same time with the property G. To express Mill's *laws of succession*, we have to build into our language some temporal notions. In connection with many laws of this type, time can be treated as discrete with the moments $t, t+1, \ldots$ The simplest laws of succession assert the existence of a nomic connection between two successive states of certain kinds of objects or systems. Assuming that these objects and their states can be described in language L_N^k, a law of succession can be expressed in the form

(103) $\quad \Box(x)(\alpha^t(x) \supset \beta^{t+1}(x)),$

where $\alpha^t(x)$ says that x satisfies the formula $\alpha(x)$ at time t, and $\beta^{t+1}(x)$ says that x satisfies $\beta(x)$ at the next moment $t+1$. If we represent the formulas $\alpha(x)$ and $\beta(x)$ as disjunctions of Q-predicates (cf. (5)), then (103) is equivalent to

(104) $\quad \Box(x) \left[\bigvee_{i \in I_\alpha} Q_i^t(x) \supset \bigvee_{i \in I_\beta} Q_i^{t+1}(x) \right]$

$\equiv \Box(x) \bigwedge_{i \in I_\alpha} \left[Q_i^t(x) \supset \bigvee_{i \in I_\beta} Q_i^{t+1}(x) \right]$

$\equiv \bigwedge_{i \in I_\alpha} \Box(x) \left[Q_i^t(x) \supset \bigvee_{i \in I_\alpha} Q_i^{t+1}(x) \right].$

It follows that all laws of succession relative to $L_N^k(\Box)$ can be expressed by disjunctions of sentences of the form

(105) $\bigwedge_{(i,j) \in T} \Diamond(\exists x)(Q_i^t(x) \& Q_j^{t+1}(x)) \&$

$\& \Box(x)\left[\bigvee_{(i,j) \in T} (Q_i^t(x) \& Q_j^{t+1}(x))\right].$

(Cf. Uchii, 1977.) (105) can again be called *a nomic constituent*. Here the index set T gives a list of all the physically possible *transformations* from one state to another.

For example, the law

$$\Box(x)[Q_1(x) \supset (Q_2(x) \vee Q_3(x))]$$

is equivalent to the disjunction of all nomic constituents of the form (105) where T contains at least one of the pairs (1, 2) and (1, 3). More generally, (104) is equivalent to the disjunction of nomic constituents where T contains at least one pair $(i, j), j \in \mathbf{I}_\beta$, for each $i \in \mathbf{I}_\alpha$.

All the ideas discussed above can be generalized from monadic to polyadic languages. The structure of the depth-d constituents in a polyadic first-order language L was given in formula (44). The corresponding depth-d, nomic constituents of $L(\Box)$ has then the following form:

$$\bigwedge_{i \in \mathbf{CT}_j^{(d-1)}} \Diamond(\exists x_1) Ct_i^{d-1}(x_1) \& \Box(x)\left[\bigvee_{i \in \mathbf{CT}_j^{(d-1)}} Ct_i^{(d-1)}(x_1)\right].$$

By similar modifications we obtain a definition of nomic attributive constituents. Then any depth-d law in $L(\Box)$ which asserts the necessary non-existence of certain kinds of individuals (or sequences of individuals) is representable as a disjunction of depth-d nomic constituents of $L(\Box)$.

Let us conclude this section by discussing the possibility of generalizing the basic ideas of inductive logic from the extensional languages L_N^k to the intensional languages $L_N^k(\Box)$ (cf. Section 10). Perhaps the most interesting question here concerns the behavior of the probabilities $P(B_i/e_n^c)$, where e_n^c is again a sentence of L_N^k describing a sample of n individuals exemplifying c different cells.

Uchii (1972, 1973) notes that all the probabilities $P(C_i)$, $P(B_j)$,

$P(e_n^c/C_i)$, $P(e_n^c/B_j)$, and $P(e_n^c)$ can be calculated as soon as the probabilities of the form $P(C_i \& B_j)$ and $P(e_n^c/C_i \& B_j)$ have been fixed. As in Section 10, let C^w be a constituent of L_N^k with the width w, and let B^w be the corresponding nomic constituent of $L_N^k(\Box)$. Note that the observation of an individual a in cell Q_i — or the sentence $Q_i(a)$ — verifies the possibility claim $\Diamond(\exists x)Q_i(x)$ and falsifies the impossibility claim $\Box \sim (\exists x)Q_i(x)$. Factual evidence statements e_n^c are, therefore, able to falsify some nomic constituents, and thereby to support the others. Moreover,

$$e_n^c \vdash B^c \supset C^c$$
$$e_n^q \vdash B^q \equiv C^q,$$

so that

$$P(B^c \& C^c/e_n^c) = P(B^c/e_n^c)P(C^c/e_n^c \& B^c) = P(B^c/e_n^c)$$
$$P(B^c/e_n^c) \leqslant P(C^c/e_n^c)$$

and

$$P(e_n^c/B^c) = P(C^c/B^c)P(e_n^c/C^c \& B^c).$$

Let P be any inductive probability measure for L_N^k which satisfies the Hintikkian condition (79), i.e., $P(C^c/e_n^c) \to 1$, when $n \to \infty$ and c is fixed. To study the behavior of $P(B^c/e_n^c)$, let us make tentatively the following assumption:

(106) $\quad P(e_n^c/C^c) = P(e_n^c/C^c \& B^c).$

(Cf. Uchii, 1972, p. 173.) (106) says that the probability of the evidence e_n^c, given knowledge about the actual constitution C^c of the universe, is not changed, if the nomic constituent B^c is added to the evidence. What $C^c \& B^c$ says over and above C^c is simply this: the cells empty by C^c are necessarily empty. According to (106), this extra knowledge does not change the probability of finding sample individuals in the exemplified cells. From (106) it follows that

$$\frac{P(B^c/e_n^c)}{P(C^c/e_n^c)} = \frac{P(B^c)P(e_n^c/B^c)}{P(C^c)P(e_n^c/C^c)}$$
$$= \frac{P(B^c)P(C^c/B^c)P(e_n^c/C^c \& B^c)}{P(C^c)P(e_n^c/C^c)}$$
$$= P(B^c/C^c).$$

Hence, (79) entails that

(107) $P(B^c/e_n^c) \to P(B^c/C^c)$, when $n \to \infty$ and c is fixed.

Further,

(108) $P(B^c/e_n^c) \to 1$, when $n \to \infty$ and c is fixed, iff $P(B^c/C^c) = 1$.

If we further assume that

(109) $P(C^w) = P(B^w)$ for all w,

then the condition $P(B^c/C^c) = 1$ entails that

$$P(C^c/B^c) = 1$$

and

$$P(e_n^c/C^c \,\&\, B^c) = P(e_n^c/B^c) = P(e_n^c/C^c).$$

Assumption (109) is not implausible, if we recall that the total number of the constituents C^w is the same as the total number of the nomic constituents B^w.

Result (107) is not surprising: if we have asymptotically become certain that C^c is the true description of the actual constitution of the universe, this conclusion is still compatible with 2^{q-c} nomic constituents B^w, $w \geqslant c$. *Claims about possible* but non-actual *worlds are*, in this sense, radically *underdetermined by observations* about the actual world. Only in the case $c = q$, where all possibilities have been realized, the distinction between actual and possible collapses, and we have $P(B^q/e_n^q) = P(C^q/e_n^q) = 1$.

Uchii (1972, 1973) has formulated systems of inductive logic which are based upon the assumptions (106), (109), and

(110) $P(B^w/C^w) = P(C^w/B^w) = 1$ for all w.

By (108), these systems satisfy the theorem

(111) $P(B^c/e_n^c) \to 1$, when $n \to \infty$ and c is fixed.

Nomic constituents, and thereby all nomic generalizations, are thus asymptotically confirmable to the degree 1 on the basis of observational evidence.

In his treatment of 'causal modalities', Uchii assumes that, for languages L_N^k without indexical predicates, a world Ω' is physically possible relative to world Ω if and only if all the Q-predicates exemplified in Ω' are already exemplified in Ω (Uchii, 1973). It follows that the strongest true universal statement becomes also physically necessary. This also means that all possible kinds of individuals already exist in the actual world — or exist there with probability one (Uchii, 1972). These systems, therefore, satisfy conditions of the form:

$$P((\exists x)a(x), \Diamond(\exists x)a(x)) = 1,$$

and, of course, (110).

These assumptions are very strong: they are in fact versions of the notorious metaphysical doctrine known as the *Principle of Plenitude*. This principle says that all genuine possibilities will be realized in the actual history of the world. This assumption is highly questionable: the point of departure for the new modal theory of Duns Scotus was to affirm the existence of physical possibilities which will never be actualized. Indeed, if 'possibly true' is equivalent to 'sometimes actually true', there is no need to develop a modal logic: the logic of physical necessity reduces to temporal logic, and the distinction between lawlike and accidental generalization collapses. For these reasons, the Principle of Plenitude — and the assumption (110) — will be rejected in this work.[28]

Is there any way of saving the result (111) without assuming dubious metaphysical ideas like the Principle of Plenitude? A positive answer to this question is defensible, if we reinterpret the whole situation to which inductive logic is applied. Virtually all systems of inductive logic are based upon the (often tacit) presupposition that the true constitution of the universe is fixed, and the aim of the evidence is to help to identify this pre-existing truth. In other words, the true constituent is assumed to be independent of our activity of sampling and examining individuals from the universe. It follows from this assumption, e.g., that $P(Q_j(a)/C_i) = 0$ if $j \notin \mathbf{CT}_i$.

In connection with lawlike statements, this static assumption is problematic. The whole point of *experimentation* — as opposed to passive observation — is that the investigator "manipulates" nature: by his planned interventions, he brings about courses of events which would not have occurred otherwise, and he prevents events which would have occurred otherwise.[29] Through his own activity, he thus realizes some physical possibilities which were not actual and would have remained unactualized without his action. Moreover, the only way of confirming laws of nature is grounded upon this activity of experimentation.[30]

From this dynamic perspective, it is not very interesting which constituent C_i of L_N^k happens to be true at a given moment of time t. If C_i is true now, we may be able to design an experiment which realizes a new, until now unactualized possibility $Q_j(x)$, $j \notin \mathbf{CT}_i$. Hence, the true constituent of L_N^k is a function of the experiments that have been made by us. In such situations, our real aim is to find out the true nomic constituent of $L_N^k(\Box)$: the permanent features of the universe are precisely those claims about physical possibility and impossibility that such a nomic constituent contains.

If the perspective outlined above is adopted, nothing prevents us from applying Hintikka's system of inductive logic or its generalizations directly to the *nomic* constituents B_i instead of the ordinary constituents C_i. In other words, we forget about the ordinary constituents of L_N^k — they express only accidental and temporally changing features of the universe — and rewrite all the assumptions and results of Section 10 by replacing C^w with the corresponding B^w. It follows immediately that the desired result (111) is provable: if the evidence e_n^c is gathered through active experimention, which tries systematically to go through all physical possibilities, then $P(B^c/e_n^c)$ grows towards the value 1, when n increases without limit and c remains fixed. In this way, all the nice features of the Hintikka-type system of inductive logic can be transported to situations involving nomic constituents and laws of nature.

CHAPTER 3

QUANTITIES, STATE SPACES, AND LAWS

It is well-known that first-order languages can contain, in addition to monadic and polyadic predicates, *n*-place function symbols for any *n*. Still, many of the central concepts introduced in Chapter 2 (*Q*-predicates, constituents etc.) are primarily applicable to formal languages containing qualitative predicates. To be sure, Carnap's notion of a monadic conceptual system leaves room for attribute spaces which are based upon some quantity (cf. Chapter 2.2). However, Carnap wished to avoid uncountable families of predicates, since that would lead to an uncountable number of *Q*-predicates as well. Therefore, he assumed that the value space of a real-valued quantity is partitioned into at most a denumerably infinite number of intervals. A more direct and more elegant way of treating quantitative concepts and laws — the so-called *state space conception of theories* — is given in this chapter. We shall see that the idea of a quantitative state space is in fact obtainable as a limiting case of Carnapian discrete conceptual systems.

3.1. QUANTITIES AND METRIZATION

Many branches of science — from physics and astronomy to biology, physiology, economics, and social psychology — use mathematical theories with quantitative concepts. Many laws of nature express functional relations between quantities. Mathematics seems, therefore, to be an indispensable part of the language of science. But how is it possible that mathematics — an abstract creation of man — is applicable to reality?[1]

The Pythagorean-Platonist tradition has a straightforward answer to this question: the world itself *is* quantitative. The founders of modern natural science accepted this view which may be called *mathematical realism*. Galileo claimed that the Book of Nature is written in the language of geometry. In effect, he distinguished objective quantities belonging to reality and subjective qualities belonging to the mind of an observer — anticipating thus John Locke's division between primary and secondary qualities. For Newton, the physical world was composed of bodies with real quantitative properties — such as their instantial position, velocity, acceleration, and mass — and forces between such bodies. Similarly, Descartes took a quantitative property, viz. extension, to be the primary attribute of the material substance.

The view opposite to mathematical realism usually relies on radical empiricism. Berkeley made the objection against Newton's mathematical conception of space that the points of the real line do not really exist, since they cannot be observed. Berkeley's phenomenalism assumes that *esse est percipi*. Similarly, later positivists and operationalists have argued that physical quantities do not really have exact values, since they cannot be measured with arbitrarily great accuracy: to be is to be the result of an observation or the value of a measurement. The instrumentalists, like Pierre Duhem, accept the use of real-valued quantities as fictions that are useful for some scientific purposes, but insist at the same time that laws involving sharp values of quantities are idealizations which are neither true nor false. In other words, such exact laws are, so to say, too sharp to fit with the imprecise reality.[2] This kind of view is defended also by Husserl's phenomenology: in criticizing Galileo's mathematical realism Husserl argued that modern mathematical physics describes an idealized construction which should not be mistaken with the true reality, i.e., with the 'life world' of everyday experience (cf. Gurwitch, 1967).

The problem for the mathematical realists is to find a place for qualities in the world of quantities. The problem for the phenomenalists and the phenomenologists is to account for the role of quantities in the world of qualities.

A critical scientific realist typically thinks that — to use Sellars' terms — the 'scientific image' of the world is ontologically but not epistemologically prior to the 'manifest image' of everyday experience (cf. Pitt, 1981). Thus, if the scientific image (or its ultimate limit) is ontologically committed to the existence of quantitative properties, then quantities are real. However, a scientific realist need not necessarily be a mathematical realist as well: he may hold that the ultimate description of reality (if such description exists at all) is to be given in a 'physicalistic' language with qualitative or classificatory predicates — so that the language of quantities is only a method for representing the qualitative properties and relations by real numbers.[3]

As long as we recognize the different positions, these metaphysical

QUANTITIES, STATE SPACES, AND LAWS

issues need not be solved in this work. The theory of truthlikeness can be developed both to qualitative and quantitative languages.

For our purposes, it is important to observe that the modern theory of measurement, as developed by Dana Scott and Patrick Suppes in the late 1950s, has shown how to build a bridge from qualitative and comparative concepts to quantities.[4] The fact that mathematics can be applied to reality is not a mystery any more — and needs no metaphysical justification. It is now possible to study, in a precise manner, the conditions that comparative concepts have to satisfy in order to be representable (uniquely or up to some class of transformations) by quantities. Whether these conditions hold in actual cases becomes then a factual question which is open to empirical investigation.

To illustrate these ideas, let **E** be a class of objects (of a certain kind), and let S be a binary relation in **E**. Here S is assumed to be the interpretation of some comparative concept, such as 'is heavier than', 'is longer than', 'is warmer than', 'is louder than', 'is more probable than' or 'is more valuable than'. Then a Representation Theorem tells that certain conditions on the system (**E**, S) are sufficient (or necessary and sufficient) for the existence of a real-valued function $f_s: \mathbf{E} \to \mathbb{R}$ which preserves the relation S:

(1) $(x, y) \in S$ iff $f_s(x) > f_s(y)$

holds for all $x, y \in \mathbf{E}$. A Uniqueness Theorem tells to what extent the function f_s is unique: the *scale* of f_s depends on the class of transformations which can be applied to the values of f_s without violating condition (1). If the structure of the system (**E**, S) is more complicated, then the representing function f_s has to satisfy additional conditions besides (1). For example, if there is an operation $\circ$ of putting two objects of **E** together, then we may want f_s to satisfy the additivity requirement

$$f_s(x \circ y) = f_s(x) + f_s(y)$$

for all $x, y \in \mathbf{E}$.

The process of introducing function $f_s: \mathbf{E} \to \mathbb{R}$ on the basis of $S \subseteq \mathbf{E} \times \mathbf{E}$ is called (fundamental) *metrization*. Function f_s is called the *quantity* determined by relation S, and its values $f_s(x)$ for $x \in \mathbf{E}$ express *quantitative properties* of the objects in **E**. For example, $f_s(x)$ may be the mass, length, temperature, loudness, probability, or utility of x. All these quantities are metrizable at least with respect to the interval scale, i.e., their values are unique up a positive linear transformation $h(z) = az + b$ $(a > 0)$, and some with respect to the ratio scale, i.e., up to transformations of the form $h(z) = az$ $(a > 0)$.

The actual determination of the value of function f_s for some object x in **E** is called the *measurement* of $f_s(x)$. Thus, measurement is a methodological procedure which employs various kinds of devices and instruments, while metrization is a method of concept formation.

Actual measurements never yield values which are determined exactly, i.e., with the accuracy of a real number. Hence, empirical data obtained by measurement are always imprecise to some extent. Empiricist instrumentalists conclude from this fact that the exact values do not exist. For example, Ludwig (1981) says that imprecision in physical measurement has an unknown limit, but belief in "infinitely high precision" is false. In this view, repeated measurements may give us more and more precise results, but they do not approach the correct sharp value of the quantity — since there is no such value.

A scientific realist answers to this view in the following way: we have to distinguish between the existence of the value $f_s(x)$ and our ability to determine it by measurement. If the conditions in the Representation Theorem for f_s are *true*, then the exact value of $f_s(x)$ exists — independently of our knowledge of this value which is always relative to the best instruments that we have so far been able to invent. On the other hand, it may happen that these conditions are highly *idealized* and, hence, not satisfied by the actual objects. For example, conditions guaranteeing the existence of subjective probability distributions and utility functions are satisfied only by ideally rational agents. Even though the sharp quantitative values do not really exist in those cases, it may nevertheless be rational for many purposes — both cognitive and practical — to make the idealized assumption of their existence.

3.2. FROM CONCEPTUAL SYSTEMS TO STATE SPACES

Consider again a monadic conceptual system with $k \geq 1$ families of predicates

$$\mathcal{M}_1 = \{M_i^1 \mid i = 1, \ldots, n_1\}$$
$$\vdots$$
$$\mathcal{M}_k = \{M_i^k \mid i = 1, \ldots, n_k\}.$$

(See Chapter 2.2.) Assume now that within each family $\mathcal{M}_j$, $j =$

1, ..., k, it is possible to speak of the relative resemblance or similarity between its predicates. In Chapter 1.3., we have already discussed the proposal that the corresponding degrees of similarity can be expressed in a quantitative way. So let us assume that a distance function $d_j: \mathcal{M}_j \to \mathbb{R}$ is defined for each family $\mathcal{M}_j$ (cf. Carnap, 1980).

For example, in the family of the color predicates, the distance function should express the fact that orange is more similar to yellow than to blue.

If $\mathcal{M}_j$ is the family of lengths, defined by

$$M_i^j(x) = \text{the length of } x \text{ is between } i-1 \text{ and } i \text{ cm,}$$

then distance between predicates $M_i^j(x)$ and $M_m^j(x)$ is $|m - i|$. More generally, if $\mathcal{M}_j$ is obtained as a partition of the real axis $\mathbb{R}$ (or some closed subinterval of $\mathbb{R}$), the distance between two elements of $\mathcal{M}_j$ is simply the distance in $\mathbb{R}$ between the midpoints of the corresponding intervals.

If the family $\mathcal{M}_j$ contains only two elements, i.e., $\mathcal{M}_j = \{M_j, \sim M_j\}$, then we can stipulate that the distance between them is 1:

(2) $d_j(M_j, \sim M_j) = 1$.

Let **Q** be the set of the Q-predicates $Q_{j_1 j_2 \ldots j_k}(x)$ relative to families $\mathcal{M}_1$, ..., $\mathcal{M}_k$ (cf. formula (2.8)). Then the distance between two Q-predicates $Q_{j_1 \ldots j_k}$ and $Q_{m_1 \ldots m_k}$ can be defined by using the Minkowski metrics

(3) $\left(\sum_{i=1}^{k} d_i(M_{j_i}^i, M_{m_i}^i)^p \right)^{1/p}$

(Cf. Example 1.2.) The definition (3) can be generalized by introducing a weight $\xi_i > 0$ for each family $\mathcal{M}_i$:

(4) $\left(\sum_{i=1}^{k} \xi_i d_i(M_{j_i}^i, M_{m_i}^i)^p \right)^{1/p}$.

(Cf. Chapter 1.4.) As special cases of (3), we obtain the city-block metric ($p = 1$), the Euclidian metric ($p = 2$), and the Tchebycheff metric ($p = \infty$).

If all the families $\mathcal{M}_i$ are considered as equally relevant, then it is reasonable to choose $\xi_i = 1/k$ for $i = 1, \ldots, k$, and p as relatively small ($p = 1$ or $p = 2$) in (4). This can be illustrated in the special case, where each $\mathcal{M}_i$ is a dichotomy, i.e., $\mathcal{M}_i = \{M_i, \sim M_i\}$, $i = 1, \ldots, k$. Then, by (2), the city-block distance between two Q-predicates Q_j and Q_m of the form

$$(\pm)M_1(x) \& \cdots \& (\pm)M_k(x)$$

(cf. formula (2.2)) is

(5) $\varepsilon(Q_j, Q_m)$ = the number of the different signs (+ or −) in the corresponding conjuncts of Q_j and Q_m, divided by k.

If a Q-predicate is represented as a k-sequence of 1's and 0's (1 corresponding to positive and 0 to negative elements), then (5) defines also the normalized Hamming distance between the Q-predicates (cf. Example 1.6). The normalized Euclidean distance Q_j and Q_m is then

(6) $\sqrt{\varepsilon(Q_j, Q_m)}$.

If each $\mathcal{M}_i$ consists of countable number of real-valued intervals M_j^i with the midpoints $\bar{x}_j^i$, then the distance (3) reduces to

(7) $\left(\sum_{i=1}^{k} |\bar{x}_{j_i}^i, \bar{x}_{m_i}^i|^p \right)^{1/p}$.

With $p = 2$, (7) gives the ordinary Euclidean distance d in $\mathbb{R}^n$ between the points $(\bar{x}_{j_1}^1, \ldots, \bar{x}_{j_k}^k)$ and $(\bar{x}_{m_1}^1, \ldots, \bar{x}_{m_k}^k)$. In this case, the set $\mathbf{Q}$ of Q-predicates with its metric is isomorphic to a countable subset of the Euclidean space $\mathbb{R}^n$.

Assume then that the countable partition corresponding to $\mathcal{M}_i$ is made finer and finer, so that in the limit the intervals degenerate to single real numbers. Then family $\mathcal{M}_i$ is replaced by the real axis $\mathbb{R}$, i.e., with the value space of the quantity — call it h_i — which determines $\mathcal{M}_i$. When this procedure is applied to all families $\mathcal{M}_i$, $i = 1, \ldots, k$, the discrete set $\mathbf{Q}$ of Q-predicates is replaced by the k-dimensional Euclidean space $\mathbb{R}^k$. The uncountable set $\mathbb{R}^k$ with its Euclidean metric is then called the *state space* generated by the quantities h_i, $i = 1, \ldots, k$.[5]

A more direct way of defining the state space $\mathbb{R}^k$ is the following. Let $h_1, \ldots, h_k$ be real-valued quantities defined for objects of the same kind $\mathbf{E}$. The state of an object $x \in \mathbf{E}$ can be expressed by the h-tuple

QUANTITIES, STATE SPACES, AND LAWS 109

$(h_1(x), \ldots, h_k(x))$. The set of all such h-tuples is the k-fold Cartesian product $\mathbb{R}^k$ of $\mathbb{R}$ with itself. With the Euclidean metric $\mathbb{R}^k$ becomes a metric state space.

If we start from a monadic language L with k one-place function symbols $\bar{h}_i$, $i = 1, \ldots, k$, and a name $\bar{r}$ for each real number $r \in \mathbb{R}$, then we can define an uncountable number of Q-predicates of the form

(8) $\quad \bar{h}_1(x) = \bar{r}_1 \& \cdots \& \bar{h}_k(x) = \bar{r}_k.$

If the Q-predicate (8) is replaced by the k-tuple

(9) $\quad (r_1, \ldots, r_k),$

then the set of the Q-predicates of L is transformed to the set $\mathbb{R}^k$. When we are dealing with quantities, it is more convenient to deal directly with the familiar mathematical space $\mathbb{R}^k$ instead of the more complicated class of linguistic entities of the form (8).

Indeed, it is a common practice in mathematical physics — at least since the time of Ludwig Boltzmann — to represent physical theories by means of equations within mathematically defined state spaces. For example, in classical mechanics and statistical mechanics the state of a system of n particles is represented as a point in the "phase space" $\mathbb{R}^{6n}$, where both the place and the impulse of each particle is given in three co-ordinates. The state space may, of course, be also a finite subspace of the Euclidean space — or any normed linear space. In particular, Quantum Mechanics can be formulated with states which are vectors in an infinite-dimensional Hilbert space. Bas van Fraassen (1970, 1972, 1980) has argued that the state space conception provides a general framework for analysing the structure and semantics of scientific theories.[6]

We have thus seen that Carnapian discrete state spaces — i.e., sets of Q-predicates with a metric — can be directly generalized to a conception of a metric state space generated by quantities. The next three sections show how different kinds of quantitative laws can be formulated by using the state space approach.

3.3. LAWS OF COEXISTENCE

Laws of coexistence express nomic connections between properties. Paradigmatic examples of such laws are the following: 'All ravens are black', 'Grass is green', 'Copper conducts electricity'. In Chapter 2.11, we argued that typical laws of coexistence can be expressed as disjunctions of nomic constituents of the form

$$(10) \quad \bigwedge_{Q_j \in \mathbf{CT}} \Diamond (\exists x) Q_j(x) \,\&\, \Box (x) \left[\bigvee_{Q_j \in \mathbf{CT}} Q_j(x) \right],$$

where Q_j's are Q-predicates in a monadic language L. As Q-predicates are defined by conjunctions of the basic predicates of the language L, a nomic constituent of the form (10) tells *which combinations of properties are physically possible and which are not*. The index set **CT** describes the region of the physically possible within the discrete state space **Q** of all Q-predicates.

Assume then that **Q** is the k-dimensional Euclidean space $\mathbb{R}^k$ determined by k quantities $h_i \colon \mathbf{E} \to \mathbb{R}$ ($i = 1, \ldots, k$). The strongest laws of coexistence again tell which combinations of the values $h_1(x)$, ..., $h_k(x)$ are physically possible and which are not. In other words, quantitative laws of coexistence typically specify the *region of the physically possible states* in **Q** (cf. van Fraassen, 1970). Normally they are formulated by means of a function $f \colon \mathbb{R}^k \to \mathbb{R}$ which expresses the permissible combinations of the values of the quantities $h_1, \ldots, h_k$:

$$(11) \quad f(h_1(x), \ldots, h_k(x)) = 0.$$

If for all $z_1, \ldots, z_{k-1} \in \mathbb{R}$ there is one and only one value $z_k \in \mathbb{R}$ such that $f(z_1, z_2, \ldots, z_k) = 0$, then the equation (11) can be solved with respect to h_k, i.e., there is a function $g \colon \mathbb{R}^{k-1} \to \mathbb{R}$ such that

$$(12) \quad h_k(x) = g(h_1(x), \ldots, h_{k-1}(x)).$$

If g is a continuous function, the equation (12) defines a surface in the space $\mathbb{R}^k$. If $k = 2$, this surface reduces to a curve in $\mathbb{R}^2$. (See Figure 1.)

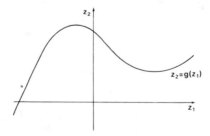

Fig. 1. A law of coexistence in $\mathbb{R}^2$.

QUANTITIES, STATE SPACES, AND LAWS

As an example of a quantitative law of coexistence one can mention Boyle's law which tells how the pressure $p(x)$, volume $V(x)$, and absolute temperature $T(x)$ of a gas x depend on each other. It can be written in the form

$$(13) \quad \frac{p(x)V(x)}{R} - T(x) = 0$$

or

$$p(x) = \frac{RT(x)}{V(x)},$$

where R is a constant. Equation (13) defines a surface in the three-dimensional space generated by the quantities V, T, and p. This surface is often represented as a family of curves in the (V, p)-space for constant values of T (see Figure 2).

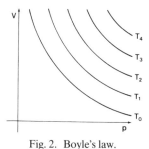

Fig. 2. Boyle's law.

Let $\mathbf{F}$ be the region of points $Q \in \mathbb{R}^k$ which satisfies the equation $f(z_1, \ldots, z_k) = 0$:

$$\mathbf{F} = \{Q \in \mathbb{R}^k \mid f(Q) = 0\}.$$

Then the law (11) could be written, in a formal language with function symbols $\bar{f}, \bar{h}_1, \ldots, \bar{h}_k$, by

$$(14) \quad \bigwedge_{Q \in \mathbf{F}} \Diamond (\exists x \in E) \left[(\bar{h}_1(x), \ldots, \bar{h}_k(x)) \in Q \right] \&$$

$$\& \bigwedge_{Q \notin \mathbf{F}} \Box \sim (\exists x \in E) \left[(\bar{h}_1(x), \ldots, \bar{h}_k(x)) \in Q \right].$$

If (11) can be written in the form (12), statement (14) entails that

(15) $\quad \Box(x)\,[x \in E\ \&\ \bar{h}_1(x) = \bar{r}_1\ \&\ \cdots\ \&\ \bar{h}_{k-1}(x) = \bar{r}_{k-1} \supset$
$\supset \bar{h}_k(x) = \tilde{g}(\bar{r}_1, \ldots, \bar{r}_{k-1})].$

For example, it is physically necessary that, for any gas x with volume V_0 and absolute temperature T_0, $p(x) = RV_0/T_0$. But the law also entails that all the values in $\mathbf{F} \subseteq \mathbb{R}^k$ are physically possible. Thus, law (11) could not be expressed in terms of ordinary constituents with existential quantifiers: it is quite accidental what volumes and temperatures the gases in the actual world happen to realize. Law (11) is not an extensional statement about a Humean regularity, but an assertion about the physically possible combinations of certain quantitative properties.

The above discussion can be generalized in one important direction: state space $\mathbf{Q}$ can be extended with n-ary quantities, for $n \geqslant 2$. The force $F(x, y)$ by which an object $x \in \mathbf{E}$ attracts another object $y \in \mathbf{E}$ is an example of a two-place quantity, i.e., $F: \mathbf{E} \times \mathbf{E} \to \mathbb{R}$. The Law of Gravitation

(16) $\quad F(x, y) = G \cdot \dfrac{m(x)m(y)}{r(x, y)^2}$

where G is the constant of gravitation, $m(x)$ and $m(y)$ are the masses of x and y, respectively, and $r(x, y)$ the distance of objects x and y in $\mathbb{R}^3$, is an example of a law of coexistence involving two-place quantities.

In a formal language, n-ary quantities are represented by n-ary function symbols. The step from 1-place to many-place function symbols corresponds to the step from monadic to polyadic languages. In these richer languages, nomic constituents have to be relativized to some quantificational depth (cf. Chapter 2.7).

The *intended models* of law (11) are structures of the form

(17) $\quad \Omega = \langle D, (g_1(x))_{x \in D}, \ldots, (g_k(x))_{x \in D} \rangle,$

where D is a non-empty domain of actual or possible objects of kind $\mathbf{E}$, and the values $g_1(x), \ldots, g_k(x)$, for $x \in D$, satisfy the equation (11). In particular, a model of the form (17) may have only one object in its domain:

(18) $\quad \Omega_a = \langle \{a\}, g_1(a), \ldots, g_k(a) \rangle.$

It is now possible to compare our notions with the structuralist concep-

QUANTITIES, STATE SPACES, AND LAWS 113

tion of a theory (cf. Chapter 2.9).[7] Let $\mathbf{M}_p$ be the class of all structures of the form (17), and let $\mathbf{M} \subseteq \mathbf{M}_p$ be the subclass which satisfies the equation $f(g_1(x), \ldots, g_k(x)) = 0$ for $x \in D$. Let $\mathbf{J} \subseteq \mathbf{M}_p$ be the class of those structures to which the scientific community intends to apply the law (11). Then $\mathbf{K} = \langle \mathbf{M}_p, \mathbf{M} \rangle$ is the *core* of a Sneedian theory-element $\langle \mathbf{K}, \mathbf{J} \rangle$, and $\mathbf{J}$ is its set of *intended applications*. The *empirical claim* of $\langle \mathbf{K}, \mathbf{J} \rangle$ is that all intended applications are models:

(19) $\mathbf{J} \subseteq \mathbf{M}$.

In the structuralist approach, it is further assumed that set $\mathbf{J}$ is defined through paradigmatic exemplars: $\mathbf{J}$ includes all the elements of a set $\mathbf{J}_0$ and other "sufficiently similar" structures. On the other hand, if $\mathbf{J}$ can be explicitly defined by the condition

(20) $(x)((\Omega_x \in \mathbf{J}) \equiv Ex)$,

then the claim (19) becomes essentially equivalent to our formulation of the law (11).

Let us finally note that many law statements of the form (12) or (15) are *idealized*.[8] It may very well happen that, in the given state space $\mathbf{Q}$ generated by $h_1, \ldots, h_k$, there is no function $g: \mathbb{R}^{k-1} \to \mathbb{R}$ such that (12) holds in all intended applications. In such cases, $\mathbf{Q}$ does not include all the factors that are *nomically relevant* to the values of h_k. By enriching $\mathbf{Q}$ with new quantities $w_1, \ldots, w_n$, we obtain a new state space $\mathbf{Q}^n$, and it may be possible to find in $\mathbf{Q}^n$ a correct law of the form

(21) $h_k(x) = g_n(h_1(x), \ldots, h_{k-1}(x), w_1(x), \ldots, w_n(x))$.

This new law may also show that the old law (12) holds under the *counterfactual* assumption that the influence of the factors $w_1, \ldots, w_n$ is neglible:

(22) $\Box(x)(w_1(x) = 0 \ \& \cdots \& \ w_n(x) = 0 \supset$
$\supset h_k(x) = g(h_1(x), \ldots, h_{k-1}(x)))$.

A law of type (22), which explicitly mentions counterfactual assumptions, are called *idealizational*: other laws are called *factual*. The process of arising from factual laws relative to $\mathbf{Q}$ towards factual laws relative to $\mathbf{Q}^n$ via the idealizational laws is called *concretization* by Nowak (1980) and *factualization* by Krajewski (1977).

For example, Boyle's law (13) is idealized, since it does not account for the influence of two additional factors: the intermolecular attractive forces ($a > 0$), and the finite size of the molecules ($b > 0$). Van der Waals' law includes these factors:

$$(23) \quad \left(p(x) + \frac{a}{V(x)^2}\right)(V(x) - b) = RT(x).$$

This result shows that Boyle's law holds on the counterfactual assumptions that $a = 0$ and $b = 0$, i.e., the following idealizational law is derivable from (23):

$$(24) \quad \Box(x)\left(a(x) = 0 \text{ \& } b(x) = 0 \supset p(x) = \frac{RV(x)}{T(x)}\right).$$

3.4. LAWS OF SUCCESSION

Laws of succession express nomic connections between the successive states of a system. For discrete state spaces **Q**, such laws can be expressed by nomic constituents of the form

$$\bigwedge_{(i,j) \in T} \Box(\exists x)(Q_i^t(x) \text{ \& } Q_j^{t+1}(x)) \text{ \&}$$

$$\text{\& } \Box(x)\left[\bigvee_{(i,j) \in T} (Q_i^t(x) \text{ \& } Q_j^{t+1}(x))\right].$$

Such sentences tell *which state transformations are physically possible and which not*.

Assume now that **Q** is the state space defined by the quantities $h_1, \ldots, h_k$. Let t be the continuous time variable taking values in $\mathbb{R}$. Then $h_i(x, t)$ is the value of the quantity h_i for object $x \in \mathbf{E}$ at time t. (In other words, we assume that the quantities h_i are functions $h_i: \mathbf{E} \times \mathbb{R} \to \mathbb{R}$.) In this case, the laws of succession correspond to *dynamical laws* which tell how the state of the system x at time t depends on time t and some (physically possible) initial state at time t_0:

$$(25) \quad (h_1(x, t), \ldots, h_k(x, t)) = F(t, h_1(x, t_0), \ldots, h_k(x, t_0)).$$

Equation (25) allows us to calculate, for each initial state at time t_0, how the system x evolves in time after t_0. Thus, F in (25) is a *transformation function* $F: \mathbb{R} \times \mathbf{Q} \to \mathbf{Q}$ which describes all the *physically possible trajectories* relative to space $\mathbf{Q}$ (cf. van Fraassen, 1970). These trajectories can be represented as curves in the space $\mathbf{Q}$.

Let $\mathbf{T}$ be the set of all physically possible trajectories relative to law (25), i.e.,

(26) $\quad \mathbf{T} = \{\langle F(t, Q) | t \geq t_0 \rangle | Q \text{ is a physically possible state in } \mathbb{R}^k\}$.

Then (25) says that all the trajectories in $\mathbf{T}$ are physically possible and all the other trajectories are physically impossible. If we wish, this statement can be expressed as a nomic sentence in a formal language (cf. (14)). The intended models of (25) are structures of the form

(27) $\quad \Omega = \langle D, (h_1(x, t))_{x \in D, t \geq t_0}, \ldots, (h_k(x, t))_{x \in D, t \geq t_0} \rangle$,

where D is a set of actual or possible objects of kind $\mathbf{E}$ (which exist through time $t \geq t_0$), and the values $h_1(x, t), \ldots, h_k(x, t)$ satisfy (25) for all $x \in D$ and $t \geq t_0$. Again, dynamic laws of the form (25) may be idealized, and can be concretized by the introduction of new factors.

In many cases, dynamic laws of the form (25) are obtained as solutions of systems of equations which involve the "state variables" h_i and their first- or second-order derivatives with respect to time t. Let us illustrate this with a familiar example from classical mechanics. Assume that a ball with mass m is thrown at time 0 and at point $(0, 0) \in \mathbb{R}^2$ with the velocity v_0 to the direction which forms the angle α with the x-axis (Figure 3). Assume further that the gravitation of the earth gives

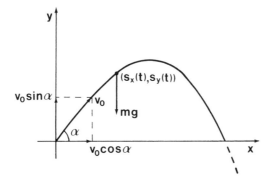

Fig. 3. Ballistic curve.

to the ball a constant acceleration g downward the y-axis, and that no other forces are operative. The position $(s_x(t), s_y(t))$ of the ball at time $t \geq 0$ is obtained by solving Newton's equations

$$m \frac{d^2 s_x(t)}{dt^2} = 0$$

$$m \frac{d^2 s_y(t)}{dt^2} = -mg$$

with the initial conditions

$$v_x(0) = v_0 \cos \alpha$$
$$v_y(0) = v_0 \sin \alpha,$$

where the velocities are defined by

$$\frac{ds_x(t)}{dt} = v_x(t)$$

$$\frac{ds_y(t)}{dt} = v_y(t).$$

The result is given by the ballistic equations

(28) $\quad s_x(t) = t v_0 \cos \alpha$

$\quad s_y(t) = t v_0 \sin \alpha - \frac{gt^2}{2}.$

By choosing $\alpha = -\pi/2$ and $v_0 = 0$, equations (28) give us as a special case Galileo's law of free fall:

(29) $\quad s_y(t) = -\frac{gt^2}{2}.$

Let us mention one way in which the ballistic equations (28) can be concretisized. One of their idealizations is the assumption that the resistance of air is zero. Assume now that the force due to the resistance of air is proportional to the velocity v of the ball, i.e., $-\beta v$

QUANTITIES, STATE SPACES, AND LAWS

where $\beta > 0$ is a constant. By solving the differential equations

$$m \frac{d^2 s_x(t)}{dt^2} = -\beta v \cos \alpha = -\beta \frac{ds_x(t)}{dt}$$

$$m \frac{d^2 s_y(t)}{dt^2} = -\beta v \sin \alpha - mg = -\beta \frac{ds_y(t)}{dt} - mg$$

we obtain

(30) $\quad s_x(t) = \frac{mv_0 \cos \alpha}{\beta} (1 - e^{-\beta t/m})$

$\quad s_y(t) = -\frac{mg}{\beta} t + \left(\frac{m^2 g}{\beta^2} + \frac{mv_0 \sin \alpha}{\beta} \right) (1 - e^{-\beta t/m}).$

While equations (28) define a parabolic curve for the projectile, equations (30) show that the curve has the line

$$x = \frac{mv_0 \cos \alpha}{\beta}$$

as its asymptote (when $t \to \infty$). (See Figure 4.) Moreover, it can be shown that the ballistic equations (30) approach equations (28) as their limit, when the resistance coefficient $\beta \to 0$. But, of course, laws (30) are still idealized with respect to many further factors.

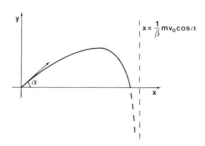

Fig. 4. Ballistic curve with resistance of air.

3.5. PROBABILISTIC LAWS

The interpretations of the concept of probability can be divided into epistemic and physical. In Chapter 2.11, we have already discussed *epistemic* interpretations where the probability $P(h/e)$ of hypothesis h given evidence e is taken to express quantitatively a rational *degree of belief* in the truth of h on the basis of e.

The *physical* interpretation is relevant in cases where the concept of probability occurs in laws of nature. According to the analysis known as the *single-case propensity interpretation*, physical probabilities are quantitative properties of certain kinds of systems or set-ups x, and their numerical values express to what extent the occurrence of certain kinds of events is determined in certain kinds of situations.[9] In this sense, the probability $P(G(x)/F(x))$ of an event of kind G on any trial of kind F is the *degree of physical possibility* of G given F. More specifically, the statement

(31) $P(G(x)/F(x)) = r$

says that the set-up x has a disposition or propensity of strength r to produce an outcome of kind G in any trial of kind F. If $rf_n(G/F)$ is the relative frequency of events G in a sequence of n independent trials of kind F on set-up x, it follows from the Strong Law of Large Numbers that, given (31),

(32) $P(rf_n(G/F) \underset{n \to \infty}{\to} r) = 1.$

The deterministic laws discussed above in Sections 3 and 4 are limiting cases of probabilistic laws. The propensity interpretation is thus in conflict with the Humean view which defines probabilistic laws in terms of actual or hypothetical relative frequencies. It is also stronger than those 'pragmatist' approaches which assign only some assertability conditons to probabilistic statements: for a 'propensitist', probabilistic laws are true or false just like the determinist ones.[10]

The attempt to formulate an explicit semantics for lawlike probabilistic statements involves problems which cannot be solved within this work. It is sufficient for my purposes to show how probabilistic laws can be obtained, in a natural way, as generalizations of the laws of coexistence and succession discussed in the preceding sections.[11]

To formulate *probabilistic laws of coexistence*, assume that **E** is a class of set-ups which are subjected to trials of a certain kind. Let **Q** be the discrete state space of Q-predicates which describes the possible results of these trials. Let P be a probability measure on the space **Q**, i.e., $P: \mathbf{Q} \to \mathbb{R}$ assigns a non-negative probability $p_i = P(Q_i)$ to each predicate Q_i in **Q**, where

$$\sum_{Q_i \in \mathbf{Q}} p_i = 1.$$

The structure of the statement

(33) $\quad \bigwedge_{Q_i \in \mathbf{Q}} [P(Q_i(x)) = p_i]$

corresponds to the 'deterministic' nomic constituent (10), when the operator of physical possibility $\Diamond$ is replaced by P, and P is applied to the open formula $Q_i(x)$ instead of the sentence $(\exists x)Q_i(x)$. Note that, by (33), state Q_i is physically possible if and only if $p_i > 0$.

If G and H are any complex predicates definable in the language for **Q**, then the representation (2.5) allows us to calculate the probability $P(G(x)/H(x))$ from the assignment (33):

$$P(G(x)/H(x)) = P(G(x) \,\&\, H(x))/P(H(x))$$

$$= P\left(\bigvee_{i \in I_{G\&H}} Q_i(x)\right) \Big/ P\left(\bigvee_{i \in I_H} Q_i(x)\right)$$

$$= \sum_{i \in I_{G\&H}} p_i \Big/ \sum_{i \in I_H} p_i.$$

A probabilistic law of the form (33) may also be dynamic in the sense that its states are relative to time t. For example, if **E** is a class of atoms of a certain radioactive substance which are not subjected to any external influences, then the state space $\mathbf{Q}(t)$ contains two time-relative states:

$Q_1(x, t) =$ atom x decays within the time-interval $[0, t]$.
$Q_2(x, t) =$ atom x does not decay within the time-interval $[0, t]$.

Then the probabilities of these states are given by the following exponential law:

(34) $\quad P(Q_1(x, t)) = 1 - e^{-\lambda t}$,

where λ is a constant. By the Law of the Large Numbers, (34) entails that, with probability approaching one, a half of a large population of similar atoms has decayed within the time $T_{\frac{1}{2}} = (\ln 2)/\lambda$. This value can be calculated from the equation

$$1 - e^{-\lambda T_{\frac{1}{2}}} = \tfrac{1}{2}.$$

For example, the 'half-life' of radon atoms is 3.82 days.

If $\mathbf{Q}$ is a non-discrete state space defined by k quantities, then the probability assignment (33) has to be replaced by a probability measure over $\mathbf{Q}$. Mathematicians have recently shown how probability measures can be defined in many-dimensional spaces, such as $\mathbb{R}^k$ or a Hilbert space, and in other metric spaces (see Parthasarathy, 1967). If measure P has a density function f_p, then the probability of a region $C \subseteq \mathbf{Q}$ can be calculated by the integral

$$\int_C f_p(x) \, dx.$$

In this sense, a probabilistic law of coexistence for $\mathbf{Q}$ allows us to determine the degree of physical possibility for any measurable region of states.

Probabilistic laws of succession are expressed in terms of *transition probabilities*. For discrete state spaces, the nomic constituent (2.105) is replaced by a matrix of probabilities

(35) $\quad \| P(Q^{t+1}(x)/Q^t(x)) \|_{\substack{i=1,\ldots,t \\ j=1,\ldots,t}} = \| p_{j/i} \|$

which, for each $i = 1, \ldots, t$, gives the probability $p_{j/i}$ of the next state ($j = 1, \ldots, t$). Here

$$\sum_{j=1}^{t} p_{j/i} = 1 \quad \text{for each } i = 1, \ldots, t.$$

In fact, (35) defines a discrete *Markov chain*, since the probability of the state at time $t + 1$ depends only on the preceding state at time t. In

QUANTITIES, STATE SPACES, AND LAWS

more complex probabilistic laws of succession, the whole prehistory of the system may influence the probabilities of future states.

Finally, let **Q** be a non-discrete state space describing the state of a system evolving in time. Let $\mathbf{Q}^t$ be the state space at time $t \geqslant t_0$, and let

$$\mathbf{Q}^{<t} = \bigcap_{t_0 \leqslant u < t} Q^u$$

$$\mathbf{Q}^{\leqslant t} = \bigcap_{t_0 \leqslant u \leqslant t} Q^u$$

$$\mathbf{Q}^{\infty} = \bigcap_{t_0 \leqslant u} Q^u$$

The elements of $\mathbf{Q}^{\infty}$ are then infinite sequences of states of the form

$$\langle Q_t \in \mathbf{Q} \mid t \geqslant t_0 \rangle.$$

Such sequences — formally functions from $[t_0, \infty)$ to **Q** — are *possible trajectories*. Laws of succession relative to **Q** are then expressed by a probability measure P on $\mathbf{Q}^{\infty}$. Such a measure determines, and is uniquely determined by, a system of probability measures on $\mathbf{Q}^{\leqslant t}$, for each $t \geqslant t_0$. Measure P on $\mathbf{Q}^{\infty}$ is a direct generalization of nomic constituents corresponding to laws of the form (25): it gives the *degrees of physical possibility for different kinds of trajectories* of a stochastic process. Moreover, P determines conditional probabilities for the state at time t given the history of the process before t:

(36) $P(Q_t / \langle Q_u \mid t_0 \leqslant u < t \rangle) =$
$\quad P(\langle Q_t \mid t_0 \leqslant u \leqslant t \rangle \cap (\mathbf{Q}^{\infty} - \mathbf{Q}^{\leqslant t})) /$
$\quad P(\langle Q_t \mid t_0 \leqslant u < t \rangle \cap (\mathbf{Q}^{\infty} - \mathbf{Q}^{<t})).$

(36) expresses a distribution of generalized transition probabilities for time-continuous stochastic processes.[12]

CHAPTER 4

COGNITIVE PROBLEMS, TRUTH, AND INFORMATION

Truth and information are usually thought to be, in some sense, the main ingredients of the concept of truthlikeness. This idea also characterizes the intended application of this notion. There are many situations — both in everyday life and in science — where we wish to find true information about the world. Requests for such information are expressed by questions, and answers to them are sought for within systematic inquiries. Truthlikeness is then a criterion for evaluating the answers to cognitive problems.[1] In this chapter, we discuss the main philosophical theories of truth, indicating the reasons for preferring a 'realist' correspondence account, and introduce the basic concepts of semantic information theory.

4.1. OPEN AND CLOSED QUESTIONS

Hintikka (1976) characterizes *questions* as requests of information. For example, WH-questions (i.e., which-, who-, where-, when-, and many what-questions) of the form

(1) Which x is such that $\alpha(x)$?

are represented by the imperative

(2) Bring it about that I know which x is such that $\alpha(x)$.

The epistemic state of affairs following the imperative operator 'Bring it about that' in (2), i.e.,

(3) I know which x is such that $\alpha(x)$,

is called the *desideratum* of (1), and the sentence

(4) $(\exists x)\alpha(x)$

expresses the *presupposition* of (1).

COGNITIVE PROBLEMS, TRUTH, AND INFORMATION 123

A similar analysis applies to propositional questions, such as why-questions

(5)   Why is it the case that h?,

with the desideratum 'I know why h' and presupposition h, and disjunctive questions

(6) h_1 or h_2 or $\cdots$ or h_n?

Here (6) expresses a *unique alternative whether-question* if its presupposition is

(7) $\bigvee_{i=1}^{n} h_i \ \& \ \bigwedge_{\substack{i,j=1 \\ i \neq j}}^{n} \sim (h_i \ \& \ h_j)$

and desideratum

(8) $\bigvee_{i=1}^{n}$ I know that h_i.

Further, (6) expresses a *complete list whether-question* if its presupposition is

$\bigvee_{i=1}^{n} h_i$

and desideratum

(9) $\bigwedge_{i=1}^{n} (h_i \supset$ I know that $h_i)$.

Yes-or-no-questions are special cases of whether-questions: the question

(10)   Is it the case that h?

can be analysed as

> Bring it about that I know whether h or $\sim h$,

so that its desideratum is

> I know that h or I know that $\sim h$.

A *potential answer* to a WH-question (1) is any singular term of the appropriate type. (Question: Who stole my watch? Answer: Mr. Smith. Question: Where was I born? Answer: Helsinki.) On this basis, WH-questions may be divided into *restricted* and *unrestricted*. The former have the form

(11)   Which x in class X is such that $\alpha(x)$?

i.e., the class X where the referent of a potential answer 'a' has to belong is specified already in the question. The latter include no such specifications. It is clear that restricted WH-questions are in principle reducible to whether-questions (possibly with infinitely many disjuncts).

For disjunctive questions, the problem of answerhood has a very simple solution: the potential answers to a unique alternative whether-question (6) are the sentences $h_1, \ldots, h_n$. They are thus paradigmatic examples of *closed questions*: in general, a question is said to be closed if we are in a position to give an effective method of enumerating or describing in advance its potential answers. In other words, the formulation of a closed question already in effect includes a specification of its potential answers.[2]

The situation is different with *open questions*. Examples of them include unrestricted WH-questions ('What are the laws of heredity among living organisms?'), why-questions ('Why did the World War I begin?'), how possible-questions ('How is life on earth possible?'), and narrative what-questions ('What happened in your childhood?'). A series of sentences can give a more and more detailed answer to such an open question, but there is no natural point to stop and to say that the answer is now complete. By contrast, for closed questions the notion of a *complete* potential answer is well defined: for example, the answers 'Nine' and 'Nine, but only two of them are closer to the sun than the earth is' are both correct replies to the question 'What is the number of planets in our solar system?', but the former is already complete, so that the latter contains *irrelevant information*.

The distinction between open and closed questions is important for our understanding of the nature of human knowledge-seeking activities. Cognitive problems arise in many different kinds of situations — they range from questions concerning singular features of our natural and social environment to abstract scientific research problems. But typically they are formulated and solved within a context which supplies the relevant presuppositions. Such contexts are associated with general long-term tasks or programmes for studying certain parts or fragments of the world from a certain perspective.[3] These studies or inquiries attempt to satisfy requests of certain kind of information about specific domains of objects or events within the 'worlds' of nature, consciousness, culture, society, or abstract man-made entities. They can thus be regarded as systematic attempts to give answers to open questions.

The most effective strategy of approaching open questions — which define perhaps 'unended quests' related to different scientific disciplines and research programmes — is to restructure them into a net of closed questions. Progress in an open-ended inquiry can be made by finding complete answers to closed questions. One way of fixing closed problems is to tentatively adopt a conceptual framework and to use it for the description of the object domain. This move helps us to make WH-questions restricted (cf. Section 2). It may transform open why-questions to closed what-questions: if a list of possible causes of death is specified, then the open question 'Why did Mr X die?' is replaced by the restricted problem 'What cause in the given list was responsible to the death of Mr X?'. The open-ended character of inquiries is reflected in the freedom that an investigator has in changing or enriching his conceptual system (cf. Section 2).

According to the argument given in this section, closed questions have a special role in human cognitive activities. The next section is devoted to the most fundamental type of such questions, viz. unique alternative whether-questions.

Finally, note that if we assume the so-called classical definition of knowledge, viz. *knowledge* as *justified true belief*, then the following holds:

(12)   I know that $h_i \supset h_i$.

Hence, the desideratum of a unique alternative whether-question (6) is not satisfied unless we have hit upon the true answer. In this sense, questions are not merely requests of information — but requests of *true information*. What notions of truth and information are involved here will be discussed below in Sections 3 and 5.

4.2. COGNITIVE PROBLEMS

A non-empty set $\mathbf{B} = \{h_i \mid i \in \mathbf{I}\}$ of statements is a *P-set* relative to b, if

$$b \vdash \bigvee_{i \in \mathbf{I}} h_i$$

(13) $\quad b \vdash \sim(h_i \,\&\, h_j) \quad$ for all $i \neq j, i, j \in \mathbf{I}$
$\quad\quad\;\;$ not $b \vdash \sim h_i \quad$ for all $i \in \mathbf{I}$.

A *P*-set $\mathbf{B}$ is thus a finite or infinite set of alternatives which are, relative to background knowledge b, mutually exclusive and jointly exhaustive. If b is true, this means that there is one and only one true element in $\mathbf{B}$.[4]

For simplicity, let us assume for a while that the background knowledge b is empty. This restriction is not essential: if $\mathbf{B} = \{h_i \mid i \in \mathbf{I}\}$ is a *P*-set, then the set

$$\mathbf{B}_b = \{h_i \in \mathbf{B} \mid \text{not } b \vdash \sim h_i\}$$

is a *P*-set relative to b. We shall also identify logically equivalent sentences in this section. (Hence, $h = h'$ iff $\vdash h \equiv h'$.)

A *P*-set of the form $\mathbf{0} = \{T\}$, where T is a tautology, is called *trivial*.

The *disjunctive closure* $D(\mathbf{B})$ of a *P*-set $\mathbf{B}$ is the set of all non-empty disjunctions of members of $\mathbf{B}$:

(14) $\quad D(\mathbf{B}) = \left\{ \bigvee_{i \in \mathbf{I}_h} h_i \mid \emptyset \neq \mathbf{I}_h \subseteq \mathbf{I} \right\}.$

The disjunction

$$h = \bigvee_{i \in \mathbf{I}_h} h_i$$

is called the *B-normal form* of h, and $\mathbf{I}_h$ is the *index set* of h relative to

COGNITIVE PROBLEMS, TRUTH, AND INFORMATION

B. The cardinality of $D(\mathbf{B})$ is the number of non-empty subsets of **B**:

$$|D(\mathbf{B})| = 2^{|\mathbf{I}|} - 1.$$

Further, for two *P*-sets **B** and **B**′:

(15) $\quad \mathbf{B} \subseteq D(\mathbf{B})$
$\quad\quad\; \mathbf{0} \subseteq D(\mathbf{B})$
$\quad\quad\; D(\mathbf{0}) = \mathbf{0}$
$\quad\quad\; D(D(\mathbf{B})) = D(\mathbf{B})$
$\quad\quad\; \mathbf{B} \subseteq D(\mathbf{B}')$ iff $D(\mathbf{B}) \subseteq D(\mathbf{B}')$.

The inclusion relation $\subseteq$ between the disjunctive closures of *P*-sets induces a partial ordering $\leq$ on the class of *P*-sets. When $\mathbf{B} \subseteq D(\mathbf{B}')$ holds (cf. (15)), **B**′ is said to be at least as *fine* as **B**. This is denoted by $\mathbf{B} \leq \mathbf{B}'$. If $\mathbf{B} \leq \mathbf{B}'$ and $\mathbf{B} \neq \mathbf{B}'$, then **B**′ is said to be *finer* than **B**. If $\mathbf{B} \leq \mathbf{B}'$, then each element of $D(\mathbf{B})$ has a **B**′-normal form, i.e., can be expressed as a disjunction of elements of **B**′. Trivially, $\mathbf{0} \leq \mathbf{B}$ for all *P*-sets **B**. Moreover, $\leq$ is a partial ordering (reflexive, transitive, and antisymmetric) on the class of *P*-sets.

The *combination* $\mathbf{B} \oplus \mathbf{B}'$ of two *P*-sets $\mathbf{B} = \{h_i | i \in \mathbf{I}\}$ and $\mathbf{B}' = \{h'_j | j \in \mathbf{J}\}$ can be defined by

(16) $\quad \mathbf{B} \oplus \mathbf{B}' = \{h_i \,\&\, h'_j | i \in \mathbf{I}, j \in \mathbf{J}\} - \{\perp\},$

where $\perp$ is a contradiction. Then

(17) $\quad \mathbf{B} \oplus \mathbf{B}' = \mathbf{B}' \oplus \mathbf{B}$
$\quad\quad\; \mathbf{B} \oplus \mathbf{0} = \mathbf{0} \oplus \mathbf{B} = \mathbf{B}$
$\quad\quad\; \mathbf{B} \leq \mathbf{B} \oplus \mathbf{B}', \mathbf{B}' \leq \mathbf{B} \oplus \mathbf{B}'$
$\quad\quad\; D(\mathbf{B}) \cup D(\mathbf{B}') \subseteq D(\mathbf{B} \oplus \mathbf{B}')$
$\quad\quad\; \mathbf{B} \leq \mathbf{B}'$ iff $\mathbf{B} \oplus \mathbf{B}' = \mathbf{B}'$
$\quad\quad\; \mathbf{B} \oplus \mathbf{B}' = \sup\{\mathbf{B}, \mathbf{B}'\}.$

To prove the last assertion, assume that $\mathbf{B} \leq C$ and $\mathbf{B}' \leq C$ for some *P*-set C. Then $\mathbf{B} \subseteq D(C)$ and $\mathbf{B}' \subseteq D(C)$. An element $h_i \,\&\, h'_j$ of $\mathbf{B} \oplus \mathbf{B}'$ can, therefore, be represented as a conjunction of two disjunctions

$$h_i \,\&\, h'_j = \bigvee_{k \in K_1} c_k \,\&\, \bigvee_{k \in K_2} c_k$$

where $c_k \in C$ for all $k \in K_1 \cup K_2$. As this conjunction is an element of $D(C)$, it follows that $\mathbf{B} \oplus \mathbf{B}' \subseteq D(C)$. Hence, $\mathbf{B} \oplus \mathbf{B}'$ is the least upper bound of $\mathbf{B}$ and $\mathbf{B}'$ (relative to the ordering $\leq$).

The *meet* $\mathbf{B} \times \mathbf{B}'$ of two P-sets $\mathbf{B}$ and $\mathbf{B}'$ can be defined by

$$(18) \quad \mathbf{B} \times \mathbf{B}' = \left\{ g \in D(\mathbf{B}) \cap D(\mathbf{B}') \,\middle|\, \bigvee_{i \in G} h_i \notin D(\mathbf{B}') \text{ for all proper subsets } G \text{ of } \mathbf{I}_g \right\}.$$

In other words, $\mathbf{B} \times \mathbf{B}'$ contains those common elements of $D(\mathbf{B})$ and $D(\mathbf{B}')$ which have a minimal number of disjuncts in their normal form. This guarantees that $\mathbf{B} \times \mathbf{B}'$ is a P-set. Two non-trivial P-sets are *isolated* if the only common element of $D(\mathbf{B})$ and $D(\mathbf{B}')$ is a tautology T, i.e., if $\mathbf{B} \times \mathbf{B}' = \mathbf{0}$. We may have $\mathbf{B} \times \mathbf{B}' = \mathbf{0}$ even if $\mathbf{B} \cap \mathbf{B}' = \emptyset$. The meet operation has the following properties:

$$(19) \quad \begin{aligned} &\mathbf{B} \times \mathbf{B}' = \mathbf{B}' \times \mathbf{B} \\ &\mathbf{B} \times \mathbf{0} = \mathbf{0} \times \mathbf{B} = \mathbf{0} \\ &\mathbf{B} \times \mathbf{B}' \leq \mathbf{B}, \mathbf{B} \times \mathbf{B}' \leq \mathbf{B}' \\ &\mathbf{B} \cap \mathbf{B}' \subseteq \mathbf{B} \times \mathbf{B}' \\ &D(\mathbf{B}) \cap D(\mathbf{B}') = D(\mathbf{B} \times \mathbf{B}') \\ &\mathbf{B} \leq \mathbf{B}' \text{ iff } \mathbf{B} \times \mathbf{B}' = \mathbf{B} \\ &\mathbf{B} \times \mathbf{B}' = \inf\{\mathbf{B}, \mathbf{B}'\}. \end{aligned}$$

To prove the last assertion, assume that $C \leq \mathbf{B}$ and $C \leq \mathbf{B}'$ for a P-set C. Then $C \subseteq D(\mathbf{B})$ and $C \subseteq D(\mathbf{B}')$, so that $C \subseteq D(\mathbf{B}) \cap D(\mathbf{B}') \subseteq D(\mathbf{B} \times \mathbf{B}')$ and $C \leq \mathbf{B} \times \mathbf{B}'$. Hence, $\mathbf{B} \times \mathbf{B}'$ is the greatest lower bound of $\mathbf{B}$ and $\mathbf{B}'$.

We have seen that the class of P-sets is a *lattice* with respect to the ordering $\leq$ with $\times$ as the infimum operation, $\oplus$ as the supremum operation, and $\mathbf{0}$ as the null element.

If $\mathbf{F} = \{h_i \mid i \in \mathbf{I}_0\}$ is a non-empty proper subset of a P-set $\mathbf{B} = \{h_i \mid i \in \mathbf{I}\}$, the *completion* $C(\mathbf{F})$ of $\mathbf{F}$ is the set

$$(20) \quad C(\mathbf{F}) = \mathbf{F} \cup \left\{ \sim \bigvee_{i \in \mathbf{I}_0} h_i \right\}.$$

For example, if $\mathbf{B} = \{h_1, \ldots, h_n\}$ and $\mathbf{F} = \{h_1, \ldots, h_m\}$, where $1 \leq m < n$, then $C(\mathbf{F}) = \{h_1, \ldots, h_m, \sim(h_1 \vee \cdots \vee h_m)\}$. $C(\mathbf{F})$ is thus a P-set such that $C(\mathbf{F}) \leqslant \mathbf{B}$:

(21)   If $\emptyset \neq \mathbf{F} \subset \mathbf{B}$, then $C(\mathbf{F}) \leqslant \mathbf{B}$.

If $\mathbf{B}$ and $\mathbf{B}'$ are two P-sets such that $\mathbf{B} \cap \mathbf{B}' \neq \emptyset$, then $C(\mathbf{B} \cap \mathbf{B}') \leqslant \mathbf{B} \times \mathbf{B}'$.

Following the terminology of Section 1, each P-set $\mathbf{B} = \{h_i | i \in \mathbf{I}\}$ corresponds to a unique alternative whether-question:

(22)   Which element of $\mathbf{B}$ is true?

We shall say that each question of the form (22) is a *cognitive problem*. Its *complete potential answers* are the sentences $h_i \in \mathbf{B}$. While non-tautological disjunctions of elements of $\mathbf{B}$ do not give complete answers to question (22), still they reduce some uncertainty about the correct answer. For this reason, it is natural to regard the elements of $D(\mathbf{B})$ as *partial* potential answers to the cognitive problem (22). A partial answer g is (logically) *stronger* than another partial answer g', if $g \vdash g'$. The weakest partial answer is the disjunction of all elements of $\mathbf{B}$, i.e., a tautology T. In this case, the answerer *suspends judgment* between the alternatives in $\mathbf{B}$. More generally, if

$$g = \bigvee_{i \in \mathbf{I}_g} h_i$$

is the strongest available (acceptable) answer, then judgment is suspended between h_i, $i \in \mathbf{I}_g$.[5]

By the definition of P-sets, there is one and only one element of $\mathbf{B}$ which is *true*. Let us denote it by h_*. For the trivial P-set $\mathbf{0} = \{T\}$, the only true element is of course the tautology T. For non-trivial P-sets, h_* is a factual sentence (cf. Chapter 2.1). Thus, the *complete correct answer* to problem $\mathbf{B}$ is given by h_*. An answer $g \in D(\mathbf{B})$ is a *partial correct* answer, if and only if $h_* \vdash g$, i.e., h_* belongs to the $\mathbf{B}$-normal form of g.

If $\mathbf{B}$ is a P-set relative to background knowledge b, then the existence of a true alternative h_* in $\mathbf{B}$ is guaranteed by the truth of b. However, if the assumption b is false, and thereby all the elements of $\mathbf{B}$ are false, then there is no true element in $\mathbf{B}$. In this case, the aim of the cognitive problem $\mathbf{B}$ is to find the *least false* element (or elements) of $\mathbf{B}$. We shall see in Chapter 6.9 how this concept can be made precise.

If $\mathbf{B} \leqslant \mathbf{B}'$ holds for two P-sets $\mathbf{B}$ and $\mathbf{B}'$, then each complete answer to problem **B** is a partial answer to problem **B**'. We can, therefore, say that **B** is a *subproblem* of **B**' just in case **B** ⩽ **B**'. There is a natural one-to-one correspondence between the subproblems of a cognitive problem **B** and whether-questions which may be asked in the problem-situation **B**. For example, if $h \in$ **B**, then the question

(23)   Is it the case that h?

corresponds to the problem $\{h, \sim h\}$ ⩽ **B**. If **F** ⊂ **B** is non-empty, then the question

(24)   Which element (if any) of **F** is true?

corresponds to the problem $C(\mathbf{F})$ ⩽ **B**.

If **B** and **B**' are two problems, then the meet **B** × **B**' is the largest common subproblem of **B** and **B**': its partial answers are partial answers to both **B** and **B**', and conversely. The combination **B** ⊕ **B**' of **B** and **B**' is the least problem which contains both **B** and **B**' as subproblems. Each complete answer to **B** ⊕ **B**' entails a complete answer to **B** and a complete answer to **B**', but there are partial answers to **B** ⊕ **B**' which do not entail partial answers to **B** or **B**'.

For two isolated non-trivial *P*-sets **B** and **B**', the correct answers h_* and h'_* are logically independent of each other. On the other hand, the true answer is preserved in the subproblem relation in the following sense:

(25)   If **B** ⩽ **B**', then $h'_* \vdash h_*$, i.e., the correct answer to **B**' is logically at least as strong as the correct answer to **B**.

In other words, if **B** ⩽ **B**', then h'_* belongs to the **B**'-normal form of h_*.

Let us consider some examples of cognitive problems which will be discussed in the later chapters. We shall see that the introduction of a language is a particularly effective way of defining systematically cognitive problem situations. A *P*-set **B** is said to be a *P*-set in language L, if all the elements of **B** are sentences of L.

EXAMPLE 1. Let $\mathscr{Z}_N^k$ be the set of all state descriptions of a monadic first-order language L_N^k (see Chapter 2.3). Then $\mathscr{Z}_N^k$ is a *P*-set, and

$$D(\mathscr{Z}_N^k) = \text{the set of all consistent sentences of } L_N^k.$$

COGNITIVE PROBLEMS, TRUTH, AND INFORMATION   131

Hence, $\mathscr{Z}_N^k$ is the finest P-set in language L_N^k:

$\mathbf{B} \leqslant \mathscr{Z}_N^k$ for all P-sets $\mathbf{B}$ in L_N^k.

EXAMPLE 2. Let $\mathscr{S}_N^k$ be the set of all structure descriptions of language L_N^k (see Chapter 2.4). Then $\mathscr{S}_N^k$ is a P-set.

EXAMPLE 3. Let Γ_N^k be the set of all constituents of language L_N^k (see Chapter 2.5). Then Γ_N^k is a P-set, and

$D(\Gamma_N^k) =$ the set of all consistent generalizations of L_N^k.

Moreover,

(26) $\Gamma_N^k \leqslant \mathscr{S}_N^k \leqslant \mathscr{Z}_N^k$.

EXAMPLE 4. Let $\Gamma_L^{(d)}$ be the set of all depth-d constituents of a polyadic first-order language L (see Chapter 2.7). Then $\Gamma_L^{(d)}$ is a P-set for all $d = 1, 2, \ldots$, and

$D(\Gamma_L^{(d)}) =$ the set of all depth-d generalizations in L.

$\bigcup_{d < \omega} D(\Gamma_L^{(d)}) =$ the set of all generalizations in L.

Moreover, for all finite depths d and d':

(27) $\Gamma_L^{(d)} \leqslant \Gamma_L^{(d')}$ if $d < d'$.

Hence,

$\Gamma_L^{(1)} \leqslant \Gamma_L^{(2)} \leqslant \cdots \leqslant \Gamma_L^{(d)} \leqslant \cdots$

EXAMPLE 5. Let $\mathscr{N}_N^k$ be the set of all nomic constituents of the monadic intensional language $L_N^k(\square)$ (see Chapter 2.11). Then $\mathscr{N}_N^k$ is a P-set, and

$D(\mathscr{N}_N^k) =$ the set of all genuine law statements in $L_N^k(\square)$.

If Γ_N^k is the set of the constituents of L_N^k, then the combination of problems Γ_N^k and $\mathscr{N}_N^k$ is the problem

$\Gamma_N^k \oplus \mathscr{N}_N^k = \{ C_i \ \& \ B_j | \ C_i \in \Gamma_N^k, B_j \in \mathscr{N}_N^k \}$.

EXAMPLE 6. Let L and L' be two monadic languages with the non-logical vocabularies λ and λ', respectively. Assume that L is a sublanguage of L', i.e., $\lambda \subseteq \lambda'$. Then the set $\mathscr{Z}$ of state descriptions of L is a subproblem of the set $\mathscr{Z}'$ of state descriptions of L':

$$\mathscr{Z} \leqslant \mathscr{Z}'.$$

EXAMPLE 7. Let L and L' be two polyadic languages. If L is a sublanguage of L', then

$$\Gamma_L^{(d)} \leqslant \Gamma_{L'}^{(d)} \quad \text{for all } d = 1, 2, \ldots.$$

EXAMPLE 8. Let L_1 and L_2 be two monadic languages with the vocabularies λ_1 and λ_2. Let L be the common extension of L_1 and L_2, with the vocabulary $\lambda = \lambda_1 \cup \lambda_2$, and let L_0 be the common sublanguage of L_1 and L_2, with $\lambda_0 = \lambda_1 \cap \lambda_2$. Then the relations between the corresponding sets of constituents is illustrated in Figure 1.

Fig. 1. Subproblem relations between sets of constituents.

EXAMPLE 9. Let **M** be the class of all structures for the polyadic first-order language L. Let $\mathbf{M}^* = \mathbf{M}/\equiv$ be the family of the equivalence classes of **M** with respect to the relation $\equiv$ of elementary equivalence between L-structures. Then each class in $\mathbf{M}^*$ corresponds to a complete theory Σ in L. (See Chapter 2.9.) The class of complete theories in L, i.e.,

$$\mathscr{CT}_L = \{\text{Th}(\Omega) \mid \Omega \in \mathbf{M}\},$$

is thus a P-set. By (2.63), each element of this set is axiomatizable by a monotone sequence of constituents in L. The disjunctive closure of $\mathscr{CT}_L$ includes all first-order theories in language L.

COGNITIVE PROBLEMS, TRUTH, AND INFORMATION   133

EXAMPLE 10. Let θ_* be the unknown value of a real-valued parameter. Then the set of statements

$$\{\theta_* = \theta \mid \theta \in \mathbb{R}\}$$

is a *P*-set (relative to the background assumption that $\theta_* \in \mathbb{R}$). There is a natural one-to-one mapping from this *P*-set to $\mathbb{R}$, so that $\mathbb{R}$ itself can be regarded as a *P*-set which defines the *estimation problem* for the real-valued parameter θ_*. The complete answers to this problem are called *point estimates* of θ_*. The partial answers in $D(\mathbb{R})$ would include all subsets of $\mathbb{R}$, but usually only the subintervals of $\mathbb{R}$ are considered as interesting elements of $D(\mathbb{R})$ (cf. Chapter 12.5). In this case, we have a problem of *interval estimation*. (This example can be generalized to cases where θ_* belongs to $\mathbb{R}^n$ or any other linear vector space.)

EXAMPLE 11. Let $h_1: E \to \mathbb{R}$ and $h_2: E \to \mathbb{R}$ be two real valued quantities, and let the background assumption b say that there is a linear function which expresses the connection between the values of h_1 and h_2. Then the class of statements

$$\{\bar{h}_2(x) = a\bar{h}_1(x) + c \mid a \in \mathbb{R}, c \in \mathbb{R}\}$$

is a *P*-set relative to b. There is a natural one-to-one correspondence between this set and the set $\{(a, c) \mid a \in \mathbb{R}, c \in \mathbb{R}\}$, i.e., $\mathbb{R}^2$. Thus, the cognitive problem associated with this *P*-set can be reduced to the problem of estimating an unknown parameter (a_*, c_*) in $\mathbb{R}^2$ (cf. Example 10).

EXAMPLE 12. Let $h_i: E \to \mathbb{R}$, $i = 1, \ldots, k$, be real-valued quantities, and let the background assumption b say that there is a continuous function $f_*: \mathbb{R}^{n-1} \to \mathbb{R}$ which expresses how the values of h_n depend on the values of $h_1, \ldots, h_{n-1}$. Then the cognitive problem associated with the determination of f_* is one-to-one correlated with the class of continuous function from $\mathbb{R}^{n-1}$ to $\mathbb{R}$, i.e.,

$$\{f: \mathbb{R}^{n-1} \to \mathbb{R} \mid f \text{ is continuous}\}.$$

This class represents then a *P*-set relative to the assumption b.

EXAMPLE 13. Let the background assumption b say that there exists a probabilistic law, expressible by a probability measure P_* on an algebra $\mathscr{A}$ of subsets of a set X. Then the class of all probability measures on $\mathscr{A}$, i.e.,

$$\{P: \mathscr{A} \to \mathbb{R} \mid P \text{ is a probability measure}\}$$

represents a P-set relative to b.

4.3. TRUTH

In Chapter 2, the notion of truth was defined for formal first-order languages by following Tarski's theory. It is, however, a matter of philosophical controversy whether the Tarskian approach is viable — or even applicable to natural and scientific languages. We shall see later that Tarski's theory serves very well our purposes in developing a theory of truthlikeness. A more general defence of his theory is given in this section.[6]

Let us start from the *redundancy theory*, with its variations (Ramsey, Strawson), which is not properly a theory about truth at all, but rather an attempt to show how one can do without the concept of truth. This doctrine claims that talk about truth can always be eliminated by the schema

(T) 'a' is true iff a.

Even if this were correct, it would not give us grounds for eliminating talk about truthlikeness: we do not seem to have anything which could be placed as condition X in the following schema:

(28) 'a' is truthlike iff X.

However, the redundancy theory is plausible in certain types of contexts, such as

(29) a knows that a $\equiv$ a knows that 'a' is true
a believes that a $\equiv$ a believes that 'a' is true,

even though it does not give a general procedure for eliminating the predicate 'true'.

Traditionally the most important theories of truth have been the correspondence theory, the coherence theory, and the various versions of the pragmatist theory. The so-called evidence theory (Descartes, Brentano), if it ever was intended as an account of truth, can best be viewed as a variant of the pragmatist theory.

According to the *correspondence theory*, truth is a relation between a belief and reality, or between a judgment and the world. This view goes back at least to Aristotle. The scholastics expressed it by saying: *veritas est adaequatio intellectus et rei*. In this account, the bearers of truth have been taken to be sentences, statements, judgments, propositions, beliefs, or ideas; their truth consists in their being in the relation of 'correspondence' with reality, world, or facts. A popular way of expressing this view is the following: a statement is a description of a 'possible state of affairs'; it is true if this state of affairs is 'actual' or exists in the 'real world', i.e., if it expresses a 'fact'; otherwise it is false.

It is not obvious that notions like 'state of affairs', 'fact', 'world', and 'reality' are better understood than the concept of truth. Moreover, the fundamental difficulty for the correspondence theory is to specify what it means to say that a statement 'corresponds' to reality. We shall argue below that Tarski's model-theoretic account of truth, proposed originally in the 1930s, is an adequate explication of this idea.

The *coherence theory* has been supported by some Hegelian idealists and by some logical empiricists (Otto Neurath). It claims that a judgment cannot 'correspond' to any extra-linguistic reality: truth has to be defined in terms of relations that judgments bear to each other. Thus, a judgment is true if it forms a coherent system with other judgments. As coherence admits of degrees, some coherence theorists (like F. H. Bradley) have talked about 'degrees of truth'.[7]

Rescher (1973) has recently defended the coherence theory as a criterion of truth. Most philosophers would agree, however, that coherence (understood as some sort of consistency) is not an adequate definition of truth. In ordinary logic, the condition that a sentence α is compatible with a consistent set Σ of true sentences is not sufficient to guarantee the truth of α. Still, this condition is *necessary* for the truth of α, since the inconsistency of α with Σ would guarantee that the negation $\sim \alpha$ of α, and hence the falsity of α, is entailed by Σ. When a sentence is defined to be true if it is consistent with Σ, it may happen that both α and $\sim \alpha$ are true — which violates the Law of Contradiction. On the other hand, if a sentence is defined to be true if it is entailed by Σ, then it may happen that neither α nor $\sim \alpha$ is true — which violates the Law of the Excluded Middle. To avoid these unattractive possibilities, we should require that Σ is a maximally large consistent set of sentences — something like a 'maximal proposition' in the sense of Freeman and Daniels (1978). But this means that Σ is a complete consistent theory (cf. Chapter 2.9), and the 'coherence' of α with Σ only means that α belongs to Σ. In other words, the 'maximal' set Σ is simply an enumeration of all the 'true' sentences, and the coherence theory of truth collapses to the unilluminating claim that a statement is 'true' if it belongs to the list of 'true' statements.

The correspondence theory makes a sharp distinction between a definition of truth (i.e., what it means for a statement to be true) and the criteria for recognizing truth (i.e., what the best indicators for the truth of a statement are): the former belongs to *semantics*, the latter to *epistemology*. Truth is not generally a manifest property of statements: some claims about reality may in fact be true even if we are not in the position to recognizing their truth.[8] Therefore, a *realist* who accepts the correspondence theory has to acknowledge the possibility that some truths transcend our cognitive capabilities: there are true statements about reality which are now unknown to us and *may* remain unknown to us forever.

The *pragmatists* think otherwise: in their view, it is not meaningful to speak of truth and reality as divorced from human practical and cognitive activities. Hence, reality *an sich* is replaced by reality *für uns*, and truth is replaced by truth-as-known-by-us. While the realists usually follow Plato in defining genuine knowledge in terms of truth (*episteme* is justified true belief), the pragmatists turn this definition upside down and define truth in terms of the results of human knowledge-seeking — and thereby intentionally blur the distinction between semantics and epistemology. Thus, the pragmatist strategy is to identify truth with the contents of human knowledge in its final or ideal state: true means the same as 'proved' (mathematical intuitionism), 'verified' (verificationism), 'warrantedly assertable' (Dewey), 'successful' or 'workable' in practice, the ideal limit of scientific inquiry, "rationally acceptable under ideal conditions" (Putnam), or the ideal consensus reached in 'free' or 'undistorted' human communication (Habermas). This conception of truth has recently become fashionable through the writings of Michael Dummett (1978), Nicholas Rescher (1977), Hilary Putnam (1978, 1981), Richard Rorty (1982), and Jürgen Habermas (1982).

Most contemporary pragmatists have been inspired by Charles Peirce's characterization of truth as the ideal limiting opinion of the scientific community. For example, Rorty interprets Peirce as suggesting that "we can make no sense of the notion that the view which can survive all objections might be false" (Rorty, 1982, p. 165). Similarly,

Putnam argues that it is not "intelligible" to suppose that "an 'ideal' theory (from a pragmatic point of view) might *really* be false" (Putnam, 1978, p. 126). In my view, the only way of making these claims convincing is to assume that phrases like 'all objections' and 'ideal theory' refer to situations where the scientific community has access to all *true* statements about the world. Without such access to at least some truths about reality, there is no guarantee that a community of investigators reaches the correct solution even to the simplest cognitive problems — whatever standards of discourse rationality we may impose on the communication between its members. In other words, it seems that the pragmatist attempt to define truth in terms of ideal consensus is valid only if this definition is circular in the sense that it already presupposes a realist notion of truth.

I have argued elsewhere that it is a mistake to regard Peirce as suggesting a consensus theory of truth (cf. Niiniluoto, 1980a).[9] Peirce was an epistemological realist and optimist who argued that the continued use of the scientific method would ultimately bring *with probability one* the scientific community to a final opinion which is in correspondence with reality. From this viewpoint, the mistake of the pragmatist interpreters of Peirce is to seek an *analytic* connection between truth and scientific inquiry: truth is 'epistemic' only in the sense that the method of science *factually* has the ability to produce true results in the long run. Therefore, it is crucially important to distinguish the objective or non-epistemic concept of truth (truth as correspondence between statements and facts, independently of our knowledge) and the epistemic surrogates of truth (coherence, consensus, assertability, acceptability, etc.)

In my view, Tarski's semantic definition of truth gives us a satisfactory non-epistemic definition of truth — it specifies the meaning of the concept of truth rather than a procedure for recognizing particular truths. Let us defend this approach against some popular objections.

Some philosophers argue against Tarski's semantic treatment that truth should be viewed as a property of non-linguistic entities (such as propositions) rather than of linguistic entities (such as sentences). This objection overlooks the fact that Tarski's theory assigns a truth value to a sentence only relative to an interpretation function V from language L to L-structures, so that in effect it deals with *interpreted sentences*, not merely with uninterpreted strings of symbols. We have seen already in Chapter 2.1 how Tarski's extensional semantics can be complemented by the intensional notions of property and proposition. The property designated by a predicate M is a function which assigns to M its extension in all possible world — and thereby also in the *actual world*. Hence, it also determines the interpretation function V_a relative to which *factual truth* is defined.

Some philosophers question the relation of Tarski's semantic theory of truth to the correspondence theory. Susan Haack claims that "Tarski did not regard himself as giving a version of the correspondence theory" (Haack, 1978, p. 114). However, this is clearly contradicted by Tarski's own statements: he wants to give a "precise expression" of "intuitions" which adhere to the "classical Aristotelean conception of truth". In Carnap's words, Tarski is *explicating* the vague formulations of the "correspondence theory" which claims that "the truth of a sentence consists in its agreement with (or correspondence to) reality".

The status and nature of Tarski's theory has been obscured by his own "material adequacy condition" for definitions of truth. According to this condition, a theory of truth should entail all equivalences of the form

(T) S is true iff α,

where α is a sentence in the object-language (or, more precisely, its homophonic translation in the meta-language) and S is the name of this sentence in the meta-language. This is the same *T*-equivalence which was written above as

(T)   'α' is true iff α.

Some philosophers tend to think that there is nothing more in Tarski's theory of truth than this simple 'disquotational schema'. If this were the case, Tarski's approach could hardly be viewed as a correspondence theory at all, since it would be based upon a correlation between *two languages* (i.e., the object-language and the meta-language) rather than between *language and world*. Haack is therefore right in saying that the condition (T) itself is "neutral between correspondence and other definitions" (*ibid.*). However, the main idea of the definition sketched above was the stipulation that the interpretation function V_a establishes a correlation between linguistic entities and non-linguistic entities in the world. Through this function V_a, linguistic formulas are satisfied, and linguistic sentences are made factually true, by entities and facts in the actual world.[10]

COGNITIVE PROBLEMS, TRUTH, AND INFORMATION 139

If quantification over propositional variables is allowed, it is indeed plausible to claim that the generalization

(30) (α) ('α' is true iff α)

is analytically true: it is a consequence of the definitions of truth and of the quotation marks. Consider, however, some instance of the T-equivalence, e.g.

(31)   'Russell is a logician' is true iff Russell is a logician.

Is this also an analytical consequence of the definition of truth? The answer to this question depends on the way in which (31) is interpreted. There are two different readings of (31) which can be represented more formally as follows:

(32)   Sentence '$M(a)$' is true relative to V iff the denotation $V(a)$ of a belongs to the extension $V(M)$ of M.

(33)   Sentence '$M(a)$' is true iff $V(a) \in V(M)$.

Here (32) is indeed a direct consequence of Tarski's definition of truth — and, hence, analytic. In contrast, the truth of (33) presupposes that $V(a)$ is the denotation of the name 'a' and $V(M)$ is the extension of the predicate 'M'. These presuppositions correspond to the assumptions

(34)   'Russell' refers to Bertrand Russell

(35)   'Logician' refers to logicians,

which are not analytically true: (34) and (35) are contingent descriptions of the way in which the (English) linguistic community has agreed to use the terms 'Russell' and 'logician'.[11]

The equivalence (31) should be a consequence of the definition of truth only if it is interpreted as a claim of the form (32) — but not of the form (33). Another way of expressing this conclusion is the following: to define the concept of factual truth, it is not sufficient to consider conditions of the form

Sentence '$M(a)$' is factually true iff the denotation of a (whatever it is) belongs to the extension of M (whatever it is), where the interpretation function V_a is mentioned only *de dicto*. Instead, we have to use a condition where V_a is *de re*:

Sentence '$M(a)$' is factually true
$\equiv$ '$M(a)$' is true relative to V_a
$\equiv V_a(a) \in V(M)$.

In an influential article, Hartry Field (1972) has argued that Tarski's theory should be understood as reducing the concept of truth to the semantic, language-world relation of denotation (for names, predicates, and function symbols), and he claims that the relation of denotation should be explicated in *physicalistic* terms. As a physicalist, Field rejects 'semanticalism', i.e., the view that there are 'irreducibly semantic facts'. While Donald Davidson thinks that Tarski's program (for him, primarily, Convention (T)) does not need the concept of reference at all, many critics of Tarski have followed Field in associating the correspondence theory of truth with physicalism.[12] In my view, such a reductionism is not needed in the theory of truth: relative to any conventional way that we choose to link parts of language with parts of the world (i.e., function V_a), Tarski's definition gives us a reasonable notion of truth. In contrast to Field's view, the choice of the interpretation function (which connects language and reality) is by no means a physical fact: rather it is based upon a *human convention*.

Human languages are symbolic — rather than indexical or iconic in Peirce's sense. The meanings of words — and the connection between language and world — are based upon human conventions, not upon any 'natural' relations. This remark is compatible with the fact that the reference of many terms can be fixed in many different ways — sometimes by employing such natural relations as causality. For example, we may accept the convention that a certain terms refers to those entities which cause such and such manifest effect. This conventionality of meanings is also the reason why statements like (34) and (35) express contingent 'semantical facts'. We may thus accept a *consensus theory of meaning* — without giving up the correspondence theory of truth.

This conclusion can be expressed also in the following way. When we know the meaning of a sentence, we thereby know its *truth conditions*, i.e., we know what the world would be like if this sentence were true in it. But knowledge of truth condition does not entail knowledge about *truth value*. This is analogous to the fact that a function may be defined without knowing its value for some particular arguments. The meaning of a sentence is determined by a stipulation within the linguistic community, but the factual truth of a sentence is ultimately determined by the actual world — rather than the ultimate consensus of a community of investigators.

We have defended above the view that truth is conceptually non-epistemic — without accepting the picture that Putnam calls *metaphysical realism*. The correspondence theory of truth does not presuppose that the world is 'ready-made', carved into 'pieces' or into individuals and essential properties independently of man. There is no privileged language for describing the world. Different conceptual systems — e.g. our ordinary language in everyday life and the various scientific frameworks — constitute the world in different ways into 'objects' or 'individuals', and they always select for consideration only some of the properties of these objects. In this sense, our ontology in practice is a reflection of the choice of the language.

However, these remarks create no problems to the Tarskian version of the correspondence theory, since it does not presuppose a ready-made world. According to the model-theoretic definition, the truth or falsity of a sentence in language L depends on the *structure* of the world relative to the descriptive vocabulary of L. True sentences are not in a relation of correspondence to a non-structured reality, but rather to a structure consisting of objects with some properties and relations. Each such L-structure represents an aspect or a fragment of the actual world, and there are as many L-structures representing the actual world as there are different languages L. In this sense, we may say that 'states of affairs' are relative to languages which are used to describe them. But it does not follow that facts depend on our knowledge or that truth is in any sense 'epistemic': as soon as a language L is given, with predicates designating some properties, it is up to the world — not to us — to 'decide' what sentences of L are factually true. For example, as soon as the meanings of the terms 'Helsinki', 'Vienna', and 'is north of' are fixed, it follows that Helsinki is north of Vienna — quite independently of any opinions that we may have concerning this fact.

The world can be conceptualized in many different ways. Each conceptual system so to say picks out its own facts from something — let us call it THE WORLD — which is not yet conceptualized or carved into pieces. In this sense, the world is not 'ready-made'. It does not follow, however, that the world is completely 'plastic' or 'malleable'

142   CHAPTER FOUR

into any form we like — just like a litre of water can be put into a container of any shape (with fixed volume). THE WORLD has 'factuality' in the sense that it is able to resist our will. This conclusion can be expressed in the following way. Let L be a semantically determinate conceptual system, i.e., a language where the basic concepts and their combination have well-defined meanings. Then the function which correlates the linguistic expressions of L with the possible worlds has unique values (denotations, extensions, truth-values) in each possible world. These assignments give us the family of all L-structures — they represent all the well-defined possible states of affairs expressive in L. The factuality of THE WORLD means now that one and only one of these L-structures fits THE WORLD. This structure, call it Ω_L^*, is partly relative to language L, but still it is 'chosen' by THE WORLD from all L-structures. Thus, Ω_L^* is a fragment of THE WORLD, which as it were contains everything of the actual world that can be 'seen' through the framework L.

Putnam's argument against realism is based on the false presupposition that we can impose any structure we like on to THE WORLD. He starts from a first-order theory T_I which is 'ideal' in some pragmatic sense, picks up a model Ω of T_I which has the same cardinality as THE WORLD, and then defines a bijective mapping f from Ω onto THE WORLD. It follows that the theory T_I is true in the structure Ω_W which is the isomorphic image of Ω via the function f. (See Figure 2.) Therefore, Putnam concludes, the ideal theory T_I could not be false in THE WORLD. One problem with this argument is that it goes through for any consistent first-order theory T_I. More seriously, if L is the language of theory T_I, then there is no guarantee that the imposed structure Ω_W coincides with the structure Ω_L^*.[13]

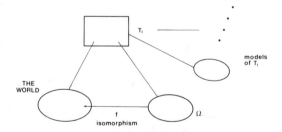

Fig. 2. Putnam's argument against realism.

COGNITIVE PROBLEMS, TRUTH, AND INFORMATION   143

Metaphysical realism assumes that there is some ideal privileged language L_{id} such that

(36)   THE WORLD = $\Omega^*_{L_{id}}$.

We don't need anything like (36) in the theory of truthlikeness.[14] The problem of truthlikeness arises in connection with cognitive problems. If the alternative solutions are expressed in language L, then our focus is the structure Ω^*_L. To have a cognitive problem relative to L means that Ω^*_L is at least partly unknown to us, but we want to find out what Ω^*_L is like. Even if unknown to us, Ω^*_L is uniquely determined by L and THE WORLD, and Ω^*_L in turn uniquely determines the truth values of the sentences of L through Tarski's definition. Our cognitive aim is to find out what these truth values are.

Sometimes it may turn out that the truths expressible in language L are not cognitively interesting — for example, they may fail to contain interesting universal generalizations or they may lack connections to other parts of our knowledge. In such cases, language L may be changed or enriched to another language L' — and our focus shifts from Ω^*_L to $\Omega^*_{L'}$. It is important to note, however, that the relations of languages L and L' are reflected in connections between the structures Ω^*_L and $\Omega^*_{L'}$. In particular, if L' is a meaning-preserving extension of L, then $\Omega^*_{L'}$ is an *expansion* of Ω^*_L in the model-theoretic sense: it just adds to Ω^*_L the interpretations of the new vocabulary in $\lambda' - \lambda$. In some cases the new language may also introduce new kinds of individuals which have relations to old individuals — for example, the individuals in the domain of Ω^*_L may be compositions or collections of new entities in the domain of $\Omega^*_{L'}$.

Thus, for a given cognitive problem **B**, there may exist a 'practically ideal' language L which is descriptively as complete as desired, i.e., λ contains all those terms that are factually relevant for the problem **B** (cf. Niiniluoto, 1984b, Ch. 5). But, equally well, conceptual enrichment helps us to formulate finer cognitive problems (cf. Section 2) and to penetrate deeper into the secrets of THE WORLD.

4.4. VAGUENESS

One familiar objection to the Tarskian correspondence theory remains to be discussed. It is the claim that Tarski's definition applies only — or at best — to formal languages. Thus, Haack claims that "Tarski, of course, is thoroughly sceptical about the applicability of his theory to natural languages" (Haack, 1978, p. 120). This is not correct: Tarski (1944) says that his solution to the definition of truth is "rigorous" for "languages whose structure has been exactly specified" and "approximate" for natural languages. To deal with natural languages, their portions that we are interested in should be replaced by a syntactically well-specified language which differs from the given language "as little as possible". Moreover, Tarski is very optimistic concerning the possibilities of using his semantic theory to the language of empirical science.

Natural language may fail to be 'exactly specified' syntactically and semantically. Grammatically ill-formed sentences are not a problem for the Tarskian program, since there is no reason to associate truth values to them. If a sentence is *syntactically ambiguous* between two or more readings, Tarski's approach allows us to assign a truth value to each such reading separately. Syntactical ambiguity is not important for the theory of truthlikeness, either, since it should be in any case eliminated from the formulations of cognitive problems.

More interesting and serious problems for classical semantics is created by the fact that human languages often fail to be semantically determinate: they contain different kinds of *vague expressions*, including predicates ('big', 'bald', 'dead', etc.), modifiers ('almost', 'approximately', etc.), and quantifiers ('most', 'few', etc.).

In some cases, the use of vague expressions is based essentially on epistemic reasons: while our language L is semantically indeterminate, there exists a semantically determinate extension L' of L. Even though L' gives sharp boundaries to the use of the expressions of L, language L may still be used for practical reasons — e.g. if it is too expensive or difficult to measure the concepts of L'. The relation of our ordinary language and the scientific languages is often of this kind: such vague qualitative terms as 'bright', 'warm' or 'fish' can be replaced by exact scientific concepts. Carnap calls this method of proceeding from vague expressions to their exact counterparts *explication*.[15]

Epistemic vagueness does not present interesting problems to the theory of truth. A vague sentence of language L turns out to be a disjunction of sharp sentences in the determinate extension L', and its truth-conditions can be expressed by a Tarskian truth definition for L'.

However, sometimes the boundaries drawn by the method of explication seem quite arbitrary — e.g. 'a is bald if a has less than 20 hairs in his head'; 'Most A's are B iff at least 90 per cent of A's are B'.

COGNITIVE PROBLEMS, TRUTH, AND INFORMATION   145

Some expressions resist explication, because they are inherently or semantically vague.[16] In such cases, there are no precise characterizations of their meanings which would fix uniquely their interpretations in the actual world. Suppose, for example, that a qualitative predicate M is introduced by the so-called method of paradigmatic examples: M is said to apply to certain paradigm cases $a_1, \ldots, a_n$, pointed out by ostension, and also to all the other cases which are sufficiently similar to some of $a_1, \ldots, a_n$. To explicate this concept M, assume that a measure sim for the degrees of similarity between objects in universe U is defined (cf. Chapter 1.3) and then fix the minimum degree of similarity, say q, required of the members of the extension of M. In other words, M is taken to apply to the members of the class

(37) $\{x \in U | \max_{i=1,\ldots,n} \text{sim}(x, a_i) \geq q\}$.

Trivially, a_i belongs to this set for all $i = 1, \ldots, n$. If the similarity between the exemplars a_i is great, then the set defined by (37) forms a cluster of similar individuals. The arbitrary feature of (37) is the fixation of the value q. Equally well we might claim that the extension of M does not have any exactly specified boundaries at all: besides clear cases of M, and clear cases of $\sim M$, there are also indeterminate boundary cases. A more refined approach is obtained by assuming that the extension of M is what Zadeh calls a *fuzzy set* (cf. Example 1.7).[17] Then each object x of universe U has a degree $K_M(x) \in [0, 1]$ in which it belongs to the extension of M. Moreover, if the function sim is available, it is natural to take function K_M to be proportional to the value

$\max_{i=1,\ldots,n} \text{sim}(x, a_i)$.

Similar remarks apply to quantitative concepts. While for a mathematical realist quantities help us to formulate an exactly true description of the mathematical reality (cf. Chapter 3.1), it has to be admitted in any case that quantities are frequently used in idealized situations, where the necessary conditions for their existence are not exactly satisfied. Then these quantities do not really have sharp values in the actual world. This conclusion is also defended by those empiricists and operationalists who think that the reality itself is vague or indeterminate with respect to mathematical descriptions. In this case, it is often assumed that the sharp value of a quantity h for object x is in reality represented by an interval in $\mathbb{R}$ (e.g. the theory of 'indeterminate probabilities' takes probabilities to be intervals of real numbers rather than real numbers). More generally, the sharp value $h(x) \in \mathbb{R}$ may be replaced by a fuzzy membership function on $\mathbb{R}$.

A common feature of genuinely semantic vagueness is the following. Let L be a language, and let $\mathbf{M}_L$ be the class of all L-structures in the standard sense. If L is semantically indeterminate, there is no unique structure Ω_L^* in $\mathbf{M}_L$ which represents the actual world relative to L (cf. Section 3). Rather, there are several L-structures which correspond to the actual world. There are two ways of making this idea more precise: Ω_L^* is replaced (i) by a non-empty set $\Theta_L \subseteq \mathbf{M}_L^*$ of L-structures, or (ii) by a fuzzy set of L-structures. Here (i) is a special case of (ii).[18]

To illustrate these alternatives, let L be a simple language with one individual constant $\mathbf{a}$ and one-place function symbol $\bar{h}$. If L is semantically determinate, the actual world is represented by a L-structure of the form

$$\Omega_L^* = \langle \{a\}, r \rangle,$$

where $r \in \mathbb{R}$, which makes the statement '$\bar{h}(\mathbf{a}) = \mathbf{r}$' and its logical consequences factually true. If L is semantically indeterminate, then alternative (i) takes the actual world to be represented by a set

(38) $\quad \Theta_L^* = \{\langle \{a\}, z \rangle \mid r_0 \leq z \leq r_1\},$

where $r_0, r_1 \in \mathbb{R}$. In the alternative (ii), the set Θ_L^* is replaced by a fuzzy subset of $\mathbf{M}_L$, i.e., by a function

(39) $\quad K_*: M_L \to [0, 1].$

What conception of truth is applicable in the case of semantical indeterminacy? One possibility is to introduce of new truth value ('indeterminate') besides 'true' and 'false'; this leads to *three-valued logics*. In *infinite-valued logics*, a sentence may have as a truth value any real number in $[0, 1]$.[19] In the standard Lukasiewicz logic, the truth values of complex sentences are defined by the following conditions:

(40) $\quad \begin{aligned} v(\sim \alpha) &= 1 - v(\alpha) \\ v(\alpha \wedge \beta) &= \min\{v(\alpha), v(\beta)\} \\ v(\alpha \vee \beta) &= \max\{v(\alpha), v(\beta)\} \\ v(\alpha \supset \beta) &= \min\{1, 1 - v(\alpha) + v(\beta)\}. \end{aligned}$

These conditions can then be written as rules for determining the fuzzy extensions of atomic formulas with vague predicates M:

$$K_{M(x)}(a) = v(M(a)),$$

and for complex formulas:

$$K_{\sim a(x)}(a) = 1 - K_{a(x)}(a).$$

This approach leads us directly to (39), if we define K_* by

(41) $K_*(\Omega) = v(z_\Omega)$ for all $\Omega \in \mathbf{M}_L$,

where z_Ω is the state description of L which is true in Ω.

The alternative illustrated by (38) does not allow us to speak of 'degrees of truth'. But there is a very natural generalization of the standard notion of truth to this case. Recall that a sentence α of a semantically determinate language L is factually true if and only if it is true in structure Ω_L^*, i.e.,

(42) α is true   iff $\Omega_L^* \vDash \alpha$
                          iff $\Omega_L^* \in \text{Mod}(\alpha)$
                          iff $\alpha \in \text{Th}(\Omega_L^*)$
                          iff $\text{Th}(\Omega_L^*) \vdash \alpha,$

where

$$\text{Th}(\Omega_L^*) = \{\beta \mid \Omega_L^* \vDash \beta\}.$$

If the actual world is represented by a set Θ_L^* of L-structures, then in analogy with (42) we may define

(43) α is true   iff $\text{Th}(\Theta_L^*) \vdash \alpha$
                          iff $\alpha \in \text{Th}(\Theta_L^*)$
                          iff $\Theta_L^* \subseteq \text{Mod}(\alpha)$
                          iff $\Omega \vDash \alpha$ for all $\Omega \in \Theta_L^*.$

Hence, truth in this extended sense means truth in all the structures Ω within the set Θ_L^*.[20]

4.5. SEMANTIC INFORMATION

Karl Popper's *Logik der Forschung* (1934) introduced to the philosophy of science the idea of the information content of scientific theories. Popper argued against what he called 'inductivists' that the aim of science cannot be highly probable theories. Good theories, he urged, should be falsifiable and, hence, they should exclude many empirical statements. Falsifiable theories are thus bold, informative — and improbable. The information content and the (logical) probability of a theory are, therefore, inversely related to each other.[21]

Popper did not develop a measure for the amount of information, but he defined two concepts of 'content' for scientific hypotheses. The *logical content* of a hypothesis h is the class of its non-tautological logical consequences

(44) $\{g \mid h \vdash g \text{ and not } \vdash g\}$.

The *empirical content* of a consistent statement h is the class of its potential falsifiers:

(45) $\{e \mid e \text{ is a basic empirical sentence and } e \vdash \sim h\}$.

The first systematic study of the concept of semantic information was given by Rudolf Carnap and Yehoshua Bar-Hillel in their 'An Outline of a Theory of Semantic Information' (1952). They consider the sentences h of a monadic first-order language L_N^k. After rejecting $\{g \text{ in } L_N^k \mid h \vdash g\}$ and Popper's logical content (44), they defined the *information content* Cont(h) of h as the class of the content elements entailed by h, where a *content element* in L_N^k is a negation of a state description in L_N^k (cf. Chapter 2.3). Thus,

(46)   Cont(h) = $\{\sim s \mid s \in \mathscr{Z} \text{ and } h \vdash \sim s\}$.

Cont(h) defined by (46) is obviously a 'mirror image' of the range $R(\sim h)$ of the negation of h, since

$$\begin{aligned}\text{Cont}(h) &= \{\sim s \mid s \in \mathscr{Z} \text{ and } s \vdash \sim h\} \\ &= \{\sim s \mid s \in R(\sim h)\}.\end{aligned}$$

In fact, Bar-Hillel (1952) defines the information content of h directly as

(47) $R(\sim h) = \{s \in \mathscr{Z} \mid s \vdash \sim h\}$

instead of (46). If Popper's 'basic sentences' could be identified with the state descriptions, then (47) would be equivalent to his notion of empirical content (45).

Let $\mathscr{CE}$ be the class of the content elements of L_N^k. Then, by (46) and (2.17), Cont(h) has the following properties:

(48) (a) Cont(h') $\subseteq$ Cont(h) iff $\vdash h \supset h'$
  (b) Cont(h) = $\emptyset$ iff $\vdash h$
  (c) Cont(h) = $\mathscr{CE}$ iff $\vdash \sim h$
  (d) Cont(s) = $\mathscr{CE} - \{\sim s\}$ if $s \in \mathscr{Z}$
  (e) Cont($\sim s$) = $\{s\}$ if $s \in \mathscr{Z}$
  (f) Cont($\sim h$) = $\mathscr{CE}$ − Cont(h)
  (g) Cont($h \vee h'$) = Cont(h) $\cap$ Cont(h')
  (h) Cont($h \& h'$) = Cont(h) $\cup$ Cont(h')
  (i) Cont(h) $\cap$ Cont(h') = $\emptyset$ iff $\vdash h \vee h'$.

Here (a) says that information content covaries with logical strength. The content of logical truths is minimal, by (b), and that of contradiction maximal, by (c). Among factual statements in L_N^k, state descriptions are maximally informative, by (d), and content elements minimally informative, by (e).

The set of state description $\mathscr{Z}$ for language L_N^k is a cognitive problem in the sense of Section 2. Thus, definition (46) can be immediately generalized to any cognitive problem $\mathbf{B} = \{h_i | i \in I\}$: for all partial answers $g \in D(\mathbf{B})$,

(49) Cont$_\mathbf{B}(g) = \{\sim h_i | h_i \in \mathbf{B}$ and $g \vdash \sim h_i\}$.

Cont$_\mathbf{B}(g)$ contains the negations of all those complete potential answers in $\mathbf{B}$ that g excludes.

The weak point of these definitions is the fact that the information contents of two statements turn out in most cases to be *incomparable* with respect to the inclusion relation $\subseteq$. The advantage of introducing quantitative measures for the *amount* or *degree of information* is the possibility of comparing any pair of hypotheses (in a given language) with respect to their informativeness.

A simple solution, favoured by Isaac Levi (1967a), to defining the degree of information is the following. Let $\mathbf{B} = \{h_i | i \in I\}$ be a finite cognitive problem. As each $h_i \in \mathbf{B}$ represents a complete potential answer, we may regard them as equally informative (from the viewpoint of the problem $\mathbf{B}$). Further, the informativeness of a partial answer

$$g = \bigvee_{i \in I_g} h_i \in D(\mathbf{B})$$

depends on the number of complete answers that it allows, i.e., $|I_g|$, or, equally well, on the number of the elements of **B** that it excludes, i.e., $|I| - |I_g|$. This idea leads to the definition

(50) $\quad \mathrm{cont}_u(g) = (|I| - |I_g|)/|I| = 1 - |I_g|/|I|$

for all $g \in D(\mathbf{B})$. (The reason for the notation cont_u will become clear later.) If a contradiction g is treated as a member of $D(\mathbf{B})$ with $|I_g| = 0$, this measure has the properties

(51) $\quad \mathrm{cont}_u(g) = 0 \quad \text{if} \quad \vdash g$
$\quad \mathrm{cont}_u(h_i) = (|I| - 1)/|I| \quad \text{if} \quad h_i \in \mathbf{B}$
$\quad \mathrm{cont}_u(g) = 1 \quad \text{if} \quad \vdash \sim g.$

However, as contradictions are not really answers to any cognitive problem — they entail all the complete answers at the same time — we might modify definition (50) so that it applies only to the members of $D(\mathbf{B})$:

(52) $\quad \mathrm{cont}_u^+(g) = (|I| - |I_g|)/(|I| - 1).$

This measure has then the properties

(53) $\quad \mathrm{cont}_u^+(g) = 0 \quad \text{if} \quad \vdash g$
$\quad \mathrm{cont}_u^+(h_i) = 1 \quad \text{if} \quad h_i \in \mathbf{B}.$

Definitions (50) and (52) presuppose that the elements h_i of **B** are equally informative. This assumption can be contested in two different ways. First, suppose that the investigator has an epistemic probability measure P on the set **B**, so that $P(h_i)$ is his degree of belief in the truth of $h_i \in \mathbf{B}$. Then the *surprise value* of an answer h_i to the cognitive problem **B** will be for him low if $P(h_i)$ is great and high if $P(h_i)$ is small. If the degree of information of h_i is assumed to reflect its surprise value, then this degree should be defined so that it is inversely related to the values of the probability measure P.

Secondly, it can be argued that, for many cognitive problems, some complete answers in **B** give more *substantial information* than others. For example, assume that **B** is the class of monadic constituents C^w for a language L with q Q-predicates. Then the constituent C^q, which claims that all Q-predicates are instantiated, is 'weaker' than the other constituents in the following sense: if C^q is true, then there are no true universal generalizations expressible in L. More generally, if $w < v$, then C^w entails that there are more true universal generalizations than

COGNITIVE PROBLEMS, TRUTH, AND INFORMATION 151

C^v does. For an investigator, who is interested in finding true laws about the world, the news that C^w is true is 'better' than the news that C^v is the case. In this sense, it may be argued that

(54) C^w is substantially more informative than C^v iff $w < v$.

This notion of information is again inversely related to probability, since most systems of inductive logic for language L contain the assumption that

(55) $P(C^w) \leqslant P(C^v)$ iff $w < v$.

(See (2.77).)

Carnap and Bar-Hillel (1952) proposed that the surprise value, or the unexpectedness, of a sentence h in L_N^k is defined by the logarithmic measure

(56) $\inf(h) = -\log P(h)$,

where P is c^+ or c^* of Carnap's inductive logic. (log is the logarithm function with base 2.) More generally, h may be a sentence in a finite language L and P a regular probability measure for L. Then the inf-measure has the following properties:

(57) (a) $0 \leqslant \inf(h) \leqslant \infty$
   (b) $\inf(h) = 0$ iff $\vdash h$
   (c) $\inf(h) = \infty$ iff $\vdash \sim h$
   (d) $\inf(h) \geqslant \inf(h')$ if $\text{Cont}(h') \subseteq \text{Cont}(h)$
   (e) $\inf(h) \geqslant \inf(h')$ if $\vdash h \supset h'$
   (f) $\inf(h) > \inf(h')$ if $\vdash h \supset h'$ and $\nvdash h' \supset h$
   (g) $\inf(h) = \inf(h')$ if $\vdash h \equiv h'$
   (h) $\inf(h \lor h') \leqslant \inf(h) \leqslant \inf(h \ \& \ h')$
   (i) $\inf(h \ \& \ h') = \inf(h) + \inf(h')$ if $P(h \ \& \ h') = P(h)P(h')$.

(Cf. Chapter 2.10.)

(56) is formally similar to the measure which is used in the communication theory, developed by Claude Shannon in 1948. However, Shannon's concept of information is purely syntactical, independent of the semantic content or meaning of a message — $P(h)$ measures only the stable relative frequency in which the 'message' h is transmitted through a communication channel. Even in cases, where the messages happen to be meaningful sentences, Shannon's information fails to satisfy condition (e), if the logically stronger sentence h is used more frequently in the channel than the weaker sentence h'.

The characteristic feature of the inf-measure is condition (i) which says that information is additive over conjunctions of probabilistically independent sentences. In fact, (56) is known to be the only non-negative function which satisfies condition (i).

Carnap's and Bar-Hillel's proposal for the amount of substantial information of sentence h is

(58) $\text{cont}(h) = 1 - P(h)$.

(See also Popper, 1954.) If P is a regular probability measure for language L, then the cont-measure has the following properties:

(59) (a) $\text{cont}(h) = P(\sim h)$
 (b) $\text{cont}(\sim h) = P(h)$
 (c) $0 \leqslant \text{cont}(h) \leqslant 1$
 (d) $\text{cont}(h) = 0$ iff $\vdash h$
 (e) $\text{cont}(h) = 1$ iff $\vdash \sim h$
 (f) $\text{cont}(h) \geqslant \text{cont}(h')$ if $\text{Cont}(h') \subseteq \text{Cont}(h)$
 (g) $\text{cont}(h) \geqslant \text{cont}(h')$ if $\vdash h \supset h'$
 (h) $\text{cont}(h) > \text{cont}(h')$ if $\vdash h \supset h'$ and $\nvdash h' \supset h$
 (i) $\text{cont}(h) \equiv \text{cont}(h')$ if $\vdash h \equiv h'$
 (j) $\text{cont}(h \lor h') \leqslant \text{cont}(h) \leqslant \text{cont}(h \& h')$
 (k) $\text{cont}(h \lor h') = \text{cont}(h) + \text{cont}(h') - \text{cont}(h \& h')$
 (l) $\text{cont}(h \& h') = \text{cont}(h) + \text{cont}(h') - \text{cont}(h \lor h')$
 (m) $\text{cont}(h \& h') = \text{cont}(h) + \text{cont}(h')$ if $\vdash h \lor h'$
 (n) $\text{cont}(h) = 1 - 2^{-\text{inf}(h)}$.

The main difference to inf is in condition (m): substantial information content is additive over conjunctions $h \& h'$ such that $h \lor h'$ is logically true. The condition $\vdash h \lor h'$ essentially says that h and h' do not have any common informative content (see (48) (i)): there are no factual sentences which are entailed by both h and h'.

If P is a uniform probability measure on a finite cognitive problem $\mathbf{B} = \{h_i \mid i \in I\}$, i.e.,

$$P(h_i) = 1/|I|,$$

then the definition (58) leads to

$$\text{cont}(h_i) = 1 - (1/|I|),$$

which is precisely cont_u defined in (50). Thus, cont_u is a special case of the cont-measure for equally probable alternative hypotheses.

Functions inf and cont are called measures of *semantic information* for an obvious reason: their value for a sentence h in language L depends on how many possible states of affairs, describable in L, the sentence h excludes. On the other hand, the probability measure P indicates the weight given to such states of affairs. As Hintikka (1970a) has emphasized, there is no purely semantical way of choosing this measure P — rather P serves as a context-dependent, pragmatic boundary condition for the application of inf and cont. In this sense, there is no sharp distinction between semantic and pragmatic measures of information. This does not mean that functions inf and cont need to be purely subjective: we have seen that, in many contexts, one can at least restrict the choice of P by rationality conditions.

Carnap and Bar-Hillel point out that inf and cont can be regarded as measures of information for an idealized rational investigator who is logically omniscient. This assumption is reflected in the results (57) (b)—(g) and (59) (d)—(i). If these measures are applied to polyadic first-order languages L, their values are not effectively calculable, since there is no algorithm for identifying the inconsistent constituents (cf. Chapter 2.8). Therefore, Hintikka (1970b) distinguishes the concepts of *surface information* and *depth information* in polyadic languages. The surface information of a generalization g at depth d is zero if and only if g is a surface tautology at depth d, i.e., the distributive normal form of its negation $\sim g$ at depth d contains only trivially inconsistent constituents. Non-trivial logical truths at depth d contain, therefore, a positive amount of surface information. Their depth information, obtained as a limit when $d \to \infty$, is nevertheless zero.

Following Hintikka (1968a), we may define two measures of *incremental information* by

(60) $\inf_{add}(h/e) = \inf(h \ \& \ e) - \inf(e)$
$\text{cont}_{add}(h/e) = \text{cont}(h \ \& \ e) - \text{cont}(e)$.

They tell how much information h adds to the information already contained in e. Then

(61) (a) $\inf_{add}(h/e) = -\log P(h/e)$
 (b) $\inf_{add}(h/e) = 0$ iff $\vdash e \supset h$
 (c) $\inf_{add}(h/e) = \inf(h)$ if $P(h \ \& \ e) = P(h)P(e)$.

(62) (a) $\text{cont}_{add}(h/e) = \text{cont}(e \supset h)$
 (b) $\text{cont}_{add}(h/e) = P(e \ \& \sim h)$
 (c) $\text{cont}_{add}(h/e) = 0$ iff $\vdash e \supset h$
 (d) $\text{cont}_{add}(h/e) = \text{cont}(h)$ iff $\vdash h \lor e$.

Measures of *conditional information* tell how informative a hypothesis h is in a situation where the truth of another sentence e is known:

(63) $\inf_{\text{cond}}(h/e) = -\log P(h/e)$
         $\text{cont}_{\text{cond}}(h/e) = 1 - P(h/e).$

Then $\inf_{\text{cond}}$ turns out to be the same as $\inf_{\text{add}}$, and

(64) $\text{cont}_{\text{add}}(h/e) = P(e)\text{cont}_{\text{cond}}(h/e).$

Further, $\text{cont}_{\text{cond}}$ behaves in certain respects like $\inf_{\text{add}}$:

(65) $\text{cont}_{\text{cond}}(h/e) = \text{cont}(h)$ iff $P(h \& e) = P(h)P(e).$

Measures of *transmitted information* tell how much the 'uncertainty' of h is reduced when e is learned, or how much substantial information e carries about the subject matter of h:

(66) $\text{transinf}(h/e) \quad = \inf(h) - \inf(h/e)$
         $\text{transcont}_{\text{add}}(h/e) = \text{cont}(h) - \text{cont}_{\text{add}}(h/e)$
         $\text{transcont}_{\text{cond}}(h/e) = \text{cont}(h) - \text{cont}_{\text{cond}}(h/e).$

(See Hintikka, 1968a.) Hence,

(67) $\text{transinf}(h/e) \quad = \log P(h/e) - \log P(h)$
         $\text{transcont}_{\text{add}}(h/e) = 1 - P(e \vee h)$
         $\text{transcont}_{\text{cond}}(h/e) = P(h/e) - P(h).$

(68) $-\infty \leq \text{transinf}(h/e) < \infty$
         $0 \leq \text{transcont}_{\text{add}}(h/e) \leq 1$
         $-1 < \text{transcont}_{\text{cond}}(h/e) < 1.$

These measures have the value 0, if either e or h is a tautology. Moreover,

(69)   If $\vdash e \supset h$, then
         $\text{transinf}(h/e) = \inf(h)$
         $\text{transcont}_{\text{add}}(h/e) = \text{transcont}_{\text{cond}}(h/e) = \text{cont}(h).$

(70) $\text{transinf}(h/e) > 0$ iff $P(h/e) > P(h)$
         $\text{transcont}_{\text{add}}(h/e) > 0$ iff $\nvdash e \vee h$
         $\text{transcont}_{\text{cond}}(h/e) > 0$ iff $P(h/e) > P(h)$

Thus, e transmits some positive information about h, in the sense of transinf and $\text{transcont}_{\text{cond}}$, if e is positively relevant to h, and, in the sense of transcont$_{add}$, if e and h have some common information content.

For our purposes, it is crucially important to observe that *truth and information are independent concepts*. This is not always the case in natural languages. A person is 'well-informed' if he or she knows much — and thereby is aware of many truths. When a general 'informs' his staff about the situation, he is supposed to tell what he at least believes to be true. In contrast, the 'information' processed by a computer can contain any coded data which may be, if meaningful statements at all, either true or false.

In the sense of this section, a sentence is informative if it says much — quite independently of the question whether it tells truths or blatant falsities. Sentences can be roughly divided into four classes:

(I)   true and informative
(II) true and uninformative
(III) false and informative
(IV) false and uninformative.

Tautologies are extreme examples of class (II): they are logically true, but their information content is empty. Contradictions are extreme examples of class (III): their information content is maximal, but they are logically false.

In a cognitive problem $\mathbf{B} = \{h_i | i \in I\}$, all complete answers h_i are informative — in some cases even equally informative. The aim of an investigator is to find the correct complete answer h_*. This sentence is *the most informative true sentence* in the set $D(\mathbf{B})$ of all answers. But, as h_* is unknown, the demands of information and probable truth are pulling in opposite directions. In Isaac Levi's suggestive worlds, in our desire for information we have to gamble with truth. This tension between truth and information is also a major theme in the theory of truthlikeness which will be developed in the next chapters.

However, the classification (I)–(IV) is too rough for our purposes: especially we wish to distinguish the seriousness of errors in class (III). We shall thus see that the concepts of truth and information have to be supplemented with a concept of similarity (cf. Chapter 1) in order to develop a viable theory of truthlikeness.

CHAPTER 5

THE CONCEPT OF TRUTHLIKENESS

After the preparations in Chapters 1—4, we are now ready to attack the main problem of this book, viz. the concept of truthlikeness. This chapter discusses the historical background and the intuitive motivation underlying this concept. Some of the frequent confusions associated with this explicandum are explained by distinguishing it from a number of related concepts, such as probability, partial truth, degree of truth, and approximate truth. An exposition and criticism of Karl Popper's qualitative and quantitative notions of verisimilitude is also given. The programme for explicating truthlikeness that I myself advocate is then formulated in general terms in the next two chapters.

5.1. TRUTH, ERROR, AND FALLIBILISM

The concept of truthlikeness has an intimate connection to the *fallibilist* tradition in epistemology which takes seriously the fact that human beings are liable to error in their knowledge-seeking activities — just as the slogan *errare humanum est* says. According to Charles S. Peirce, who introduced the term, fallibilism claims that "people cannot attain absolute certainty concerning questions of fact" (1.149). In other characterizations, Peirce refers also to the indeterminacy of knowledge:

Fallibilism is the doctrine that our knowledge is never absolute but always swims, as it were, in a continuum of uncertainty and of indeterminacy. (1.711.)

Still, the supporters of this view wish to maintain a 'realist' concept of knowledge and the idea of cognitive progress. The concept of truthlikeness is needed for this purpose — which distinguishes the fallibilists from the scepticists (who deny the possibility of knowledge) and the instrumentalists (who attempt to analyse science without the concepts of truth and knowledge).

The doctrine of fallibilism obviously has many different variants. First, one has to distinguish between the possibility and the actuality of error. Thus, the *weak* form of fallibilism claims only that human knowledge, as far we are able to tell, *may* always be in error. True beliefs are possible, and perhaps we have already reached them in many cases, but we cannot be certain of that in any specific case. Strong forms of fallibilism claim that human knowledge *is* always erroneous: truth is a limit which can be approached but never reached by us.[1] Secondly, the history of epistemology shows that also the concept of error is ambiguous between several alternatives. It is, therefore, useful to review briefly the development of fallibilism and its main rivals.

According to Plato, we can love only something which we miss. Love for wisdom (*philosophia*) therefore presupposes lack of knowledge, but not that sort of resigned ignorance which has given up the search for truth (*Lysis* 218a, *Symposium* 204a,b). In spite of this dynamic feature of Plato's concept of philosophy, the dominant view of Ancient and Medieval epistemology was a static one. Plato's *episteme* was defined as justified and true knowledge about the unchanging realm of ideas. For the followers of Aristotle, *scientia* consisted of necessary truths which were demonstrated through scientific syllogisms. Science did not aim to search for new truths, but rather attempted to organize the old and established propositions in their natural or proper order. It was the possession — rather than the pursuit — of knowledge that characterized rationality as the essence of man.

A vivid illustration of this ideal of man is given in the dialogue *Contra Academicos*, written by St. Augustine in the years 385—386 (see Augustine, 1950). Book One is devoted to the problem: Is the mere search for truth sufficient for happiness? Licentius, a follower of Cicero, defends the view that "a man is happy if he seeks for truth, even though he should never be able to find it" (*ibid.*, p. 43). The opponents of Licentius point out that "only the wise man could be happy", and

the wise man ought to be perfect, and ... he who was still trying to find out what truth was, was not perfect, and, on that account, was not happy. (*Ibid.*, p. 62.)

Licentius argues further that

wisdom for man consisted in the search for truth from which, on the account of the resulting peace of spirit, happiness would follow. (*Ibid.*, p. 61.)

He associates this view with the scepticist doctrine of the Old Academy: a man who only seeks for truth, cannot fall into "the approbation as true of what is not true":

he who approves of nothing, cannot approve of what is not true. Therefore, he cannot be in error. (*Ibid.*, p. 48.)

Augustine's reply is that "the wise man has sure possession of wisdom". Therefore, not only he errs who assents to what is not true, but also he who fails to assent to wisdom.

The same theme was discussed, among others, by Thomas Aquinas who regarded error as "the monster of the mind" (Byrne, 1968, p. 69). As man's intellect is naturally directed to truth, error or false assertion is "the evil of the intellect", something to be avoided and overcome. The ultimate end of man, and the road to his happiness, is certain knowledge of all things, i.e.,

to become like God by actualizing in himself that knowledge always possessed by God but possessed by man only potentially. (*Ibid.*, p. 79.)

Similar ideas can be found, e.g., in Spinoza's *Ethics*. Following Descartes, Spinoza thought that all known truths are at the same time certain: "he, who has a true idea, simultaneously knows that he has a true idea" (*Prop.* xliii). In particular, "the human mind has an adequate knowledge of the external and infinite essence of God" (*Prop.* xlvii). Therefore, error or 'falsity' arises only from lack of knowledge and from confusion:

Falsity consists in the privation of knowledge, which inadequate, fragmentary, or confused ideas involve. (*Prop.* xxxv.)

The infallibilist quest for certainty, and the demonstrative ideal of science, withheld its strong position among the rationalists and the empiricists until the eighteenth and nineteenth century. As a countermove, the Greek scepticism was revived in the Western countries in the sixteenth century (cf. Popkin, 1960), and received support from David Hume's criticism of induction in 1748. Following this tradition, W. K. Clifford argued in 1886 that "it is wrong always, everywhere, and for everyone, to believe anything upon insufficient evidence".

The two views discussed above — the *infallibilist* picture of *episteme* or *scientia* consisting of demonstrated and incorrigible truths, and the *scepticist* view of the non-existence or the impossibility of knowledge — leave no room for anything like gradation of truth. They share the conviction that error is an evil which prevents man from reaching his perfection or happiness. However, the concept of *error*, as used in these discussions, covered at least three kinds of situations (see Figure 1) which are easily confused with each other:

THE CONCEPT OF TRUTHLIKENESS

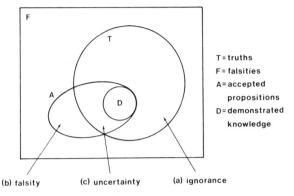

Fig. 1. Three kinds of errors.

(a) error as *ignorance*: the failure to know some truth which could be known,
(b) error as *falsity*: the acceptance of an untrue proposition,
(c) error as *uncertainty* (rash judgment): the acceptance of a true proposition on insufficient grounds.

It is easy to see that errors of kind (b) and (c) can be avoided by two strategies: by accepting nothing (scepticism), or by accepting only demonstrated truths (infallibilism). Both of these alternatives are problematic, if we realize that the class D can contain at most analytical truths of logic and mathematics: they commit the error (a) of ignorance with respect to *all* factual truths in T.

In developing a middle position between infallibilism and scepticism, a fallibilist may be concerned with the inevitable uncertainty of human knowledge. Instead of regarding uncertainty as an 'error', he may try to construct systems for non-demonstrative inference and measures for the epistemic uncertainty of propositions relative to available evidence. This is the line taken in the modern theory of inductive inference and in the systems of inductive probabilities (cf. Chapter 2.10). The Bayesian school of induction assumes that there is an 'epistemic probability' $P(h/e)$ for each scientific hypothesis h given evidence e, which measures the rational degree of belief in the truth of h on e. Thus, knowledge of truth is possible — and even probable in certain situations. It is interesting to note, however, that of the leading inductivists Carnap in fact follows the advice of Ancient sceptics: influenced by Hume, Carnap thinks that a rational man never 'accepts' scientific hypotheses as true.

In this sense, Carnap is a non-sceptical instrumentalist — rather than a fallibilist who is willing to employ a concept of 'knowledge' for the historically evolving body of accepted propositions. Other Bayesians, like Isaac Levi and Jaakko Hintikka, have proposed rules for the tentative acceptance of hypotheses upon inconclusive evidence.[2]

Another fallibilist line of thought is more concerned with errors of type (a) and (b), i.e., with the incompleteness and the falsity of human knowledge. The concepts of truthlikeness, partial truth, and degree of truth arise primarily in this context. In the next sections, I shall make some comments on the history of this complex of ideas.

5.2. PROBABILITY AND VERISIMILITUDE

In many Western languages, the concept of 'probability' is etymologically derived from the concept of 'truth' and either 'likeness' or 'appearance'. This is not only the case with the Latin *verisimilitudo* and its English counterparts 'verisimilitude', 'verisimilar', and 'likely', but also with the Russian 'veroyatnost', the German 'Wahrscheinlichkeit', the Swedish 'sannolikhet' (meaning literally 'likeness to the truth'), and the Finnish 'todennäköisyys' (meaning 'truth-looking' as in the construction 'hyvännäköinen' for 'good-looking').

The Greek concept of probability was developed primarily in the Stoic and sceptic discussions about belief (*pistin*), that is, about the fallible and fluctuating attitudes that Plato called *doxa* and later Latin writers *opinio*. The concept *pithanos* referred to a view which is 'probable' in the sense of plausible or persuasive. This word was translated to Latin by Cicero (106—43 B.C.) as *probabilis* and *verisimilis*. Another Greek concept *eikos* was used by Aristotle for "that which occurs in most cases".

Xenophanes, in the sixth century B.C., argued that no man has known or will know "certain truth", but he conjectured that his own views are *eioikota*. Popper (1963) translates this as 'like the truth', and Feyerabend (1984a) as 'truth-seeming'. Popper argues that Xenophanes uses this word in the objective sense of 'similarity to truth' rather than in the subjective sense of 'plausibility'. Therefore, he regards Xenophanes as a precursor of his own 'critical rationalism'.

While the textual evidence for interpreting Xenophanes is scanty and hardly conclusive, a more interesting and clearer position was developed in the New Academy, sometimes called the Third Academy, by Carneades (c. 214–129 B.C.). According to his view, the wise man avoids the crime of error by refraining from assent altogether, but still in matters concerning practical action he can follow the 'probability' (*pithanotes*). Cicero, who supported the doctrine of the New Academy, used the Latin terms *probabilitas* and *verisimilitudo* in this connection. Thus, we may say that the idea of 'verisimilitude' was originally proposed as a reply to the argument 'He who does not assent to anything, does nothing'. Augustine criticized this view, in *Contra Academicos*, in two ways. First, he claims that the followers of Carneades and Cicero commit an error, since they approve of what-is-like-truth and that what-is-like-truth is not true (i.e., an error of type (b)). (See Augustine, 1950, p. 147.) Secondly, he argues that it is absurd to rely on what-is-like-truth without assuming that truth is — and can be — known:

"I beg you", said I, "give me your best attention. If a man who never saw your father himself, yet, on seeing your brother, asserted that he was like your father, would you not think that he was mentally affected or talking at random?" ... "It is obvious that in the same way are your Academics to be laughed at, since they say that in practical matters they follow 'what-is-like-truth', although actually they do not know what truth itself is." (*Ibid.*, pp. 82, 85.)

Carneades is often taken to suggest that we use 'probability', in the sense of uncertain truth, as the guide of life. For example, Jeffrey (1984) draws a line from Carneades and Cicero via Montaigne and Hume to de Finetti. In this interpretation, Carneades is a forerunner of Bayesianism or of decision theory under uncertainty, who would recommend us to act on those assumptions which we believe and hope to be true or which yield the largest gains in terms of probabilistic expectation. However, in Augustine's interpretation — which is most interesting for the theory of truthlikeness — Carneades rather claimed that we have to act on something which falls short of truth and which we do not accept as true. This view is precisely the same as Popper's "solution to the pragmatic problem of induction" which suggests that we use verisimilitude as the guide of life without accepting any hypothesis as true or even probably true (see Chapter 12.4 below).

The Ancient philosophers had some grasp of the existence of grades of probability and verisimilitude. This idea was further discussed by Medieval discussions about *opinio* resulting from 'dialectical' disputations (as opposed to 'apodeictic' or demonstrative argumentation). According to Byrne's interpretation, Thomas Aquinas, who regarded error as an evil of the intellect, thought that some opinions are 'more evil' and, hence, 'more false' than other opinions which are 'less evil', 'more good', or 'more true' (Byrne, 1968, pp. 77—78). There was also a tendency to think that in a dialectical disputation both sides may contain some amount of truth — one 'more' than the other (*ibid.*, p. 159). However, for Thomas, the phrase 'more true' means in effect 'probably true' (*ibid.*, pp. 155—156). Thus, the Medieval notion of *probabilitas* was not a measure of the degree of truth or truthlikeness, but it indicated the number of the authorities and the strength of their arguments in favour of an opinion. Thomas acknowledged that, in some situations where genuine demonstrated knowledge is impossible, we are entitled to approve an opinion on the basis of inconclusive arguments. The concept of probability thereby had such connotations as argument and proof (*probatio*), the 'probity' of authorities, and the 'approval' of opiniotative knowledge (*ibid.*, pp. 144, 188).

The founders of the probability calculus in the seventeenth century used freely both terms *probabilitas* and *verisimilitudo*, but the doxastic interpretation continued to have a strong influence. Thus, Immanuel Kant regarded probability and verisimilitude as two different kinds of opinions:

> By probability is to be understood a holding-to-be-true out of insufficient reasons, which, however, bear a greater proportion to the sufficient ones that the reasons of the opposite. By this explanation we distinguish probability (*probabilitas*) from *verisimilitude* (*verisimilitudo*), a holding-to-be-true out of insufficient reasons so far as these are greater than the reasons of the opposite. (Kant, 1974, p. 89.)

Probability is an objectively valid "approximation to certainty" (i.e., ratio of the insufficient reasons to the sufficient ones), while verisimilitude is only subjectively valid "magnitude of persuasion" (i.e., ratio of the insufficient reasons for a claim to those for the opposite claim). Peirce, too, used the term 'verisimilar' in the epistemic sense indicating a special kind of relation between theory and evidence:

> I will now give an idea of what I mean by *likely* or *verisimilar*. I call that theory *likely* which is not yet proved but is supported by such evidence that if the rest of the conceivable possible evidence should turn out upon examination to be of a *similar* character, the theory would be conclusively proved. (2.663)

Popper's (1963) proposal to use the term 'verisimilitude' for an objective, non-epistemic concept indicating 'closeness to the truth' runs thus against the tradition. Popper's great merit is his recognition that *probability and truthlikeness are two quite distinct concepts* — no matter how the latter concept is explicated. Their difference is highlighted in the following observation: if reliable evidence e contradicts a hypothesis h, then the epistemic probability $P(h/e)$ of h given e is zero; yet, h may be highly truthlike, since false theories (even theories known to be false) may be 'close' to the truth.

The historians of probability have generally failed to appreciate the distinction between probability and verisimilitude. An amusing example of this can be found in Isaac Todhunter's classical work *A History of the Mathematical Theory of Probability* (1865). Todhunter tells that the gentlemen in Florence once raised the following problem in their "learned conversations":

A horse is really worth a hundred crowns, one person estimated it at ten crowns and another at a thousand; which of the two made the more extravagant estimate?

When consulted on this matter, Galileo pronounced the two estimates to be equally extravagant, "because the ratio of a thousand to a hundred is the same as the ratio of a hundred to ten". A priest called Nozzolini instead claimed that the higher estimate is more mistaken, since "the excess of a thousand above a hundred is greater than that of a hundred above ten".

Todhunter does not find the discussion about this issue to be "of any scientific interest or value". He suggests that the gentlemen — who already had renounced such 'frivolities' as attention to ladies, the stables, and excessive gaming — "might have investigated questions of greater moment than that which is here brought under our notice". We can recognize the Galileo—Nozzolini controversy as an early attempt to measure quantitatively the size of the error (of type (b)) of rival false point estimates of a real-valued parameter (cf. Chapter 8). It is no wonder that Todhunter does not find this discussion relevant to the history of probability theory, since it concerns the notion of 'closeness to the truth'.

However, one area — described in detail in Todhunter's excellent book — where the probability calculus in fact had an early contact with the idea of truthlikeness is the *theory of errors* within astronomical observations, from Thomas Simpson (1757), J. L. Lagrange (1770—73), and Daniel Bernoulli (1778) to P. S. Laplace's and Denis Poisson's classical formulations in the early nineteenth century. This theory, which lead to the discovery of the normal (Gaussian) distribution and to the concept of 'probable error', studied the following type of question: given the true value of a quantity, a device for measuring this value, and a probability distribution of errors within a single measurement, find the probability that the mean result of n observations shall lie between assigned limits around the true value. In other words, this theory is concerned with the idea of the probability of making an error of a given size (cf. Chapter 7.4 below).

5.3. APPROACH TO THE TRUTH

How can a finite being like man have knowledge of the infinite God? This theological question has had an important role in epistemology: in the tradition of objective idealism, there is a tendency of identifying the 'absolutely infinite being' (this is Spinoza's definition of God) with the whole of the spiritual reality — or with Truth in Capital 'T'. For example, this is how G. W. F. Hegel characterized the object of logic:

> The objects of philosophy, it is true, are upon the whole the same as those of religion. In both the object is Truth, in that supreme sense in which God and God only is the Truth.
>
> The first question is: What is the object of our science? The simplest and most intelligible answer to this question is that Truth is the object of logic. Truth is a noble word, and the thing is nobler still. So long as man is sound at heart and in spirit, the search for truth must awake all the enthusiasm of his nature. But immediately there steps in the objection — are *we* able to know the truth? There seems to be a disproportion between finite beings like ourselves and the truth which is absolute: and doubts suggest themselves whether there is any bridge between the finite and the infinite. God is truth: how shall we know Him? (Hegel, 1975, pp. 3, 26.)

On the other hand, the philosophical materialists can interpret the infinite as the mind-independent, material world with its unlimited complexity, i.e., with its 'inexhaustible' variety of attributes and their levels. This is precisely the strategy of such materialist followers of Hegel as Friedrich Engels and V. I. Lenin.

The Thomist solution to the possibility of human knowledge of infinite God was based on the concept of analogy. Another trend is the so-called negative theology which claims that we can only know what God is *not* like. This view is more interesting for our purposes, since it — and its materialist versions — have provided a significant motivation to the idea that human knowledge may make gradual 'approach' to the infinite truth.

Nicholaus Cusanus, in *De docta ignorantia* (1440), combined negative theology with the thesis that the opposites coincide at infinity — just as a circle with infinite radius is also a straight line (see Cusanus, 1954). The main lesson of "learned ignorance" is that the "infinite as infinite" or "the absolute truth" is "unknown" or "beyond our grasp". While truth itself does not have degrees, "absolute truth enlightens the darkness of our ignorance", so that we can make approach towards truth.

> From the self-evident fact that there is no gradation from infinite to finite, it is clear that the simple maximum is not to be found where we meet degrees of more or less; for such degrees are finite, whereas the simple maximum is necessarily infinite. It is manifest, therefore, that when anything other than the simple maximum is given, it will always be possible to find something greater.

> A finite intellect, therefore, cannot by means of comparison reach the absolute truth of things. Being by nature indivisible, truth excludes 'more' or 'less', so that nothing but truth itself can be the exact measure of truth ... In consequence, our intellect, which is not the truth, never grasps the truth with such precision that it could not be comprehended with infinitely greater precision. The relationship of our intellect to the truth is like that of a polygon to a circle; the resemblance to the circle grows with the multiplication of the angles of the polygon; but apart from its being reduced to identity with the circle, no multiplication, even if it were infinite, of its angles will make the polygon equal the circle.

Cusanus proceeds to give an account of a comparative concept 'truer' for propositions about God:

> Sacred ignorance has taught us that God is ineffable, because He is infinitely greater than anything that words can express. So true is this that it is by the process of elimination and the use of negative propositions that we come nearer the truth about Him.

> From this it is clear how in theology negative propositions are true and affirmative ones inadequate; and that of the negative ones those are truer which eliminate greater imperfections from the infinitely Perfect. It is truer, for example, to deny that God is a stone than to deny that He is life or intelligence ... In affirmative propositions the contrary holds good: It is truer to assert that God is intelligence and life than to assert that He is earth, stone or anything material.

Cusanus' model for the progress of human knowledge as endless approach towards the truth is based upon a geometrical analogy, essentially similar to the method of exhaustion of Eudoxos and Archimedes. Later philosophers and scientists — from Robert Boyle, Robert Hooke, and G. W. Leibniz in the seventeenth century, David Hartley, Joseph Priestley, and Georges LeSage in the eighteenth century, to C. S. Peirce and Ernst Mach in the nineteenth century — extended the analogy from geometry to algebraic and analytical examples of convergence.[3]

Quoting Francis Bacon's remark that "truth does more easily emerge out of error than confusion", Boyle argued that the temporary use of false conjectures in science is legitimate — comparable to the arithmetical method of *regula falsi* for solving linear equations. This method proceeds as follows: to solve the equation

$$ax + b = 0$$

replace the unknown x with a conjectural solution x_0 and calculate the value of

$$ax_0 + b.$$

If the result is c, then the correct solution can be calculated from x_0, a, and c:

$$x = x_0 - c/a.$$

Iterative methods for solving equations, first developed in the seventeenth century, give even more dramatic models for 'convergence toward the truth': given an equation $f(x) = 0$, we can systematically construct an infinite sequence x_i, $i = 0, 1, 2, \ldots$ of conjectures such that

$$\lim_{i \to \infty} x_i = x$$

in the ordinary mathematical sense. A good illustration is given by Newton's tangent method (Figure 2) which, for sufficiently smooth functions f, converges quite rapidly to the correct solution.

The idea of approach to the truth had special prominence in the latter half of the nineteenth century, the main impulses coming from Hegel's dynamic metaphysics and from evolutionism. This can be seen in the work of Peirce, the founder of American pragmatism, Engels and Lenin, the classics of dialectical materialism, and F. H. Bradley, the leading representative of idealist Hegelians in England.

Peirce was so impressed by the idea of convergence towards truth

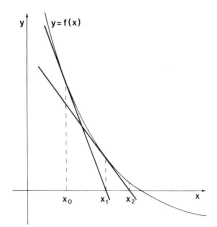

Fig. 2. Newton's tangent method.

that, in his classical essays in 1877, he made it the corner-stone of his theory of truth and reality:

The opinion which is fated to be ultimately agreed to by all who investigate, is what we mean by the truth, and the object represented in this opinion is the real. (5.407)

Peirce argued that truth and reality can be characterized as ultimate limits of scientific inquiry, conducted by "a COMMUNITY, without definite limits, and capable of a definite increase of knowledge". The characteristic feature of the method of science is that the "real things" influence, through causal interaction, the opinions of the members of the scientific community, so that

different minds may set out with most antagonist views, but the process of investigation carries them by force outside of themselves to one and the same conclusion. (5.407)

Peirce also attempted to prove this general claim in the special case of induction, leading from a sample to a statistical generalization about the population. In the end he realized, however, that even in the most favourable situations convergence towards the truth can be guaranteed only *with probability one*. (Cf. Niiniluoto, 1980a, 1984b.)

For Engels, the "revolutionary character" of Hegel's philosophy lies in the fact it "dissolves all conceptions of final, absolute truth":

Truth lay now in the process of cognition itself, in the long historical development of science, which mounts from lower to ever higher levels of knowledge without ever reaching, by discovering so-called absolute truth, a point at which it can proceed no further. (Engels, 1946, pp. 13–14.)

In *Anti-Dühring* (Part I, Ch. IX), Engels argued — in a way strongly resembling Peirce — that the limitations of individual human beings can be "solved" only in "the endless succession of generations of mankind": absolute truth cannot be realized "except through an endless eternity of human existence". In his comments on K. W. Nägeli in *The Dialectics of Nature*, Engels defended the knowability of the infinite:

Just as the infinity of knowable matter is composed of the purely finite things, so the infinity of the thought which knows the absolute is composed of an infinite number of finite human minds, working side by side and successively at this infinite knowledge, committing practical and theoretical blunders, setting out from erroneous, one-sided, and false premises, pursuing false, tortous, and uncertain paths, and often not even finding what is right when they run their noses against it (Priestley). The cognition of the infinite is therefore beset with double difficulty and from its very nature can only take place in an infinite asymptotic progress. And that fully suffices us in order to be able to say: the infinite is just as much knowable as unknowable, and that is all that we need. (Engels, 1976, pp. 234–235.)

These ideas are also reflected in Lenin's remarks on absolute and relative truth in *Materialism and Empirio-Criticism*, written in 1908 against Kantian agnosticism and Machian positivism. Like Engels and Peirce, Lenin argued against Kant's unknowable things-in-themselves: there are things "outside of us", but they can become "things-for-us", when our knowledge grows.

In the theory of knowledge, ... we must think dialectically ... determine how *knowledge* emerges from *ignorance*, how incomplete, inexact knowledge becomes more complete and more exact. (Lenin, 1927, p. 99.)

Human thought then by its nature is capable of giving, and does give, absolute truth, which is compounded of a sum-total of relative truths. Each step in the development of science adds new grains to the sum of absolute truth, but the limits of the truth of each scientific proposition are relative, now expanding, now shrinking with the growth of knowledge. (*Ibid.*, pp. 133–134.)

In his notes 'On Dialectics' (1916), Lenin characterized the identity or coincidence of opposites as the essence of dialectics. He concludes with Cusanus-type geometrical analogies:

Dialectics as a *Living*, many-sided knowledge (with the number of sides eternally increasing) with an infinite number of shadings of every sort of approach and approximation to reality (with a philosophical system growing into a whole out of each side) ... Human knowledge is not (or does not follow) a straight line, but a curve, which endlessly approximates to a series of circles, a spiral. (*Ibid.*, pp. 381—382.)

A further variant of the tradition of Cusanus can be found in Ch. XXIV, 'Degrees of Truth and Reality', of Bradley's *Appearance and Reality* (1893). Bradley himself says that here he is "perhaps even more than elsewhere, indebted to Hegel". He starts by pointing out that the distinction of more and less belongs only to the "world of appearance":

The Absolute, considered as such has of course no degrees; for it is perfect, and there can be no more or less in perfection. (Bradley, 1930, p. 318.)

But our judgements are always 'relative' and 'imperfect' — they "never reach as far as perfect truth". Hence, "truth and error, measured by the Absolute, must each be subject always to degree" (*ibid.*, p. 321). This degree depends on how much 'alteration' it would take to "convert" a relative truth to a complete, internally harmonious and all-inclusive reality.

Hence to be more or less true, and to be more or less real, is to be separated by an interval, smaller or greater, from all-inclusiveness or self-consistency. Of two given appearances the one more wide, or more harmonious, is more real. It approaches nearer to a single, all-containing, individuality. To remedy its imperfections, in other words, we should have to make a smaller alteration. The truth and the fact, which to be converted into the Absolute, would require less rearrangement and addition, is more real and truer. And this is what we mean by degrees of reality and truth. (*Ibid.*, pp. 322—323.)

In the Appendix, Bradley attempts to answer an objection raised by many critics:

If all appearances are equally contradictory, all are equally incapable of aiding us to get nearer to the ultimate nature of Reality.

Bradley pleads to the old analogy between error and evil: there is no 'hopeless puzzle' here, since the similar problem for evil and for good have been solved a long time ago:

But suppose that in theology I say that all men before God, and measured by him, are equally sinful — does that preclude me from also holding that one is worse or better than another? (*Ibid.*, p. 495.)

Bradley's reply is hardly adequate, but in a sense he is right. There is no problem in saying, e.g., that two finite natural numbers m and n are infinitely far from the least infinite ordinal ω (since $\omega - m = \omega - n = \omega$ for all $m < \omega$ and $n < \omega$), but still we may have $m > n$. We can say that a monotonely increasing sequence of natural numbers a_i, $i = 1, 2, \ldots$, grows towards infinity ω, if for any natural number n there is i_0 such that $a_i > n$ for $i \geq i_0$. We may also have monotonely increasing infinite sequences of finite sets such that their limit (i.e., union) is infinite: for example, if $A_n = \{0, 1, \ldots, n\}$ for each $n < \omega$, then $A_0 \subseteq A_1 \subseteq \ldots$ and

$$\omega = \bigcup_{n \in \omega} A_n.$$

There are thus perfectly meaningful numerical and set-theoretical analogies for saying that one finite thing is 'closer' to the infinite than another. There need not be any hopeless puzzles for a theory of degrees of truth in this direction.

The 'dialectics of finite and infinite' seems to be a pseudo-problem also in the sense that human knowledge is by no means restricted to the finite. Every sentence has an infinite number of deductive consequences. If scientific theories are defined as deductively closed sets of sentences (cf. Chapter 2.9), then each theory is an infinite set. Mathematicians and scientists have now learned to handle even such theories which have an infinite recursive set of axioms.

These remarks do not, however, reduce the interest of the idea of convergence to the truth. Regardless of metaphysical issues about the infinity of the world, many human tasks are 'infinite' in obvious ways. For example, to estimate a real-valued parameter requires knowledge with infinite precision (cf. Chapter 3.1). It is also possible to approximate finite objects with infinite convergent sequences, not only in the relatively trivial sense in which the infinite sequence.

$$4, 3, 2, 1, 0, 0, 0, \ldots$$

approaches 0 as its limit, but also in the sense that infinite sums of non-zero numbers may have a finite value, as shown by the infinite geometrical series

$$\sum_{n < \omega} q^n = \lim_{n \to \infty} (1 - q^n)/(1 - q) = 1/(1 - q) \quad (0 \leq q < 1).$$

Popper's falsificationism, as developed in *Logik der Forschung* (1934), and complemented with the theory of verisimilitude in *Conjectures and Refutations* (1963), can be regarded as a modern secular version of Cusanian negative theology. For Popper, the so-called scientific knowledge consists of bold conjectures which the scientists put to severe tests. A scientific hypothesis can be refuted, but never accepted as true, or as probable, in a test. We may assert only negative propositions about reality. Still, through the elimination of false conjectures we shall come nearer to the truth about the world — some conjectural theories are more truthlike than others.

During the half millenium separating Cusanus and Popper, there were hardly any serious attempts to develop exact mathematical or logical accounts of the notion of degree of truth or verisimilitude. (Reichenbach is an exception, as we see in Section 4.) The contrast to the theory of probability is striking in this respect. W. V. O. Quine could still in 1960 argue against Peirce that

> there is a faulty use of numerical analogy in speaking of a limit of theories, since the notion of limit depends on that of 'nearer than', which is defined for numbers and not for theories. (Quine, 1960, p. 23.)

Part of the work to be done in the next chapters certainly is concerned with the concept of 'nearer than' for scientific statements. However, first we must consider even a more fundamental question: if a theory is said to be 'nearer to the truth' than another, what is meant by the phrase 'the truth'? Do we mean some particular truth that we happen to be interested in? Or perhaps 'the whole truth'? Or known and certain truth? Or do we, like Hegel, mean the noble thing, i.e., reality itself, to which 'the true' is sometimes taken to refer? These alternatives have a direct connection to the classification of 'errors' in Section 1: approach to the whole truth is connected with reduction of the incompleteness of our knowledge (error of type (a)); approach to a given true statement through false ones is connected with the reduction of falsity (error of type (b)); approach to certified truth is associated with the reduction of uncertainty (error of type (c)). We shall see in the next section how these distinctions help to clarify some confusions in earlier discussions about the nearness to the truth.

5.4. TRUTH: PARTS AND DEGREES

Following the well-known slogan of Heracleitos, 'All is in flux', Hegel's dialectics claims that

> everything finite, instead of being stable and ultimate, is rather changeable and transient. (Hegel, 1975, p. 118.)

Peirce based his doctrine of 'synechism' (see 6.169) on this 'principle of continuity': no things are absolute but "swim in continua" (1.711).

If this principle is applied to the idea of absolute or final truth, we obtain a dynamic conception of knowledge (cf. the quotation from Engels in the preceding section). A more radical application is directed to the distinction between truth and falsity. Thus, Peirce sometimes claimed that all knowledge is indeterminate (1.711). Bradley argued that

> there cannot for metaphysics be, in short, any hard and absolute distinction between truths and falsehoods (Bradley, 1930, p. 323.),

but every judgment is partly true and partly false. Similarly, Engels claimed that

> that which is recognized now as true has also its latent false side which will later manifest itself, just as that which is now regarded as false has also its true side by virtue of which it could previously be regarded as true. (Engels, 1946, p. 45.)

On closer inspection, the motivation for these views about partial truth and partial falsity turns out to be rather obscure. Let us start from Bradley's difficulties with the concept of 'error':

> Error is without any question a dangerous subject, and the chief difficulty is as follows. We cannot, on the one hand, accept anything between non-existence and reality, while, on the other hand, error obstinately refuses to be either. It persistently attempts to maintain a third position, which appears nowhere to exist, and yet somehow is occupied. In false appearance there is something attributed to the real which does not belong to it. But if the appearance is not real, then it is not false appearance, because it is nothing. On the other hand, if it is false, it must therefore be true reality, for it is something which is. And this dilemma at first sight seems insoluble. (Bradley, 1930, pp. 164–165.)

Bradley's solution to this dilemma is to claim that errors have a place in reality, because they in fact contain some truth:

Error *is* truth, it is partial truth, that is false only because partial and left incomplete. The Absolute *has* without subtraction all those qualities, and it has every arrangement which we seem to confer upon it by our mere mistake. The only mistake lies in our failure to give also the complement. The reality owns the discordance and the discrepancy of false appearance; but it possesses also much else in which this jarring character is swallowed up and dissolved in fuller harmony. (*Ibid.*, pp. 169—170.)

For example, if a subject a seems to possess two incompatible qualities P and Q, it may turn out the 'real subject' is $a + b$, rather than mere a, and that a possesses P and b possesses Q.

It is clear that the roots of Bradley's difficulties with the concept of error are in his idealism: if we distinguish mental acts, their propositional contents, and the external facts to which propositions refer, Bradley's dilemma disappears. More specifically, he systematically fails to make a distinction between judgment as a mental act and what is judged (see Ewing, 1934, p. 208). When a false proposition is asserted, the act of judgment is real, even though the content of the judgment does not fit with reality.

Bradley's thesis that all judgments are partly false is directly influenced by Hegel's claim that an 'immediate qualitative judgment', "however correct it may be, cannot contain truth" (Hegel, 1975, p. 237). Hegel preserves here the concept of 'truth' to 'notional judgments' which state the 'coincidence' of an object with its 'notion'. This means in effect that a judgment of the form 'a is P' is true only if P is the essence of a. Findlay says that here 'true' means something like 'profundity' or 'importance' in ordinary usage (Findlay, 1966, p. 232). On the other hand, Bradley seems to be mislead by the fact that Hegel's logic does not make a distinction between the copula 'is' as sign of predication and as sign of identity (cf. *ibid.*, p. 231): he says that "any categorical judgment must be false", since "the subject and the predicate, in the end, cannot either *be* the other" (Bradley, 1930, p. 319). (Cf. Ewing, 1934, p. 223.)

A more interesting premiss in Bradley's argumentation is his holism: every judgment is "conditional" in the sense that its truth depends on an infinite number of conditions. Bradley seems to think that these presuppositions are in some way contained in the judgment (cf. Ewing, 1934, p. 221), and as the infinity of these conditions cannot be known, our knowledge is always incomplete. It does not follow, however, that every judgment is partly false (cf. also Haack, 1980). A judgment may quite well be true, even if no one knows the truth of its presuppositions.

Moreover, we can still make a distinction between the *falsity* and the *incompleteness* of a proposition. It is correct to say that a proposition is false if and only if it is inconsistent with the whole truth (i.e., with the totality of all true propositions). It does not follow that, in order to be true, a proposition has to contain the whole truth. Bradley's doctrine of degrees of truth thus confuses the question whether a proposition h is *wholly* (totally) *true* with the question whether h is the *whole truth* (Woozley, 1949, p. 157; cf. Ewing, 1934).

The distinction between two kinds of errors — falsity and incompleteness — is also relevant the evaluation of Engels' thesis that every truth has its 'latent false side'. When Engels, in *Ludwig Feuerbach*, says that the arrival at 'absolute truth' would mean the end of world history (Engels, 1946, p. 17), he obviously means that the whole edifice of human knowledge would have reached its final state. Indeed, as Lenin pointed out to A. Bogdanov, Engels says in *Anti-Dühring* that such single propositions as 'Napoleon died on May 5, 1825' express 'eternal truths'. Bogdanov, who claimed that Marxism denies the "unconditional objectivity of any truth whatsoever", confuses two questions:

(1) Is there such a thing as objective truth . . . ? (2) If so, can human ideas, which give expression to objective truth, express it all at one time, as a whole, unconditionally, absolutely, or only approximately, relatively? (Lenin, 1927, p. 120.)

On this interpretation, in referring to the 'false side' of a true proposition, Engels in fact means only its incompleteness, its failure to express the whole truth.

After giving a conclusive criticism of Bradley's doctrine of degrees of truth in *Idealism: A Critical Study* (1934), A. C. Ewing makes the important remark that nevertheless the false propositions "do not all diverge equally from truth". If we think that a mountain is 20 000 feet high, while it is really 21 000 feet, we are both theoretically and practically in a better position than if we estimate it to be only 5000 feet high.

There may be topics where the best we can do is not to arrive at propositions that are true but to arrive at propositions that are as little false as possible. Such false propositions are not only of practical but of theoretical value . . . Some false propositions may be far more worth asserting, may be of far more importance for developing a less incorrect view of reality or parts of reality than hosts of true propositions of less range and significance. (Ewing, 1934, pp. 226—227.)

While Ewing has in mind primarily metaphysical and theological propositions, like 'God is a person' (cf. Cusanus once again!), the content of his remarks anticipates remarkably well the basic thesis of Popper's philosophy of science in the 1960s.

The above discussion leads us to some systematic conclusions. We can start by distinguishing, wholly within the classical bivalent logic, two different senses of 'partial truth'. First, a statement h is *partly true*, if h is, in some sense, a composite proposition and a non-empty *part of h is true* (cf. Haack, 1980, p. 8). Similarly, h is *totally true*, if all of h is true. Hence, if h is totally true, it is true; if h is partly but not totally true, it is false. If no part of h is true, it is *totally false*. These definitions, illustrated in Figure 3, can be made precise only by telling what a 'part' of a proposition is. For example, as each false statement (even a totally false one) entails true statements, does this mean that all falsities are partly true? If h is true and g is false, is the truth $h \vee g$ totally or partly true? A further problem arises from the fact that this concept fails to be invariant under logical equivalence: h is logically equivalent to $h \& t$, where t is an arbitrary tautology, but clearly $h \& t$ has more true 'parts' than h. If this remedy cannot be repaired, the notion of 'partly true' is hardly interesting.

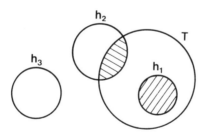

Fig. 3. h_1 is totally true, h_2 partly true, h_3 totally false.

Secondly, a true statement h may be said to be 'partial' if it is incomplete or short of the whole truth. A witness in court, who does not lie but conceals some relevant facts, tells a 'half-truth' precisely in this sense. Thus, h is a *partial truth*, if it, in some sense, gives or expresses a non-empty *part of the whole truth*. If h expresses all of the truth, it is the *whole truth*, and if h does not give any part of the whole truth, it is a *pure falsity*. These definitions presuppose that partial truths are true, and pure falsities are false — it would be odd to say that some false statement is a (partial) truth. On the other hand, false propositions can give 'information about the truth' (cf. Hilpinen, 1976). We may, therefore, extend the notion of partial truth to cover all kinds of propositions: a statement h (true or false) has a positive *degree of partial truth*, if h gives a non-empty part of the whole truth. (See Figures 4 and 5.) It follows that, for true propositions, the degree of partial truth should depend on their information content (cf. Chapter 4.5).

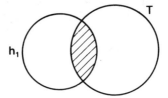

Fig. 4. h_1 has a positive degree of partial truth.

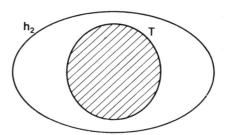

Fig. 5. h_2 has a maximal degree of partial truth.

If h is factually (non-tautologically) true, it has a positive degree of partial truth. In contrast, logical truths, due their tautological or uninformative character (cf. Chapter 4.5), do not tell anything about the truth. If h is false but partly true, it will have a positive degree of partial truth.[4] Contradictions will have a maximum degree of partial truth, since they entail all statements whatsoever — and, hence, the whole truth as well.

Thirdly, still in the two-valued logic, we may say that a statement is *almost true, nearly true*, or *approximately true*, indicating thereby that it is false but close to being true. For example, an estimate of the height of a mountain may be 'close to the truth' in this sense (cf. Ewing's example). This concept presupposes some metric which allows us to say that an almost true false proposition makes a 'small error' or that its 'distance from the truth' is small. A numerical value expressing this distance could be called the 'degree of nearness to the truth', the 'degree of approximate truth', or shortly the *degree of truth*. Then all true propositions would have the maximal degree of truth, but the degree of truth of false propositions would depend on the size of the error that they make.

A quantitative statement h is *approximately true*, if it is almost true, i.e., sufficiently close to being true (see Figure 6). There is, however, another way of speaking about 'approximations to the truth'. Sometimes a statement is said to be an *approximate truth*, if it is true but not very sharp — it covers the truth but only by making a relatively weak or uninformative claim (see Figure 7). In this case, finding a better 'approximation' means the introduction of a sharper or stronger true statement.[5]

Fig. 6. h is approximately true.

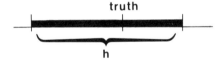

Fig. 7. h is an approximate truth.

The difference between partial truth and nearness to the truth can be summarized in the following principles:

(1) Tautologies have a minimal degree of partial truth but a maximal degree of nearness to the truth.

(2) Contradictions have a maximal degree of partial truth but a low degree of nearness to the truth.

CHAPTER FIVE

(3) All true statements have the same degree of nearness to the truth, but their degrees of partial truth vary depending on their information content.

(4) If h has a positive degree of partial truth, its negation $\sim h$ cannot be the whole truth. If h has a non-maximal degree of nearness to the truth, its negation $\sim h$ has the maximal degree of nearness to the truth.

The first part of (4) follows from the fact that the information about the truth in h cannot be included in $\sim h$. The second part follows from the fact that the negation of a false statement is true.

The concept of 'degree of truth' is used also in connection with the many-valued and fuzzy logics, which replace the two classical truth values 1 (true) and 0 (false) with the continuum [0, 1] (or its fuzzy subsets). We have already seen in Chapter 4.4 how the idea of fuzziness can arise in connection with vague predicates and continuous quantities. As the latter constituted perhaps the most significant motivation for Peirce's synechism, it might be argued that the semantics of vagueness is the best modern explication of Hegelian principles of continuity. This claim is of course controversial. Haack (1980) has given a survey of the alleged linguistic evidence for introducing fuzzy truth values, and she is able to point out troubles in the introduction of 'linguistic truth values' by means of 'hedges' like 'very' and 'more or less'. For example, if K is the membership function of the fuzzy extension of the predicate 'tall', and the value $K(a)$ is high, i.e., a is clearly tall or very tall, then Zadeh (1975) says that the sentence 'a is tall' is *very true*. However, Haack argues, the modifier 'very' behaves in different ways in front of the adjectives 'true' and 'tall'. Further, there is no general agreement about the truth tables for continuous truth values (see (4.40) for one suggestion). In any case, the widely accepted principle

$$v(\sim h) = 1 - v(h)$$

indicates a clear difference to the rule (4) about the degree of nearness to the truth: if $v(h) > 0$, then $v(\sim h) < 1$. With respect to negation — but not in other respects (cf. also Chapter 6.4) — continuous truth values behave like degrees of partial truth.

In this connection, it is in order to emphasize again that the theory of truthlikeness which will be developed in the later chapters can be formulated without departing from the classical two-valued logic. However, we shall also extend our treatment to semantically indeterminate languages (see Chapter 6.8).

We shall see in Chapter 6.4 how the concepts of partial truth and nearness to the truth can be defined in a precise and satisfactory way. Sections 6 and 7 also shows how Popper's notion of truthlikeness attempts to combine these concepts into one comparative or quantitative notion. In the next section, we shall briefly evaluate some other attempts to explicate 'degrees of truth'.[6]

5.5. DEGREES OF TRUTH: ATTEMPTED DEFINITIONS

Hans Reichenbach constructed, in his *Wahrscheinlichkeitslehre* (1935), a simple model for continuous degrees of truth. He called this a 'logic of an individual verifiability'. Assume that a marksman makes the statement $h = $ 'I shall hit the center'. After the shot we measure the distance r of the hit from the center (see Figure 8). Then

(5) $v(h) = 1/(1 + r)$,

which varies between 0 and 1, is the 'truth value' of h. In this case, we may also say that 'h holds to the degree $1/(1 + r)$'. In other words, $v(h)$ can be interpreted as the 'degree of truth' of statement h (see Reichenbach, 1949, p. 390).

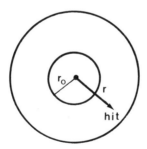

Fig. 8. Reichenbach's model for degrees of truth.

Reichenbach's model corresponds directly to our notion degree of nearness to the truth, since $v(h)$ measures the closeness of the claim h from the truth. It could be interpreted in the framework of two-valued logic — perhaps with the addition that h is said to be approximately or almost true if $v(h)$ is sufficiently close to 1. Reichenbach himself uses the model to obtain an interpretation of many-valued logic, i.e., a logic with continuous truth values, from which two- and three-valued logics can be obtained by introducing 'demarcation values', such as

(6) $\quad h$ is $\begin{cases} \text{true, if } r \leq r_0 \\ \text{false, if } r > r_0 \end{cases}$

or

(7) $\quad h$ is $\begin{cases} \text{true,} & \text{if } r \leq r_0 \\ \text{false,} & \text{if } r \geq r_1 \\ \text{indeterminate, if } r_0 < r < r_1. \end{cases}$

(See also Reichenbach, 1938, §36.) Further, he suggests that we may construct a hierarchy of metalanguages, where statements about truth values (e.g., '$v(h) = a$') have themselves continuous truth values.

Reichenbach also proposed a 'probability logic', where the classical truth value of a singular sentence h is replaced by its 'weight' $P(h)$; it is numerically identical with the relative frequency in which the 'posits' of form h are true in an infinite sequence of predictive inferences (see Reichenbach, 1949, §81; Reichenbach, 1938, §35). The 'truth values' of this logic then satisfy the probabilistic formulas

(8) $\quad P(h \vee g) = P(h) + P(g) - P(h \& g)$
$P(\sim h) = 1 - P(h)$
$P(h \supset g) = 1 - P(h) + P(h \& g)$
$P(h \equiv g) = 1 - P(h) - P(g) + 2P(h \& g).$

The motivation for this 'logic of weights' comes from Reichenbach's frequency interpretation of probability and from his verifiability theory of meaning. I cannot find any reason to assume that the probabilistic weights of single cases have anything to do with 'truth values' or 'degrees of truth'; instead, they may indicate a rational degree of belief in the truth of a predictive statement.

In *The Myth of Simplicity* (1963), Ch. 8, Mario Bunge defines a concept of 'partial truth' by a function V which takes values in the interval $[-1, 1]$. Miller (1977) has shown that Bunge's axioms for negation and conjunction lead to a disastrous result: unless logically equivalent propositions may have different degrees of partial truth, the theory permits at most the degrees 1 and −1. It is also difficult to see what the explicandum of Bunge's definition is. As Hilpinen (1976) notes, the condition

(9) $V(\sim h) = -V(h)$

may be plausible for a measure of the acceptability or plausibility of h (if h is highly plausible, its negation is highly implausible), but not for the degree of nearness to the truth. If h is 'almost true' but not strictly true (i.e., $0 \ll V(h) < 1$), then $\sim h$ is not 'almost true', as Bunge says, but rather true (cf. (4)). It may be added that our concept of partial truth satisfies something like (9) (cf. (4) again). Still, Bunge's concept cannot be our partial truth, since his V satisfies the conditions

(10) $V(h) = 1$, if $\vdash h$
        $V(h) = -1$, if $\vdash \sim h$,

which are characteristic to the degrees of nearness to the truth (cf. (1) and (2)).

Bunge's new system of 'degrees of truth', in *Treatise on Basic Philosophy 2* (1974), is based on a function V which has the range $[0, 1]$ and formally satisfies the properties (8) of a probability measure. Bunge, who does not accept the concept of probability of a statement, distinguishes sharply the concepts 'degree of truth' and 'probability', but he does not give justification for thinking that degrees of truth satisfy the same principles as probability. He does not explicitly define the function V in any special case, and the interpretation of 'degrees of truth' remains as problematic as earlier. The suggested readings

$V(h) = 1$ h is true
$0 \ll V(h) < 1$ h is approximately true
$\frac{1}{2} < V(h) < 1$ h is partially true
$0 < V(h) \ll 1$ h is nearly false
$V(h) = 0$ h is false

(*ibid.*, p. 117) do not help very much, since (outside fuzzy logics) a measure for the degree of truth should satisfy

(11) h is false iff $0 \leq V(h) < 1$.

(Cf. Hilpinen, 1976.) In particular, the phrase 'nearly false' does not make much sense in this connection, for Bunge is not willing to change classical logic to a many-valued one (*ibid.*, p. 106). And again, unless one is dealing with fuzzy logic, the condition

$$V(\sim h) = 1 - V(h)$$

might hold for partial truth, but then the requirements

$$V(\text{tautology}) = 1$$
$$V(\text{contradiction}) = 0$$

should be reversed: contradictions have a maximal degree of partial truth.

Nicholas Rescher proposes, in *The Coherence Theory of Truth* (1973), two measures for 'degrees of truth'. Rescher distinguishes carefully this attempt from probability and from what we have called partial truth (*ibid.*, p. 198). The degree of truth of a statement depends on the extent of its 'coherence' with an inconsistent 'data-base' Σ. Thus, let Σ be an inconsistent set of statements, and let $MC(\Sigma)$ be the class of the maximally consistent subsets of Σ. Then statement h is an 'inevitable consequence' of Σ, if $\Gamma \vdash h$ for all Γ in $MC(\Sigma)$, and a weak consequence of Σ, if $\Gamma \vdash h$ for some Γ in $MC(\Sigma)$. Let

$$q(h) = \text{the proportion of } \Gamma \text{ in } MC(\Sigma) \text{ such that } \Gamma \vdash h$$

and

(12) $\Delta(h) = q(h) - q(\sim h).$

(*Ibid.*, pp. 358—360.) Rescher notes that this Δ-measure has some common properties with Bunge's (1963) function V:

$$-1 \leqslant \Delta(h) \leqslant 1$$
$$\Delta(h) = 1, \quad \text{if } \vdash h$$
$$\Delta(h) = -1, \text{if } \vdash \sim h$$
$$\Delta(\sim h) = -\Delta(h).$$

Therefore, it is again difficult to see what Rescher's explicandum has to do with gradation of truth. He himself admits that the Δ-measure presupposes a 'precriterial' perspective upon truth, where it is not yet settled how well a proposition 'fits' into a given context. It is thus more natural to regard Rescher's Δ as a measure of the *plausibility* of a proposition, given an inconsistent body of data, rather than as a degree of truth (cf. Hilpinen, 1976).

5.6. POPPER'S QUALITATIVE THEORY OF TRUTHLIKENESS

Popper agrees with Kant that truth is a 'regulative principle' of science: while there are no general criteria for recognizing truth, "there are criteria of progress toward the truth" (Popper, 1963, p. 226). However, "we want more than mere truth: what we look for is *interesting truth*" or informative truth (*ibid.*, p. 229). Thus, the concept of 'verisimilitude' or 'truthlikeness' should be

> so defined that maximum verisimilitude would be achieved only by a theory which is not only true, but completely comprehensively true: if it corresponds to *all* facts, as it were, and, of course, only to *real* facts. (*Ibid.*, p. 234.)

This idea of a better approximation to the truth should not be confused with probability:

> Verisimilitude ... represents the idea of approaching comprehensive truth. It thus combines truth and content while probability combines truth with lack of content. (*Ibid.*, p. 237.)

Moreover, it is a semantic notion, like truth and content, not an epistemological or epistemic idea (*ibid.*, p. 234). Still, it should be applicable to the comparison of "theories which are *at best* approximations" — "theories of which we actually know that they cannot be true" (*ibid.*, p. 235). Indeed,

> Search for verisimilitude is a clearer and a more realistic aim than the search for truth, ... while we can never have sufficiently good arguments in the empirical sciences for claiming that we have actually reached the truth, we can have strong and reasonably good arguments for claiming that we may have made progress towards the truth. (Popper, 1972, pp. 57–58.)

While all appraisals of truthlikeness are fallible, Popper suggests that his 'degrees of corroboration' can serve as indicators of degrees of verisimilitude (*ibid.*, p. 103).

For Popper, the concept of truthlikeness expresses "the idea of a degree of better (or worse) correspondence to truth". While his theory is committed to Tarski's version of the correspondence account of truth, we shall see that it also has an interesting relation to the classical coherentist theories of degrees of truth.

Let T be the class of all true statements (in Tarski's sense) in some first-order language L. Thus, T represents 'the whole truth' (with respect to L).[7] Let F be the class of false statements in L. As we are dealing with classical two-valued logic, $\mathbf{T} \cap \mathbf{F} = \emptyset$, and $\mathbf{T} \cup \mathbf{F}$ is the whole class of the sentences of L. Moreover, $\mathbf{T}$ is consistent and complete in L, since, for each sentence h in L, either $h \in \mathbf{T}$ or $\sim h \in \mathbf{T}$ but not both. For any statement h in L, the logical content of h is given by its consequence class

$$Cn(h) = \{g \text{ in } L \mid h \vdash g\}.$$

If $\text{Mod}(h)$ is the class of L-structures that are models of L, i.e.,

$$\text{Mod}(h) = \{\Omega \mid \Omega \vDash h\}$$

then

(13) $h \vdash g$ iff $Cn(g) \subseteq Cn(h)$ iff $\text{Mod}(h) \subseteq \text{Mod}(g)$.

If **Taut** is the class of tautologies in L, then

(14)   **Taut** $\subseteq Cn(h)$ for all h in L.

As true statements have only true consequences, we have

(15)   If h is true, then $Cn(h) \subseteq \mathbf{T}$.

As false statements have both true and false consequences, we have also

(16)   If h is false, then $Cn(h) \cap \mathbf{F} \neq \emptyset$
and $(Cn(h) \cap \mathbf{T}) -$ **Taut** $\neq \emptyset$.

Let us say that the class of all true consequences of h in L is its *truth content* $Ct_T(h)$, and the class of all false consequences of h in L is its *falsity content* $Ct_F(h)$:

(17) $Ct_T(h) = Cn(h) \cap \mathbf{T}$
$Ct_F(h) = Cn(h) \cap \mathbf{F}$.

Hence,

(18) $Cn(h) = Ct_T(h) \cup Ct_F(h)$
$Ct_T(h) \cap Ct_F(h) = \emptyset$
Taut $\subseteq Ct_T(h)$
$Ct_F(h) = \emptyset$,   if h is true
$Ct_T(h) = Cn(h)$, if h is true
$Ct_T(h) =$ **Taut**,   if $\vdash h$
$Ct_T(h) = \mathbf{T}$,   if $\vdash \sim h$
$Ct_F(h) = \mathbf{F}$,   if $\vdash \sim h$.

(See Figure 9.)

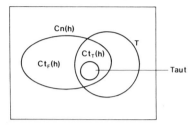

Fig. 9. The truth content and the falsity content of h.

If the class **T** is finitely axiomatizable, let t_* be the finite conjunction of its axioms. Then $\mathbf{T} = Cn(t_*)$. In this case, the truth content of h corresponds to the common content of h and t_* (cf. Chapter 4.4); it is the consequence class of the sentence

(19) $h_T =_{df} h \vee t_*$,

since

$Cn(h_T) = Cn(h \vee t_*) = Cn(h) \cap Cn(t_*) = Cn(h) \cap \mathbf{T} = Ct_T(h)$.

Every statement h entails its truth content in the sense that

(20) $h \vdash h_T$.

Moreover, h_T is the logically *strongest true statement* in L entailed by h. The falsity content $Ct_F(h)$ of h is not a consequence class of any sentence, since it is not closed under deduction.

Class **T** is always a theory, since it is a deductively closed set of sentences in L. However, it need not be (recursively) axiomatizable — this is the case e.g. in arithmetic, as Gödel's incompleteness theorems shows. If h itself is a theory **A**, the truth content h_T of h is the intersection $\mathbf{A} \cap \mathbf{T}$ of the theories **A** and **T** in Tarski's sense (Popper, 1972, pp. 49, 330; cf. Chapter 2.9).

The sentences in $\mathbf{T} - Ct_T(h)$, i.e., true sentences of L not entailed by h, represent the 'errors' of h in the sense of ignorance or incompleteness; the sentences in $Ct_F(h)$ correspond to the 'errors' of h in the sense of falsity (cf. Figure 1 in Section 1). It is, therefore, natural to require that for a highly truthlike statement both of these classes are small. If this is the case, then intuitively h is *close to the whole truth* **T**. This is the idea of 'closeness' or 'similarity' to the truth that Popper attempted to capture in his qualitative definition of verisimilitude: assuming that the truth contents and the falsity contents of statements h_1 and h_2 (in L) are comparable with respect to inclusion, h_2 is *more truthlike* than h_1 if and only if

(21) (a) $Ct_T(h_1) \subset Ct_T(h_2)$ and $Ct_F(h_2) \subseteq Ct_F(h_1)$, or
(b) $Ct_T(h_1) \subseteq Ct_T(h_2)$ and $Ct_F(h_2) \subset Ct_F(h_1)$.

(See Popper, 1963, p. 233; Popper, 1972, p. 52.) This is equivalent to each of the requirements

(22) $Ct_T(h_1) \subseteq Ct_T(h_2)$ and $Ct_F(h_2) \subseteq Ct_F(h_1)$, and $Cn(h_1) \neq Cn(h_2)$.

(23) $Ct_F(h_2) \cup (\mathbf{T} - Ct_T(h_2)) \subset Ct_F(h_1) \cup (\mathbf{T} - Ct_T(h_1))$

(24) $Cn(h_2)\Delta\mathbf{T} \subset Cn(h_1)\Delta\mathbf{T}$.

Thus, h_2 is more truthlike than h_1 if and only if the symmetric difference between $Cn(h_2)$ and **T** is properly included in the symmetric difference between $Cn(h_1)$ and **T** (see Figure 10).

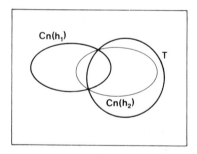

Fig. 10. Popper's comparative notion of truthlikeness.

Definition (21) entails immediately two important results:

(25)   If h is false, its truth content $Ct_T(h)$ is more truthlike than h.

(26)   If h_1 and h_2 are true, then h_2 is more truthlike than h_1 iff $h_2 \vdash h_1$.

Principle (26) says that, for *true* statements, truthlikeness varies directly with logical strength (see Figure 11).

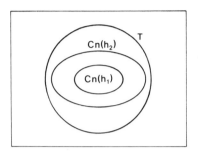

Fig. 11. h_2 is more truthlike than h_1.

Popper (1966) proved that comparison of truth contents can often be replaced by comparison of logical contents:

(27)   If **T** is unaxiomatizable, then
    (a) $Ct_T(h_1) = Ct_T((h_2)$ iff $Cn(h_1) = Cn(h_2)$
    (b) $Ct_T(h_1) \subset Ct_T(h_2)$ iff $h_1 \vdash h_2$ and $h_2 \nvdash h_1$.

This theorem can be generalized to the case where h_1 and h_2 are arbitrary axiomatizable theories (see Miller, 1974a, Th.5, Cor.2). Popper hoped that this result would allow us to measure verisimilitude by the formula

    Content minus falsity content

instead of

    Truth content minus falsity content.

In other words, the theory with the greater logical content will also be "the one with the greater verisimilitude unless its falsity content is also greater" (Popper, 1972, p. 53). However, this hope collapses in a dramatic way, as David Miller and Pavel Tichý noted independently of each other in 1972—73 (see Miller, 1974a; Tichý, 1974). It turns out that the comparability of the truth contents guarantees also the comparability of falsity contents, but unfortunately in a way which trivializes the definition (21).

Let **A** and **B** be two theories, and assume that **B** is more truthlike than **A**. Then, by (21), **A** ∩ **T** ⊆ **B** ∩ **T**. We have now four possible cases.

(i) Both **A** and **B** are true. Then **A** ∩ **F** = **B** ∩ **F** = ∅ and **A** ⊆ **B**. In this case, **B** ⊢ **A**.

(ii) **A** is false and **B** is true. Then ∅ = **B** ∩ **F** ⊂ **A** ∩ **F** and **A** ∩ **T** ⊆ **B** ∩ **T** = **B**. In this case, **A** ∩ **T** ⊆ **B** or **B** ⊢ **A** ∩ **T**.

(iii) **A** is true and **B** is false. Then ∅ = **A** ∩ **F** ⊂ **B** ∩ **F** and **A** ∩ **T** ⊆ **B** ∩ **T**, which contradicts the assumption that **B** is more truthlike than **A**.

(iv) Both **A** and **B** are false. The following theorem (cf. Miller, 1974a, Th.1.3; Tichý, 1974, Prop.2.4; Harris, 1974, Th.4.1) shows that this case too leads to a contradiction.

(28) If **A** and **B** are false theories, then
A ∩ **T** ⊆ **B** ∩ **T**   iff   **A** ⊆ **B**   iff   **A** ∩ **F** ⊆ **B** ∩ **F**.

To prove (28), assume first that **A** ∩ **T** ⊆ **B** ∩ **T**. Since **B** is false, there is a sentence $b \in$ **B** ∩ **F**, so that $\sim b \in$ **T**. If now $a \in$ **A**, then $a \vee \sim b \in$ **A** ∩ **T** ⊆ **B** ∩ **T**. Hence, both $b \supset a$ and b belong to **B** so that $a \in$ **B** by the deductive closure of **B**. Trivially, **A** ⊆ **B** entails **A** ∩ **F** ⊆ **B** ∩ **F**. Secondly, assume that **A** is false and **A** ∩ **F** ⊆ **B** ∩ **F**. Then there is a sentence $c \in$ **A** ∩ **F**. If now $a \in$ **A**, then $c \in$ **F** entails that $a \& c \in$ **F**, and $c \in$ **A** entails that $a \& c \in$ **A**. Hence, $a \& c \in$ **A** ∩ **F** ⊆ **B** ∩ **F** ⊆ **B**, so that $a \in$ **B**. Again, **A** ⊆ **B** trivially entails that **A** ∩ **T** ⊆ **B** ∩ **T**.

These observations can be summarized in the following theorem:

(29) If theory **B** is more truthlike than theory **A** in Popper's sense, then one of the following conditions holds:
  (i) **A** and **B** are true, and **B** entails **A**
  (ii) **A** is false, **B** is true, and **B** entails the truth content **A** ∩ **T** of **A**.

Definition (21) does not allow us to compare the verisimilitude of two false theories:

(30) If **B** is more truthlike than **A**, then **B** must be true.[8]

This result means that the borderline between the sets **T** and **F** works as it were a mirror: if you add something to the truth content of a false theory, you also add something to its falsity content. Similarly, if you reduce the falsity content of a theory, you also thereby reduce its truth content. The only way to push a false theory towards the set **T** (cf. Figure 10) is, so to say, to squeeze it inside entirely (cf. Figure 12).

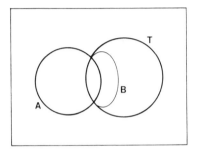

Fig. 12. B is more truthlike than A.

If **T** is unaxiomatizable and **B** is true, then condition **B** = **B** ∩ **T** ⊇ **A** ∩ **T** entails **B** ⊇ **A** by (27), so that **A** too must be true. This excludes the possibility (ii) of Theorem (29), illustrated in Figure 12. Hence,

(31) Assume that **T** is unaxiomatizable, and that **A** and **B** are axiomatizable theories. If **B** is more truthlike than **A**, then both **A** and **B** are true.

(See Miller, 1974a, Th.5, Cor.1.)

Results (30) and (31) clearly show that — in spite of its initial plausibility — the definition (21) is inadequate for its original purpose. John Harris (1974) has shown that they can be generalized to cases where the comparison theory **T** is replaced by an arbitrary (not necessarily complete) theory. Martin Hyland's proposal for reformulating Popper's definition by replacing the whole theories by their 'bases' (i.e., independent sets of axioms) was refuted by Miller (1974b), who showed that this suggestion fails to define an ordering. Peter Mott (1978) has explored the idea that, in order to be more truthlike than theory **B**, a theory **A** should preserve the 'short theorems' of **B** and add new 'short truths'. His definition allows us to compare at least some false theories; it satisfies (26) but unfortunately not (25). Chris

Mortensen (1978) has shown that the proof of (30) is not valid any more, if we replace classical logic by relevance logic, but later he has pointed out that the Popperian definition leads to other serious troubles in a large class of non-classical logics (Mortensen, 1983).[9]

Popper's first reaction to the Miller–Tichý-theorem (30) was the suggestion that one might define truthlikeness just in terms of the truth content — and forget about the trouble-making falsity content (cf. Tichý, 1974, p. 157; Harris, 1974, p. 165; Miller, 1978a, p. 415). As Tichý (1974) pointed out, on this definition it would be "child's play to increase the verisimilitude of any false theory": just add to theory **A** any false sentence h which does not follow from **A**! Indeed, we know by (28) that the truth contents of false statements depend only on their logical contents. Hence,

(32)   If h_1 and h_2 are false, then $Ct_T(h_1) \subseteq Ct_T(h_2)$ iff $h_2 \vdash h_1$.

Moreover, by (27), the same holds for all statements (true or false) if **T** is unaxiomatizable. Further, by (18), $Ct_T(h) = \mathbf{T}$ for a contradiction h, so that logically false sentences would have maximal truthlikeness in the proposed sense. For these reasons, the concept of truthlikeness should not satisfy the principle

(33)   If h_1 and h_2 are false, then h_2 is more truthlike than h_1 in case $h_2 \vdash h_1$.

(33) is a reasonable adequacy condition for the concept of partial truth rather than truthlikeness: adding falsities to a given theory increases its *coverage* of the truth **T**, but not its *closeness* to **T**.

In his comments on the Miller–Tichý-result, Popper (1976) is impressed by the observation that his original definition (21) can be motivated by a "general definition of distance between theories". He refers to Mazurkiewicz's probabilistic distance function

(34) $P(\mathbf{A} \triangle \mathbf{B})$

between two theories **A** and **B** in the Boolean algebra of deductive systems. (See (1.22).) If theories are considered as sets of sentences, the qualitative (or Boolean) distance between theory **A** and truth **T** can be defined by $\mathbf{A} \triangle \mathbf{T}$, so that its complement, the 'verisimilitude' of **A**, can be defined as the set

(35) $(\mathbf{A} \cap \mathbf{T}) \cup (\mathbf{F} - \mathbf{A} \cap \mathbf{F})$.

If $\mathbf{A} = Cn(h)$, (35) reduces to

(36) $Ct_T(h) \cup (\mathbf{F} - Ct_F(h))$.

Comparison of two statements h_1 and h_2 by means of (36) leads directly to (24) which is equivalent to Popper's definition (21).

Miller (1978a) has explored the alternative where theories are considered as elements of the Lindenbaum algebra $\mathscr{A}_L$ of language L (cf. Chapter 2.1). He formulates a monotonicity condition in order to guarantee that any Boolean valued distance function on algebra $\mathscr{A}_L$ is identical with the symmetric difference Δ. Assuming that the truth is finitely axiomatizable, and thus representable by an element t_* of $\mathscr{A}_L$, Miller shows that $a \Delta t_* < b \Delta t_*$ holds for a and b in $\mathscr{A}_L$ if and only if either (i) $b < a$, and a and b have the same truth value, or (ii) $b \leq a \cup t_*$, and b is true and a is false. This means that the definition of comparative truthlikeness in terms of $a \Delta t_*$ allows both of the cases of theorem (29) and also the case corresponding to (33). In other words, somewhat reluctantly, Miller is lead to the 'less amusing' result that "the stronger of two false theories is the closer to the truth" (*ibid.*, p. 423). If $\mathbf{T}$ is unaxiomatizable, Miller's consideration of autometric operations in 'Brouwerian algebras' leads him to the unintuitive result that "the distance from the truth of an axiomatizable theory is equal to its truth content" (*ibid.*, p. 427). This means that "approach to the truth via axiomatizable theories" turns out to be "quite independent of where the truth in truth is".

Miller's third proposal is to replace theories $\mathbf{A}$ by their classes of models Mod($\mathbf{A}$), and then define the distance of theory $\mathbf{A}$ from the truth $\mathbf{T}$ by

(37)   Mod($\mathbf{A}$) Δ Mod($\mathbf{T}$).

(*Ibid.*, p. 428.) This definition has also been proposed by Kuipers (1983) for the general case, where $\mathbf{T}$ need not be complete. If $\mathbf{T}$ is complete, as we have assumed in this chapter, then the falsity of theory $\mathbf{A}$ means that $\mathbf{A} \cap \mathbf{T}$ is inconsistent and Mod($\mathbf{A}$) $\cap$ Mod($\mathbf{T}$) = $\emptyset$. Hence, Mod($\mathbf{A}$) Δ Mod($\mathbf{T}$) = Mod($\mathbf{A}$) $\cup$ Mod($\mathbf{T}$), and

(38)   If $\mathbf{A}$ and $\mathbf{B}$ are false theories and $\mathbf{B} \vdash \mathbf{A}$,
then Mod($\mathbf{B}$) Δ Mod($\mathbf{T}$) $\subseteq$ Mod($\mathbf{A}$) Δ Mod($\mathbf{T}$).

In this case, definition (37) leads again to the unsatisfactory result (33).

Perhaps (33) is acceptable after all? Miller proposes an interesting geometrical model to support this idea: if **T** is in the north pole, and if the complete theories are scattered within a short distance of **T**, then from many parts of the globe we can get closer to the terminus **T** just by heading north — i.e., towards every complete theory or towards greater content. The crucial assumption of this model is that all complete theories are in fact, roughly, at the same short distance from the north pole. This assumption does not hold in those cases where the strongest answers to our cognitive problems are located at different distances from the strongest correct answer. The fundamental weakness of all definitions of truthlikeness — among them Popper's original proposal and Miller's reformulations of it — that are based on the symmetric difference $Cn(h) \Delta \mathbf{T}$ is simply the fact that they don't pay any attention to the underlying metric structure of the space of complete answers. It is not the *size* of the falsity content that matters, but rather the distance of its elements from the truth. The weakness of Popper's and Miller's qualitative theories of verisimilitude is due to the fact that they in effect only reflect adequately the idea of partial truth: they fail to handle false theories, because they don't include tools for measuring the degrees of 'nearness to the truth'. This claim can be further supported by considering Popper's proposal for a quantitative measure of verisimilitude.

5.7. QUANTITATIVE MEASURES OF VERISIMILITUDE

Popper's qualitative concept of truthlikeness presupposes a very strong comparability condition between truth contents and falsity contents. Already for this reason it excludes the possibility of comparing most pairs of statements for their closeness to the truth. For example, it does not say anything about two true statements h_1 and h_2 such that $h_1 \not\vdash h_2$ and $h_2 \not\vdash h_1$. To guarantee the comparability of all statements, Popper introduced a quantitative measure for the *degree of verisimilitude* $Vs(h)$ of h (Popper, 1963, pp. 392–397). Given Vs, we may then say that h_2 is more truthlike than h_1 if and only if $Vs(h_2) > Vs(h_1)$.[10]

First, the *degree of the truth content* of h is defined as the measure of the information content of the truth content h_T:

(39) $\quad ct_T(h) = \text{cont}(h_T) = 1 - P(h_T) = 1 - P(h \vee t_*)$

where P is a regular probability measure for language L. (We assume here that the truth set **T** is axiomatized by sentence t_* in L.) Then

(40) $0 \leq ct_T(h) \leq \text{cont}(t_*)$.
$ct_T(h) = \text{cont}(h)$,   if h is true.
$ct_T(h) = 0$,   if $\vdash h$.
$ct_T(h) = \text{cont}(t_*)$,   if $\vdash \sim h$.
$ct_T(h_T) = ct_T(h)$.

$ct_T(h)$ is, therefore, a measure of degree of partial truth in the sense of Section 4, since it satisfies the adequacy conditions (1)–(3) for degrees of partial truth. It is also interesting to note that $ct_T(h)$ is equal to the amount of substantial information that h transmits about t_* in the sense of (4.67):

(41) $ct_T(h) = \text{transcont}_{\text{add}}(t_*/h)$.

Secondly, the *degree of the falsity content* of h is defined as

(42) $ct_F(h) = 1 - P(h/h_T)$.

In other words, $ct_F(h)$ is the conditional content of h relative to h_T:

(43) $ct_F(h) = \text{cont}_{\text{cond}}(h/h_T)$

(see Chapter 4.5). Popper considers also the alternative definition

(44) $\text{cont}(h_T \supset h) = 1 - P(h_T \supset h)$,

which is equivalent to the incremental content of h relative to h_T:

$$\text{cont}_{\text{add}}(h/h_T) = \text{cont}(h \,\&\, h_T) - \text{cont}(h_T)$$
$$= P(h_T) - P(h)$$
$$= P(h \lor t_*) - P(h),$$

which would lead to the nice-looking result

$$ct_T(h) + ct_F(h) = 1 - P(h_T) + P(h_T) - P(h)$$
$$= \text{cont}(h).$$

Popper rejects definition (44), since it would give the result that $ct_F(\text{contradiction}) = P(t_*)$ which may be smaller than $ct_T(\text{contradiction}) = 1 - P(t_*)$. A better reason to reject (44) is the fact that it would give the same measure of falsity content $P(t_*)$ to all false statements h. Definition (43) implies instead that the falsity content of h increases when $P(h)$ decreases or cont(h) increases:

(45) If h is false, then $ct_F(h) = P(t_*)/(P(h) + P(t_*))$.

The measure $ct_F(h)$ has the following properties

(46) $0 \leq ct_F(h) \leq \text{cont}(h)$.
$ct_F(h) = 0$, if h is true.
$0 < ct_F(h) \leq 1$, if h is false.
$ct_F(h) = 1$, if $\vdash \sim h$.

A simple way of measuring degrees of verisimilitude is the difference between truth content and falsity content:

$$ct_T(h) - ct_F(h) = 1 - P(h_T) - (1 - P(h/h_T))$$
$$= P(h/h_T) - P(h_T).$$

If h is true, this difference equals cont(h). Therefore, if $Vs(h)$ is measured by normalizing this difference with any of the factors cont(h), $ct_T(h)$ or $ct_T(h) + ct_F(h)$, we obtain the result:

If h is true, $Vs(h) = 1$.

As Popper correctly notes, this would make Vs a measure of 'degree of truth-value' rather than truthlikeness (cf. *ibid.*, p. 397). Therefore, his proposal is the following function

(47) $Vs(h) = (P(h/h_T) - P(h_T))/(P(h/h_T) + P(h_T))$,

which is identical with the normalization

$$(ct_T(h) - ct_F(h))/(2 - (ct_T(h) + ct_F(h))).$$

Hence,

(48) $-1 \leq Vs(h) \leq 1$
$Vs(h) = 0$, if $\vdash h$
$Vs(h) = -1$, if $\vdash \sim h$.

Moreover,

(49) $Vs(g) \gtreqless Vs(h)$ iff
$Ct_T(g) - P(h_T)Ct_F(g) \gtreqless Ct_T(h) - P(g_T)Ct_F(h)$.

(50) If h is true, then
$Vs(h) = (1 - P(h))/(1 + P(h)) = (\text{cont}(h))/(1 + P(h))$.

(51) If h is false, then
$$Vs(h) = (P(h) - (P(h) + P(t_*))^2)/(P(h) + (P(h) + P(t_*))^2).$$

It follows from (49) and (50) that Vs has several nice properties:

(52) $Vs(h) \leqslant V(t_*)$ for all h.

(53) $Vs(h) < V(h_T)$, if h is false.

(54) For true statements h, $Vs(h)$ varies directly with $\text{cont}(h)$.

(55) If h and g are true and $g \vdash h$, then $Vs(h) \leqslant Vs(g)$.

(52) says that the whole truth t_* has the maximal degree of truthlikeness. (53) and (55) correspond directly with the results (25) and (26). Theorem (54) follows from the fact that, by (50), $Vs(h)$ is a monotomically decreasing function of $P(h)$ for a true h. (55) is of course a direct consequence of (54).

Measure Vs avoids the Miller–Tichý-result (30), since it allows for comparisons of false hypotheses. However, it gives intuitively wrong results in most of these comparisons. Result (51) shows that, for false h, $Vs(h)$ is a monotonely non-increasing function of the argument $P(h)$, if and only if

(56) $P(h) \geqslant P(t_*)$.

If the truth t_* is 'interesting' in Popper's own sense, $P(t_*)$ is small, and condition (56) can be expected to hold for most consistent hypotheses h. For example, if t_* is an element of a cognitive problem **B** with equally probable potential complete answers, then (56) holds for all h in $D(\mathbf{B})$ (see Chapter 4.2). Thus,

(57) Let h be a consistent false statement. As long as $P(h) \geqslant P(t_*)$ holds, $Vs(h)$ varies directly with $\text{cont}(h)$.

Hence, Vs satisfies, for most false statements h_1 and h_2, the undesirable result (33) which claims that truthlikeness increases with logical strength. This can be seen directly by calculating the values $Vs(h_1)$ and $Vs(h_2)$, as shown in Niiniluoto (1978b), p. 294. The same criticism applies also to the measures of verisimilitude that Popper elsewhere proposed as alternatives to Vs:

(58) $(\text{cont}(h))/(2 - \text{cont}(h))$ (Popper, 1966, p. 353.)

(59) $\begin{cases} \text{cont}(h) + P(t_*), \text{ if } h \text{ is true} \\ \text{cont}(h) - P(t_*), \text{ if } h \text{ is false} \end{cases}$ (Popper, 1976, p. 153.)

If g is true and h is false, then $Vs(g) > Vs(h)$ if and only if

(60) $P(g) < P(h) + 2P(t_*) + P(t_*)^2/P(h)$

(cf. Niiniluoto, 1978b, p. 295). This condition is satisfied, if g is sufficiently bold. It is natural that, in the case where h and g are equally informative, the true one wins:

(61)   If g is true, h is false, and $\text{cont}(h) = \text{cont}(g)$, then $Vs(g) > Vs(h)$.

The right side of (60) is minimized by choosing $P(h) = P(t_*)$. Therefore,

(62)   If g is true and if $P(g) < 4P(t_*)$, then $Vs(g) > Vs(h)$ for all false statements h.

To illustrate (62), assume that g is of the form $t_* \lor g'$, where g' is a false statement such that $P(g') < 3P(t_*)$. Then, no matter how strange claim g' may be, there is no way coming closer to the truth than g by means of false statements. As a general result this may seem strange, if we take seriously the idea that sometimes false statements may be more truthlike than some true ones.

Let $\mathbf{B} = \{h_i \mid i \in I\}$ be a finite cognitive problem such that all hypotheses h_i are equally probable, i.e., $P(h_i) = 1/w$, where $w = |I|$. In this case, the true element h_* in $\mathbf{B}$ corresponds to the sentence t_* expressing the 'whole truth'. Then formulas (50) and (51) reduce to

$$Vs(h_i) = (w - 1)/(w + 1), \text{ if } h_i \text{ is true}$$
$$= (w - 4)/(w + 4), \text{ if } h_i \text{ is false}$$

(cf. Niiniluoto, 1978b, p. 295). In this case, all false complete answers h_i are *equally distant* from the truth. Verisimilitude is thus independent on where truth in fact is.

The source for the difficulties of Popper's measure Vs can be found in the fact that Vs — and the underlying measure of falsity content — does not reflect the different *distances* in which the different maximally informative statements are from the true one t_*. There is no reason to

THE CONCEPT OF TRUTHLIKENESS   197

assume generally that the probability measure P would express such distances (cf., however, Miller, 1978a). Just like the qualitative Boolean distances and their numerical versions, Vs fails to account for the underlying metric structure of the cognitive problems. Therefore, it is no wonder that it does not give us an adequate explication of closeness to the truth.

Popper has, in fact, observed that his measure Vs should not be applied in those situations where "we have a kind of measure of the *distance* of our guesses from the truth" (Popper, 1963, p. 397). For these cases, he proposes that one should change the definition (19) of h_T so that h_T contains as its disjuncts h, t_*, and the guesses between them. This suggestion is *ad hoc* — and it is not easy to say what it amounts to in the general case. In any case, Popper has not developed this idea further. It will be our task in the next chapter to show how the distance between our guesses gives the key to the logical problem of truthlikeness.

To conclude this chapter, we may note how the same problem with distance arises also in connection with a measure of 'degree of truth' proposed by Håkan Törnebohm (1976). Törnebohm's measure

(63) $v(h) = (\log P(h) - \log P(h/h_T))/\log P(h)$

is equal to the ratio

$(\text{transinf}(h/h_T))/\text{inf}(h) = \log P(h_T)/\log P(h)$.

It is a measure of degree of truth, rather than truthlikeness, since $v(h) = 1$ iff h is true. Again, measure v fails to make a difference between the seriousness of mistakes (cf. Niiniluoto, 1978b, p. 296). If P is a uniform probability over a cognitive problem $\mathbf{B} = \{h_i \mid i \in I\}$, all false complete answers h_i have the same degree of truth:

$v(h_i) = \log(2/w)/\log(1/w)$.

If h_i and h_j ($i \neq j$) are *any* false elements of $\mathbf{B}$, and $\mathbf{w} = |I| \geq 6$, we have also that $v(h_i \vee h_j) > v(h_i)$ which is clearly against intuitive expectations concerning degrees of truth.

CHAPTER 6

THE SIMILARITY APPROACH TO TRUTHLIKENESS

The similarity approach to truthlikeness was discovered in 1974, in two different forms and independently of each other, by Risto Hilpinen within possible worlds semantics (see Hilpinen, 1976) and Pavel Tichý within propositional logic (see Tichý, 1974). The convergence of their apparently different ideas, and the application of the general approach to monadic first-order logic, was given in my paper for the 1975 International Congress for Logic, Methodology, and Philosophy of Science in London, Ontario (see Niiniluoto, 1977b). Later extensions cover full first-order logic (Tichý, 1976, 1978; Niiniluoto, 1978a, b; Tuomela, 1978; Oddie, 1979), higher-order and intensional logics (Oddie, 1982; Niiniluoto, 1983b), probabilistic hypotheses (Rosenkrantz, 1980), and quantitative languages (Niiniluoto, 1982b, c, 1986a, c; Festa, 1986). The technical and philosophical controversies surrounding this approach have ranged from disagreements about the details in working out the programme (cf. Oddie, 1981; Niiniluoto, 1982a) to more general challenges to the whole enterprise (cf. Miller, 1976, 1978; Popper, 1976; Urbach, 1983).[1]

In this chapter, I formulate the basic ideas of the similarity approach in a general framework which is provided by the concept of a cognitive problem (cf. Chapter 4.2). If g is a potential partial answer to problem **B**, its degree of truthlikeness depends on the similarity between the states of affairs allowed by g and the true state expressed by the true element h_* of **B**. For this definition, two functions have to be specified: the *similarity function*, which expresses the distances between the elements of **B**, and a *reduction function*, which extends the similarity function from complete answers in **B** to partial answers in $D(\mathbf{B})$. The choice of the reduction function — which determines the general properties of the concept of truthlikeness — is discussed in Sections 3—5. Section 7 develops a further generalization of the reduction function to a measure of the distance between statements in $D(\mathbf{B})$. This function is applied in Section 8 to measure the distance of a statement from indefinite truth. Section 9 deals with the case where a cognitive problem is defined relative to a false presupposition. Chapter 7 gives a general outline of a method, first proposed in Niiniluoto (1977b) for monadic languages, for estimating degrees of truthlikeness on the basis of evidence. In the later chapters, the abstract treatment of cognitive problems is interpreted in different kinds of knowledge-seeking contexts by specifying the 'targets' that represent the 'whole truth' (cf. Section 2). The problem of defining the similarity function is treated separately for each such specific context.

6.1. SPHERES OF SIMILARITY

In his paper to the Warsaw conference on Formal Methodology in 1974, Hilpinen analysed the concepts of approximate truth and truthlikeness by means of possible worlds semantics. Following the idea of David Lewis (1973), he employed, as an undefined primitive notion, a comparative concept of *similarity between possible worlds*. This concept can be expressed, without assuming quantitative degrees of similarity, by a system $\mathcal{N}_u$ of *nested spheres of similarity* around each possible world u. The elements S of $\mathcal{N}_u$ are sets of possible worlds such that

(1) $u \in S$ for all S (weak centering)

(2) $S \subseteq S'$ or $S' \subseteq S$ for all S, S' (nesting).

Hilpinen assumed that there is a smallest set S_0 in $\mathcal{N}_u$ such that $S_0 \neq \{u\}$, and the largest set S_1. A world v is *distant* from u if v does not belong to any sphere in $\mathcal{N}_u$.

The system $\mathcal{N}_u$ then allows us to define a comparative notion $v \geq_u w$ (v is *at least as close to u as w is*) in the following way: $v \in S$ for all spheres S in $\mathcal{N}_u$ such that $w \in S$. In other words, the smallest sphere in $\mathcal{N}_u$ to which v belongs indicates the closeness of v to u. It is clear that the relation $\geq_u$ is reflexive, transitive, and connected in the class of all worlds. The relations of being *closer to u* ($>_u$) and *equally close to u* ($\sim_u$) can then be defined by

(3) $v >_u w$ iff $v \geq_u w$ and not $w \geq_u v$

(4) $v \sim_u w$ iff $v \geq_u w$ and $w \geq_u v$.

In particular, all distant worlds are equally close to u.

If a proposition h is identified with its 'truth set' Mod(h), i.e., the set of possible worlds in which h is true, then the closeness of h to world u depends on the location of the set Mod(h) in the system of spheres $\mathcal{N}_u$

around u (see Figure 1). Proposition h *allows* all the possible worlds in Mod(h) and *excludes* all the possible worlds outside Mod(h). If u is chosen to be the *actual world*, then the closeness of Mod(h) to u can be used to explicate the ideas of closeness to the truth and truthlikeness. Hilpinen says that h is

(5)   *approximately true (almost true)*   iff   Mod(h) ∩ $S_0 \neq \emptyset$,
    clearly true   iff $S_0 \subseteq$ Mod(h),
    completely false   iff   Mod(h) ∩ $S_1 = \emptyset$,
    trivially true   iff   ∼h is completely false.

Thus, h is almost true if h is true in some world minimally close to u, clearly true if h is true in all worlds minimally close to u, and trivially true if h is true in all non-distant worlds relative to u.

Let $\mathbf{E}_u(h)$ be the set of spheres in $\mathcal{N}_u$ which do not intersect Mod(h):

(6) $\mathbf{E}_u(h) = \{S \in \mathcal{N}_u | \text{Mod}(h) \cap S = \emptyset\}$.

Then the small size of $\mathbf{E}_u(h)$ is an indication that h is near to the truth u in the sense that h allows possible worlds which are close to the world u. Thus,

(7) h is at least as *near to the truth* as g
     iff $\mathbf{E}_u(h) \subseteq \mathbf{E}_u(g)$.

The motivation for this definition can be seen also in the following results:

(8) $\mathbf{E}_u(h) = \emptyset$ iff h is approximately true
     $\mathbf{E}_u(h) = \emptyset$ if h is logically true
     $\mathbf{E}_u(h) = \mathcal{N}_u$ if h is completely false
     $\mathbf{E}_u(h) = \mathcal{N}_u$ if h is logically false
     $\mathbf{E}_u(h \lor g) = \mathbf{E}_u(h) \cap \mathbf{E}_u(g)$
     $\mathbf{E}(g) \subseteq \mathbf{E}(h)$ if h is logically stronger than g
     $\mathbf{E}_u(h) \subseteq \mathbf{E}_u(h \land g)$
     $\mathbf{E}_u(h) = \emptyset$ or $\mathbf{E}_u(\sim h) = \emptyset$.

$\mathbf{E}_u(h)$ is thus clearly a qualitative counterpart to the concept of degree of truth. A slight modification of Hilpinen's formulation would make it even better in this respect: if condition (1) is replaced by the stronger centering condition

(9) $\{u\} \in \mathcal{N}_u$,

THE SIMILARITY APPROACH TO TRUTHLIKENESS   201

so that $\{u\}$ is the smallest element of $\mathcal{N}_u$, then the first result of (8) would be replaced by

(10) $\mathbf{E}_u(h) = \emptyset$ iff h is true
        $\mathbf{E}_u(h) = \{u\}$ iff h is almost but not strictly true.

In contrast to (8), (10) allows us to say that true propositions are nearer to the truth than false but approximately true ones.

Let $\mathbf{I}_u(h)$ be the set of the spheres in $\mathcal{N}_u$ which include $\mathrm{Mod}(h)$:

(11) $\mathbf{I}_u(h) = \{S \in \mathcal{N}_u \mid \mathrm{Mod}(h) \subseteq S\}$.

Then the large size of $\mathbf{I}_u(h)$ indicates that h contains information about the truth u, since then h is effective in excluding outside $\mathrm{Mod}(h)$ possible worlds which are far from u. Thus,

(12) h is at least as *informative about the truth* as g
        iff $\mathbf{I}_u(g) \subseteq \mathbf{I}_u(h)$.

Hence,

(13) $\mathbf{I}_u(h) = \emptyset$ if h is logically true
        $\mathbf{I}_u(h) = \emptyset$ if h is consistent and completely false
        $\mathbf{I}_u(h) = \mathcal{N}_u$ if h is logically false
        $\mathbf{I}_u(h \vee g) = \mathbf{I}_u(h) \cap \mathbf{I}_u(g)$
        $\mathbf{I}_u(g) \subseteq \mathbf{I}_u(h)$ if h is logically stronger than g
        $\mathbf{I}_u(h) \subseteq \mathbf{I}_u(h \wedge g)$
        $\mathbf{I}_u(h) = \emptyset$ or $\mathbf{I}_u(\sim h) = \emptyset$.

Hilpinen modifies the definition (11) so that $\mathbf{I}_u(h) = \emptyset$ for contradictions.

In order to be informative about the truth u, proposition h has to be informative in the absolute sense of Chapter 4.4, i.e., $\mathrm{Mod}(h)$ has to be a small set. However, not all propositions with high information content are also informative about the truth u — for this purpose, the whole set $\mathrm{Mod}(h)$ has to be located close to the world u. The difference between these two concepts corresponds to Popper's distinction between logical content and truth content: as Hilpinen notes, definition (12) corresponds closely to Popper's concept of truth content (except that Popper does not have a concept of similarity).

Hilpinen defines a comparative concept of truthlikeness as follows:

(14) h is at least as *truthlike* as g
iff $\mathbf{E}_u(h) \subseteq \mathbf{E}_u(g)$ and $\mathbf{I}_u(g) \subseteq \mathbf{I}_u(h)$.

Further, h is *more truthlike* than g iff h is at least as truthlike as g but g is not at least as truthlike as h. If the condition given in (14) does not hold, propositions h and g are *incomparable* with respect to truthlikeness. A typical situation where h is more truthlike than g is illustrated in Figure 1.

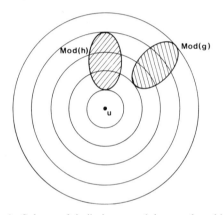

Fig. 1. Spheres of similarity around the actual world u.

Definition (14) has the decisive advantage that it is free from the troubles that Miller and Tichý raised over Popper's concept of verisimilitude (see Chapter 5.5); it makes some false propositions comparable with respect to truthlikeness. However, it is a weak theory in the sense that it makes many pairs of rival propositions incomparable. Hilpinen was quite aware of that — he pointed out that "any *numerical* measure of truthlikeness must involve some mechanism by which the two main components of truthlikeness can be balanced against each other" (*ibid.*, p. 32). This suggestion was worked out in my London Ontario paper in 1975 (see Niiniluoto, 1977b; cf. Sections 3 and 4 below), where I employed a quantitative distance function d to represent degrees of similarity. With function d, the system of spheres around world u can be defined by 'balls' of radius r:

$$S_r = \{v \mid d(u, v) \leq r\},$$

and then one can proceed to give a quantitative definition of degrees of truthlikeness.

The use of a similarity metric is not only a matter of technical convenience, but it also helps us to get rid of some unintuitively strong features of definition (14):

(15) Proposition h is at least as truthlike as a tautology iff h is approximately true.

(16) If h is not approximately true and g is approximately true, then h cannot be at least as truthlike as g.

If we make modifications (9) and (10), then principles (15)–(16) follow from definition (14) with 'approximate true' replaced by 'true'. Thus, only true propositions have the possibility to beat tautologies, and, *given any true proposition h, no false proposition can be at least as truthlike as h*. This is clearly in conflict with the requirement that some false but informative propositions may be more truthlike than some true but uninformative propositions.

There is also another motivation for dealing with similarity metrics in the theory of truthlikeness. It was argued in Chapter 4.3 that 'truth in the actual world' always means in effect Tarskian truth in some fragment of the actual world. If h is a contingent statement in a semantically determinate language L, then h is factually true if and only if $\Omega_L^* \vDash h$, where Ω_L^* is the fragment of THE WORLD relative to L. This idea has to be taken seriously also in the theory of truthlikeness. In defining the degree of truthlikeness of a proposition h, our 'goal' or 'target' is not THE WORLD — the actual world in its infinite variety — but only that part of it which exhibits the relevant features of the cognitive problem situation. If the relevant language in which h is expressed is L, then — as a first approximation at least — we may choose as our target the L-structure Ω_L^*.

This argument suggests that, instead of dealing with the class of all possible worlds, we should restrict our attention to the class $\mathbf{M}_L$ of L-structures for some first-order language L. Moreover, there is no reason to be satisfied with the notion of similarity as an undefined primitive notion. So the general problem underlying the theory of truthlikeness can be formulated as follows: *find a reasonable similarity metric in the space* $\mathbf{M}_L$ (cf. Niiniluoto, 1978a). We know such metrics exist: $\mathbf{M}_L/\equiv$ (i.e., $\mathbf{M}_L$ with elementarily equivalent structures identified) is metrizable, since, as the Stone space of the Lindenbaum algebra $\mathscr{A}_L$ of L (cf. Chapter 2.9), it is a compact Hausdorff space (cf. Chapter 1.2).[2] The problem is to find such a metric or at least a distance function — and construct a definition of degrees of truthlikeness on its basis.

6.2. TARGETS

My first reaction to Hilpinen's similarity approach, already during the 1974 congress in Warsaw, was to suggest a linguistic reformulation to it: replace possible worlds by *maximally strong descriptions of possible worlds*, i.e., Hintikka's constituents, and then specify explicitly a distance measure between constituents.[3] At least for the case of monadic constituents, it seemed to be easy to define such a metric (cf. Niiniluoto, 1977b). In working out the extension of this approach to polyadic constituents, in my paper for the Jyväskylä Logic Conference in 1976, I found out that a solution to this problem gives us also a metric on the space $\mathbf{M}_L/\equiv$ (Niiniluoto, 1978a; cf. Chapter 10.3 below).

It turned out that the linguistic version of the similarity approach had been first suggested by Tichý (1974), who formulated it only for propositional logic — but hinted at the extension to first-order logic *via* Hintikka's theory of distributive normal forms.

Let PL be a system of propositional logic with the atomic sentences $p_1, \ldots, p_n$ and the ordinary connectives $\sim$, $\wedge$, $\vee$, $\supset$, and $\equiv$. Then each sentence h of PL can be expressed as a finite disjunction of conjunctions of the form

(17) $(\pm)p_1 \& (\pm)p_2 \& \cdots \& (\pm)p_n.$

This disjunction is known as the full *disjunctive normal form* of h. Let us call conjunctions (17) the constituents of PL, and define their distance as the number of atomic sentences negated in one constituent but unnegated in the other. Then Tichý's proposal is to define the verisimilitude of h as the arithmetical mean of the distances between the true constituent and the constituents in the normal form of h (*ibid.*, p. 159). The same definition, apparently without knowledge of Tichý, was later rediscovered by H. Vetter (1977).

A constituent of PL is a description of a possible world — or a maximally strong statement within the conceptual resources of the language of PL. It is, indeed, the most rudimentary type of description of a 'world' or a state of affairs that we can imagine. Its weakness in this respect is due to the fact that in propositional logic there is no language-world-correlation in the same way as in the semantics of first-order languages. Propositional logic does not exhibit the inner structure of statements, and thereby it treats states of affairs as unanalysed wholes — as if 'facts' were the basic elements of the world. In contrast with predicate logic, propositional logic does not attempt to link linguistic expressions to such elements of reality as individuals, properties, and relations (cf. the remarks on realism in Chapter 4.3). As the world can be divided into facts by sentences in indefinitely many ways, it is no wonder that Tichý's definition of truthlikeness in propositional logic has been claimed to be oversensitive to translation (see Miller, 1975). This problem of linguistic invariance will be discussed in more general setting in Chapter 13.2 below.

We have thus seen that Hilpinen and Tichý approached the similarity view from two extremes: the former from possible worlds, and the latter from the weakest kinds of descriptions of possible worlds. It is easy to observe that there is an infinite variety of alternatives between these extremes. In the first place, we have to replace the actual world by its fragment Ω_L^*, which is relativized to a given set of descriptive vocabulary λ. This of course does not mean that the language L is fixed once and for all times — rather the choice of λ indicates so to say the *descriptive depth* at which we consider the world at a given time. By enriching the vocabulary λ, we can expand the structure Ω_L^* and focus on ever larger fragments of the actual world (see Chapter 4.3). In explicating the concept of truthlikeness, our target cannot be 'the whole truth' about the actual world, if this is meant to include all truths about the world (cf. Popper, 1972, p. 124). It is not even clear that such ideas of 'the whole truth' and 'closeness to the whole truth' are meaningful at all. Therefore, a measure of truthlikeness should reflect the distance of a statement from the world at some given descriptive depth (cf. Niiniluoto, 1977b, pp. 124—127). The fact that this descriptive depth can be chosen in different ways, brings in an interesting dynamic feature to the theory of truthlikeness.

When the descriptive vocabulary λ, and thereby the corresponding structure Ω_L^*, has been chosen, we can still choose the *logical depth* of the descriptions of Ω_L^*. In fact, no linguistic statement h can capture the structure Ω_L^* uniquely — so that $\text{Mod}(h) = \{\Omega_L^*\}$. The strongest descriptions — Scott sentences in infinitary logic, infinitely deep constituents in the Hintikka—Rantala languages, and categorical theories in higher-order logic — are able to specify a first-order structure in $\mathbf{M}_L$ up to isomorphism $\approx$, so that they in effect focus on the elements of $\mathbf{M}_L/\approx$ rather than $\mathbf{M}_L$. The state descriptions of L can do 'almost' the same (cf. Chapter 2.3). Complete theories in L are able to specify elements in the space $\mathbf{M}_L/\equiv$, i.e., structures up to elementary equivalence $\equiv$. Constituents of depth d in L do the same up to d-elementary equivalence $\equiv_d$; with increasing quantificational depth d, sequences of constituents give more and more fine finite approximations of the elements of $\mathbf{M}_L/\equiv$. Monadic constituents tell only what kinds of individuals one can find in the world Ω_L, while structure descriptions tell also how many individuals exemplify these different kinds. Some levels of this hierarchy of logical depth are included in Figures 2 and 3.

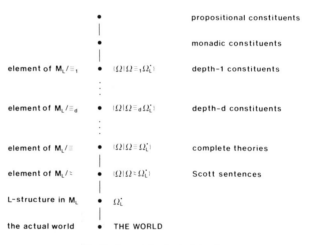

Fig. 2. Increasing logical depth.

The basic strategy of the similarity approach can now be expressed in the following way. The degree of truthlikeness of a statement g does not measure its distance from the actual world in all its variety, but rather *from the most informative true description of the world*. This description h_* is called the *target* of the definition (cf. Figure 4). The choice of the target is restricted by the requirement that it has to be *factually true* in the Tarskian correspondence-theoretical sense. It is also relative to the descriptive depth, indicated by the choice of the

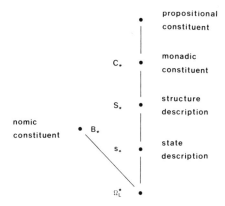

Fig. 3. Levels of logical depth for a monadic language L.

language L with vocabulary λ, and to the logical depth, indicated by the type of sentence in L that expresses all the information that is considered to be relevant to truth-*likeness*.

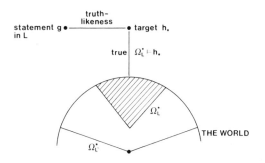

Fig. 4. Target $h_* =$ the most informative true statement in L.

Different choices of the target sentence h_* give us different concepts of truthlikeness: for example,

closeness to the true state description s_*
closeness to the true monadic constituent C_*
closeness to the true nomic constituent B_*
closeness to the complete theory $\mathrm{Th}(\Omega_L^*)$
closeness to the strongest true law of succession

define concepts that differ from each other. Therefore, the question naturally arises whether some of these conceptions is 'better' than the others. Is there some way of choosing the 'best' target for the purpose of the theory of truthlikeness?

The view that is taken in this book gives a negative answer to this question (cf. Niiniluoto, 1983b). It seems to me that all the targets which lead to natural similarity metrics are satisfactory from a logical point of view. There are no rules of logic which would dictate the selection of the same target for all situations. The choice of the target sentence reflects our context-dependent cognitive interests: the language L indicates the aspects of reality that we are interested in, and the logical type of the target in L tells what sort information is taken to be relevant. For example, even if it is the case that

$$s_* \vdash S_* \vdash C_*,$$

the extra logical strength in the state description s_* vis-à-vis the monadic constituent C_* may be irrelevant to our cognitive interests. If we are interested in the properties of specific individuals in the actual world, then s_* is a natural target. But, for example, in the context of science only the invariant — general and lawlike — features of the world may be taken to be relevant goals of inquiry. In this spirit, it was argued in Niiniluoto (1977b) that the theory of truthlikeness for scientific hypotheses should be built on their distances to the true constituent of some first-order language. Similarly, one could argue that the proper goal of scientific theorizing is the true nomic constituent (cf. Niiniluoto, 1983b).

The idea that the choice of the relevant target h_* is a function of cognitive interests can be expressed by the concept of cognitive problem (cf. Chapter 4.2). The fixing of a conceptual framework L, and one of its sentences as the target h_*, is tantamount to formulating a cognitive problem $\mathbf{B} = \{h_i | i \in I\}$ such that h_* is the complete correct answer to $\mathbf{B}$. The other elements of $\mathbf{B}$ are precisely the other sentences of L of the same logical type: if h_* is s_*, then $\mathbf{B}$ is the set of all state descriptions in L; if h_* is C_*, then $\mathbf{B}$ is the set of all monadic constituents in L; if h_* specifies the value of a real-valued quantity, then $\mathbf{B}$ is the set of all real numbers (or the range of the possible values of this quantity), etc. And, as will be seen in Chapters 7—11, it is precisely the fact that all the elements of the cognitive problem $\mathbf{B}$ have the same logical type, reflected systematically also in their syntactic form, which helps us to define a similarity metric for various kinds of cognitive situations.

Instead of attempting to restrict the variety of possible types of targets, this book develops a flexible theory of truthlikeness which is able to handle all of them along the lines of the linguistic similarity approach. If the reader is not satisfied with this multiplicity of notions of truthlikeness — in spite of the fact that they are all treated by the same method — he or she is invited to pick up a favourite among the different targets and limit attention to that one.

6.3. DISTANCE ON COGNITIVE PROBLEMS

Let $\mathbf{B} = \{h_i | i \in I\}$ be a cognitive problem, i.e., a finite or infinite collection of mutually exclusive and jointly exhaustive statements. Problem $\mathbf{B}$ is said to be *discrete*, if $\mathbf{B}$ is finite or denumerably infinite, and *continuous*, if the index set I of $\mathbf{B}$ is a subspace of some uncountably infinite space (such as $\mathbb{R}^n$).

Assume that a two-place real-valued function $\Delta: \mathbf{B} \times \mathbf{B} \to \mathbb{R}$ is defined on $\mathbf{B}$, so that

(18) $\quad \Delta_{ij} = \Delta(h_i, h_j)$

expresses the *distance* between h_i and h_j in $\mathbf{B}$. More precisely, Δ_{ij} is the distance of h_j from h_i. We assume that Δ is at least a semimetric on $\mathbf{B}$:

$\Delta_{ij} \geq 0 \quad$ for all $i, j \in I$
$\Delta_{ij} = 0 \quad$ iff $i = j$.

It is often convenient to assume that Δ is normalized, so that

$\Delta_{ij} \leq 1 \quad$ for all $i, j \in I$.

This assumption can be made without loss of generality: if Δ takes values from 0 to infinity, then

(19) $\quad \Delta'_{ij} = \Delta_{ij}/(\Delta_{ij} + 1)$

varies from 0 to 1. Another way of cutting down indefinitely large values of Δ is by truncation: if Δ varies from 0 to infinity, choose some sufficiently high value $\delta \in \mathbb{R}$ and let

(20) $\quad \Delta'_{ij} = \Delta_{ij}/\delta,$ if $\Delta_{ij} \leq \delta$
$\phantom{(20) \quad \Delta'_{ij}} = 1, \quad$ if $\Delta_{ij} > \delta.$

In many cases, Δ is a genuine metric, i.e., it is also symmetric ($\Delta_{ij} = \Delta_{ji}$ for all $i, j \in I$) and satisfies the triangle inequality.

When Δ is normalized, the *degree of similarity* between h_i and h_j in **B** is defined by

(21) $\quad \text{sim}(h_i, h_j) = 1 - \Delta_{ij}.$

Then

$$0 \leq \text{sim}(h_i, h_j) \leq 1$$
$$\text{sim}(h_i, h_j) = 1 \quad \text{iff} \quad i = j.$$

Note that if Δ' is defined from Δ by (19), then

(22) $\quad \text{sim}(h_i, h_j) = 1 - \Delta'_{ij} = 1 - \Delta_{ij}/(\Delta_{ij} + 1) = 1/(\Delta_{ij} + 1)$

(cf. (1.55) and Reichenbach's formula (5.5)).

The function Δ defines, relative to each $i \in I$, an ordering $\leq_i$ of **B**:

(23) $\quad h_j \leq_i h_k \quad \text{iff} \quad \Delta_{ij} \leq \Delta_{ik} \quad \text{for } j, k \in I.$

The ordering $\leq_i$ is a comparative similarity relation in the sense of Section 1, and it defines a system of spheres of similarity around h_i (cf. Figure 5). The relation of *equidistance* from h_i,

(24) $\quad h_j \sim_i h_k \quad \text{iff} \quad \Delta_{ij} = \Delta_{ik} \quad \text{for } j, k \in I,$

is an equivalence relation in **B**. Note that the Δ-measure is not defined for contradictions g, since their **B**-normal form is empty, and the empty set $I_g = \emptyset$ does not have any definite location in the spheres around any h_i. In this sense, contradictions are more distant from any h_i than any consistent statement in $D(\mathbf{B})$.

We shall say that the ordering $\leq_i$ is *finite*, if $\mathbf{B}/\sim_i$ is finite, that is, there is only a finite number of equivalence classes of $\sim_i$. (It is of course possible that **B** is infinite and $\mathbf{B}/\sim_i$ is finite.) The ordering $\leq_i$ is *dense*, if for all h_j and h_k in **B** such that $\Delta_{ij} < \Delta_{ik}$ there is a h_m in **B** such that $\Delta_{ij} < \Delta_{im} < \Delta_{ik}$.

An element h_j of **B** is called a Δ-*complement* of h_i, if $\Delta_{ij} = \max \Delta_{ik}$, $k \in I$. In general, h_i may have many (or none) Δ-complements in **B**. If each $h_i \in \mathbf{B}$ has a unique Δ-complement h_j in **B**, such that $\Delta_{ij} = 1$, the system $(\mathbf{B}, \Delta)$ is said to be Δ-*complemented*.

The orderings $\leq_i$ and $\leq_j$, for $i, j \in I$, are Δ-*isomorphic*, if there is a bijective mapping $f: I \to I$ such that $f(i) = j$ and $\Delta_{ik} = \Delta_{jf(k)}$ for all $k \in I$. The system $(\mathbf{B}, \Delta)$ is *structurally symmetric*, if all the orderings

$\leq_i$, $i \in I$, are Δ-isomorphic with each other. This requirement means that every element h_i of **B** has around it essentially the same system of spheres.[4] The condition that the function Δ is symmetric is not sufficient to guarantee the structural symmetry of (**B**, Δ). This can be seen from the example, where $\mathbf{B} = \{h_x | x \in [0, 1] \subseteq \mathbb{R}\}$, and $\Delta(h_x, h_y) = |x - y|$ for $x, y \in \mathbf{B}$. The trivial metric

(25) $\quad \Delta_{ij} = 0 \quad \text{if} \quad i = j$
$\quad\quad\quad\quad = 1 \quad \text{if} \quad i \neq j$

is always structurally symmetric.[5]

If **B** is finite (i.e., $|I| < \omega$), then, for each $i \in I$, the *average distance* of the elements of **B** from h_i is defined by

(26) $\quad \operatorname{av}(i, \mathbf{B}) = \dfrac{1}{|I|} \sum_{j \in I} \Delta_{ij}.$

This concept can be generalized to infinite discrete problems **B** by ordering the elements of **B** in the increasing distance from *i*, and by setting

(27) $\quad \operatorname{av}(i, \mathbf{B}) = \lim_{n \to \infty} \operatorname{av}(i, {}^i\mathbf{B}^n) = \lim_{n \to \infty} (1/|{}^i\mathbf{B}^n|) \sum_{h_j \in {}^i\mathbf{B}^n} \Delta_{ij}$

where ${}^i\mathbf{B}^n$ is the set of the first n elements of **B** in this ordering. If the similarity relation $\leq_i$ is finite, so that Δ_{ij} takes only the values $\delta_0, \ldots, \delta_r$, let $p_0, \ldots, p_r$ be the proportions of elements of **B** with the distances $\delta_0, \ldots, \delta_r$, respectively, from h_i. In other words, $p_0, \ldots, p_r$ are the relative sizes of the equivalence classes in the set $\mathbf{B}/\sim_i$. Then (27) gives the same result as

(28) $\quad \sum_{s=0}^{r} p_s \delta_s.$

Finally, if **B** is a continuous problem, then the definition (26) is replaced by the integral

(29) $\quad \operatorname{av}(i, \mathbf{B}) = \dfrac{\int_I \Delta(h_i, h_x)\, dx}{\int_I dx} = \dfrac{\int_I \Delta_{ix}\, dx}{\int_I dx}.$

If the denominator of (29) is infinite, then the value of $\operatorname{av}(i, \mathbf{B})$ has to be defined by a limiting process in analogy with (27); for example, we may consider I as the limit of balls $I_{ir} = \{j | \Delta_{ij} \leq r\}$ around i, when $r \to \infty$.

If $(\mathbf{B}, \Delta)$ is structurally symmetric, then the value av$(i, \mathbf{B})$ is the same for all $i \in I$. This constant is then denoted by av$(\mathbf{B})$. We shall say that $(\mathbf{B}, \Delta)$ is *balanced*, if av$(\mathbf{B}) = \frac{1}{2}$.

Let g be a partial answer to problem $\mathbf{B}$, i.e., $g \in D(\mathbf{B})$, so that g has a $\mathbf{B}$-normal form

$$g = \bigvee_{j \in \mathbf{I}_g} h_j.$$

Then g can be represented as the non-empty set $\{h_j | j \in \mathbf{I}_g\}$ in the ordered set $(\mathbf{B}, \leq_i)$ (cf. Figure 5). It is then natural to say that the distance of g from the element $h_i \in \mathbf{B}$ can be defined as the *distance of the set $\mathbf{I}_g$ from the center i.*

Formally, our problem is to extend the function $\Delta: \mathbf{B} \times \mathbf{B} \to \mathbb{R}$ to a function $\Delta: \mathbf{B} \times D(\mathbf{B}) \to \mathbb{R}$, such that $\Delta(h_i, g)$ expresses the distance of $g \in D(\mathbf{B})$ from $h_i \in \mathbf{B}$. A natural adequacy condition is the requirement that the ordering $\leq_i$ defined by Δ on $D(\mathbf{B})$, i.e.,

(30) $g \leq_i g'$ iff $\Delta(h_i, g) \leq \Delta(h_i, g')$,

coincides with the original ordering $\leq_i$ on $\mathbf{B}$ (cf. (23)). In other words, the following should hold for discrete problems:

(31)   Assume that $\mathbf{I}_g = \{j\}$ and $\mathbf{I}_{g'} = \{k\}$. Then $\Delta(h_i, g) \leq \Delta(h_i, g')$ iff $\Delta_{ij} \leq \Delta_{ik}$.

A simple way — but not the only one — to guarantee (31) is to choose the extended Δ so that it coincides with the original Δ on $\mathbf{B} \times \mathbf{B}$:

(32) $\Delta(h_i, g) = \Delta_{ij}$, if $\mathbf{I}_g = \{j\}$.

It is also natural to assume that $\Delta(h_i, g)$ is a function of the distances Δ_{ij}, $j \in \mathbf{I}_g$, between h_i and the elements h_j of the $\mathbf{B}$-normal form of g. A function which expresses the connection between the extended Δ and the values Δ_{ij} is called *the reduction function* red:

(33) $\Delta(h_i, g) = \text{red}(\langle \Delta_{ij} | j \in \mathbf{I}_g \rangle)$.

Condition (32) requires that red$(\langle \Delta_{ij} \rangle) = \Delta_{ij}$.

Condition (31) need not hold generally for continuous problems $\mathbf{B}$. Let us say that $g \in D(\mathbf{B})$ is a *sharp* hypothesis, if $\mathbf{I}_g \subseteq I$ has the

THE SIMILARITY APPROACH TO TRUTHLIKENESS   213

measure zero in the space I. Then, in particular, all *point* hypotheses h_x, $x \in I$, are sharp. If the distance $\Delta(h_i, g)$ is assumed to depend on the values Δ_{ix}, $x \in \mathbf{I}_g$, through integrals over $\mathbf{I}_g$, e.g.,

$$\int_{\mathbf{I}_g} \Delta_{ix}\, dx$$

and

$$\int_{\mathbf{I}_g} dx,$$

then it may happen that $\Delta(h_i, g)$ is undefined for all sharp g, since these integrals vanish when $\mathbf{I}_g$ has the measure zero. Alternatively, it may happen that $\Delta(h_i, g)$ has the same value (zero) for all sharp g. In both cases, condition (31) is violated.

Two choices are open to us in this situation. First, we may limit our attention to those elements of $D(\mathbf{B})$ which are not sharp. Then we may require that (31) holds, when g and g' are finite balls I_{jr} and I_{kr} ($r > 0$) around h_j and h_k, respectively. This is not always satisfactory, since it forces us to give up the goal of finding the true complete answer to the continuous problem **B** — in other words, to change the original cognitive problem.[6] Secondly, even if $\Delta(h_i, g)$ may be formally undefined for sharp g, we may decide to extend $\Delta(h_i, g)$ to the case $g = h_j$ as follows:

(34)   If $\mathbf{I}_g = \{j\}$, then $\Delta(h_i, g) = \lim_{r \to 0} \Delta(h_i, I_{jr})$,

where $I_{jr} = \{k \in I \mid \Delta_{jk} \leq r\}$ is a finite ball around h_j. If $\Delta(h_i, g)$, defined by (34), coincides with Δ_{ij}, so that (32) is satisfied, we shall say that the distance function $\Delta: \mathbf{B} \times D(\mathbf{B}) \to \mathbb{R}$ is *homogeneous*.

The topological measures for the distance between a point and a set give us two reduction functions which satisfy (32) and (33) for discrete and continuous problems:

(35) $\Delta_{\inf}(h_i, g) = \inf_{j \in \mathbf{I}_g} \Delta(h_i, h_j) = \inf_{j \in \mathbf{I}_g} \Delta_{ij}$

(36) $\Delta_{\sup}(h_i, g) = \sup_{j \in \mathbf{I}_g} \Delta(h_i, h_j) = \sup_{j \in \mathbf{I}_g} \Delta_{ij}.$

(See formulas (1.23) and (1.25).) If $\leq_i$ is dense, it may happen that $\Delta_{\inf}(h_i, g) = 0$ even if $i \notin \mathbf{I}_g$, and $\Delta_{\sup}(h_i, g) = 1$ even if $\mathbf{I}_g$ does not contain a Δ-complement of h_i. If the order $\leq_i$ is finite, these possibilities are excluded. Then definitions (35) and (36) can be written in the form

(37) $\quad \Delta_{\min}(h_i, g) = \min_{j \in \mathbf{I}_g} \Delta_{ij}$

(38) $\quad \Delta_{\max}(h_i, g) = \max_{j \in \mathbf{I}_g} \Delta_{ij}.$

$\Delta_{\min}$ measures the *minimum distance* of g from h_i, and $\Delta_{\max}$ the corresponding *maximum distance* (see Figure 5).

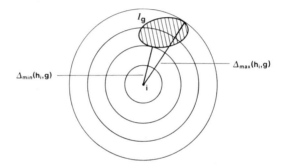

Fig. 5. Functions $\Delta_{\min}$ and $\Delta_{\max}$.

When **B** is finite, two further reduction functions are the following:

(39) $\quad \Delta_{\text{av}}(h_i, g) = \dfrac{1}{|\mathbf{I}_g|} \sum_{j \in \mathbf{I}_g} \Delta_{ij}$

(40) $\quad \Delta_{\text{sum}}(h_i, g) = \dfrac{\sum_{j \in \mathbf{I}_g} \Delta_{ij}}{\sum_{j \in I} \Delta_{ij}} = \dfrac{1}{|I|\,\text{av}(i, \mathbf{B})} \sum_{j \in \mathbf{I}_g} \Delta_{ij}.$

Here Δ_{av} gives the *arithmetical mean* of the distances between h_i and the elements h_j in the **B**-normal form of g, and Δ_{sum} the *normalized sum* of the corresponding distances.

If **B** is a continuous problem, (39) and (40) correspond formally to

$$(41) \quad \Delta_{av}(h_i, g) = \frac{\int_{I_g} \Delta_{ix} \, dx}{\int_{I_g} dx}$$

$$(42) \quad \Delta_{sum}(h_i, g) = \frac{\int_{I_g} \Delta_{ix} \, dx}{\int_{I} \Delta_{ix} \, dx}.$$

If the denominators of (39)–(42) are infinite, we have to use again a limiting process to define their values. However, (40) and (42) can be used for comparative purposes even in those cases where their denominators are infinite: for example, (40) allows us to say that g' is more distant from h_i than g if and only if

$$\sum_{i \in I_g} \Delta_{ij} < \sum_{i \in I_{g'}} \Delta_{ij}.$$

Measure Δ_{av} satisfies the homogeneity condition (32) for finite problems. In continuous problems, if the values of Δ_{av} are extended to point hypotheses by the definition (34), Δ_{av} is again homogeneous. Measure Δ_{sum} behaves differently. For finite **B**,

$$\Delta_{sum}(h_i, h_j) = \frac{\Delta_{ij}}{|I| \, av(i, \mathbf{B})}$$

so that Δ_{sum} satisfies (31) but not (32). However, for continuous problems, the limiting process (34) gives the result that

$$\Delta_{sum}(h_i, h_j) = 0 \text{ for all } h_j \in \mathbf{B}.$$

Hence, Δ_{sum} is not homogeneous. More generally, (42) implies that $\Delta_{sum}(h_i, g) = 0$ for a sharp hypothesis g independently of the location of I_g with respect to i.

Δ_{sum} can be combined with homogeneous distance functions, so that interesting reduction functions are obtained. Such combinations are not 'pure', since they contain some weights as extra-logical parameters. Of special interest are the following *min-max* and *min-sum* functions:

216   CHAPTER SIX

(43) $\quad \Delta_{mm}^{\gamma}(h_i, g) = \gamma \Delta_{min}(h_i, g) + (1 - \gamma)\Delta_{max}(h_i, g)$

(44) $\quad \Delta_{ms}^{\gamma\gamma'}(h_i, g) = \gamma \Delta_{min}(h_i, g) + \gamma' \Delta_{sum}(h_i, g),$

where $0 \leq \gamma \leq 1$ and $0 \leq \gamma' \leq 1$. If $\leq_i$ is dense, the functions Δ_{min} and Δ_{max} in (43) and (44) should be replaced by Δ_{inf} and Δ_{sup}. Hence,

$$\Delta_{mm}^{1} = \Delta_{ms}^{10} = \Delta_{min}$$
$$\Delta_{mm}^{0} = \Delta_{max}$$
$$\Delta_{ms}^{01} = \Delta_{sum}$$
$$\Delta_{ms}^{11} = \Delta_{min} + \Delta_{sum}$$
$$\Delta_{mm}^{\frac{1}{2}} = \tfrac{1}{2}(\Delta_{min} + \Delta_{max})$$
$$\Delta_{ms}^{\frac{1}{2}\frac{1}{2}} = \tfrac{1}{2}(\Delta_{min} + \Delta_{sum})$$
$$\Delta_{ms}^{\gamma(1-\gamma)} = \gamma \Delta_{min} + (1 - \gamma)\Delta_{sum}.$$

When $0 < \gamma \leq 1$ and $0 < \gamma' \leq 1$, the value of γ (relative to γ') indicates the relative weight given to the minimum distance *vis-à-vis* the maximum distance (Δ_{max}) or the normalized sum (Δ_{sum}).

Measure Δ_{mm}^{γ} is homogeneous for all γ. Measure $\Delta_{ms}^{\gamma\gamma'}$, for $\gamma > 0$, satisfies the adequacy condition (31) even for continuous problems: if $\mathbf{I}_g = \{j\}$, then

$$\lim_{r \to 0} \Delta_{ms}^{\gamma\gamma'}(h_i, I_{jr}) = \gamma \cdot \Delta_{ij} + \gamma' \cdot 0 = \gamma \Delta_{ij}.$$

In the definition (44), it is important to use the normalized Δ_{sum}-function (40), since Δ_{min} also varies between 0 and 1.[7] However, if Δ itself is not normalized, as it may happen in many continuous problems, then we could replace Δ_{sum} in (44) by the unnormalized sum-function: for discrete problems,

(45) $\quad \Delta_{ms}^{\gamma\gamma'}(h_i, g) = \gamma \Delta_{min}(h_i, g) + \gamma' \sum_{j \in I_g} \Delta_{ij}.$

and for continuous problems,

$$\Delta_{ms}^{\gamma\gamma'}(h_i, g) = \gamma \Delta_{min}(h_i, g) + \gamma' \int_{I_g} \Delta_{ix}\, dx.$$

These functions can be normalized by (19).

We shall not investigate the properties of the distance measures directly, since it is more interesting to study the extended similarity functions on $\mathbf{B} \times D(\mathbf{B})$ that they define through the formula

THE SIMILARITY APPROACH TO TRUTHLIKENESS 217

$1 - \Delta(h_i, g)$, especially in the case where h_i is chosen to be the true element h_* of **B**. These concepts of 'closeness to the truth' are studied in the next section.

6.4. CLOSENESS TO THE TRUTH

Let $\mathbf{B} = \{h_i | i \in I\}$ be a cognitive problem, and let h_* be the true element of **B**. If Δ is a distance function on **B**, the distances of $h_j \in \mathbf{B}$ from the truth h_* are denoted by

$$\Delta_{*j} = \Delta(h_*, h_j).$$

The extended distance function $\Delta: \mathbf{B} \times D(\mathbf{B}) \to \mathbb{R}$ expresses then the distances of $g \in D(\mathbf{B})$ from the truth in terms of the distances Δ_{*j}, $j \in I_g$:

$$\Delta(h_*, g) = \text{red}(\{\Delta_{*j} | j \in I_g\}).$$

As the function Δ is assumed to be normalized, the *closeness of* $g \in D(\mathbf{B})$ *to* $h_i \in \mathbf{B}$ can be defined

(46) $\quad M(g, h_i) = 1 - \Delta(h_i, g).$

Thus, M is a function $M: D(\mathbf{B}) \times \mathbf{B} \to \mathbb{R}$ which satisfies

$$0 \leq M(g, h_i) \leq 1.$$

By choosing h_i as the target sentence h_*, we obtain a measure for the *closeness* of g *to the truth*:

(47) $\quad M(g, h_*) = 1 - \Delta(h_*, g).$

Then g' is *at least as close to the truth* as g if and only if $M(g', h_*) \geq M(g, h_*)$, and g' is *closer to the truth* than g if and only if $M(g', h_*) > M(g, h_*)$.

By substituting different definitions of Δ in (46), a number of alternative measures $M: D(\mathbf{B}) \times \mathbf{B} \to \mathbb{R}$ are obtained:

(48) $\quad M_{\min}(g, h_i) = 1 - \Delta_{\min}(h_i, g)$
$\quad\quad\quad M_{\max}(g, h_i) = 1 - \Delta_{\max}(h_i, g)$
$\quad\quad\quad M_{\text{av}}(g, h_i) = 1 - \Delta_{\text{av}}(h_i, g)$
$\quad\quad\quad M_{\text{sum}}(g, h_i) = 1 - \Delta_{\text{sum}}(h_i, g)$
$\quad\quad\quad M_{\text{mm}}^{\gamma}(g, h_i) = 1 - \Delta_{\text{mm}}^{\gamma}(h_i, g)$
$\quad\quad\quad M_{\text{ms}}^{\gamma\gamma'}(g, h_i) = 1 - \Delta_{\text{ms}}^{\gamma\gamma'}(h_i, g).$

Measures (48) then give six different definitions for the closeness to the truth:

(49) $M_{\min}(g, h_*) = 1 - \Delta_{\min}(h_*, g)$
$M_{\max}(g, h_*) = 1 - \Delta_{\max}(h_*, g)$
$M_{av}(g, h_*) = 1 - \Delta_{av}(h_*, g)$
$M_{sum}(g, h_*) = 1 - \Delta_{sum}(h_*, g)$
$M_{mm}^{\gamma}(g, h_*) = 1 - \Delta_{mm}^{\gamma}(h_*, g)$
$M_{ms}^{\gamma\gamma'}(g, h_*) = 1 - \Delta_{ms}^{\gamma\gamma'}(h_*, g)$.

Is any of the measures (49) a reasonable explicate for degrees of truthlikeness? We shall discuss in the next two sections M_{av}, M_{mm}^{γ}, and $M_{ms}^{\gamma\gamma'}$ as candidates for a definition of truthlikeness. In this section, it will be seen that $M_{\min}$, $M_{\max}$, and M_{sum} express concepts of closeness to the truth that can be clearly distinguished from truthlikeness.

The basic properties of the *minimum-measure* $M_{\min}$ are given in the next theorems.

(50) $M_{\min}(g, h_i) = 1$ iff $i \in \mathbf{I}_g$
$M_{\min}(g, h_i) = 1$ if g is a tautology
$M_{\min}(h_j, h_i) = 1 - \Delta_{ij} = \text{sim}(h_i, h_j)$
$M_{\min}(g, h_i)$ is minimal if g is a Δ-complement of h_i
$M_{\min}(g \vee g', h_i) = \max\{M_{\min}(g, h_i), M_{\min}(g', h_i)\}$
$M_{\min}(g, h_i) \leq M_{\min}(g', h_i)$ if g is logically stronger than g'
$M_{\min}(g \& g', h_i) \leq \min\{M_{\min}(g, h_i), M_{\min}(g', h_i)\}$
$M_{\min}(\sim g, h_i) = 1$ iff $M_{\min}(g, h_i) < 1$.

(51) $M_{\min}(g, h_*) = 1$ iff g is true
$M_{\min}(g, h_*) = 1$ if g is a tautology
$0 \leq M_{\min}(g, h_*) < 1$ iff g is false
$M_{\min}(h_j, h_*) = 1 - \Delta_{*j} = \text{sim}(h_*, h_j)$
$M_{\min}(g, h_*)$ is minimal if g contains only Δ-complements of h_*
$M_{\min}(g, h_*) \leq M_{\min}(g', h_*)$ if $g \vdash g'$.

If $(\mathbf{B}, \Delta)$ is Δ-complemented, then the minimal value mentioned in (50) and (51) is equal to 0.

Theorem (51) shows that $M_{\min}(g, h_*)$ is an explicate of the notion of *degree of truth* or degree of truth value (cf. Chapter 5.3). It measures quantitatively *nearness to the truth* (or closeness to being true) in the sense of Hilpinen's definition (7). All the true statements g in $D(\mathbf{B})$ have the same value 1, and the value for a false statement depends on the closest guess to the truth h_* contained in its normal form. In other words, $M_{\min}(g, h_*)$ is high if g includes elements of $\mathbf{B}$ which are near to the target t_*. (50) also shows that $M_{\min}$ is truth-functional with respect to disjunction $\vee$, but not with respect to conjunction. Indeed, it may quite well happen that two false statements g and g' in $D(\mathbf{B})$ are both very near to the truth, but still $g \& g'$ is a contradiction.

As a measure of nearness to the truth, $M_{\min}$ is *insensitive to information content* in the following sense: as long as the best guess of g remains the same, any number of worse guesses may be added to g without changing the $M_{\min}$-value. In particular,

(52)   If g is any false statement in $D(\mathbf{B})$,
then $M_{\min}(h_* \vee g, h_*) = M_{\min}(h_*, h_*) = 1$.

Already this shows that $M_{\min}$ does not explicate the concept of truthlikeness — which attempts to combine the idea of truth and information.

The basic properties of the *maximum-measure* $M_{\max}$ are given in the next theorems.

(53) $M_{\max}(g, h_i) = 1$ iff $g = h_i$
$M_{\max}(h_j, h_i) = 1 - \Delta_{ij} = \text{sim}(h_i, h_j)$
$M_{\max}(g, h_i)$ is minimal if g contains a Δ-complement of h_i
$M_{\max}(g, h_i)$ is minimal if g is a tautology
$M_{\max}(g \vee g', h_i) = \min\{M_{\max}(g, h_i), M_{\max}(g', h_i)\}$
$M_{\max}(g, h_i) \geq M_{\max}(g', h_i)$ if g is logically stronger than g'
$M_{\max}(g \& g', h_i) \geq \max\{M_{\max}(g, h_i), M_{\max}(g', h_i)\}$, if $g \& g'$ is consistent
$M_{\max}(g, h_i)$ is minimal or $M_{\max}(\sim g, h_i)$ is minimal.

(54) $M_{\max}(g, h_*) = 1$ iff $g = h_*$
$M_{\max}(h_j, h_*) = 1 - \Delta_{*j} = \text{sim}(h_*, h_j)$
$M_{\max}(g, h_*)$ is minimal if g contains a Δ-complement of h_*
$M_{\max}(g, h_*)$ is minimal if g is a tautology
$M_{\max}(g, h_*) \geq M_{\max}(g', h_*)$ if $g \vdash g'$.

If $(\mathbf{B}, \Delta)$ is Δ-complemented, then minimal values mentioned in (53) and (54) are equal to 0.

Theorem (54) shows that $M_{max}(g, h_*)$ shares some properties with the notion of *degree of partial truth* in the sense of Chapter 5. It has a value less than 1 if and only if g falls short of the 'whole truth' h_*. However, now the whole truth is not represented by the *set* of all truths, but instead by a single element h_* of the cognitive problem **B**. Therefore, the idea of partial truth has to be built into the similarity function sim: the value $sim(h_*, h_i)$ is high if and only if h_i in some sense 'agrees' with a large part of h_*. If this condition holds, $M_{max}(g, h_*)$ serves as an explicate of partial truth.

At the same time, $M_{max}(g, h_*)$ is a measure of the *relief from error* that g gives us. Recall that the suspension of judgment relative to problem **B** is represented by a tautology — and a tautology allows, besides the truth h_*, all the false elements of **B**. If g is not a tautology, the value $M_{max}(g, h_*)$ depends on the false elements of **B** that g excludes. More precisely, it depends on the worst possibility that g still allows. Thus, $M_{max}(g, h_*)$ measures quantitatively *information about the truth* in the sense of Hilpinen's definition (12). Thus, maximal information about the truth h_* can be given only by h_* itself. Minimal information about h_* is given by tautologies and by the Δ-complements of h_*. In general, a statement may have a high degree of truth, but a low degree of information about the truth — e.g., tautologies. However, as $M_{min}(g, h_*) \geq M_{max}(g, h_*)$ for all $g \in D(\mathbf{B})$, a statement with high degree of information about the truth will also have a high degree of truth. Again, M_{max} is functional with respect to disjunction but not to conjunction.

As a measure of relief from error, M_{max} is *insensitive to truth values* in the following sense: as long as the worst guess of g remains the same, better guesses may be added to g without changing the M_{max}-value. In particular,

(55)   If g is any false statement in $D(\mathbf{B})$,
then $M_{max}(h_* \vee g, h_*) = M_{max}(g, h_*)$.

This result says that a false g and its truth content $h_* \vee g$ receive the same M_{max}-value. Already this shows that M_{max} is not an explicate of the concept of truthlikeness (cf. condition (5.25)). Another feature of M_{max}, which distinguishes it from truthlikeness, is the following:

(56)   If g and g' are any false statements in $D(\mathbf{B})$ such that $g \vdash g'$,
then $M_{max}(g, h_*) \geq M_{max}(g', h_*)$.

(Cf. condition (5.33).)

The basic properties of the *sum-measure* M_{sum} for finite problems **B** are given in the following theorems:

(57) If **B** is discrete, $M_{sum}(g, h_i) = \dfrac{\sum\limits_{j \notin I_g} \Delta_{ij}}{\sum\limits_{j \in I} \Delta_{ij}}$.

(58) Assume that **B** is finite.
$M_{sum}(g, h_i) = 1$ iff $g = h_i$
$M_{sum}(h_j, h_i) = 1 - \Delta_{ij}/(|I| \operatorname{av}(i, \mathbf{B}))$
$M_{sum}(g, h_i) = 0$ iff $I - \{i\} \subseteq I_g$
$M_{sum}(g, h_i) = 0$ if g is a tautology
$M_{sum}(g \vee g', h_i) = M_{sum}(g, h_i) + M_{sum}(g', h_i) -$
$\quad - M_{sum}(g \& g', h_i)$
$M_{sum}(g, h_i) \geq M_{sum}(g', h_i)$ if g is logically stronger than g'
$M_{sum}(g \& g', h_i) \geq \max\{M_{sum}(g, h_i), M_{sum}(g', h_i)\}$, if $g \& g'$ is consistent
$M_{sum}(\sim g, h_i) = 1 - M_{sum}(g, h_i)$.

(59) Assume that **B** is finite.
$M_{sum}(g, h_*) = 1$ iff $g = h_*$
$M_{sum}(h_j, h_*) = 1 - \Delta_{*j}/(|I| \operatorname{av}(*, \mathbf{B}))$
$M_{sum}(g, h_*) = 0$ iff g contains all false statements $h_j \in \mathbf{B}$
$M_{sum}(g, h_*) = 0$ if g is a tautology
$M_{sum}(g, h_*) \geq M_{sum}(g', h_*)$ if $g \vdash g'$.

For disjunction, conjunction, and negation, M_{sum} satisfies formally similar conditions as probability. However, like M_{max} but unlike probability, M_{sum} increases with logical strength.

M_{sum} is not a measure of truthlikeness, since it satisfies conditions corresponding to (55) and (56). It may be regarded as a measure for the degree of *information about the truth*. Its main difference to M_{max} is the minimality condition: $M_{max}(g, h_*) = 0$ (or minimal) if g allows for a situation which is the worst possible guess about the truth h_*, while $M_{sum}(g, h_*) = 0$ if g allows for all false guesses. This means that M_{sum} gives credit to g for *all* the false elements of **B** that g excludes (cf.

(57)). In this sense, $M_{sum}(g, h_*)$ is more sensitive than $M_{max}(g, h_*)$ as a measure for the degree of relief from error. This can be seen also from the following results:

(60)   If $\Delta_{*i} = \Delta_{*j} > 0, i \neq j$,
then $M_{max}(h_i \vee h_j, h_*) = M_{max}(h_i, h_*)$

(61)   If $\Delta_{*i} = \Delta_{*j} > 0, i \neq j$,
then $M_{sum}(h_i \vee h_j, h_*) < M_{sum}(h_i, h_*)$.

For continuous problems **B**, we get the results:

(62)   If **B** is continuous,

$$M_{sum}(g, h_i) = \frac{\int_{I-\mathbf{I}_g} \Delta_{ix}\,dx}{\int_I \Delta_{ix}\,dx}.$$

(63)   Assume that **B** is continuous.
$M_{sum}(g, h_i) = 1$ iff $\mathbf{I}_g$ has the measure zero.
$M_{sum}(g, h_i) = 0$ iff $I - \mathbf{I}_g$ has the measure zero.

(64)   Assume that **B** is continuous.
$M_{sum}(g, h_*) = 1$ iff g is sharp.
$M_{sum}(g, h_*) = 0$ iff $\sim g$ is sharp.

Theorem (64) shows that, in the extreme cases, $M_{sum}(g, h_*)$ reduces to a measure of *absolute* information content which is independent where the truth h_* is. If neither g nor $\sim g$ is a sharp hypothesis, then $M_{sum}(g, h_*)$ behaves in analogy with the results (58) and (59).

6.5. DEGREES OF TRUTHLIKENESS

The combined *min-max-measure* M^γ_{mm}, i.e.,

(65) $M^\gamma_{mm}(g, h_i) = 1 - \Delta^\gamma_{mm}(h_i, g)$
$= 1 - \gamma\Delta_{min}(h_i, g) - (1-\gamma)\Delta_{max}(h_i, g)$
$= \gamma(1 - \Delta_{min}(h_i, g)) + (1-\gamma)(1 - \Delta_{max}(h_i, g))$
$= \gamma M_{min}(g, h_i) + (1-\gamma) M_{max}(g, h_i)$

(66) $M^\gamma_{mm}(g, h_*) = \gamma M_{min}(g, h_*) + (1-\gamma) M_{max}(g, h_*)$,

THE SIMILARITY APPROACH TO TRUTHLIKENESS   223

where $0 < \gamma < 1$, was recommended as a measure of truthlikeness in Niiniluoto (1977b). (66) gives the weight $\gamma > 0$ to the degree-of-truth-factor $M_{\min}(g, h_*)$ and the weight $1 - \gamma > 0$ to the information-about-the-truth-factor $M_{\max}(g, h_*)$. The basic properties of M^{γ}_{mm} are the following:

(67) $M^{\gamma}_{mm}(g, h_i) = 1$ iff $g = h_i$
       $M^{\gamma}_{mm}(h_j, h_i) = 1 - \Delta_{ij} = \text{sim}(h_i, h_j)$
       $M^{\gamma}_{mm}(g, h_i)$ is minimal iff g contains only Δ-complements of h_i
       $M^{\gamma}_{mm}(g, h_i) = \gamma + (1 - \gamma)M_{\max}(g, h_i)$ if $i \in \mathbf{I}_g$
       $M^{\gamma}_{mm}(g \vee g', h_i) = \gamma \max\{M_{\min}(g, h_i), M_{\min}(g', h_i)\} +$
       $\qquad + (1 - \gamma)\min\{M_{\max}(g, h_i), M_{\max}(g', h_i)\}.$

(68)   Assume that $(\mathbf{B}, \Delta)$ is Δ-complemented.
       $M^{\gamma}_{mm}(g, h_i) = 0$ iff g is the Δ-complement of h_i
       $M^{\gamma}_{mm}(g, h_i) = \gamma$ if g is a tautology
       $M^{\frac{1}{2}}_{mm}(g, h_i) = \frac{1}{2}$ if g is a tautology.

(69) $M^{\gamma}_{mm}(g, h_*) = 1$ iff $g = h_*$
       $M^{\gamma}_{mm}(h_j, h_*) = 1 - \Delta_{*j} = \text{sim}(h_*, h_j)$
       $M^{\gamma}_{mm}(g, h_*)$ is minimal iff g contains only Δ-complements of h_*
       $M^{\gamma}_{mm}(g, h_*) = \gamma + (1 - \gamma)M_{\max}(g, h_*)$ if g is true.

Measure M^{γ}_{mm} has several nice properties. (54) and (69) together show that truthlikeness covaries with logical strength among true statements:

(70)   If g and g' are true and $g \vdash g'$, then
       $M^{\gamma}_{mm}(g, h_*) \geq M^{\gamma}_{mm}(g', h_*).$

(See Figure 6.) Hence, a tautology has the lowest degree of truthlikeness among true statements. The same principle does not hold for false statements:

(71)   There are false g and g' such that $g \vdash g'$
       but $M^{\gamma}_{mm}(g, h_*) < M^{\gamma}_{mm}(g', h_*).$

(See Figure 7.)

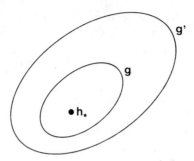

Fig. 6. g is more truthlike than g' by measure M^γ_{mm}.

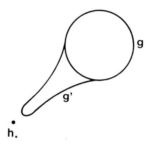

Fig. 7. g' is more truthlike than g by measure M^γ_{mm}.

If g is a false partial answer, M^γ_{mm} implies that g can be made better in two ways: by adding to g a new disjunct which is closer to h_* than the best guess in g, or by eliminating from g the worst guess. This is expressed in the following result:

(72) $\quad M^\gamma_{mm}(g \vee h_j, h_*) > M(g, h_*)$ iff $\Delta_{*j} < \Delta_{\min}(h_*, g)$
$\qquad\qquad\qquad\qquad\quad = \quad$ iff $\Delta_{\min}(h_*, g) \leq \Delta_{*j} \leq \Delta_{\max}(h_*, g)$
$\qquad\qquad\qquad\qquad\quad < \quad$ iff $\Delta_{*j} > \Delta_{\max}(h_*, g)$.

This entails, besides (71), that M^γ_{mm} satisfies the truth content principle:

(73) $\quad$ If g is false, then $M^\gamma_{mm}(h_* \vee g, h_*) > M^\gamma_{mm}(g, h_*)$.

If γ is not too close to 1, M^γ_{mm} satisfies also the condition that some false statements are more truthlike than some true statements (see Figure 8). With (70), this means that some false statements may be

THE SIMILARITY APPROACH TO TRUTHLIKENESS

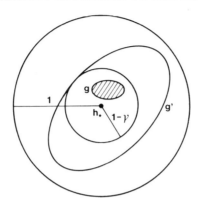

Fig. 8. g is more truthlike than g' by measure M^γ_{mm}.

more truthlike than tautologies:

(74) When $\vdash g'$, there is a false statement g such that
$M^\gamma_{mm}(g, h_*) > M^\gamma_{mm}(g', h_*)$.

If Δ defines a finite order relation, condition (74) holds if

$$1 - \gamma \geqq \min_{j \neq *} \Delta_{*j}.$$

As an answer to the cognitive problem **B**, a tautology is equivalent to the suspension of judgment. (70) thus implies that *all non-tautological true answers are better than the suspension of judgment*. (74) in turn says that *some false answers are better than no answer at all*. For a Δ-complemented $(\mathbf{B}, \Delta)$, the result (68) gives a methodological interpretation to the parameter γ: the value of γ is the *minimum treshold* that the degree of truthlikeness of a partial answer g has to *exceed in order to be worth giving*. We can thus say that those complete answers h_i are *misleading* which are situated outside the circle of radius $1 - \gamma$, i.e., $\Delta_{*i} > 1 - \gamma$ (cf. Figure 8).

In spite of its many desirable properties, the min-max-measure M^γ_{mm} has a weakness: since $M^\gamma_{mm}(g, h_*)$ depends only on the minimum and the maximum distances between h_* and $\mathbf{I}_g$, it is insensitive to variations inside these two extremes (see (72)). For this reason, it is not able to make a difference between 'fat' and 'thin' hypotheses:

(75) If $\Delta_{*i} = \Delta_{*j} > 0, i \neq j$,
then $M^\gamma_{mm}(h_i \vee h_j, h_*) = M^\gamma_{mm}(h_i, h_*)$.

It also violates the following strengthening of (70):

(76) If g and g' are true, $g \vdash g'$, and $g' \nvdash g$,
then $M^\gamma_{mm}(g, h_*) > M^\gamma_{mm}(g', h_*)$.

In this sense, it may be claimed that M^γ_{mm} is not sufficiently sensitive to information content.

If M_{max} is replaced by the measure cont^+_u (or, more generally, cont) in (66), the resulting *min-cont-measure*

(77) $M^\gamma_{mc}(g, h_*) = \gamma M_{min}(g, h_*) + (1 - \gamma)\text{cont}^+_u(g)$

shares many properties with M^γ_{mm}:

(78) $M^\gamma_{mc}(g, h_*) = 1$ iff $g = h_*$
$M^\gamma_{mc}(h_i, h_*) = \gamma(1 - \Delta_{*i}) + (1 - \gamma) = 1 - \gamma\Delta_{*i}$.
$M^\gamma_{mc}(g, h_*) = \gamma + (1 - \gamma)\text{cont}^+_u(g)$ if g is true
$M^\gamma_{mc}(g, h_*) = \gamma$ if g is a tautology.

M^γ_{mc} has the advantage over M^γ_{mm} that it makes a difference between 'fat' and 'thin' statements (cf. principles (75) and (76)). However, M^γ_{mc} is not satisfactory, since it is too insensitive to distances from the truth (see Niiniluoto, 1978b, p. 292). For example, when $i \neq *$,

$$M^\gamma_{mc}(h_* \vee h_i, h_*) = \gamma + (1 - \gamma) \cdot \frac{|I| - 2}{|I| - 1}$$

is entirely independent of the distance Δ_{*i}. Further, if h_1 and h_2 are close to h_*, while h_3 is very far from h_*,

$$M^\gamma_{mc}(h_* \vee h_1 \vee h_2, h_*) = \gamma + (1 - \gamma) \cdot \frac{|I| - 3}{|I| - 1}$$
$$< M^\gamma_{mc}(h_* \vee h_3, h_*).$$

The following combination of M^γ_{mm} and M^γ_{mc} was proposed in Niiniluoto (1978b), p. 292:

(79) $M^\gamma_{mmc}(g, h_*) = \gamma M_{min}(g, h_*) + (1 - \gamma)\text{cont}^+_u(g)M_{max}(g, h_*)$.

This *min-max-cont-measure* takes into account the minimum and the

THE SIMILARITY APPROACH TO TRUTHLIKENESS 227

maximum distances between $\mathbf{I}_g$ and h_*, and the size of $\mathbf{I}_g$. Thus, assuming $(\mathbf{B}, \Delta)$ to be Δ-complemented,

(80) $M^\gamma_{mmc}(g, h_*) = 1$ iff $g = h_*$
$M^\gamma_{mmc}(h_i, h_*) = 1 - \gamma \Delta_{*i}$
$M^\gamma_{mmc}(g, h_*) = \gamma + (1 - \gamma)\text{cont}^+_u(g) M_{max}(g, h_*)$ if g is true
$M^\gamma_{mmc}(g, h_*) = \gamma$ if g is a tautology
$M^\gamma_{mmc}(g, h_*) = 0$ iff g is the Δ-complement of h_*.

The basic difference between M^γ_{mm} and M^γ_{mmc} can be seen by comparing (72) and the following result:

(81)   Assume $j \notin \mathbf{I}_g$. Then
$M^\gamma_{mmc}(g \vee h_j, h_*) < M^\gamma_{mmc}(g, h_*)$ if $\Delta_{*j} \geq \Delta_{min}(h_*, g)$.

(81) says that it is always disadvantageous to add to g a new disjunct h_j, unless h_j is better than the best guess in g. It follows immediately that M^γ_{mmc} satisfies (70), (76), and

(82)   If $\Delta_{*i} = \Delta_{*j} > 0, i \neq j$,
then $M^\gamma_{mmc}(h_i \vee h_j, h_*) < M^\gamma_{mmc}(h_i, h_*)$.

Moreover,

(83) $M^\gamma_{mmc}(g \vee h_j, h_*) > M^\gamma_{mmc}(g, h_*)$ if $\Delta_{*j} < \Delta_{min}(h_*, g)$ and γ is sufficiently high.

In particular, M^γ_{mmc} satisfies the truth content principle (cf. (73)) if and only if

$$\frac{\gamma}{1 - \gamma} > \frac{1 - \Delta_{max}(g, h_*)}{(|I| - 1)\Delta_{min}(g, h_*)}$$

(cf. Niiniluoto, 1978b, p. 293).

Measure M^γ_{mmc} has still a weakness, which is illustrated by the following result:

(84)   Assume $\Delta_{*1} < \Delta_{*i} < \Delta_{*2}$. Then
$M^\gamma_{mmc}(h_1 \vee h_i \vee h_2, h_*) = \gamma(1 - \Delta_{*1}) + (1 - \gamma) \cdot$
$\cdot \frac{|I| - 3}{|I| - 1} \cdot (1 - \Delta_{*2})$.

Thus, $M^\gamma_{mmc}(h_1 \vee h_i \vee h_2, h_*)$ is independent of the distance Δ_{*i} — it may vary from Δ_{*1} to Δ_{*2} without any effect in the degree of truthlikeness. More generally, let g and g' be two statements such that

$$\Delta_{\min}(h_*, g) = \Delta_{\min}(h_*, g')$$
$$\Delta_{\max}(h_*, g) = \Delta_{\max}(h_*, g')$$
$$|\mathbf{I}_g| = |\mathbf{I}_{g'}|.$$

Assume that g is ovate and g' obovate (see Figure 9). Then intuitively g should be closer to the truth than g', but still $M^\gamma_{\text{mmc}}(g, h_*) = M^\gamma_{\text{mmc}}(g', h_*)$.

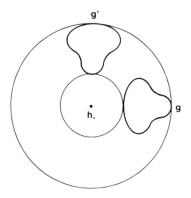

Fig. 9. g is more truthlike than g'.

This weakness of M^γ_{mmc} is obviously a consequence of the fact that $M^\gamma_{\text{mmc}}(g, h_*)$ depends only on the *number* of disjuncts in g between the best and the worst guess, not on their *distances* from h_*. A simple modification of M^γ_{mmc}, which corrects it in this respect, leads to the weighted *min-sum-measure* $M^{\gamma\gamma'}_{\text{ms}}$

(85) $\quad M^{\gamma\gamma'}_{\text{ms}}(g, h_i) = 1 - \Delta^{\gamma\gamma'}_{\text{ms}}(h_i, g)$
$\quad\quad\quad\quad\quad\quad = 1 - \gamma\Delta_{\min}(h_i, g) - \gamma'\Delta_{\text{sum}}(h_i, g)$
$\quad\quad\quad\quad\quad\quad = 1 - \gamma - \gamma' + \gamma M_{\min}(g, h_i) + \gamma' M_{\text{sum}}(g, h_i)$

(86) $\quad M^{\gamma\gamma'}_{\text{ms}}(g, h_*) = 1 - \gamma - \gamma' + \gamma M_{\min}(g, h_*) + \gamma' M_{\text{sum}}(g, h_*)$,

where $0 < \gamma \leq 1$ and $0 < \gamma' \leq 1$. In the special case where $\gamma' = 1 - \gamma$, $M^{\gamma\gamma'}_{\text{ms}}$ is denoted by M^γ_{ms}:

$$M^\gamma_{\text{ms}} = M^{\gamma(1-\gamma)}_{\text{ms}}.$$

THE SIMILARITY APPROACH TO TRUTHLIKENESS 229

When **B** is finite, the basic properties of this measure are the following:

(87) $M_{ms}^{\gamma\gamma'}(g, h_i) = 1$ iff $g = h_i$
$M_{ms}^{\gamma\gamma'}(h_j, h_i) = 1 - \gamma\Delta_{ij} - \gamma'\Delta_{ij}/(|I|\text{av}(i, \mathbf{B}))$
$M_{ms}^{\gamma\gamma'}(g, h_i) = 1 - \gamma' + \gamma' M_{\text{sum}}(g, h_i)$ if $i \in \mathbf{I}_g$
$M_{ms}^{\gamma\gamma'}(g, h_i) = 1 - \gamma'$ if g is a tautology
$M_{ms}^{\gamma\gamma'}(g \vee g', h_i) = 1 - \gamma - \gamma' + \gamma\max\{M_{\min}(g, h_i),$
$M_{\min}(g', h_i)\} + (1 - \gamma)[M_{\text{sum}}(g, h_i) +$
$+ M_{\text{sum}}(g', h_i) - M_{\text{sum}}(g \& g', h_i)].$

(88) $M_{ms}^{\gamma\gamma'}(g, h_*) = 1$ iff $g = h_*$
$M_{ms}^{\gamma\gamma'}(h_j, h_*) = 1 - (\gamma + \gamma'/(|I|\text{av}(*, \mathbf{B})))\Delta_{*j}$
$M_{ms}^{\gamma\gamma'}(g, h_*) = 1 - \gamma' + \gamma' M_{\text{sum}}(g, h_*)$ if g is true
$M_{ms}^{\gamma\gamma'}(g, h_*) = 1 - \gamma'$ if g is a tautology.

Furthermore,

$M_{ms}^{\gamma}(g, h_i) = \gamma + (1 - \gamma)M_{\text{sum}}(g, h_i)$ if $i \in \mathbf{I}_g$
$M_{ms}^{\gamma}(g, h_i) = \gamma$ if g is a tautology.

If $(\mathbf{B}, \Delta)$ is balanced, so that $\text{av}(i, \mathbf{B}) = \frac{1}{2}$, γ and γ' could be chosen so that

$$M_{ms}^{\gamma\gamma'}(h_i, h_j) = 1 - \Delta_{ij}.$$

This equation holds by (87) if

$$\gamma + \frac{2\gamma'}{|I|} = 1.$$

When **B** is continuous, $M_{ms}^{\gamma\gamma'}$ behaves essentially in the same way, if attention is restricted to statements $g \in D(\mathbf{B})$ such that $\mathbf{I}_g$ has a non-zero measure. For the case where $\mathbf{I}_g$ has the measure 0, M_{ms}^{γ} is essentially similar to the min-cont-measure M_{mc}^{γ} (see (77)).

Note that when $\Delta_{ms}^{\gamma\gamma'}(h_i, g)$ is the non-normalized min-sum-distance (45), then $M_{ms}^{\gamma\gamma'}(g, h_i)$ has to be defined instead of (85) by

(85') $$\frac{1}{1 + \gamma\Delta_{\min}(h_i, g) + \gamma' \sum_{j \in \mathbf{I}_g} \Delta_{ij}}.$$

This function has the properties

(88') $M_{ms}^{\gamma\gamma'}(g, h_*) = 1$ iff $g = h_*$
$M_{ms}^{\gamma\gamma'}(h_j, h_*) = 1/(1 + (\gamma + \gamma')\Delta_{*j})$

$$M_{ms}^{\gamma\gamma'}(g, h_*) = 1 \bigg/ \left(1 + \gamma' \sum_{j \in I_g} \Delta_{ij}\right) \text{ if } g \text{ is true}$$

$$M_{ms}^{\gamma\gamma'}(g, h_*) = \frac{1}{1 + \gamma'|I|\text{av}(*, \mathbf{B})} \text{ if } g \text{ is a tautology.}$$

It follows from (87) and (88) that the min-sum-measure $M_{ms}^{\gamma\gamma'}$ satisfies all the desiderable properties of truthlikeness. For true statements, $M_{ms}^{\gamma\gamma'}(g, h_*)$ depends on $M_{sum}(g, h_*)$ — and, hence, covaries with logical strength. $M_{ms}^{\gamma\gamma'}$ also accounts for the difference between fat and thin statements, and between obovate and ovate statements. $M_{ms}^{\gamma\gamma'}$ agrees with M_{mmc}^{γ} that it is never advantageous to add to g a new disjunct h_j which is worse than the best guess of g:

(89) Assume $j \notin \mathbf{I}_g$. Then
$M_{ms}^{\gamma\gamma'}(g \vee h_j, h_*) < M_{ms}^{\gamma\gamma'}(g, h_*)$ if $\Delta_{*j} \geq \Delta_{min}(h_*, g)$.

However, if the weight γ of the truth-factor M_{min} is sufficiently great relative to γ', a false statement g can be made better by increasing its degree of truth in spite of the corresponding loss in information content:

(90) Assume $\Delta_{*j} < \Delta_{*1} = \Delta_{min}(h_*, g)$. Then
$M_{ms}^{\gamma\gamma'}(g \vee h_j, h_*) > M_{ms}^{\gamma\gamma'}(g, h_*)$

iff $\gamma(\Delta_{*1} - \Delta_{*j}) > \dfrac{\gamma'\Delta_{*j}}{|I|\text{av}(*, \mathbf{B})}$.

The condition (90) holds for all γ and γ', if g is false (i.e., $\Delta_{*1} > 0$) and h_j is true (i.e., $\Delta_{*j} = 0$). (90) thus entails the truth content principle for all γ and γ'. For most problems $\mathbf{B}$, the condition (90) is satisfied already for relatively small values of γ, as can be seen by writing it in the form

$$\Delta_{*j} < \frac{\gamma|I|\text{av}(*, \mathbf{B})}{\gamma|I|\text{av}(*, \mathbf{B}) + \gamma'} \Delta_{*1}.$$

(Recall that $|I|\text{av}(*, \mathbf{B})$ is the sum of all distances $\Delta_{*j}, j \in I$, so that it

THE SIMILARITY APPROACH TO TRUTHLIKENESS 231

is usually a large number relative to γ, γ', Δ_{*1}, and Δ_{*j}.) If the similarity ordering $\leq_*$ on **B** is finite, and δ is the minimum distance between two different elements of **B**, define

$$\gamma_0 = \frac{1-\delta}{\delta |I| \text{av}(*, \mathbf{B})}$$

Then, by choosing $\gamma/\gamma' > \gamma_0$, the condition (90) holds for all g and h_j.

If g is a false partial answer, and $\gamma/\gamma' > \gamma_0$, g can be made more truthlike by eliminating from it disjuncts which are worse than the best guess in g. However, the elimination of the best guess from g leads to a decrease in truthlikeness. For this reason, truthlikeness does not covary with logical strength among false statements.

Which statement in $D(\mathbf{B})$ is minimally truthlike according to $M_{\text{ms}}^{\gamma\gamma'}$? The answer to this question depends on the values of γ and γ'. Note first that the disjunction of all false answers is less truthlike than a tautology, since its degree is less than

$$1 - \delta\gamma - \gamma' < 1 - \gamma'.$$

A complete answer h_j is less truthlike than a tautology — i.e., 'misleading' in the sense defined on p. 225 — if and only if

(91) $$\Delta_{*j} > \frac{\gamma'}{\gamma + \frac{\gamma'}{|I|\text{av}(*, \mathbf{B})}}.$$

The right side of (91) is always larger than γ'; it is larger than 1, so that no h_j is misleading, if

(92) $$\gamma < \gamma' \left(1 - \frac{1}{|I|\text{av}(*, \mathbf{B})}\right).$$

In other words, if γ is sufficiently small (roughly less than γ'), every complete answer to the problem **B** is better than the suspension of judgment. If γ is chosen to be larger than the right side of (92), there are misleading potential complete answers in **B**. For example, if $\gamma = \frac{1}{2}$ and $\gamma' = \frac{1}{4}$, then h_j is misleading if and only if

$$\Delta_{*j} > \frac{|I|\text{av}(*, \mathbf{B})}{1 + 3|I|\text{av}(*, \mathbf{B})}.$$

This is a good reason for choosing γ to be relatively large — but not so large that all errors are misleading and M10 fails (cf. (74)). If Δ is balanced and $|I|$ is relatively large, we can approximately state that *the ratio γ'/γ expresses the radius of the circle of non-misleading complete answers.*

Note that when γ' approaches 0, the right side of (92) becomes arbitrarily close to 0. It follows that, for sufficiently large values of γ, the minimality conditions for $M_{\text{ms}}^{\gamma\gamma'}$ and M_{mmc}^{γ} agree with each other:

(93)   When γ is sufficiently large relative to γ', $M_{\text{ms}}^{\gamma\gamma'}(g, h_*)$ is minimal iff g is the disjunction of all Δ-complements of h_*.

The measure $M_{\text{ms}}^{\frac{1}{2}\frac{1}{2}}$ was proposed as a special case of a measure for theory-distance in my paper for the Salzburg congress in 1983 (see Niiniluoto, 1986a). Measure $M_{\text{ms}}^{\gamma\gamma'}$ was first proposed in my paper for the PSA meeting in 1984 (see Niiniluoto, 1986b). We have now seen that it is able to satisfy all the intuitively appealing principles for the quantitative concept of truthlikeness.

It follows that also some other combinations of the Δ_{min}- and Δ_{sum}-functions share these nice properties. For example, it will turn out (cf. Chapter 12.5) that in some contexts it is convenient to use generalizations of the definitions (85) and (85'),

(94) $1 - \gamma \Delta_{\text{min}}(h_i, g)^\mu - \gamma' \Delta_{\text{sum}}(h_i, g)$

and

(95) $\dfrac{1}{1 + \gamma \Delta_{\text{min}}(h_i, g)^\mu + \gamma' \sum\limits_{j \in I_g} \Delta_{ij}}$,

where $\mu \geq 1$ is a natural number.

6.6. COMPARISON WITH THE TICHÝ–ODDIE APPROACH

Our previous discussion in Chapters 5 and 6 has suggested a number of adequacy conditions which an explicate of the concept of truthlikeness should satisfy. We shall state these conditions for a measure Tr: $D(\mathbf{B}) \times \mathbf{B} \to \mathbb{R}$ which defines the degree of truthlikeness $\text{Tr}(g, h_*)$ for a statement $g \in D(\mathbf{B})$ relative to the target $h_* \in \mathbf{B}$. Further, a non-trivial distance function $\Delta: \mathbf{B} \times \mathbf{B} \to \mathbb{R}$ between the elements of $\mathbf{B}$ is assumed to be given.

(M1)   (*Range*) $0 \leq \text{Tr}(g, h_*) \leq 1$.

(M2)   (*Target*) $\text{Tr}(g, h_*) = 1$ iff $g = h_*$.

(M3) (*Non-triviality*) All true statements do not have the same degree of truthlikeness; all false statements do not have the same degree of truthlikeness.

(M4) (*Truth and logical strength*) Among true statements, truthlikeness covaries with logical strength.
 (a) If g and g' are true statements and $g \vdash g'$, then $\text{Tr}(g', h_*) \leq \text{Tr}(g, h_*)$:
 (b) If g and g' are true statements and $g \vdash g'$ and $g' \nvdash g$, then $\text{Tr}(g', h_*) < \text{Tr}(g, h_*)$.

(M5) (*Falsity and logical strength*) Among false statements, truthlikeness does not covary with logical strength; there are false statements g and g' such that $g \vdash g'$ but $\text{Tr}(g, h_*) < \text{Tr}(g', h_*)$.

(M6) (*Similarity*) $\text{Tr}(h_i, h_*) \gtreqless \text{Tr}(h_j, h_*)$ iff $\Delta_{*i} \lesseqgtr \Delta_{*j}$ for all $h_i, h_j \in \mathbf{B}$.

(M7) (*Truth content*) If g is a false statement, then $\text{Tr}(h_* \vee g, h_*) > \text{Tr}(g, h_*)$.

(M8) (*Closeness to the truth*) Assume $j \notin \mathbf{I}_g$. Then $\text{Tr}(g \vee h_j, h_*) > \text{Tr}(g, h_*)$ iff $\Delta_{*j} < \Delta_{\min}(h_*, g)$.

(M9) (*Distance from the truth*) Let $\Delta_{*1} < \Delta_{*i}$. Then $\text{Tr}(h_1 \vee h_i, h_*)$ decreases when Δ_{*i} increases.

(M10) (*Falsity may be better than truth*) Some false statements may be more truthlike than some true statements.

(M11) (*Thin better than fat*) If $\Delta_{*i} = \Delta_{*j} > 0, i \neq j$, then $\text{Tr}(h_i \vee h_j, h_*) < \text{Tr}(h_i, h_*)$.

(M12) (*Ovate better than obovate*) If $\Delta_{*1} < \Delta_{*i} < \Delta_{*2}$, then $\text{Tr}(h_1 \vee h_i \vee h_2, h_*)$ increases when Δ_{*i} decreases.

(M13) (Δ-*complement*) $\text{Tr}(g, h_*)$ is minimal, if g consists of the Δ-complements of h_*.

These conditions are by no means independent: for example, M4b entails M4a, M8 entails M5 and M7, M6 entails M3.

Popper's qualitative definition of truthlikeness stumbled on the triviality condition M3, since it made false statements incomparable. Popper's truth content measure and Miller's Boolean distance measure violate M5, and Hilpinen's comparative definition violates M10.

The success of the measures studied in Sections 4 and 5 is summarized in Table I, where $+(\gamma)$ means that the measure satisfies the given condition with some restriction on the value of γ (and γ').

TABLE I
Measures of truthlikeness

condition \ measure	M^γ_{mm}	M^γ_{mc}	M^γ_{mmc}	$M^{\gamma\gamma'}_{ms}$	M^γ_{av}
M1	+	+	+	+	+
M2	+	+	+	+	+
M3	+	+	+	+	+
M4a	+	+	+	+	−
M4b	−	+	+	+	−
M5	+	$+(\gamma)$	$+(\gamma)$	$+(\gamma)$	+
M6	+	+	+	+	+
M7	+	$+(\gamma)$	$+(\gamma)$	+	+
M8	+	$+(\gamma)$	$+(\gamma)$	$+(\gamma)$	−
M9	+	−	+	+	+
M10	$+(\gamma)$	$+(\gamma)$	$+(\gamma)$	$+(\gamma)$	+
M11	−	+	+	+	−
M12	−	−	−	+	+
M13	+	$+(\gamma)$	+	$+(\gamma)$	+

Table I includes also the *average-measure*

$$(96) \quad M_{av}(g, h_i) = 1 - \Delta_{av}(h_i, g)$$

$$= 1 - \frac{1}{|\mathbf{I}_g|} \sum_{j \in \mathbf{I}_g} \Delta_{ij}$$

$$= \frac{1}{|\mathbf{I}_g|} \sum_{j \in \mathbf{I}_g} (1 - \Delta_{ij})$$

$$= \frac{1}{|\mathbf{I}_g|} \sum_{j \in \mathbf{I}_g} \mathrm{sim}(h_i, h_j)$$

$$M_{av}(g, h_*) = \frac{1}{|\mathbf{I}_g|} \sum_{j \in \mathbf{I}_g} (1 - \Delta_{*j}).$$

THE SIMILARITY APPROACH TO TRUTHLIKENESS   235

This function was proposed by Tichý (1974) and later it has been defended in Tichý (1978) and Oddie (1979, 1982). (See also Vetter, 1977). The definition of truthlikeness by the linguistic similarity approach with the reduction function M_{av} may therefore be called the Tichý–Oddie approach.

For finite problems, the basic properties of M_{av} are the following:

(97) $M_{av}(g, h_i) = 1$ iff $g = h_i$
$M_{av}(h_j, h_i) = 1 - \Delta_{ij} = \text{sim}(h_i, h_j)$
$M_{av}(g, h_i)$ is minimal if g contains only Δ-complements of h_i

$$M_{av}(g \vee g', h_i) = \frac{|\mathbf{I}_g| M_{av}(g, h_i) + |\mathbf{I}_{g'}| M_{av}(g', h_i) - |\mathbf{I}_{g \& g'}| M_{av}(g \& g', h_i)}{|\mathbf{I}_g| + |\mathbf{I}_{g'}| - |\mathbf{I}_{g \& g'}|}$$

(98)   Assume that $(\mathbf{B}, \Delta)$ is Δ-complemented and structurally symmetric.
$M_{av}(g, h_i) = 0$ iff g is the Δ-complement of h_i.
$M_{av}(g, h_i) = \text{av}(\mathbf{B})$, if g is a tautology.
Assume that $(\mathbf{B}, \Delta)$ is also balanced. Then $M_{av}(g, h_i) = \frac{1}{2}$ if g is a tautology.

Results (97) and (98) hold then also for the special case where $h_i = h_*$.

As a measure of truthlikeness, M_{av} has many common properties with M^γ_{mm}. In particular, it satisfies the truth content principle M7, and the requirement M5 about the behaviour of truthlikeness with respect to logical strength among false statements. M_{av} and $M^{\frac{1}{2}}_{mm}$ agree with each other for all statements g which are *symmetric* with respect to their midpoint in the sense that

$$\tfrac{1}{2}(\Delta_{min}(h_*, g) + \Delta_{max}(h_*, g)) = \Delta_{av}(h_*, g).$$

In contrast to M^γ_{mm} and M^γ_{mmc}, M_{av} is able to distinguish between ovate and obovate statements (cf. Figure 8). However, in contrast to M^γ_{mmc}, M_{av} and M^γ_{mm} fail to make a difference between fat and thin statements.

There is one central feature of M_{av} which has been taken to be a sufficient reason for rejecting it as a measure of truthlikeness (see Miller, 1976; Popper, 1976; Niiniluoto, 1978b). According to M_{av}, the degree of truthlikeness of g can be increased by adding to g a new disjunct which is closer to h_* than the average distance in g:

(99) $M_{av}(g \vee h_j, h_*) \gtreqless M_{av}(g, h_*)$ iff $\Delta_{*j} \lesseqgtr \Delta_{av}(h_*, g)$.

This result violates condition M8. It follows that, in contrast with M^γ_{mm}

and M^γ_{mmc}, even many *true* statements g can be made more M_{av}-truthlike by adding to them new *false* disjuncts, i.e.,

(100) There are true statements g and g' such that $g \vdash g'$ and $M_{\text{av}}(g, h_*) < M_{\text{av}}(g', h_*)$.

M_{av} thus violates Popper's fundamental requirement M4 that among true answers to a cognitive problem logically stronger statements are more truthlike than the weaker ones.

In his criticism of Tichý, Popper presents a number of 'trivial counterexamples' to the average measure M_{av} (Popper, 1976, pp. 147–149). For example, let $\mathbf{B} = \{h_i \mid i = 1, \ldots, 8\}$, where $h_1 = h_*$ and

$$\Delta(h_*, h_1) = 0$$
$$\Delta(h_*, h_2) = \Delta(h_*, h_3) = \Delta(h_*, h_4) = \tfrac{1}{3}$$
$$\Delta(h_*, h_5) = \Delta(h_*, h_6) = \Delta(h_*, h_7) = \tfrac{2}{3}$$
$$\Delta(h_*, h_8) = 1.$$

Thus, $(\mathbf{B}, \Delta)$ is Δ-complemented. Hence,

$$\Delta_{\text{av}}(h_*, h_1 \vee h_8) = \tfrac{1}{2} > \tfrac{4}{9} = \Delta_{\text{av}}(h_*, h_1 \vee h_2 \vee h_8)$$
$$\Delta_{\text{av}}(h_*, h_2 \vee h_3) = \tfrac{1}{3} = \Delta_{\text{av}}(h_*, h_2)$$

$$\Delta_{\text{av}}(h_*, h_1 \vee h_8) = \Delta_{\text{av}}\left(h_*, \bigvee_{i=1}^{8} h_i\right) = \tfrac{1}{2} < \tfrac{5}{9} =$$

$$= \Delta_{\text{av}}(h_*, h_1 \vee h_5 \vee h_8).$$

It follows that, by M_{av}, $h_1 \vee h_2 \vee h_8$ is more truthlike than the stronger true statement $h_1 \vee h_8$. Moreover, $h_1 \vee h_5 \vee h_6$ is a true statement which is less truthlike than a tautology.

It seems that Tichý did not originally note the discrepancy between his measure M_{av} and condition M4: at least Popper tells that Tichý accepted M4 still in the summer of 1975 (see Popper, 1976, pp. 148–149). However, in the light of Popper's examples, Tichý rejected M4 but upheld M_{av}.

Is there any method for resolving such a conflict? Tichý and Oddie think that intuitive examples should be used to test general principles, and not *vice versa*. "In the absence of clear-cut examples for regulating speculation, *anything goes*." (Oddie, 1981, p. 249.) Thus, when an intuitive judgment about a singular example and a high-level principle come into conflict, the principle has to be jettisoned (*ibid.*, p. 262). I doubt, however, that such appeal to intuitions is always helpful.

When intuitions about examples conflict, who decides whose intuitions are most reliable? When I gave a counter-example to Tichý's definition, which is as clear as anything can be according to *my* intuition (see Niiniluoto, 1979b, p. 375), Oddie replied that my example is "a borderline case, and should not be used to decide between rival theories" (Oddie, 1981, p. 264). So who decides which examples are 'clear-cut' and which 'borderline cases'?

It is also doubtful that there exist absolutely reliable intuitions about 'clear-cut' examples. In science, it is not necessary to assume that observations are absolutely certain — still, we use observations to test general theories. But, *vice versa*, theories are also used to interpret and reinterpret observation. Similarly, there is a two-way traffic between intuitive examples and general principles in all problems of explication — such as the definition of the concept of 'set' in axiomatic set theory, the definition of the concept of 'confirmation' in inductive logic, and the explication of the concept of truthlikeness. I do not assume that anyone has a pre-existing intuition about truthlikeness which is absolutely certain or sufficiently rich for testing all interesting principles (Niiniluoto, 1979b, p. 372). Whatever intuitions we have concern both simple examples and general adequacy conditions. The task of explication is to find a 'reflective equilibrium' between them.[8] In other words, intuitive examples are used to test general principles and attempted definitions, but conversely principles and definitions are also needed to test and to develop our intuition.

In the case of truthlikeness — which is certainly not a concept in ordinary language — we are trying to define a new technical term which has interesting applications especially in the philosophy of science. If we go to test the intuition of some persons who have never heard about verisimilitude, we are likely to get no answer at all — or we are asked first to give some explanation *in general terms* what this concept is supposed to mean. (The same holds for the concept of probability which has become a part of the ordinary language.) This is one way of arguing that such concepts as truthlikeness are learned at least partly by describing their general properties. Therefore, intuitive judgments about single examples are not 'pure' but rather already laden with some higher-level principles. Popper's intuitions (and mine) are no doubt laden with his ideas about truth and content, Tichý's intuitions with his reliance on the average function.

One way out of this conflict would be to admit that the Popperian verisimilitude and the Tichý-Oddie notion of truthlikeness are in fact different concepts, i.e., that their *explicata* are different. In this way we have earlier distinguished from each other such concepts as partial truth, nearness to the truth, and truthlikeness. However, Tichý and Oddie themselves present their measure M_{av} as a rival to the Popperian concept, i.e., as a more satisfactory explicate of verisimilitude than any measure (such as M_{mm}^γ, M_{mc}^γ, and M_{ms}^γ) which satisfies M4. Therefore, we have to still consider the motivation that they have attempted to give for their approach.

Tichý (1978) presents three arguments in defense of his measure M_{av}. The first of them is directly concerned with Popper's example above. Tichý imagines a situation where a person, John, has to rely on alternative theories in choosing his equipment for a walk. Moreover, he assumes that John's risk in trusting in a hypothesis h_i ($i = 1, \ldots, 8$) can be measured by $\Delta(h_*, h_i)$. Then John should give preference to $h_1 \vee h_2 \vee h_8$ over $h_1 \vee h_8$, since the *average risk* involved with the former is only .44 as compared to the average risk .50 of the latter. For similar reasons, John should be indifferent between h_2 and $h_2 \vee h_3$.

The import of this argument is quite different from Tichý's intention: even if it were the case that function M_{av} serves to measure the degree of *trustworthiness* (or pragmatic preference for action) of a hypothesis, it would not follow that M_{av} is an adequate measure of *truthlikeness* (cf. Niiniluoto, 1978b, p. 288). These two concepts are clearly distinct. Tichý tends to think that a disjunction of two constituents $h_1 \vee h_2$ is something like a lottery ticket which gives us the alternatives h_1 and h_2 with equal probabilities: if we 'put our trust' to $h_1 \vee h_2$, we have to make a 'blind choice' between h_1 and h_2 (Tichý, 1978, pp. 188—191; cf. Niiniluoto, 1982a, p. 192). This idea is irrelevant, if we are dealing with answers to cognitive problems — the connection between truthlikeness and practical action is quite another question which has to be studied separately.[9]

Oddie has repeated Tichý's argument without the questionable appeal to practical action. Consider three statements:

(101) g_1: $\theta = 9$
       g_2: $\theta = 9 \vee \theta = 10000$
       g_3: $\theta = 8 \vee \theta = 9 \vee \theta = 10000$

where g_1 is true. Then Oddie says that g_3 is closer to the truth than g_2, for g_3 gives us "the opportunity of choosing the very good theory"

THE SIMILARITY APPROACH TO TRUTHLIKENESS 239

$\theta = 8$ which g_2 excludes, and in "choosing from" g_2 "one stands a fifty percent risk of a disastrous theory" (Oddie, 1981, p. 257). But, again, the idea that a disjunctive answer is a fair lottery ticket with equally probable alternatives seems to be *ad hoc*. It may in some sense describe the epistemic situation of a person who is given only a true disjunctive answer and who tries to hit the truth h_* by a 'blind choice'. But this is not what we are explicating in our theory: the concept of truthlikeness is intended to tell how close to the *given* target h_* statements in $D(\mathbf{B})$ are. Tichý and Oddie thus make an unwarranted shift from the logical problem of truthlikeness to correlated practical and epistemic problems.

Tichý's second argument for measure M_{av} is an alleged counterexample to M4: it would be 'grotesque' to maintain that theory 'Snow is white or: grass is green and the Moon is made of green cheese' is closer to the truth than 'Snow is white or grass is green'. This example can be formalized in propositional logic by letting

h = 'Snow is white'
g = 'Grass is green'
k = 'The Moon is made of green cheese'.

Then Tichý's claim is that proposition $h \vee g$ should be closer to the truth $h \& g \& \sim k$ than proposition $h \vee (g \& k)$. Indeed, these distances by Tichý's measure are 7/18 and 2/5, respectively. However, the situation is not quite clear intuitively. To see this, let us formulate a general principle to the effect that

(102)   If h and g are true generalizations and k is a false generalization, then $h \vee g$ has a greater degree of truthlikeness than $h \vee (g \& k)$.

Then it is easy to see that measure M_{av} does not generally satisfy (101), i.e., Tichý's own measure of truthlikeness sometimes gives a 'grotesque' result.

This argument is not conclusive, since Tichý can of course deny that he approves (102) (see Tichý, 1978, p. 195), but then he owes us an explanation of why $h \vee g$ in this example, but not in other similar ones, should be thought to be more truthlike than $h \vee (g \& k)$. Again, I guess that Tichý's intuition is somehow connected with idea that the choice from $h \vee (g \& k)$ may lead us to the falsity k. If this were correct, presumably the following should hold also: if h and g are true,

$h \lor (g \,\&\, k)$ becomes less and less truthlike, when the alternative $g \,\&\, k$ becomes more and more grotesque. It would follow that the least truthlike of statements of this form is $h \lor (g \,\&\, {\sim}g)$ which is logically equivalent to h. Hence, $h \lor g$ would be more truthlike than h, which is not at all plausible as a general principle. (Both h and $h \lor g$ contain the truth h_* in their normal form, but the latter contains more false disjuncts in the normal form than the former.)

Tichý's third argument in favour of M_{av} (and thus implicitly against M4) is based upon an assumed analogy between an archer and a proponent of a theory. If an archer makes m shots with his bow at some target a_*, then his performance is the better the closer the individual hits $a_1, \ldots, a_m$ are *on the average* to a_*. Similarly, a theory $h_1 \lor \ldots \lor h_m$ tries to hit h_*, so that it is the better the closer the constituents $h_1, \ldots, h_m$ are on the average to the target, i.e., to the true constituent h_*.

I find this analogy seriously misleading (Niiniluoto, 1978b, pp. 289–290). In a contest between archers, each participant will make the *same number* of shots — so that their performance can be evaluated equally well by the total sum of points or by the average number of points which they score during the contest. The target is divided into ten regions and the archer will gain points, in a scale ranging from one to ten, depending on the place where his arrow hits the target. The best result is obtained by shooting *all* the arrows to the middle of the target, i.e., to the ten points region a_*. Otherwise this region a_* does not have any special status: it is better than the other regions only in the sense that the archer scores *more* points by hitting it. Moreover, if the performance of the archer has not been maximally good, there always is hope that he can make his average result better by making further shots.

When expressed in their normal form, first-order generalizations can be viewed as offering a number of 'guesses' as to which constituent is the true one h_*. But the situation is quite different from an archery contest. Generalizations offer *different* number of guesses. A good theory should hit the truth *as soon as possible*, i.e., with as few guesses as possible; thus, the best result is obtained by hitting the truth h_* with the *first* guess (cf. M2). The true constituent h_* has a special status as a target — the difference between the right and the wrong guesses is not only a matter of degree, in the sense that *as soon as the theory has hit h_* all further additions to it become superfluous*. In the above example, suppose that we first make the guesses h_8 and h_1, i.e., we propose the theory $h_1 \lor h_8$. We have then hit upon the truth h_* by two guesses, and our task in finding the truth has been completed. There would be no point in making further guesses — in particular, the addition of guess h_2 would not make our theory better. Another way of viewing the situation is the following: if we have first made the worst possible guess h_8, then it is better to make the correct guess h_1 in the next time than to make the guess h_1 only after another wrong guess h_2.

It may be misleading to speak here about 'guessing' and 'finding the truth', since these words have epistemic connotations. If we were trying to define an epistemic notion of truthlikeness, i.e., estimates of degrees of truthlikeness relative to some evidence (see Chapter 7 below), then we could also assume that the proponent of a theory who offers a number of guesses *does not know when he has hit upon the truth*. In such a situation, it might be reasonable to make further guesses even when one has already 'found' the truth, since one need not know about his success. But we are here concerned with an objective notion of truthlikeness which is independent of all epistemic considerations, i.e., we are trying to solve the *logical*, rather than the *epistemological*, problem of truthlikeness (cf. Niiniluoto, 1977b).

Oddie has replied to this criticism that the interest of hitting the truth directly would "jettison truthlikeness as an aim altogether" (Oddie, 1981, p. 258). However, his own definition satisfies the target principle M2, and measures like M^γ_{mm} and M^γ_{ms} preserve the crucial importance of distances from the truth both among true and false statements. Oddie's argument that, in example (101), the 'very good guess' $\theta = 8$ compensates for the bad guess $\theta = 10000$ fails to observe that $\theta = 8$ may be good *relative* to $\theta = 10000$ but not so good any more relative to $\theta = 9 \lor \theta = 10000$ (cf. Niiniluoto, 1982a, p. 292). My behaviour would be odd indeed, if in a contest I should wish, *after* hitting upon the correct answer, to make some additional *false* guesses in order to compensate for my first poor guesses.

I conclude that Tichý's and Oddie's average measure for the logical concept of truthlikeness rests upon intuitions which pertain to the epistemic problem of truthlikeness. That measure M_{av} is not acceptable can be argued also indirectly by showing how a measure of truthlikeness satisfying M1—M13 can be given a very natural and appealing motivation — without confusing any epistemic ideas to this stage of discussion.

Instead of thinking of statements g in $D(\mathbf{B})$ as *disjunctions* of *admitted* guesses from $\mathbf{B}$, let us view such g as *conjunctions* of *excluded* elements of $\mathbf{B}$. A tautology does not exclude any elements of $\mathbf{B}$, so that it (i.e., suspension of judgment) corresponds to the initial situation, where each element h_i of $\mathbf{B}$ is represented by a point at the distance Δ_{*i} from the truth h_*. The archer is now replaced by a 'bomber' who tries to hit the false elements of $\mathbf{B}$, but not the true one, by throwing snow balls on the target. A hit on a false h_i in $\mathbf{B}$ means that h_i is excluded or wiped out of the picture; when h_i is hit (for the first time), the person scores Δ_{*i} points. In other words, the credit that he gets from the hit depends on the distance of h_i from h_*: the larger Δ_{*i} is, the greater the gredit. The number of throws is not limited, but if all the points in $\mathbf{B}$ are excluded the game is lost. When the person stops throwing, his total *score from exclusions* is divided by the largest possible cumulative score (i.e., the sum of all Δ_{*i}, $i \in I$). In this way, we obtain the relative score from exclusions sc_e. If the bomber has not hit the truth h_* with a ball, his *truth bonus* is one. Otherwise, we look for the closest element h_j from h_* which has not been hit. Then the truth bonus tb is $1 - \Delta_{*j}$; in order to have a high bonus, some element of $\mathbf{B}$ close to h_* should be saved from exclusion. The final result is then the weighted sum of the truth bonus tb and the score from exclusions sc_e:

$$\gamma \cdot \text{tb} + \gamma' \cdot sc_e.$$

This measure, defined by our game of excluding falsity and preserving truth, is identical with $M_{\text{ms}}^{\gamma\gamma'}$.

If the rules of the game are modified so that each exclusion of a false point gives the same score 1, and $\gamma' = 1 - \gamma$, then the measure M_{mc}^{γ} is obtained. If the original game is modified by stipulating that the exclusion of h_i gives the score Δ_{*i} if and only if all elements $h_j \in \mathbf{B}$, $j \neq i$, $\Delta_{*i} \leq \Delta_{*j}$, have already been excluded, measure M_{mm}^{γ} is obtained. These modified games are, however, less natural than the one leading us the min-sum-measure $M_{\text{ms}}^{\gamma\gamma'}$.

6.7. DISTANCE BETWEEN STATEMENTS

In the preceding sections, we have discussed various possibilities to extend the distance measure $\Delta: \mathbf{B} \times \mathbf{B} \to \mathbb{R}$ to a measure of the form $\Delta: \mathbf{B} \times D(\mathbf{B}) \to \mathbb{R}$. For many purposes (cf. Sections 8 and 9), it is

THE SIMILARITY APPROACH TO TRUTHLIKENESS   243

useful to have also an extended distance function δ between any two elements g_1 and g_2 of $D(\mathbf{B})$, i.e., a function of the form $\delta: D(\mathbf{B}) \times D(\mathbf{B}) \to \mathbb{R}$ such that

(103) $\delta(g_1, g_2) = 0$ iff $g_1 = g_2$

and

(104) $\delta(h_i, h_j) \gtreqless \delta(h_i, h_k)$ iff $\Delta_{ij} \gtreqless \Delta_{ik}$.

One way of guaranteeing (104) is to choose δ so that

(105) $\delta(h_i, h_j) = \Delta(h_i, h_j) = \Delta_{ij}$ for all $h_i, h_j \in \mathbf{B}$.

δ induces also a measure $\delta(h_i, g)$ for the distance of a statement $g \in D(\mathbf{B})$ from an element h_i of $\mathbf{B}$, and thereby also a measure for the distance $\delta(h_*, g)$ of g from the truth h_*. It is, therefore, interesting to see what candidates for a measure of truthlikeness different normalized δ-measures define through the formula

(106) $\mathrm{Tr}(g, h_*) = 1 - \delta(h_*, g)$.

But, as the δ-measures can be used for different philosophical and methodological purposes, it cannot be expected that all interesting measures δ would yield a reasonable notion of truthlikeness via (106).[10]

It is again natural to assume that $\delta(g_1, g_2)$ can be defined as a function of the distances Δ_{ij}, $i \in I_{g_1}$, $j \in I_{g_2}$, between the elements of the **B**-normal forms of g_1 and g_2, respectively. Perhaps the simplest possibility which satisfies (103) is

(107) $\delta_{\mathrm{SD}}(g_1, g_2) = \dfrac{|I_{g_1} \Delta I_{g_2}|}{|I|}$,

i.e., the relative cardinality of the symmetric difference between sets I_{g_1} and I_{g_2} (cf. Example 1.7). This measure, which is a quantitative counterpart to Miller's (1978a) Boolean-valued distance between theories, simply tells in how many elements of I the **B**-normal forms of g_1 and g_2 disagree. The elements of $I_{g_1} - I_{g_2}$ are excluded by g_2 but not g_1; the elements of $I_{g_2} - I_{g_1}$ are excluded by g_1 but not g_2 (cf. Figure 10). Hence,

$0 \leq \delta_{\mathrm{SD}}(g_1, g_2) \leq 1$
$\delta_{\mathrm{SD}}(g_1, g_2) = 0$ iff $g_1 = g_2$
$\delta_{\mathrm{SD}}(g_1, g_2) = 1$ iff $g_1 = {\sim} g_2$.

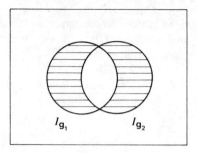

Fig. 10. Definition of $\delta_{SD}(g_1, g_2)$.

Further,

$$\delta_{SD}(h_i, g) = \frac{|\mathbf{I}_g| + 1}{|I|} \quad \text{if } i \notin \mathbf{I}_g$$

$$= \frac{|\mathbf{I}_g| - 1}{|I|} \quad \text{if } i \in \mathbf{I}_g$$

so that $1 - \delta_{SD}(h_*, g)$ is a linear function of $\text{cont}_u^+(g)$ — and, for false g, independent of where h_* is. Furthermore, if $i \neq j$,

$$\delta_{SD}(h_i, h_j) = \frac{2}{|I|}$$

so that δ_{SD} violates condition (104). As δ_{SD} is entirely independent of the underlying metric structure Δ of **B**, it is too coarce to be satisfactory. The same is true of the Mazurkiewicz metric

(108) $\quad \delta_M(g_1, g_2) = P(\mathbf{I}_{g_1} \triangle \mathbf{I}_{g_2})$

(see (1.22)).

An example of a measure which reflects the distances Δ_{ij} on **B** is the minimum distance

(109) $\quad \delta_{\min}(g_1, g_2) = \min_{\substack{i \in \mathbf{I}_{g_1} \\ j \in \mathbf{I}_{g_2}}} \Delta_{ij}$

(see (1.23)), which is an extension of $\Delta_{\min}$:

$$\delta_{\min}(h_i, g) = \Delta_{\min}(h_i, g).$$

Hence, (109) violates condition (103):

$$\delta_{\min}(g_1, g_2) = 0 \quad \text{iff} \quad \mathbf{I}_{g_1} \cap \mathbf{I}_{g_2} \neq \emptyset.$$

THE SIMILARITY APPROACH TO TRUTHLIKENESS 245

Another possibility is the Hausdorff distance

(110) $\delta_H(g_1, g_2) = \max\{\max_{i \in I_{g_1}} \Delta_{\min}(h_i, g_2), \max_{j \in I_{g_2}} \Delta_{\min}(h_j, g_1)\}$

(see (1.24)) which is an extension of $\Delta_{\max}$:

$$\delta_H(h_i, g) = \Delta_{\max}(h_i, g),$$

and therefore satisfies (103) and (105):

$0 \leq \delta_{\max}(g_1, g_2) \leq 1$
$\delta_{\max}(g_1, g_2) = 0$ iff $g_1 = g_2$
$\delta_{\max}(g_1, g_2) = 1$ iff some h_j in g_2 is a Δ-complement of some h_i in g_1.

The average of all distances between pairs of elements from g_1 and g_2, i.e.,

(111) $\delta_{av}(g_1, g_2) = \dfrac{1}{|I_{g_1}||I_{g_2}|} \sum_{j \in I_{g_1}} \sum_{j \in I_{g_2}} \Delta_{ij},$

does not satisfy (103), since

$$\delta_{av}(g, g) = \dfrac{1}{|I_g|^2} \sum_{i \in I_g} \sum_{j \in I_g} \Delta_{ij},$$

which is 0 if and only if $|I_g| = 1$. Still, it is a generalization of Δ_{av}, since

$$\delta_{av}(h_i, g) = \Delta_{av}(h_i, g).$$

Oddie (1979, p. 231), has proposed a generalization of Δ_{av} which satisfies (103), (104), and (105). Let η be a surjection of the larger of two sets I_{g_1} and I_{g_2} onto the other. Then

(112) $\delta_0(g_1, g_2) = \min_\eta \dfrac{1}{|\eta|} \sum_{\langle i,j \rangle \in \eta} \Delta_{ij},$

i.e., $\delta_0(g_1, g_2)$ is the average breadth of the linked elements $\langle i, j \rangle \in \eta$ for the most economical surjection η between the normal forms of g_1 and g_2. Oddie also requires that η has to be 'fair' in the sense that it divides the elements of one set to the other set as evenly as possible. However, the use of averages leads to unnatural results (see Figure 13 below). Also the condition of fairness has unintuitive consequences: in

Figure 11, $\delta_0(g_1, g_2)$ is relatively high, since h_2 has to be linked with h_5, but $\delta_0(g_1 \vee h_7, g_2)$ is much smaller than it. (Cf. also the discussion in Chapter 9.3.)

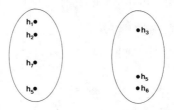

Fig. 11. $g_1 = h_1 \vee h_2 \vee h_5, g_2 = h_3 \vee h_5 \vee h_6$.

A natural method for defining δ is the following. Let us go through all the elements h_j in the **B**-normal form of g_2 and give to each h_j a 'fine' which equals its minimum distance from g_1; this 'penalty' will be zero, if h_j belongs to the **B**-normal form of g_1 as well. Similarly, we go through the elements h_i of g_1 and 'fine' them with the minimum distance of h_i from g_2. Then the sum of these penalties can be taken to measure the distance between g_1 and g_2:

(113) $\quad \delta_{ms}(g_1, g_2) = \sum_{i \in \mathbf{I}_{g_1}} \min_{j \in \mathbf{I}_{g_2}} \Delta_{ji} + \sum_{j \in \mathbf{I}_{g_2}} \min_{i \in \mathbf{I}_{g_1}} \Delta_{ji}.$

If Δ is symmetric, (113) can be written in the form

(114) $\quad \sum_{i \in \mathbf{I}_{g_1}} \Delta_{\min}(h_i, g_2) + \sum_{j \in \mathbf{I}_{g_2}} \Delta_{\min}(h_j, g_1).$

As (114) can be written also in the form

(115) $\quad \sum_{i \in \mathbf{I}_{g_1} \Delta \mathbf{I}_{g_2}} \beta_i,$

where

$$\beta_i = \Delta_{\min}(h_i, g_1) \quad \text{if } i \in \mathbf{I}_{g_2} - \mathbf{I}_{g_1}$$
$$= \Delta_{\min}(h_i, g_2) \quad \text{if } i \in \mathbf{I}_{g_1} - \mathbf{I}_{g_2},$$

measure δ_{ms} is a *weighted symmetric difference* between g_1 and g_2 with the minimum distance function $\Delta_{\min}$ determining the weights β_i (see

Figure 12). If $\delta_{ms}(g_1, g_2)$ is normalized by dividing its value by $|I|$, then the normalized measure

(116) $\quad \delta_{nms}(g_1, g_2) = \dfrac{1}{|I|} \delta_{ms}(g_1, g_2),$

which takes values in the interval $[0, 1]$, reduces to $\delta_{SD}(g_1, g_2)$ if all distances Δ_{ij} are equal to 1. In other words,

(117) $\quad$ If Δ is trivial, $\delta_{nms}(g_1, g_2) = \delta_{SD}(g_1, g_2)$.

It is then immediately clear that δ_{ms} and δ_{nms} satisfy condition (103).

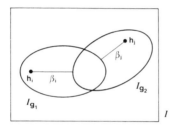

Fig. 12. Measure δ_{ms}.

Measures δ_{ms} and δ_{nms} satisfy also the condition (104), since

$$\delta_{ms}(h_i, h_j) = 2\Delta_{ij}$$
$$\delta_{nms}(h_i, h_j) = 2\Delta_{ij}/|I|.$$

The normalization

$$\tfrac{1}{2} \delta_{ms}(g_1, g_2),$$

which satisfies (105) as well, was proposed in Niiniluoto (1986a) (see formula (17)).

Functions δ_{ms} and δ_{nms} give an interesting justification to the idea that the distance of g from the truth h_* can be defined by a weighted combination of the minimum distance $\Delta_{min}(h_*, g)$ and the sum distance $\Delta_{sum}(h_*, g)$:

(118) $\quad \delta_{ms}(h_i, g) = \Delta_{min}(h_i, g) + \displaystyle\sum_{j \in I_g} \Delta_{ji}$

$\qquad\qquad = \Delta_{min}(h_i, g) + |I|\mathrm{av}(i, \mathbf{B}) \cdot \Delta_{sum}(h_i, g)$

(119) $\delta_{\text{nms}}(h_i, g) = \dfrac{1}{|I|} \Delta_{\min}(h_i, g) + \text{av}(i, \mathbf{B})\Delta_{\text{sum}}(h_i, g).$

If $(\mathbf{B}, \Delta)$ is balanced, then $\text{av}(i, \mathbf{B}) = \frac{1}{2}$ and, hence, independent of i. Then $\delta_{\text{nms}}(h_i, g)$ is identical with a special case of the min-sum distance $\Delta_{\text{ms}}^{\gamma\gamma'}(h_i, g)$:

(120) If $\gamma = 1/|I|$ and $\gamma' = \frac{1}{2}$, then
$\delta_{\text{nms}}(h_i, g) = \Delta_{\text{ms}}^{\gamma\gamma'}(h_i, g).$

However, this special case of $\Delta_{\text{ms}}^{\gamma\gamma'}$ is not among those which define a reasonable notion of truthlikeness, since here γ is small relative to γ' (cf. Section 5). In other words, δ_{nms} does not allow us to give enough weight to the $\Delta_{\min}$-factor *vis-à-vis* the Δ_{sum}-factor.

These observations suggest a more flexible generalization of (113). The 'errors' of g_2 in the sets $\mathbf{I}_{g_2} - \mathbf{I}_{g_1}$ and $\mathbf{I}_{g_1} - \mathbf{I}_{g_2}$ are, from the viewpoint of g_1, of different types: the latter are *mistaken exclusions* and the former *failures to make an exclusion*. Thus, we may give these two types of errors different weights α and α', respectively. Define the values of β_i in (115) by

$$\begin{aligned}\beta_i &= \alpha \Delta_{\min}(h_i, g_1) \quad \text{if } i \in \mathbf{I}_{g_2} - \mathbf{I}_{g_1} \\ &= \alpha' \Delta_{\min}(h_i, g_2) \quad \text{if } i \in \mathbf{I}_{g_1} - \mathbf{I}_{g_2},\end{aligned}$$

where $0 < \alpha \leq 1$ and $0 < \alpha' \leq 1$. The resulting normalized measure is denoted by $\delta_{\text{nms}}^{\alpha\alpha'}$:

(121) $\delta_{\text{nms}}^{\alpha\alpha'}(g_1, g_2) = \dfrac{\alpha}{|I|} \sum_{i \in \mathbf{I}_{g_1}} \Delta_{\min}(h_i, g_2) + \dfrac{\alpha'}{|I|} \sum_{j \in \mathbf{I}_{g_2}} \Delta_{\min}(h_j, g_1).$

Hence,

$\delta_{\text{nms}}^{11} = \delta_{\text{nms}}.$

Furthermore,

$\delta_{\text{nms}}^{\alpha\alpha'}(h_i, h_j) = (\alpha + \alpha')\Delta_{ij}/|I|$

(122) $\delta_{\text{nms}}^{\alpha\alpha'}(h_i, g) = \dfrac{\alpha}{|I|} \Delta_{\min}(h_i, g) + \alpha' \cdot \text{av}(i, \mathbf{B})\Delta_{\text{sum}}(h_i, g).$

If $(\mathbf{B}, \Delta)$ is balanced, so that $\text{av}(i, \mathbf{B}) = \frac{1}{2}$, we get by (122):

(123) If $\gamma = \alpha/|I|$ and $\gamma' = \alpha'/2$, then
$\delta_{\text{nms}}^{\alpha\alpha'}(h_i, g) = \Delta_{\text{ms}}^{\gamma\gamma'}(h_i, g).$

If now α is sufficiently large relative to α', $\delta_{\text{nms}}^{\alpha\alpha'}(h_*, g)$ satisfies the

THE SIMILARITY APPROACH TO TRUTHLIKENESS   249

adequacy conditions for a distance measure defining the notion of truthlikeness.

A basic difference between Oddie's δ_0 and δ_{nms} can be seen in Figure 11, where $\delta_0(g_1, g_2)$ is large but $\delta_{nms}(g_1, g_2)$ is small. Moreover, the addition of h_7 to g_1 increases the distance:

$$\delta_{nms}(g_1 \vee h_7, g_2) > \delta_{nms}(g_1, g_2).$$

Another way of normalizing the definition of δ_{ms} was proposed in Niiniluoto (1978b) (see μ_2 on p. 362):

$$(124) \quad \delta_{mav}(g_1, g_2) = \frac{1}{2|\mathbf{I}_{g_1}|} \sum_{i \in \mathbf{I}_{g_1}} \Delta_{min}(h_i, g_2) +$$

$$+ \frac{1}{2|\mathbf{I}_{g_2}|} \sum_{j \in \mathbf{I}_{g_2}} \Delta_{min}(h_j, g_1).$$

This measure, which satisfies (103) and (105), results from (115), when the weights β_i are defined as follows:

$$\beta_i = \frac{1}{2|\mathbf{I}_{g_2}|} \Delta_{min}(h_i, g_1) \quad \text{if } i \in \mathbf{I}_{g_2} - \mathbf{I}_{g_1}$$

$$= \frac{1}{2|\mathbf{I}_{g_1}|} \Delta_{min}(h_i, g_2) \quad \text{if } i \in \mathbf{I}_{g_1} - \mathbf{I}_{g_2}.$$

This choice is not as natural as the one leading to δ_{nms}, since here we are dealing in effect with the *average* of the minimum errors (rather than their sum). Therefore $\delta_{mav}(g_1, g_2)$ decreases, when we add to g_2 a new disjunct h_k such that $k \notin \mathbf{I}_{g_1}$ and

$$\delta_{min}(g_1, g_2) \leq \Delta_{min}(h_k, g_1) < \frac{1}{|\mathbf{I}_{g_2}|} \sum_{j \in \mathbf{I}_{g_2}} \Delta_{min}(h_j, g_1).$$

For example, in the situation described in Figure 13,

$$\delta_{SD}(g_1, g_2) < \delta_{SD}(g_1, g_2 \vee h_m)$$
$$\delta_{nms}(g_1, g_2) > \delta_{nms}(g_1, g_2 \vee h_m)$$
$$\delta_{mav}(g_1, g_2) > \delta_{mav}(g_1, g_2 \vee h_m)$$
$$\delta_0(g_1, g_2) > \delta_0(g_1, g_2 \vee h_m)$$

but

$$\delta_{nms}(g_1, g_2) < \delta_{nms}(g_1, g_2 \vee h_k)$$
$$\delta_{mav}(g_1, g_2) > \delta_{mav}(g_1, g_2 \vee h_k)$$
$$\delta_0(g_1, g_2) > \delta_0(g_1, g_2 \vee h_k).$$

This feature is reflected also in the result

(125) $\quad \delta_{mav}(h_i, g) = \tfrac{1}{2}\Delta_{min}(h_i, g) + \tfrac{1}{2}\Delta_{av}(h_i, g),$

which shows that $\delta_{mav}(h_i, g)$ is a combination of the Δ_{min}-factor and the Δ_{av}-factor. By (125),

$$\delta_{mav}(h_i, g) = \tfrac{1}{2}\Delta_{av}(h_i, g) \quad \text{if } i \in \mathbf{I}_g.$$

δ_{mav} is thus almost an extension of Tichý's distance function Δ_{av}.

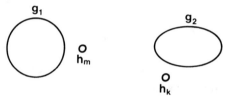

Fig. 13.

Similar results follow, if δ_{ms} is normalized by another averaging procedure:

(126) $\quad \delta'_{mav}(g_1, g_2) = \dfrac{\delta_{ms}(g_1, g_2)}{|\mathbf{I}_{g_1}| + |\mathbf{I}_{g_2}|}.$

This measure, which satisfies (105), gives the distance of g from h_i again as a combination of Δ_{min} and Δ_{av}:

(127) $\quad \delta'_{mav}(h_i, g) = \dfrac{1}{1 + |\mathbf{I}_g|} \Delta_{min}(h_i, g) + \dfrac{|\mathbf{I}_g|}{1 + |\mathbf{I}_g|} \Delta_{av}(h_i, g).$

$$= \dfrac{|I|}{1 + |\mathbf{I}_g|} \delta_{nms}(h_i, g).$$

If $i \in \mathbf{I}_g$, this reduces to

$$\dfrac{|\mathbf{I}_g|}{1 + |\mathbf{I}_g|} \Delta_{av}(h_i, g),$$

which implies that $\delta'_{mav}(h_*, g)$ does not satisfy the condition M4 for truthlikeness.

THE SIMILARITY APPROACH TO TRUTHLIKENESS 251

The weighted symmetric difference (115) can be modified by replacing the weights β_i defined by $\Delta_{\min}$ with weights defined by average distances:

$$\beta_j = \frac{1}{|\mathbf{I}_{g_1}|} \sum_{i \in \mathbf{I}_{g_1}} \Delta(h_i, h_j) \quad \text{if } j \in \mathbf{I}_{g_2} - \mathbf{I}_{g_1}$$

$$= \frac{1}{|\mathbf{I}_{g_2}|} \sum_{i \in \mathbf{I}_{g_2}} \Delta(h_i, h_j) \quad \text{if } j \in \mathbf{I}_{g_1} - \mathbf{I}_{g_2}.$$

If Δ is symmetric, these weights can be written as follows:

$$\beta_j = \Delta_{\text{av}}(h_j, g_1) \quad \text{if } j \in \mathbf{I}_{g_2} - \mathbf{I}_{g_1}$$
$$= \Delta_{\text{av}}(h_j, g_2) \quad \text{if } j \in \mathbf{I}_{g_1} - \mathbf{I}_{g_2}.$$

(see Figure 14.) This choice leads to the definition

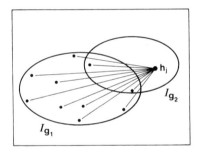

Fig. 14. Measure $\delta_{\text{avs}}(g_1, g_2)$.

$$(128) \quad \delta_{\text{avs}}(g_1, g_2) = \frac{1}{2} \sum_{i \in \mathbf{I}_{g_1} - \mathbf{I}_{g_2}} \Delta_{\text{av}}(h_i, g_2) + \frac{1}{2} \sum_{j \in \mathbf{I}_{g_2} - \mathbf{I}_{g_1}} \Delta_{\text{av}}(h_j, g_1).$$

Hence,

$$\delta_{\text{avs}}(h_i, h_j) = \Delta_{ij}$$

$$\delta_{\text{avs}}(h_i, g) = \tfrac{1}{2} \Delta_{\text{av}}(h_i, g) + \tfrac{1}{2} \sum_{j \in \mathbf{I}_g} \Delta_{ji}$$

$$= \tfrac{1}{2} \Delta_{\text{av}}(h_i, g) + \tfrac{1}{2} |I| \text{av}(i, \mathbf{B}) \cdot \Delta_{\text{sum}}(h_i, g)$$

$$(129) \qquad = \frac{(1 + |\mathbf{I}_g|)}{2} \cdot \Delta_{\text{av}}(h_i, g).$$

252 CHAPTER SIX

Again, (129) implies that $\delta_{avs}(h_*, g)$ does not satisfy the condition M4 for truthlikeness.

Let us further note that weights β_i for (115) could be defined as weighed averages of minimum and maximum distances:

$$\beta_j = \gamma \min_{i \in I_{g_1}} \Delta(h_i, h_j) + (1 - \gamma) \max_{i \in I_{g_1}} \Delta(h_i, h_j), \text{ if } j \in I_{g_2} - I_{g_1}$$

$$= \gamma \min_{i \in I_{g_2}} \Delta(h_i, h_j) + (1 - \gamma) \max_{i \in I_{g_2}} \Delta(h_i, h_j), \text{ if } j \in I_{g_1} - I_{g_2}$$

If Δ is symmetric, these weights can be written as follows

$$\beta_j = \Delta^\gamma_{mm}(h_j, g_1) \quad \text{if } j \in I_{g_2} - I_{g_1}$$
$$= \Delta^\gamma_{mm}(h_j, g_2) \quad \text{if } j \in I_{g_1} - I_{g_2}.$$

(See Figure 15.) This choice leads to the measure

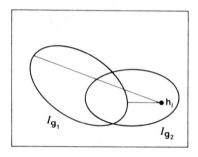

Fig. 15. Measure $\delta^\gamma_{mm}(g_1, g_2)$.

$$(130) \quad \delta^\gamma_{mm}(g_1, g_2) = \frac{1}{2} \sum_{i \in I_{g_1} - I_{g_2}} \Delta^\gamma_{mm}(h_i, g_2) +$$

$$+ \frac{1}{2} \sum_{j \in I_{g_2} - I_{g_1}} \Delta^\gamma_{mm}(h_j, g_1).$$

THE SIMILARITY APPROACH TO TRUTHLIKENESS 253

Hence,
$$\delta^\gamma_{mm}(h_i, h_j) = \Delta_{ij}$$
$$\delta^\gamma_{mm}(h_i, g) = \tfrac{1}{2}\Delta^\gamma_{mm}(h_i, g) + \tfrac{1}{2}\sum_{j \in I_g} \Delta_{ji}$$
$$= \tfrac{1}{2}\Delta^\gamma_{mm}(h_i, g) + \tfrac{1}{2}|I|\text{av}(i, \mathbf{B}) \cdot \Delta_{sum}(h_i, g).$$

Thus, $\delta^\gamma_{mm}(h_i, g)$ is a weighted combination of the min-max-factor Δ^γ_{mm} and the Δ_{sum}-factor. Moreover,

$$\frac{\delta^\gamma_{mm}(h_i, g)}{|I|/2} = (1 - \gamma)\Delta_{max}(h_i, g) + \delta^{\gamma 1}_{nms}(h_i, g).$$

Instead of (115), we could also imitate directly the definition of Δ^γ_{mm} by using Δ_{min} in the place of Δ:

(131) $\quad \delta^\gamma_{mmm}(\gamma_1, \gamma_2) = \tfrac{1}{2}(\gamma \min_{i \in I_{g_1}} \Delta_{min}(h_i, g_2) +$

$\qquad\qquad\qquad + (1 - \gamma) \max_{i \in I_{g_1}} \Delta_{min}(h_i, g_2)) +$

$\qquad\qquad\qquad + \tfrac{1}{2}(\gamma \min_{j \in I_{g_2}} \Delta_{min}(h_j, g_1) +$

$\qquad\qquad\qquad + (1 - \gamma) \max_{j \in I_{g_2}} \Delta_{min}(h_j, g_1)).$

(See Niiniluoto, 1978b, p. 322.) Hence,

$$\delta^\gamma_{mmm}(h_i, h_j) = \Delta_{ij}.$$

If Δ is symmetric,

$$\delta^\gamma_{mmm}(h_i, g) = \tfrac{1}{2}\Delta_{min}(h_i, g) + \tfrac{1}{2}(\gamma \min_{j \in I_g} \Delta_{ji} + (1 - \gamma) \max_{j \in I_g} \Delta_{ji})$$
$$= \tfrac{1}{2}\Delta_{min}(h_i, g) + \tfrac{1}{2}\Delta^\gamma_{mm}(h_i, g)$$
$$= \frac{1 + \gamma}{2} \cdot \Delta_{min}(h_i, g) + \frac{1 - \gamma}{2} \cdot \Delta_{max}(h_i, g).$$

Many definitions in this section presuppose that $\mathbf{B} = \{h_i | i \in I\}$ is a finite problem. When $|I| = \omega$, we could make similar modifications as in Section 6.3, but for continuous cognitive problems there are special troubles — created by sets of measure zero. If we try to directly generalize the definition (115) of weighted symmetric difference by replacing the sum with an integral, then δ_{ms} and all its modifications lead to the following result: when $\mathbf{H}_1$ and $\mathbf{H}_2$ are two subsets of the continuous space $\mathbf{Q} \subseteq \mathbb{R}^k$, and $\mathbf{H}_1 \triangle \mathbf{H}_2$ has the measure zero, then the distance between $\mathbf{H}_1$ and $\mathbf{H}_2$ is zero. This will hold, e.g., when $\mathbf{H}_1$ and $\mathbf{H}_2$ are two countable unions of lines in $\mathbb{R}^2$. Further, if $\mathbf{H}_1$ and $\mathbf{H}_2$ are points in $\mathbf{Q}$, their distance is zero. Thus, both conditions (103) and (104) are violated by δ_{nms} in the continuous case.

One possible way out would be to stipulate that sets of measure zero are simply ignored in continuous cognitive problems. This choice — in spite of its mathematical elegance — is not desirable for our purposes, since we are interested in many sharp statements — at least in the point hypotheses which, as the elements of space $\mathbf{Q}$, define the relevant cognitive problem $\mathbf{B}$. Therefore, we should find a measure for the distance between statements which works for a reasonable choice of sharp and non-sharp hypotheses in the class $D(\mathbf{B})$.

Assume now that $\mathbf{Q} \subseteq \mathbb{R}^k$ and Δ is the Euclidean distance on $\mathbf{Q}$. Let $\mathbf{H}_1$ and $\mathbf{H}_2$ be measurable sets in $\mathbf{Q}$. Then the distance between $\mathbf{H}_1$ and $\mathbf{H}_2$ cannot be simply

$$\alpha \int_{\mathbf{H}_1} \Delta_{\min}(Q, \mathbf{H}_2) \, dQ + \alpha' \int_{\mathbf{H}_2} \Delta_{\min}(Q, \mathbf{H}_1) \, dQ$$

(cf. (121)). This quantity has to be supplemented by a factor which is independent of the geometrical size of the symmetric difference $\mathbf{H}_1 \triangle \mathbf{H}_2$. The choice of this factor seems to depend on the cognitive problem. For example, if $\mathbf{Q} = \mathbb{R}$ and the only sharp hypotheses in $D(\mathbf{B})$ are points of $\mathbb{R}$, then it would be sufficient to add

$$\Delta_{\min}(\mathbf{H}_1, \mathbf{H}_2) = \min_{\substack{Q \in \mathbf{H}_1 \\ Q' \in \mathbf{H}_2}} \Delta(Q, Q').$$

However, if $\mathbf{Q} = \mathbb{R}^2$ and $D(\mathbf{B})$ includes lines in $\mathbb{R}^2$, then $\Delta_{\min}(\mathbf{H}_1, \mathbf{H}_2) = 0$ for any two intersecting lines $\mathbf{H}_1$ and $\mathbf{H}_2$ in $\mathbb{R}^2$. Therefore, the added factor should reflect not only the minimum distance but also the maximum distance between $\mathbf{H}_1$ and $\mathbf{H}_2$. In the general case, we may thus choose the function δ^γ_{mmm} defined by (131). Let us define the distance between $\mathbf{H}_1$ and $\mathbf{H}_2$ by

(132) $\delta^{\gamma aa'}(\mathbf{H}_1, \mathbf{H}_2) = \delta^{\gamma}_{mmm}(\mathbf{H}_1, \mathbf{H}_2) + \delta^{aa'}_{mm}(\mathbf{H}_1, \mathbf{H}_2)$

$= \gamma \Delta_{\min}(\mathbf{H}_1, \mathbf{H}_2) + \dfrac{1-\gamma}{2} \max_{Q \in \mathbf{H}_1} \Delta_{\min}(Q, \mathbf{H}_2) +$

$+ \dfrac{1-\gamma}{2} \max_{Q \in \mathbf{H}_2} \Delta_{\min}(Q, \mathbf{H}_1) +$

$+ \alpha \int_{\mathbf{H}_1} \Delta_{\min}(Q, \mathbf{H}_2) \, dQ +$

$+ \alpha' \int_{\mathbf{H}_2} \Delta_{\min}(Q, \mathbf{H}_1) \, dQ,$

where $0 < \gamma < 1$ and $\alpha > 0$, $\alpha' > 0$. Hence, when $\mathbf{H}_1$ is a point hypothesis Q,

(133) $\delta^{\gamma aa'}(Q, \mathbf{H}_2) = \dfrac{\gamma+1}{2} \Delta_{\min}(Q, \mathbf{H}_2) + \dfrac{1-\gamma}{2} \Delta_{\max}(Q, \mathbf{H}_2) +$

$+ \alpha' \int_{\mathbf{H}_2} \Delta(Q', Q) \, dQ'.$

When $\mathbf{H}_2$ is a point hypothesis Q' as well, we obtain

$\delta^{\gamma aa'}(Q, Q') = \Delta(Q, Q'),$

so that conditions (104) and (105) are satisfied. For the special case, where $\gamma = 1$, we have

$\delta^{1 aa'}(\mathbf{H}_1, \mathbf{H}_2) = \Delta_{\min}(\mathbf{H}_1, \mathbf{H}_2) +$

$+ \alpha \int_{\mathbf{H}_1} \Delta_{\min}(Q, \mathbf{H}_2) \, dQ +$

$+ \alpha' \int_{\mathbf{H}_2} \Delta_{\min}(Q, \mathbf{H}_1) \, dQ$

$\delta^{1 aa'}(Q, \mathbf{H}_2) = \Delta_{\min}(Q, \mathbf{H}_2) + \alpha' \int_{\mathbf{H}_2} \Delta(Q', Q) \, dQ'$

$\delta^{1 aa'}(Q, Q') = \Delta(Q, Q').$

This shows that $\delta^{1aa'}(Q, \mathbf{H}_2)$ is essentially the same as our non-normalized min-sum-measure $\Delta_{ms}(Q, \mathbf{H}_2)$ defined by (45), i.e., a weighted combination of a Δ_{min}-factor and a non-normalized Δ_{sum}-factor.

6.8. DISTANCE FROM INDEFINITE TRUTH

So far we have assumed that the cognitive problem **B** is formulated in a semantically determinate language L, so that the actual world is represented by a single L-structure Ω_L^*. If language L is semantically indeterminate, then the actual world has to be represented either by (i) a class Θ_L^* of L-structures, or by (ii) a fuzzy set $K_*: M \to [0, 1]$ of L-structures (see Chapter 4.4). This section shows how the concept of truthlikeness can be extended to the case of *indefinite truth*.

Assume first that the actual world relative to L corresponds to a class Θ_L^* of L-structures. Then a statement h in L is *true* if and only if $\text{Th}(\Theta_L^*) \vdash h$, where

$$\text{Th}(\Theta_L^*) = \bigcap_{\Omega \in \Theta_L^*} \text{Th}(\Omega)$$

(see (4.43)). If each theory $\text{Th}(\Omega)$ is finitely axiomatizable by some sentence A_Ω in L, then $\text{Th}(\Theta_L^*)$ is axiomatizable by the disjunction of these statements:

$$\text{Th}(\Theta_L^*) = \text{Cn}\left(\bigvee_{\Omega \in \Theta_L^*} A_\Omega\right).$$

(Cf. Chapter 2.9.) This disjunction, if it is expressible in L, is therefore the strongest true statement in L, and it is natural to take it as the *target* of a cognitive problem.

More generally, let $\mathbf{B} = \{h_i | i \in I\}$ be a P-set in an indeterminate language L. Then the strongest (and most informative) true statement is not an element of **B** but rather a disjunction of some elements h_i, $i \in I_*$, of **B**. Let us denote this disjunction by H_*:

$$H_* = \bigvee_{i \in I_*} h_i.$$

If now H_* is chosen as the target, then the problem of truthlikeness reduces to the following question: what is the distance of a statement $g \in D(\mathbf{B})$ from H_*? As H_* itself is an element of $D(\mathbf{B})$, this problem was solved already in Section 7. Let δ be the weighted symmetric difference measure (121), or its generalization (132), for the distance between two elements of $D(\mathbf{B})$. Then, relative to the target $H_* \in D(\mathbf{B})$, we stipulate that $\delta(H_*, g)$ is the *distance of g from the truth*, and

(134) $\quad \text{Tr}(g, H_*) = 1 - \delta(H_*, g)$

is the *degree of truthlikeness* of g (cf. Niiniluoto, 1986b). This proposal has the nice feature that it contains as a special case our earlier treatment of truthlikeness, where H_* contains only one disjunct h_*.

It can immediately be seen that definition (134) shares the nice properties of our earlier notion of truthlikeness (see the list M1—M13 in Section 6). Assume $\delta = \delta_{nms}^{aa'}$. Then, in particular, for a true statement g,

$$\text{Tr}(g, H_*) = 1 - \frac{a'}{|I|} \sum_{j \in I_g - I_*} \Delta_{\min}(h_j, H_*).$$

Among true statements, the most truthlike is H_* itself, i.e.,

$$\text{Tr}(H_*, H_*) = 1,$$

while tautologies are least truthlike. Further, if $j \notin I_*$, then

$$\text{Tr}(h_j \vee H_*, H_*) = 1 - \frac{a'}{|I|} \Delta_{\min}(h_j, H_*),$$

which increases when h_j comes closer to the set I_*. Again, the truth content $h_j \vee H_*$ of h_j ($j \notin I_*$) is more truthlike than h_j itself, since

$$\text{Tr}(h_j, H_*) = 1 - \frac{a}{|I|} \sum_{i \in I_*} \Delta_{ij} - \frac{a'}{|I|} \Delta_{\min}(h_j, H_*).$$

If the distance Δ on $\mathbf{B}$ is symmetric and balanced, so that $\text{av}(i, \mathbf{B}) = \frac{1}{2}$, then we have also

$$\text{Tr}(h_j, H_*) = 1 - \Delta_{ms}^{\frac{a'}{2} \frac{a}{|I|}} (h_j, H_*).$$

A feature that distinguishes indefinite truth from the case of definite truth is the possibility of having only 'partly true' statements which are not true but still entail the target H_*. If g is such a statement which is too sharp to be true, so that $\mathbf{I}_g \subseteq I_*$, then

$$\mathrm{Tr}(g, H_*) = 1 - \frac{\alpha}{|I|} \sum_{j \in I_* - \mathbf{1}_g} \Delta_{\min}(h_j, g).$$

This value can be increased by making g logically weaker. In particular, if $i \in I_*$, then

$$\mathrm{Tr}(h_i, H_*) = 1 - \frac{\alpha}{|I|} \sum_{\substack{j \in I_* \\ j \neq i}} \Delta_{ji}.$$

This value is maximized by choosing h_i so that i is located at the "center of gravity" of the set I_*.[11]

Another possibility of treating indefinite truth is to replace the structure Ω_L^* by a fuzzy set of structures, i.e., by a function $K_*: \mathbf{M} \to [0, 1]$. In this case, it is natural to allow that the alternative hypotheses are also fuzzy sets over $\mathbf{M}$. If $K: \mathbf{M} \to [0, 1]$ corresponds to such a hypothesis, one possibility of defining its distance from K_* is given by the Hamming distance

$$(135) \quad \sum_{\Omega \in \mathbf{M}} |K(\Omega) - K_*(\Omega)|$$

(cf. Example 1.7). The value of (135) can be normalized, and it may be replaced by an integral when $\mathbf{M}$ is an infinite space. This proposal is not satisfactory, since it leads to Miller's definition (5.34) in the special case, where K and K_* are ordinary sets of structures. In other words, definition (135) is not able to reflect the underlying metric structure Δ of the space $\mathbf{M}$ of L-structures.[12]

As an alternative to (135), we may suggest that a fuzzy set $K: \mathbf{M} \to [0, 1]$ is replaced by the ordinary set K^q of those elements $\Omega \in \mathbf{M}$ which have a sufficiently great ($\geq q$) value by K:

$$K^q = \{\Omega \in \mathbf{M} \mid K(\Omega) \geq q\},$$

where the treshold value q satisfies $\frac{1}{2} \leq q \leq 1$. Then the distance between the fuzzy sets K and K_* can be defined by using the distance function $\delta = \delta_{\mathrm{nms}}^{aa'}$ as above. In other words, the degree of truthlikeness of K relative to the indefinite truth K_* is

$$(136) \quad \mathrm{Tr}(K, K_*) = 1 - \delta(\mathrm{Th}(K^q), \mathrm{Th}(K_*^q)).$$

If K and K_* are ordinary sets, definition (136) reduces to our earlier suggestion (134).

6.9. COGNITIVE PROBLEMS WITH FALSE PRESUPPOSITIONS

In the preceding sections, it has been assumed that the cognitive problem **B** — or, as in Section 8, its disjunctive closure $D(\mathbf{B})$ — contains a true element. This assumption covers the cases where **B** satisfies the conditions for a P-set relative to an empty or a true presupposition b. However, if the presupposition b is false, then it may happen that all the elements of **B** are likewise false. It is clear that the theory of truthlikeness should be applicable to this situation as well. In this section, we shall show how the case of false presuppositions can be reduced to our earlier treatment.

To illustrate these ideas, assume that a scientist is working within a Kuhnian 'disciplinary matrix' or a Lakatosian 'research programme'. This means that he accepts some theory or theoretical assumptions, and allows this 'background knowledge' to generate and to define his research problems. Some philosophers would go so far as to claim that all cognitive problems in science are relative to such 'paradigmatic' assumptions. However, the background knowledge of the scientists is never incorrigible, and it may quite well be false.

In some cases, it is even *known* that the working hypothesis of our inquiry is false. Popper himself says that

ultimately, the idea of verisimilitude is most important in cases where we know that we have to work with theories which are *at best* approximations — that is to say, theories of which we actually know that they cannot be true. (Popper, 1963, p. 235.)

This is the case with laws and theories which are known to be idealized. For example, if we are interested in the functional relation between two quantities $g_1: E \to \mathbb{R}$ and $g_2: E \to \mathbb{R}$, we may start with the false but simplifying assumption that there is a linear function $f_*: \mathbb{R} \to \mathbb{R}$ such that $g_2(x) = f_*(g_1(x))$ for all $x \in E$. The cognitive problem $\mathbf{B}_b$ relative to this presupposition b is then the task of finding the 'best' element in the set of linear functions:

$$\mathbf{B}_b = \{f: \mathbb{R} \to \mathbb{R} \mid (\exists a \in \mathbb{R})(\exists b \in \mathbb{R})(z \in \mathbb{R}) \\ (f(z) = az + b)\}.$$

As b is false, no element of $\mathbf{B}_b$ gives a correct solution to the problem

(137) What is the *true* functional relation between the quantities g_1 and g_2?

Instead of (137), the assumption b leads us to the following question:

(138) Which of the elements of $\mathbf{B}_b$ is the *least false* answer to the original problem (137)?

Let us now assume that $\mathbf{B}_b = \{h_i | i \in I\}$ is a cognitive problem relative to b, where b and all statements $h_i \in \mathbf{B}$, $i \in I$, are false. How can we order the elements of $D(\mathbf{B})$ with respect to their truthlikeness? To do this, we should need a target — but the earlier definition of the target h_* as "the most informative true statement in $\mathbf{B}$" (cf. Section 2) does not help us here, since no element of $D(\mathbf{B})$ is true.

A natural solution to our problem can be given as follows. Let L be the relevant language where b and h_i's are expressed, and let Ω_L^* be the actual world relative to L. Language L is again assumed to be semantically determinate. It may be richer than the language of the statements h_i. (For example, b may falsely state that some factors are not relevant, and therefore these factors do not appear in the statements h_i.) Let Δ be a metric on the space $\mathbf{M}_L$ of L-structures (cf. Chapter 10.3), and Θ_*^b be the class of L-structures $\Omega \in \mathbf{M}_L$ which are models of b and minimally distant from Ω_L^* (see Figure 16):

$$\Theta_*^b = \{\Omega \in \mathbf{M}_L | \Omega \vDash b \text{ and } \Delta(\Omega_L^*, \Omega) \leq \Delta(\Omega_L^*, \Omega') \text{ for all } \Omega' \in \mathbf{M}_L \text{ such that } \Omega' \vDash b\}.$$

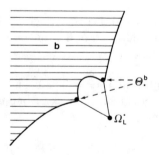

Fig. 16. The Δ-closest b-worlds to Ω_L^*.

THE SIMILARITY APPROACH TO TRUTHLIKENESS 261

For each model Ω of b, there is one and only one element $h_*(\Omega)$ of $\mathbf{B}_b$ which is true in Ω. This follows directly from the conditions

$$b \vdash \bigvee_{i \in I} h_i$$

$$b \vdash \sim(h_i \& h_j) \quad (i \neq j).$$

If $\mathbf{\Theta}_*^b$ contains only one structure Ω, we (counterfactually) pretend that Ω is the actual world, and define the *target* of $\mathbf{B}_b$ as the statement $h_*(\Omega)$. If $\mathbf{\Theta}_*^b$ contains several structures, we choose as the target the disjunction

$$(139) \quad \bigvee_{\Omega \in \mathbf{\Theta}_*^b} h_*(\Omega).$$

The degree of truthlikeness of an arbitrary statement $g \in D(\mathbf{B}_b)$ can then be defined, as in Sections 5 and 8 above, by means of the distance of g from the target.[13]

This method gives us a comparative notion of truthlikeness within the set $\mathbf{B}_b$ of false statements. The difference to our earlier treatment is illustrated in Figure 17 which should be compared with Figure 4 of Section 2. Here the target $h_*(\Omega)$ itself is not true in the actual world Ω_L^*, but still it is the most truthlike element in $\mathbf{B}_b$, since it is the most informative statement which is true in the b-world closest to Ω_L^*.

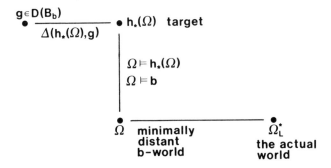

Fig. 17. Truthlikeness with a false presupposition b.

The proposal of this section can be expressed in a very perspicious way by noting that the target (139) would be true if b were true, i.e.,

$$(140) \quad b \mathbin{\Box\!\!\rightarrow} \bigvee_{\Omega \in \Theta_*^b} h_*(\Omega)$$

holds, when $\Box\!\!\rightarrow$ is the counterfactual if-then-operator of D. Lewis (1971). The chosen target is moreover the strongest of statements satisfying (140). Hence, the target for the problem $\mathbf{B}_b$ with false presupposition b is *the most informative statement* in $D(\mathbf{B}_b)$ which *would be true if b were true*.

CHAPTER 7

ESTIMATION OF TRUTHLIKENESS

The similarity approach, as developed in Chapter 6, gives a solution to the *logical* problem of truthlikeness. Just as the Tarskian or model-theoretical definition characterizes the meaning of the concept of truth by defining the truth or falsity of a given statement g in a *given* structure Ω, we have characterized the meaning of the concept of truthlikeness by defining the distance of a given statement g from a *given* target h_*.

In this chapter, we shall show how the similarity approach helps us to solve also the *epistemic* problem of truthlikeness. This problem concerns the conditions for rationally claiming, on some evidence, that a statement g is truthlike — or at least is more truthlike than some other statement — *even when the truth h_* is unknown*. Apart from some trivial cases, appraisals of the relative distances from the truth presuppose that an epistemic probability distribution on a cognitive problem B is available. In this sense, it will be argued, the problem of estimating verisimilitude is neither more nor less difficult than the traditional problem of induction.

7.1. THE EPISTEMIC PROBLEM OF TRUTHLIKENESS

The preceding chapter showed how to define the degree of truthlikeness $Tr(g, h_*)$ of a statement g relative to the given target h_*. This solution to the logical problem of verisimilitude shows that it is meaningful to say that

(1) g_1 is more truthlike than g_2,

even when g_1 and g_2 are false statements. This result is in itself an important achievement, since it guarantees — just as Popper originally intended — that talk about 'nearness' and 'approach' to the truth is perfectly significant rather than "just so much mumbo-jumbo" (Laudan, 1981, p. 32). Hence, it gives to a critical scientific realist the possibility of systematically developing and defending a theory of *cognitive progress* in terms of increasing truthlikeness (see Niiniluoto, 1984b).

While Popper has repeatedly emphasized that truthlikeness is not an 'epistemological' or 'epistemic' idea, he also suggests that "we *can* have strong and reasonably good arguments for claiming that we may have made progress toward the truth" (Popper, 1972, p. 58).

> But some of us (for example Einstein himself) sometimes wish to say such things as that we have reason to conjecture that Einstein's theory of gravity is *not true*, but that it is a *better approximation to truth* than Newton's. To be able to say such things with a good conscience seems to me a major desideratum in the methodology of the natural sciences. (*Ibid.*, p. 335.)

In *Conjectures and Refutations*, Popper gives a list of six types of cases where "we should be inclined to say" that a theory T_1 "seems — as far as we know — to correspond better to the facts" than another theory T_2 (Popper, 1963, p. 232). All of these cases have the common feature that the "empirical content" of T_1 exceeds that of T_2: given the class e of the relevant 'basic statements', i.e., empirical test statements, theory T_1 describes or explains more facts in e than T_2, or T_1 explains these facts more precisely or in more detail than T_2. While appraisals of comparative verisimilitude on some evidence are always tentative and conjectural, they are for Popper "a major desideratum" in a reasonable account of verisimilitude. In other words, in addition to statements of type (1), the theory of verisimilitude should give an analysis of judgements of the following type:

(2)   Given evidence e, it is rational to claim (or conjecture) that g_1 is more truthlike than g_2.

Popper's own solution to the epistemic problem of truthlikeness is based on the suggestion that his "degrees of corroboration" could serve as fallible indicators of verisimilitude:

> The degree of corroboration of a theory has always a temporal index: it is the degree to which the theory appears well tested at the time t. Although it is not a measure of its verisimilitude, it can be taken as an indication of how its verisimilitude *appears* at the time t, compared with another theory. Thus the degree of corroboration is a guide to the preference between two theories at a certain stage of the discussion with respect to their then apparent approximation to truth. But it only tells us that one of the theories offered *seems — in the light of the discussion —* the one nearer to truth (Popper, 1972, p. 103.)

It is clear, however, that Popper's own concept of corroboration fails in many cases to do this job. The *degree of corroboration* of a theory g

relative to evidence e is defined by the formula

(3) $$C(g, e) = \frac{P(e/g) - P(e)}{P(e/g) - P(e \& g) + P(e)}$$

(see Popper, 1959, Appendix *ix; Popper, 1963, p. 288). Hence, if evidence e refutes g, then $C(g, e)$ takes its minimum value -1. But this is not desirable, if $C(g, e)$ is intended to serve as an indicator of the degree of truthlikeness of g. If evidence e contradicts the hypothesis g, then of course the posterior probability $P(g/e)$ of g given e will have its minimum value 0, but the same principle should not hold for verisimilitude: g may be highly truthlike, even if it is known to be false. If g_1 and g_2 are both refuted by e, then $C(g_1, e) = -1$ and $C(g_2, e) = -1$, but still one of these hypotheses may be much more truthlike than another. Furthermore, as some false statements may be more truthlike than some true ones, it should be possible that, given evidence e, some refuted hypotheses appear to be more verisimilar than some unrefuted ones.

We may thus conclude that Popper has not succeeded in giving an adequate solution to the problem about judgements of type (2). In her paper for the 1975 conference at Kronberg, Noretta Koertge also noted that it is not satisfactory to "award Newton's theory and flat earth theory" the same degree of corroboration "because they are both refuted". She urged that "a good theory of the estimate of the degree of verisimilitude of a theory would be of great value to philosophers of science of all persuasion" (Koertge, 1978, p. 276).

Any attempt to estimate degrees of verisimilitude has to face a serious problem raised by *Augustine's objection*: to judge that a son resembles his father presupposes acquaintance with the father; similarly, to judge that a theory resembles the truth presupposes that the truth is already known (see Chapter 5.2). And, we may add, if we already know the true theory, why should we be any more interested in the merely truthlike ones? The whole concept of truthlikeness seems thus to be useless for any epistemological purposes.[1] A clear formulation of this classical argument was presented by F. C. S. Schiller in *Logic for Use* (1929):

We cannot assume that the road to truth runs in a straight line, that we are approaching truth at every step, and that every apparent approach is really on the road to success. ... One might as well assume that the right route up a mountain must always be one that goes straight for the summit, whereas it may only lead to the foot of an unclimbable cliff, and the true *route* may lead a long way round up a lateral *arête*. This simple consideration really disposes of the assumption that we can declare one theory truer than another, in the sense of coming nearer to absolute truth, without having previously reached the latter; and the history of the sciences fully confirms this inference by furnishing many examples of theories which have long seemed all but completely true and have then had to be discarded, while others which looked quite unpromising have in the end proved far more valuable.[2]

There is one way of trying to get clear of Augustine's objection. Even when we don't know the father, it may be a good guess for us that his son resembles him — for the simple reason that in most families sons resemble their fathers. Moreover, this is not an accidental fact, since it results from the lawful nature of the generating mechanism which transmits 'genetic information' from parents to their off-springs. Perhaps it could be argued, on similar grounds, that it is for us a good guess to claim that the results of scientific inquiry resemble the truth. Perhaps the 'mechanism' which generates scientific statements lawfully guarantees that they must be close to the truth.[3]

Unfortunately, this argument does not work: at least from our viewpoint, it puts the carriage before the horses. The generalization that many sons resemble their fathers is based on the fact that we are able to study its instances: there is a method for evaluating the similarity between persons, and we have had the opportunity of employing it for many pairs of sons and fathers.[4] In order to defend the claim that many results of scientific investigation must resemble the truth, or that science has a great propensity on generating truthlike theories, we have to be able to verify this claim in particular cases. However, there does not seem to be any way of doing this without begging the question about Augustine's objection — unless we dogmatically assume that science has in fact reached the truth in many cases, and we know it.

However, the testimony provided by the history of science is by no means unambiguous. As Schiller points out, the road to truth cannot be assumed to be a 'straight line'. Thomas Kuhn rejects the idea that "successive theories grow ever closer to, or approximate more and more closely to, the truth" by claiming that he cannot see any "coherent direction of ontological development" in the mechanical theories of Aristotle, Newton, and Einstein (Kuhn, 1970, p. 206). Hilary Putnam has seriously worried about the question of how to block the following 'meta-induction' from the history of science: just as the theoretical terms of past science did not refer, so it will turn out that no theoretical term used now refers (Putnam, 1978, p. 25). (See also Newton—Smith, 1981.) Larry Laudan (1981, 1984) likewise reminds us that some of the most successful theories in the history of science (e.g., the ether theories in the nineteenth century) were "by present lights non-referring", and concludes that the 'upward' inference from the empirical success of a theory to its 'approximate truth' is unwarranted. Laudan also rejects the suggestion that the approximate truth of the best theories can be abductively inferred as the 'best explanation' of the continuing success of science (cf. Putnam, 1978; Niiniluoto, 1984b, p. 51), since it is not at all clear what explanatory 'downward' inference leads from the approximate truth of a theory to the success of its predictions.

The force of these objections to scientific realism is hard to evaluate, since they seem to presuppose something that they themselves are challenging. To claim that science is *not* approaching towards the truth presupposes the existence of viable conceptions of convergence and distance for theories: something like our function Tr is needed to evaluate the 'direction' of theory sequences in the history of science. To study the relation between an approximately true theory and its deductive predictions, a precise definition for 'approximate truth' and 'successful prediction' are needed. To urge that some of the past theories were 'by present lights' non-referring presupposes either the validity of induction, applied at the level of meta-induction, or the validity of abductive inference to our present theories (i.e., to theories in the light of which we regard their predecessors to be non-referring). Moreover, to show that our best theories are false does not refute scientific realism: in contrast, this is precisely what a convergent realist who is also a strong fallibilist claims (cf. Chapter 5.1).

On the other hand, realism without sufficiently precise conceptual tools for discussing 'the nearness of truth' is not adequate — and has deserved much of the criticism from its opponents. Thus, the treatment of the logical concept of truthlikeness in Chapter 6 has not completed our task. It seems to me that the only satisfactory way of answering the epistemological challenges to realism is to give a solution to Augustine's objection, i.e., to find a method for tentatively estimating degrees of truthlikeness on some evidence e.

It is important to emphasize that, as Popper observes, the indicators of verisimilitude are always conjectural and corrigible. The evidence e

itself may turn out to be incorrect, and e may also be incomplete and misleading in the sense that further evidence forces us to revise our appraisals. Schiller is right in thinking that *infallible* comparisons of truthlikeness presuppose that 'absolute truth' (or target h_*) has previously been reached. Augustine is right in thinking that appraisal of verisimilitude is impossible if we are *completely* ignorant about the truth. My proposal avoids these extremes by showing how the epistemic probabilities of the rival hypotheses help us to construct a fallible and revisable indicator of their verisimilitude.[5]

7.2. ESTIMATED DEGREES OF TRUTHLIKENESS

Miller (1975) complained that "possible-world approaches like that of Hilpinen threaten not to be empiricist enough", since "there is no indication how experience could influence judgements of truthlikeness at all". However, in my linguistic formulation of the similarity approach in the 1975 Conference in London, Ontario, I showed that it is in fact quite easy to supplement the 'semantical' definition with an epistemic method for estimating degrees of truthlikeness on the basis of empirical and theoretical evidence (see Niiniluoto, 1977b, Section 6).

Let $\mathbf{x}$ be a random variable which takes the values $x_i \in \mathbb{R}$ ($i = 1, \ldots, k$) with the physical probabilities p_i ($i = 1, \ldots, k$), respectively. Then the *expected value* of $\mathbf{x}$ is defined by

$$(4) \quad \sum_{i=1}^{k} p_i x_i.$$

For example, if $\mathbf{x}$ takes the values 1 ('tails') and 0 ('heads') with the probabilities .75 and .25, then the expected value of $\mathbf{x}$ is .75.

The concept of expectation can be defined also with respect to epistemic probabilities. Let $x_* \in \mathbb{R}$ be the unknown value of a quantity, and let x_i, $i \in I$, be the possible values that x_* could have. Assume that there is, for each $i \in I$, a probability $P(x_i/e) \geq 0$ given evidence e that x_i is the correct value x_*. Then the expected value of x_* on the basis of e and P is

$$(5) \quad \sum_{i \in I} P(x_i/e) x_i.$$

If the probabilities $P(x_i/e)$ express rational degrees of belief in the claim

$x_* = x_i$, then the value of (5) is also a rational *estimate* of the unknown x_* on evidence e (cf. Carnap, 1962, Ch. IX).

Assume now that $\mathbf{B} = \{h_i \mid i \in I\}$ is a discrete cognitive problem with an unknown target h_*, and let $g \in D(\mathbf{B})$ be a partial answer. If e expresses the available evidence relevant to problem $\mathbf{B}$, what is the most reasonable estimate of the unknown degree of truthlikeness $\mathrm{Tr}(g, h_*)$ of g? Following the general idea explained above, note that the possible values of $\mathrm{Tr}(g, h_*)$ are $\mathrm{Tr}(g, h_i)$, $i \in I$. Let P be an epistemic regular probability measure defined on $\mathbf{B}$, so that

$$\sum_{i \in I} P(h_i) = 1$$

$$P(h_i \ \& \ h_j) = 0 \quad \text{if } i \neq j$$

$$P(g/e) = 1 \quad \text{if } e \vdash g.$$

Then $\mathrm{Tr}(g, h_*)$ takes the values $\mathrm{Tr}(g, h_i)$ with the probabilities $P(h_i/e)$ ($i \in I$). Hence, the *expected value* of the unknown degree $\mathrm{Tr}(g, h_*)$ on the basis of evidence e and relative to the probability measure P is

(6) $\quad \mathrm{ver}(g/e) = \sum_{i \in I} P(h_i/e) \mathrm{Tr}(g, h_i).$

The value of $\mathrm{ver}(g/e)$ can then be taken to be the *estimated degree of truthlikeness* of g on the basis of e.

If $\mathbf{B}$ is a continuous space, then measure P in (6) has to be replaced by a density function $p: \mathbf{B} \to \mathbb{R}$ and the sum operator Σ by the integral:

(7) $\quad \mathrm{ver}(g/e) = \int_{\mathbf{B}} p(x/e) \mathrm{Tr}(g, x) \, \mathrm{d}x.$

For simplicity, the results of this chapter are stated for discrete sets $\mathbf{B}$, but similar conclusions hold for (7). (See also Chapter 12.5.)

A direct way of motivating the definition (6) is the following: $\mathrm{ver}(g/e)$ is the *weighted average* of the possible values of $\mathrm{Tr}(g, h_*)$, where each of these values $\mathrm{Tr}(g, h_i)$ is weighted with the degree of belief $P(h_i/e)$ on e that h_i is true. In the same way, if $\Delta(h_i, g)$ measures the distance of $g \in D(\mathbf{B})$ from $h_i \in \mathbf{B}$, then the *expected distance of g from the truth*

h_* is definable by

(8) $\quad \text{Exp}_e \Delta(h_*, g) = \sum_{i \in I} P(h_i/e) \Delta(h_i, g).$

When $M(g, h_i)$ is defined as $1 - \Delta(h_i, g)$, where Δ is the min-sum distance $\Delta_{ms}^{\gamma\gamma'}$, then formulas (6) and (8) imply that

(9) $\quad \text{ver}(g/e) = 1 - \text{Exp}_e \Delta(h_*, g),$

since P has to satisfy

$$\sum_{i \in I} P(h_i/e) = 1.$$

Function ver: $D(\mathbf{B}) \rightarrow [0, 1]$ gives us an explicate for claims about comparative verisimilitude of the form (2):

(10) $\quad g_1$ seems to be more truthlike on evidence e than g_2 iff $\text{ver}(g_1/e) > \text{ver}(g_2/e).$

In other words, ver gives a solution to the epistemic problem of truthlikeness.[6]

Function ver gives us also a systematic tool for studying what may be called *verisimilitude kinematics* — in analogy with 'probability kinematics' (cf. Jeffrey, 1965). Assume that the degree of truthlikeness of $g \in D(\mathbf{B})$ is estimated, on evidence e, to be $\text{ver}(g/e)$. When some new evidence e' is obtained, the new estimated degree on the combined evidence $e \& e'$ is

$$\text{ver}(g/e \& e') = \sum_{i \in I} P(h_i/e \& e') \text{Tr}(g, h_i).$$

Changes in the estimated verisimilitude of a hypothesis g are thus based upon the changes of the whole probability distribution P on $\mathbf{B}$: the new probabilities $P(h_i/e \& e')$ are obtained from the old ones $P(h_i/e)$ by Bayesian conditionalization.[7] The conditions for evidence e' to *increase the estimated verisimilitude* of g relative to e (i.e., for $\text{ver}(g/e) < \text{ver}(g/e \& e')$) are thus clearly different from the conditions for e' to *confirm* g relative to e (i.e., for $P(g/e) < P(g/e \& e')$).

ESTIMATION OF TRUTHLIKENESS

Let us now study some of the general properties of the measure ver for finite cognitive problems **B**. By (6.87), we obtain:

(11) $\text{ver}(g/e) = 1 - \gamma'$, if g is a tautology.

Thus, for a tautology g, the value of $\text{ver}(g/e)$ is the same for any evidence e. Again by (6.87), the elements h_j of **B** satisfy

(12) $\text{ver}(h_j/e) = 1 - \sum_{i \in I} P(h_i/e) \left(\gamma + \dfrac{\gamma'}{|I| \text{av}(i, \mathbf{B})} \right) \Delta_{ij}.$

In particular,

If $\text{av}(i, \mathbf{B}) = 1/2$ for all $i \in I$,

then $\text{ver}(h_j/e) = 1 - \left(\gamma + \dfrac{2\gamma'}{|I|} \right) \sum_{i \in I} P(h_i/e) \Delta_{ij}.$

Formula (12) shows that $\text{ver}(h_j/e) = 1$ if and only if P concentrates the whole probability mass on that element h_i of **B** which satisfies $\Delta_{ij} = 0$, i.e.,

$\text{ver}(h_j/e) = 1$ iff $P(h_j/e) = 1.$

By (8) and (9), the same result holds more generally for all $g \in D(\mathbf{B})$, since $\Delta_{\text{ms}}(h_i, g) = 0$ if and only if $g = h_i$:

(13) $\text{ver}(g/e) = 1$ iff $g = h_i$ and $P(h_i/e) = 1.$

Result (12) shows that $\text{ver}(h_j/e)$ increases, when the probability P becomes more and more concentrated around those elements $h_i \in \mathbf{B}$ which are close to h_j:

(14) $\text{ver}(h_j/e) < \text{ver}(h_m/e)$ iff $\sum_{i \in I} P(h_i/e) \Delta_{ij} > \sum_{i \in I} P(h_i/e) \Delta_{im}.$

Further, (12) entails that all the values $\text{ver}(h_j/e)$ are equal, if the probabilities $P(h_i/e)$ are equal to each other:

(15) If $\text{av}(i, \mathbf{B}) = 1/2$ for all $i \in I$, and if $P(h_i/e) = 1/|I|$

for all $i \in I$, then $\text{ver}(h_j/e) = 1 - \left(\dfrac{\gamma}{2} + \dfrac{\gamma'}{|I|} \right).$

Thus, when $(\mathbf{B}, \Delta)$ is balanced, and P given e is a uniform probability distribution on $\mathbf{B}$, function ver cannot make any difference between the complete answers $h_j \in \mathbf{B}$. In this case, we may say that we are *completely ignorant* about the correct answer to $\mathbf{B}$. More generally, on the conditions of (15), we have for any partial answer $g \in D(\mathbf{B})$:

$$\text{Exp}_e \, \Delta(h_*, g) = \frac{1}{|I|} \sum_{i \in I} (\gamma \Delta_{\min}(h_i, g) + \gamma' \Delta_{\text{sum}}(h_i, g))$$

$$= \frac{\gamma}{|I|} \sum_{i \notin \mathbf{I}_g} \Delta_{\min}(h_i, g) + \frac{\gamma'}{|I|} \sum_{i \in I} \frac{2}{|I|} \sum_{j \in \mathbf{I}_g} \Delta_{ij}$$

$$= \frac{\gamma}{|I|} \sum_{i \notin \mathbf{I}_g} \Delta_{\min}(h_i, g) + \frac{2\gamma'}{|I|} \sum_{j \in \mathbf{I}_g} \frac{1}{|I|} \sum_{i \in I} \Delta_{ij}$$

$$= \frac{\gamma}{|I|} \sum_{i \notin \mathbf{I}_g} \Delta_{\min}(h_i, g) + \frac{2\gamma'}{|I|} \sum_{j \in \mathbf{I}_g} \frac{1}{2}$$

$$= \frac{\gamma}{|I|} \sum_{i \notin \mathbf{I}_g} \Delta_{\min}(h_i, g) + \gamma' \cdot \frac{|\mathbf{I}_g|}{|I|}$$

Hence,

(16) If $\text{av}(i, \mathbf{B}) = 1/2$ for all $i \in I$, and if $P(h_i/e) = 1/|I|$ for all $i \in I$, then for $g \in D(\mathbf{B})$

$$\text{ver}(g/e) = 1 - \frac{\gamma}{|I|} \sum_{i \notin \mathbf{I}_g} \Delta_{\min}(h_i, g) - \frac{\gamma'}{|I|} \cdot |\mathbf{I}_g|.$$

As a special case with $g = h_j$, formula (16) reduces to

$$1 - \frac{\gamma}{|I|} \sum_{i \in I} \Delta_{ij} - \frac{\gamma'}{|I|} = 1 - \gamma \cdot \text{av}(i, B) - \frac{\gamma'}{|I|}$$

$$= 1 - \frac{\gamma}{2} - \frac{\gamma'}{|I|},$$

which equals (15). When g is a tautology, (16) reduces to $1 - \gamma'$, which equals (11).

From (11) and (15) it follows that, in the case of complete ignorance, a tautology seems more truthlike than any complete answer $h_j \in \mathbf{B}$ if and only if

$$\frac{\gamma'}{\gamma} < \frac{|I|}{2(|I|-1)},$$

i.e., γ' is sufficiently small relative to γ. (Roughly, $2\gamma' < \gamma$.) Indeed, if γ' is relatively small, then the 'penalty' for low information content in formula (16) is likewise small. This can be seen also from the following calculation. Let $g \in D(\mathbf{B})$ and $j \notin \mathbf{I}_g$. Let J be the set of $h_i \in \mathbf{B}$ such that $i \notin \mathbf{I}_g$ and

$$\Delta_{ij} < \Delta_{\min}(h_i, g).$$

J is thus the set of $h_i \in \mathbf{B}$ outside g which are closer to h_j than to g. Then, for the case of complete ignorance, (16) entails that

$$\text{ver}(g/e) < \text{ver}(g \vee h_j/e) \text{ iff } \frac{\gamma}{\gamma'} < \sum_{i \in J} (\Delta_{\min}(h_i, g) - \Delta_{ij}).$$

Let us next consider the case where evidence e makes us completely *certain* that a particular hypothesis h_j is the true element of $\mathbf{B}$, i.e., $P(h_j/e) = 1$ and $P(h_i/e) = 0$ for $i \neq j$, $i \in I$. As P is a regular probability measure, this will hold if and only if $e \vdash h_j$. Then formula (6) reduces simply to $\text{Tr}(g, h_j)$:

(17) If $P(h_j/e) = 1$ for some $j \in I$, and $P(h_i/e) = 0$ for $i \neq j$, $i \in I$, then $\text{ver}(g/e) = \text{Tr}(g, h_j)$.

On the same conditions,

$$\text{Exp}_e \Delta(h_*, g) = \Delta(h_j, g).$$

In particular, if h_j is certain given e, then

$\text{ver}(h_i/e) = \text{Tr}(h_i, h_j) = 1 - \gamma \Delta_{ji} - \gamma' \Delta_{ji}/(|I| \text{av}(j, \mathbf{B}))$ for all $i \in I$
$\text{ver}(h_j/e) = \text{Tr}(h_j, h_j) = 1$
$\text{ver}(g/e) = 1 - \gamma' \Delta_{\text{sum}}(h_j, g)$, if $j \in \mathbf{I}_g$.

Furthermore, if $\text{av}(j, \mathbf{B}) = 1/2$ for all $j \in I$, we see that

$$\text{ver}(h_j/e) = 1 - \left(\gamma + \frac{2\gamma'}{|I|}\right) \Delta_{ij},$$

i.e., ver(h_j/e) is inversely proportional to the distance Δ_{ij} between h_j and the certain hypothesis h_i. It follows that ver behaves here essentially in the same way as function Tr. For example, the following results for comparative verisimilitude follow from the properties of Tr.

(18) Assume that $P(h_i/e) = 1$ for some $j \in I$.
Then ver(g_1/e) > ver(g_2/e) holds if one of the following conditions obtains:
(i) $g_1 = h_j$ and $g_2 \neq h_j$
(ii) $g_1 = g_2 \lor h_j$ and $j \notin \mathbf{I}_{g_2}$
(iii) $g_1 = h_i$, $g_2 = h_m$, and $\Delta_{ji} < \Delta_{jm}$
(iv) $j \in \mathbf{I}_{g_1}, j \in \mathbf{I}_{g_2}, g_1 \vdash g_2$, and $g_2 \nvdash g_1$.
(v) $j \in \mathbf{I}_{g_1}$, g_1 is not a tautology, and g_2 is a tautology

(Cf. Niiniluoto, 1979b, p. 252.) These results say that, among the answers $g \in D(\mathbf{B})$ entailed by the certain element h_j, the best one is h_j itself, and the worst one is a tautology. According to (ii), if g_2 is certainly false, then its "estimated truth content" $g_2 \lor h_j$ seems more truthlike than g_2.

Result (iv) highlights the difference between estimated verisimilitude and posterior probability, since it assigns a *higher* degree of verisimilitude to the logically *stronger* claim g_1 than to the weaker g_2, while probability behaves in the opposite way: if $g_1 \vdash g_2$, and $g_2 \nvdash g_1$, we have $P(g_1/e) \leq P(g_2/e)$, instead of ver(g_1/e) > ver(g_2/e). This means also that high probability is not sufficient for high estimated truthlikeness: for a tautology g, we have $P(g/e) = 1$ but ver(g/e) = $1 - \gamma'$.

Another crucial difference between probability and estimated verisimilitude follows from (17). Suppose that $i \neq j$, but Δ_{ij} is very small. Then h_i is 'almost equivalent' to the certain answer h_j, and its estimated verisimilitude ver(h_i/e) is very close to one — in spite of the fact that its probability $P(h_i/e))$ is zero. Hence,

(19) It is possible that ver(g/e) ≈ 1 but $P(g/e) = 0$.

This means, in other words, that evidence e may in fact refute a hypothesis g, but still e may indicate that g is close to the truth. Further, some hypotheses g refuted by e may seem more truthlike on e than some hypotheses entailed by e. (This is a counterpart to M10 of Chapter 6.6.)

Theorem (17) can be written also in an approximate form which

ESTIMATION OF TRUTHLIKENESS 275

does not presuppose that h_j is deducible from e:

(20) If $P(h_j/e) \approx 1$ for some $j \in I$, and $P(h_i/e) \approx 0$
for $i \neq j, i \in I$, then
$\mathrm{ver}(g/e) \approx \mathrm{Tr}(g, h_j)$.

Therefore, theorem (18) holds approximately as well. There is also an asymptotic version of these results, if the evidence e_n has a size n which may grow without limit (cf. Chapter 2.10):

(21) If $P(h_j/e_n) \to 1, n \to \infty$, for some $j \in I$, and
$P(h_i/e_n) \to 0, n \to \infty$, for $i \neq j, i \in I$, then
$\mathrm{ver}(g/e_n) \to \mathrm{Tr}(g, h_j), n \to \infty$.

Hence, in particular,

(22) If h_j is asymptotically certain on e_n, then
$\mathrm{ver}(g/e_n) \xrightarrow[n \to \infty]{} 1$ iff $g = h_j$.

A more general asymptotic result holds, when several elements of **B** have a non-zero asymptotic probability:

(23) If $P(h_1/e_n) \to p_1, \ldots, P(h_k/e_n) \to p_k$, and
$P(h_i/e_n) \to 0, n \to \infty$, for $i > k, i \in I$, then
$$\mathrm{ver}(g/e_n) \xrightarrow[n \to \infty]{} \sum_{j=1}^{k} p_j \mathrm{Tr}(g, h_j).$$

In this case, $\mathrm{ver}(g/e_n)$ is a weighted average of the closeness of g to the 'cluster' of statements $\{h_1, \ldots, h_k\}$. If again $\mathrm{av}(i, \mathbf{B}) = 1/2$ for all $i \in I$, the expected verisimilitude of the disjunction $h_1 \vee \cdots \vee h_k$ is asymptotically

(24) $$\mathrm{ver}\left(\bigvee_{j=1}^{k} h_j/e_n\right) \xrightarrow[n \to \infty]{} 1 - \frac{2\gamma'}{|I|} \sum_{i=1}^{k} \sum_{j=1}^{k} p_i \Delta_{ij}.$$

This value is relatively high, if the cluster $\{h_1, \ldots, h_k\}$ is concentrated, so that all the distances Δ_{ij} in (24) are small. If the cluster is balanced, in the sense that the sum

$$\sum_{j=1}^{k} \Delta_{ij} = a \text{ (constant) for all } i = 1, \ldots, k,$$

then the limit (24) reduces to

$$1 - \gamma' \cdot \frac{a}{|I|/2}.$$

For example, if $k = 2$, then

$$\text{ver}(h_1 \vee h_2/e_n) \xrightarrow[n \to \infty]{} 1 - \gamma' \cdot \frac{2\Delta_{12}}{|I|}.$$

If $k = |I|$, then $a = |I|/2$, and the above formula gives the expected verisimilitude $1 - \gamma'$ of a tautology.

As $\text{ver}(g/e)$ is only a fallible estimate of the 'true' degree of truthlikeness of g, there is in general no guarantee that $\text{ver}(g/e) = \text{Tr}(g, h_*)$.[8] This equation will hold, if evidence e entails one of the elements h_j in **B** and e is true, since then h_j must be h_*:

(25) If $e \vdash h_j$ for some $j \in I$, and if e is true,
then $\text{ver}(g/e) = \text{Tr}(g, h_*)$.

(Cf. (17).) When h_j becomes asymptotically certain on e_n ($n \to \infty$), then $\text{ver}(g/e_n)$ approaches the degree $\text{Tr}(g, h_*)$ if and only if $h_j = h_*$. For this condition, it is not sufficient that e_n is true, but e_n has to be 'fully informative' as well: even true evidence can be eventually misleading, if it does not exhibit all the relevant variety of the target domain (cf. Niiniluoto, 1979b, p. 250):

(26) If $P(h_j/e_n) \to 1$, $n \to \infty$, for some $j \in I$, and if e_n is true and fully informative, then $\text{ver}(g/e_n) \to \text{Tr}(g, h_*)$, when $n \to \infty$.

(Cf. Chapter 9.5.)

An important feature of the comparative evaluations of verisimilitude of type (10) is their *revisability*: even if

$$\text{ver}(g_1/e) < \text{ver}(g_2/e),$$

it may quite well happen that

$$\text{ver}(g_1/e \ \& \ e') > \text{ver}(g_2/e \ \& \ e').$$

In this case, the new evidence e' gives us a rational reason for reversing the earlier judgement that g_2 is more truthlike than g_1.

Miller (1980) has complained that the definition (6) of $\text{ver}(g/e)$ bypasses the problem about the *falsifiability* of relative judgements of truthlikeness. As the claim, on some evidence e, that g_2 is more truthlike than g_1 is an empirical conjecture, for a Popperian like Miller such a claim should be falsifiable but not verifiable. In my view, this argument bypasses the fact that the estimation of verisimilitude by ver is sensitive to evidence and leads to empirically revisable judgements of relative truthlikeness.

Moreover, Miller's insistence of the Popperian falsifiability and non-verifiability requirement is problematic also for the reason that it conflicts with the prospect of having a *connected* comparative notion of truthlikeness (such as those based on our measures Tr and ver): if we succeed in *falsifying* the claim that g_2 is more truthlike than g_1, and if the relation 'more truthlike than' is connected, then we have in fact *verified* the converse claim that g_1 is at least as truthlike as g_2 (Niiniluoto, 1982b, p. 188).

There are, indeed, some special cases where judgements of comparative truthlikeness may be in principle falsifiable and verifiable. Suppose that evidence e verifies one element h_j in **B** and thus falsifies all the others. Then, by (17), all comparative claims of verisimilitude are decided by the degrees of the form $\text{Tr}(g, h_j)$:

(27)   If evidence e verifies h_j for some $j \in I$, and if $\text{Tr}(g_2, h_j) > \text{Tr}(g_1, h_j)$, then e verifies the claim 'g_2 is more truthlike than g_1' and falsifies the claim 'g_1 is at least as truthlike as g_2'.

More generally, e may falsify some elements of **B**, and thereby verify the disjunction of the non-refuted ones. Then we obtain the result:

(28)   Evidence e verifies the claim 'g_2 is more truthlike than g_1' and falsifies the claim 'g_1 is at least as truthlike as g_2' if and only if, for some $J \subseteq I$,
(i) e verifies the disjunction $\bigvee_{i \in J} h_j$, and
(ii) $\text{Tr}(g_2, h_i) > \text{Tr}(g_1, h_i)$ for all $i \in J$.

Condition (ii) can be stated without any assumptions about the probability measure P on **B**. But, on the other hand, it is very restricted, since it requires that g_2 is *uniformly better* than g_1 with respect to all non-falsified elements $h_i \in$ **B**. This means that there will be many pairs of statements (e.g., h_i and h_j for any $i \in J$ and $j \in J$) which cannot be compared by condition (ii).

We have thus seen that a person who wishes to avoid all the probabilistic considerations involved in the definition (6) of ver may restrict his or her attention solely to falsifiable judgements of comparative verisimilitude. The price is high, however, since this move would make the concept of truthlikeness methodologically relatively uninteresting, in the sense that this notion could be applied only to two special cases: (i) one element of **B** is verified by the evidence, and (ii) several elements of **B** remain non-falsified by the evidence, and comparability is restricted to those rare situations where g_1 is uniformly better than g_2.

The virtue of function ver is that it helps us to go beyond the special cases (i) and (ii) — and to make revisable judgements of comparative truthlikeness for all interesting situations. However, this solution presupposes the existence of epistemic probability measures which express rational degrees of belief. In this precise sense, the epistemic problem of truthlikeness is equally difficult as the traditional problem of induction.[9]

7.3. PROBABLE VERISIMILITUDE

Let P be an epistemic probability measure on the cognitive problem **B**. In the preceding section, it was shown how P with the measure Tr for truthlikeness yields a function ver which can be used for estimating degrees of truthlikeness. This is not the only possibility of combining functions Tr and P, however. When P is available, we can also calculate degrees of *probable verisimilitude*.

Let $g \in D(\mathbf{B})$, and let $\varepsilon > 0$ be a small real number. Then the *ε-neighbourhood* of g in **B** is defined by

$$\mathcal{U}_\varepsilon(g) = \{i \in I \mid \Delta(h_i, g) \leq \varepsilon\}.$$

Then

$$i \in \mathcal{U}_\varepsilon(g) \quad \text{iff} \quad \operatorname{Tr}(g, h_i) \geq 1 - \varepsilon.$$

The *probability that the degree of truthlikeness of g is at least $1 - \varepsilon$ given evidence e*, is therefore

(29) $\quad P(\operatorname{Tr}(g, h_*) \geq 1 - \varepsilon / e) = P(* \in \mathcal{U}_\varepsilon(g)/e) = \sum_{i \in \mathcal{U}_\varepsilon(g)} P(h_i/e).$

ESTIMATION OF TRUTHLIKENESS 279

Let us denote this value by

$$PTr_{1-\varepsilon}(g/e).$$

Then the function $PTr_{1-\varepsilon}: D(\mathbf{B}) \to \mathbb{R}$ has the following properties:

(30) (a) $0 \leqslant PTr_{1-\varepsilon}(g/e) \leqslant 1$ for all $g \in D(\mathbf{B})$.
(b) $PTr_{1-\varepsilon}(g/e)$ is a non-decreasing function of ε.
(c) If $\varepsilon \geqslant 1$, then $PTr_{1-\varepsilon}(g/e) = 1$ for all $g \in \mathbf{B}$.
(d) When g is a tautology,
$$PTr_{1-\varepsilon}(g/e) = 1, \text{ if } \varepsilon \geqslant \gamma'$$
$$= 0, \text{ if } \varepsilon < \gamma'.$$
(e) When $j \in I$,
$$PTr_{1-\varepsilon}(h_j/e) = \sum_{i \in \mathscr{U}_\varepsilon(h_j)} P(h_i/e),$$
where
$$\mathscr{U}_\varepsilon(h_j) = \{i \in I \mid \Delta_{ij} \leqslant \varepsilon'\}$$
$$\varepsilon' = \frac{\varepsilon}{\gamma + \gamma'/|I| \operatorname{av}(i, \mathbf{B})}$$
(f) When $\varepsilon \to 0$,
$$PTr_{1-\varepsilon}(g/e) \to 0, \text{ if } g \notin \mathbf{B}$$
$$PTr_{1-\varepsilon}(h_i/e) \to P(h_i/e) \text{ for all } i \in I.$$

Assuming that $\varepsilon < \gamma'$, result (e) shows that $PTr_{1-\varepsilon}(h_j/e) = 1$ if and only if the whole probability mass 1 is concentrated on the ε-neighbourhood $\mathscr{U}_\varepsilon(h_j)$ of h_j. Hence, in particular,

(31) When $P(h_j/e) = 1$ for some $j \in I$,
$$PTr_{1-\varepsilon}(h_i/e) = 1 \quad \text{iff} \quad i \in \mathscr{U}_\varepsilon(h_j) \quad \text{iff} \quad \Delta_{ij} \leqslant \varepsilon'.$$

It is thus possible that more than one hypothesis $h_i \in \mathbf{B}$ receives the maximum value 1. Moreover, it may happen that $P(h_i/e) = 0$ but $PTr_{1-\varepsilon}(h_i/e) = 1$.[10]

There is another related concept which is obtained by replacing the min-sum distance Δ with $\Delta_{\min}$ in the definition of $\mathscr{U}_\varepsilon(g)$:

$$\mathscr{V}_\varepsilon(g) = \{i \in I \mid \Delta_{\min}(h_i, g) \leqslant \varepsilon\}.$$

When ε is small, we may say that g is *almost true* or *approximately true* if and only if $\Delta_{\min}(h_*, g) \leqslant \varepsilon$. This condition guarantees that g has a high ($\geq 1 - \varepsilon$) degree of *nearness to the truth* (cf. Chapter 5.4). The probability that g is approximately true, given evidence e, is thus

(32) $\quad P(* \in \mathscr{V}_\varepsilon(g)/e) = \sum_{i \in \mathscr{V}_\varepsilon(g)} P(h_i, e).$

Let us denote this value by

$$PA_{1-\varepsilon}(g/e).$$

Then the function $PA_{1-\varepsilon}: D(\mathbf{B}) \to \mathbb{R}$ has the following properties:

(33) (a) $0 \leq PA_{1-\varepsilon}(g/e) \leq 1$.
 (b) $PA_{1-\varepsilon}(g/e)$ is a non-decreasing function of ε.
 (c) If $\varepsilon \geq 1$, then $PA_{1-\varepsilon}(g/e) = 1$ for all $g \in \mathbf{B}$.
 (d) $PA_{1-\varepsilon}(g/e) = 1$, if g is a tautology.
 (e) When $j \in I$,
 $$PA_{1-\varepsilon}(h_j/e) = \sum_{i \in \mathscr{V}_\varepsilon(h_j)} P(h_i/e)$$
 where
 $$\mathscr{V}_\varepsilon(h_j) = \{i \in I \mid \Delta_{ij} \leq \varepsilon\}.$$
 (f) $P(g/e) \leq PA_{1-\varepsilon}(g/e)$.
 (g) When $\varepsilon \to 0$, $PA_{1-\varepsilon}(g/e) \to P(g/e)$.
 (h) When $P(h_j/e) = 1$ for some $j \in I$,
 $PA_{1-\varepsilon}(h_i/e) = 1 \quad$ iff $\quad i \in \mathscr{V}_\varepsilon(h_j) \quad$ iff $\quad \Delta_{ij} \leq \varepsilon$.
 (i) $PA_{1-\varepsilon}(g_1/e) \leq PA_{1-\varepsilon}(g_2/e)$, if $g_1 \vdash g_2$.

Here (g) says that $PA_{1-\varepsilon}$ reduces to P, when ε approaches 0, i.e., the probability that g is almost true approaches in the limit to the probability that g is strictly true. Result (h) shows again that $PA_{1-\varepsilon}(g/e)$ may be 1, even if $P(g/e) = 0$.

Result (i) indicates that $PA_{1-\varepsilon}(g/e)$ decreases, when g becomes logically stronger. This feature, which $PA_{1-\varepsilon}$ shares with ordinary probability P, distinguishes $PA_{1-\varepsilon}$ from the estimation function ver (cf. (18) (iv)).

7.4. ERRORS OF OBSERVATION

Let $\mathbf{B}$ be a cognitive problem relative to the presupposition b, and let $g \in D(\mathbf{B})$. Assume ver is defined relative to a probability measure P which satisfies the appropriate epistemic rationality conditions. If ver$(g/e) = r$, is it rational to claim on e that the degree of truthlikeness of g is r? In particular, if ver$(g/e) = 1$, is it rational to claim that g is the complete true answer to the problem $\mathbf{B}$? A positive answer to this question has at least two preconditions.[11] First, it must be rational to assume that the presupposition b is true. Otherwise all the members of $\mathbf{B}$ are false, and even the best answer in $D(\mathbf{B})$ fails to be true (cf. Chapter 6.9). Secondly, e has to represent the total available evidence relevant to the problem $\mathbf{B}$, and e has to be acceptable as true. For example, if $e \vdash h_i$, then by (17) ver$(h_i/e) = 1$; but if e is known to be false, this result gives us no warrant to claim that h_i is actually true.

In this section, I shall discuss the possibility of appraising truthlikeness in the basis of *false evidence*. The next section will consider the case of *false presuppositions*.

The standard treatments of inductive logic assume that observational errors in evidence have the probability zero (cf. Chapter 2.10). For example, if a monadic constituent C_i claims that the cells $\mathbf{CT}_i$ are non-empty and other cells are empty, then the conditional probability $P(Q_j(a_n)/C_i)$ of observing an individual a_n in a cell $Q_j \notin \mathbf{CT}_i$, which is empty given C_i, is 0. However, a similar assumption is unrealistic in many situations. Perhaps the most important of them is the case, where the evidence is obtained by measuring some quantities. Due to errors in measurement, there is a positive probability that the measuring process gives a result which deviates from the true value.

The classical *theory of errors*, developed by Laplace and other mathematicians at the turn of the nineteenth century for the systematic treatment of astronomical observations, can be understood as an attempt to make estimates of probable verisimilitude on the basis of false evidence.[12] Let x be a random variable which takes as its values the results of the measurement of an unknown real-valued quantity $\theta \in \mathbb{R}$. Then a standard assumption is to postulate that x is $N(\theta, \sigma^2)$, i.e., normally distributed around the true value θ with variance $\sigma^2 > 0$. This means that the density function $f(x/\theta)$ of the physical probability distribution for x is defined by *Gauss's curve*

$$(34) \quad \frac{1}{\sigma\sqrt{2\pi}} e^{-(x-\theta)^2/2\sigma^2},$$

where the variance σ^2 depends on the accuracy of the measuring device. If $\mathbf{x}_1, \ldots, \mathbf{x}_n$ are n independent repetitions of the measurement, and $\mathbf{y}$ is their mean value

$$\mathbf{y} = \frac{1}{n} \sum_{i=1}^{n} \mathbf{x}_i,$$

then $\mathbf{y}$ is also normally distributed with the same mean θ but with a smaller variance σ^2/n. Thus, when n grows, the probability distribution for the observed mean $\mathbf{y}$ becomes more and more concentrated around the true value θ (see Figure 1). The assumption (34) thus excludes the existence of such 'systematic errors' that would make $\mathbf{y}$ a *biased* estimate of θ.

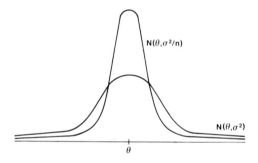

Fig. 1. Gauss's curve of errors.

Suppose then a *prior probability density* $g(\theta)$ is defined for the unknown $\theta \in \mathbb{R}$. Thus $g: \mathbb{R} \to \mathbb{R}$ is a continuous function such that, for each measurable set $A \subseteq \mathbb{R}$,

(35) $\quad g(A) = \int_A g(\theta) \, d\theta.$

is the epistemic probability of the claim $\theta \in A$. In particular,

$$\int_{-\infty}^{\infty} g(\theta) \, d\theta = 1.$$

Then the *posterior density* for θ given x is obtained by multiplying the prior probability $g(\theta)$ with the likelihood function $f(x/\theta)$: according to

ESTIMATION OF TRUTHLIKENESS 283

Bayes's formula for distributions,

(36) $\quad g(\theta/x) = \dfrac{g(\theta)f(x/\theta)}{f(x)},$

where

$$f(x) = \int_{-\infty}^{\infty} g(\theta)f(x/\theta)\,d\theta.$$

If $g(\theta)$ is $N(\mu, \sigma_1^2)$ and, as above, $f(y/\theta)$ is $N(\theta, \sigma^2/n)$, then (35) entails that $g(\theta/y)$ is $N(\mu_n, \sigma_n^2)$, where

(37) $\quad \mu_n = \left(\dfrac{\mu}{\sigma_0^2} + \dfrac{ny}{\sigma^2}\right) \bigg/ \left(\dfrac{1}{\sigma_0^2} + \dfrac{n}{\sigma^2}\right)$

$$\dfrac{1}{\sigma_n^2} = \dfrac{1}{\sigma_0^2} + \dfrac{n}{\sigma^2}.$$

If $g(\theta)$ is sufficiently flat, so that σ_0 is very large, then $g(\theta/y)$ is approximately $N(y, \sigma^2/n)$.[13] When n grows, it becomes increasingly probable that the unknown value θ is close to the observed mean y. For example,

$$g\left(-\dfrac{1{,}64\sigma}{\sqrt{n}} \leq y - \theta \leq \dfrac{1{,}64\sigma}{\sqrt{n}} \bigg/ \theta\right) = .95.$$

If the distance between θ and y is measured simply by the geometrical distance $|y - \theta|$ (cf. Chapter 1.1.), for any $\varepsilon > 0$,

(38) $\quad PA_{1-\varepsilon}(\theta = y/y) \to 1$, when $n \to \infty$,

i.e., the probability that the claim $\theta = y$ is approximately true, given evidence y, approaches 1, when the size n of y grows without limit (cf. (32)).

The classical Bayesian treatment of observational errors can thus be understood in terms of probable approximate truth. But there is also a straightforward connection to the function ver: if $g(\theta/y)$ is $N(y, \sigma^2/n)$, then the expected distance of the claim $\theta = y$ from the truth (cf. (8)) is

$$\int_{-\infty}^{\infty} g(\theta/y)|\theta - y|\,d\theta = \dfrac{2\sigma}{\sqrt{2\pi n}}.$$

This value goes to 0, when $n \to \infty$. This means that

(39) $\text{ver}(\theta = y/y) \to 1$, when $n \to \infty$,

i.e., the expected degree of verisimilitude for the claim that the observed mean y is equal to the true value θ grows towards its maximum value 1, when n grows without limit.

These results illustrate the fact that rational claims about probable verisimilitude and estimated degrees of truthlikeness can be based upon evidence which is known (with probability one) to be erroneous. Comparative judgements of verisimilitude are therefore also possible on the same basis. The legitimacy of such inferences is grounded on a factual assumption about the physical probability distribution of errors associated with a measuring device.

An important feature of the above argument is the assumption that the error curve, defined by the likelihood function $f(x|\theta)$, is symmetric around the mean value θ and decreases when $|x - \theta|$ increases. Thus, smaller errors are more probable than greater ones. When $g(\theta)$ is sufficiently 'uninformative', so that $f(x/\theta)$ dominates $g(\theta)$, it follows that the posterior distribution $g(\theta/x)$ is a decreasing function of the distance $|x - \theta|$ between θ and the observed value of x. When y is the mean of n repeated observations, then $g(\theta/y)$ is still a decreasing function of the distance $|y - \theta|$, and becomes more and more concentrated around the mean y when $n \to \infty$. This fact leads to the results (38) and (39), guaranteeing a high probable verisimilitude and estimated truthlikeness for the claim that $\theta = y$.

It should be added that these results can be proved under very general conditions: the Central Limit Theorem of probability calculus guarantees that, for sufficiently large values of n, the mean of n independent, identically distributed random variables is approximately normal. The basic idea — the probability of evidence decreases when its distance from the truth grows — can be generalized from point estimation to other kinds of cognitive problems as well (cf. Chapter 4.2).

7.5. COUNTERFACTUAL PRESUPPOSITIONS AND APPROXIMATE VALIDITY

In Section 4, we discussed a cognitive problem **B**, where evidence e is known to deviate from the truth, and the likelihood $P(e/h_i)$ decreases with the distance between the statements e and h_i. However, it was assumed the true hypothesis h_* is an element of **B**, and it is the evidence e which contains errors (due to the inevitable inaccuracy of measurement). Another way of viewing such examples is to assume that the evidence e is correct, but all the relevant hypotheses in **B** are false. This is typically the case when the inquiry with **B** is based upon a methodological assumption b that is in fact false. For example, b may be a counterfactual idealizing condition.

When the cognitive problem **B** is based on such a false presupposition b, each hypothesis h_i in **B** is known to be false, and any rational probability measure P, defined on **B** relative to our total evidence e_0, gives the value 0 to the elements of **B**.[14] However, the condition $P(h_i/e_0) = 0$ for all $i \in I$ entails, by (6), that $\text{ver}(g/e_0) = 0$ for all $g \in D(\mathbf{B})$. Therefore, the definition of ver cannot be applied for rational estimation of truthlikeness in this situation.

Some philosophers have developed notions of confirmation which are intended to apply to rival theories which are known to be strictly speaking false. For example, in his 'tempered' version of Bayesianism, Abner Shimony reinterprets the epistemic probabilities of the form $P(h/e)$ as "rational degrees of commitment":

> This suggests that a person whose belief in the literal truth of a general proposition h, given evidence e, is extremely small may nevertheless have a nonnegligible credence that h is related to the truth in the following way: (i) within the domain of current experimentation h yields almost the same observational predictions as the true theory; (ii) the concepts of the true theory are generalizations or more complete realizations of those of h; (iii) among the currently formulated theories competing with h, there is none which better satisfies conditions (i) and (ii). (Shimony, 1970, pp. 94—95.)

This idea allows Shimony to assign a positive prior degree of commitment to each 'seriously proposed' hypothesis h, even if the degree of belief in h (in the ordinary sense) is zero or 'extremely small'.

Shimony's condition (iii) excludes the true theory from the set **B** of relevant hypotheses. However, if **B** is expanded to a finer cognitive problem **B**′ which contains the true target h_*, then Shimony's degrees of commitment presumably behave like the function PA of probable approximate truth. Therefore, their properties also differ from those of estimated degrees of verisimilitude.

To illustrate another idea, suppose that we study the ballistic behaviour of projectiles near the surface of the earth, and we make the counterfactual assumption b that the resistance of air has no influence. The relevant hypotheses in **B** are then quantitative laws which describe the path of a projectile in terms of its initial position, initial velocity, time, and gravitation (see (3.28)). How can we test these alternatives by data e, obtained by observing projectiles in the air? The obvious problem is that this evidence e describes movements under the real conditions with resisting air, not under the idealizing assumption b.

One possibility is to say that evidence e indirectly confirms a hypothesis h_i in **B**, if e confirms the concretization of h'_i of h_i, where h'_i is obtained from h_i by eliminating the false assumption b (see (3.30)). Here we obviously need some theory which tells how the influence of air is to be added to h_i. Alternatively, we may follow the suggestion of Suppes (1962) and try — again using some theoretical background assumptions — to substract the influence of air from the data e. Let e' be what the data e would have been in the idealized situation b. Then e indirectly confirms h_i, if e' confirms h_i. (See Figure 2.)[15]

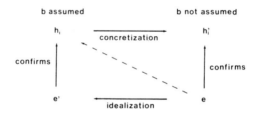

Fig. 2. Indirect confirmation of an idealized hypothesis h_i.

In spite of the fact that $P(h_i/e \ \& \ {\sim}b) = 0$, we may have $P(h'_i/e \ \& \ {\sim}b) > 0$ and $P(h_i/e' \ \& \ b) > 0$. In other words, idealized hypotheses can be confirmed by idealized evidence, and factual hypotheses by factual evidence. It follows that we also have $\mathrm{ver}(h'_i/e \ \& \ {\sim}b) > 0$ and $\mathrm{ver}(h_i/e' \ \& \ b) > 0$. The first alternative means that an idealized cognitive problem **B**, which does not contain any true hypothesis, is replaced by a concretized problem **B'**, which contains a true target. The second alternative means that the problem **B** is studied relative to data e' which tell what the actual evidence e would have been in an b-world (cf. Chapter 6.9). The latter method allows us to make comparative judgements of the estimated truthlikeness of the idealized hypotheses in $D(\mathbf{B})$:

(40) $g_1 \in D(\mathbf{B})$ seems more truthlike on e than $g_2 \in D(\mathbf{B})$
iff $\mathrm{ver}(g_1/e' \ \& \ b) > \mathrm{ver}(g_2/e' \ \& \ b)$.

ESTIMATION OF TRUTHLIKENESS 287

This suggestion for applying the function ver to cognitive problems with a false presupposition works, when we have a method for transforming e to e' or h_i to h'_i. But in many cases we may lack a theory that would allow us to do this, and we may fail even to know what precisely the idealizing assumptions b are. Perhaps the set **B** has been chosen primarily on the basis of simplicity and manageability. It is therefore important to find some other way of ordering the elements of **B** with respect to their presumed truthlikeness.

One natural method, which relies on the idea of *distance from evidence* (rather than expected distance from truth), is the following. Suppose that it is possible to define a function D which measures the distance $D(h_i, e)$ between a hypothesis $h_i \in \mathbf{B}$ and the evidence statement e. Function D thus expresses the 'fit' between h_i and e. If $D(h_i, e) = 0$, h_i is said to be *valid* with respect to e. If $D(h_i, e)$ is small ($< \varepsilon$ for some preassigned treshold value $\varepsilon > 0$), then h_i is *approximately valid*. When evidence e is taken to be sufficiently representative about the truth, the elements of **B** can ordered by their D-distance from e:

(41) Evaluate on evidence e that $h_i \in \mathbf{B}$ is more truthlike than $h_j \in \mathbf{B}$ if and only if $D(h_i, e) < D(h_j, e)$.

It is reasonable to expect that rule (40), wherever applicable, leads to the same result as (41).

The definition $D(h_i, e)$ does not reduce to the measures $\delta(g_1, g_2)$, $g_1 \in D(\mathbf{B})$, $g_2 \in D(\mathbf{B})$, studied in Chapter 6.7, since evidence statement e — or at least its observational part — typically does not belong to $D(\mathbf{B})$. For example, when **B** consists of constituents, observational evidence e normally is a finite conjunction of singular sentences; when **B** is a set of quantitative laws, e is a set of data obtained from measurement. I shall discuss the definition of D in special cases in the later chapters,[16] but here it is sufficient to illustrate the general idea by a historically important example.

Let $h_i: E \to \mathbb{R}$ $(i = 1, \ldots, k)$ be real-valued quantities, and let **B** be a class of functions $g: \mathbb{R}^{k-1} \to \mathbb{R}$ which express a lawful connection between the values of $h_1, \ldots, h_{k-1}, h_k$:

(42) $h_k(x) = g(h_1(x), \ldots, h_{k-1}(x))$.

Here **B** may be restricted, e.g., by the counterfactual assumption b that g is a linear function (see Example 4.11). Let evidence e consist of a finite set of data $\{\langle h_1(a_j), \ldots, h_k(a_j)\rangle | j = 1, \ldots, n\}$, obtained by measuring the quantities $h_1, \ldots, h_k$ for n objects $a_1, \ldots, a_n$. Then the distance between the hypothesis (42) and the evidence e can be defined as a specialization of the L_p-metrics between functions (see formulas (1.8) and (1.9)):

$$(43) \quad \left(\sum_{i=1}^{n} |g(h_1(a_i), \ldots, h_{k-1}(a_i)) - h_k(a_i)|^p \right)^{1/p}.$$

(Cf. Niiniluoto, 1986a, formula (32).) (43) is then the Minkowski-distance between the observed values $h_k(a_i)$ and the theoretical values calculated by the function g. For $p = 2$, (43) gives the traditional formula for the least square difference

$$\left(\sum_{i=1}^{n} (g(h_1(a_i), \ldots, h_{k-1}(a_i)) - h_k(a_i))^2 \right)^{\frac{1}{2}}$$

between the law (42) and evidence e. For $p = \infty$, formula (43) reduces to the Tchebycheff distance

$$\max_{i=1,\ldots,n} |g(h_1(a_i), \ldots, h_{k-1}(a_i)) - h_k(a_i)|$$

(cf. Patryas, 1977). When $k = 2$, these definitions yield the requirements that the linear function $g(x) = ch_1(x) + d$ should be chosen so that the distance

$$\sum_{i=1}^{n} (ch_1(a_i) + d - h_2(a_i))^2$$

or the distance

$$\max_{i=1,\ldots,n} |ch_1(a_i) + d - h_2(a_i)|$$

is minimized (see Figure 3). The linear function chosen from **B** by these criteria is then the one that has the greatest approximate validity relative to e. It is indeed the hypothesis that we judge, on the basis of evidence e, to be the *least false* element of **B**. Stated in other terms, it is the hypothesis in **B** of which we can rationally claim on e that it *would be true* if the assumption b were true.

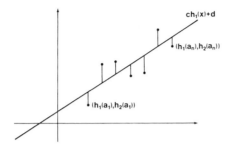

Fig. 3. Linear regression function.

In a more complicated methodological situation, the points $(h_1(a_i), \ldots, h_k(a_i))$ are replaced by observed means of measurements with a multidimensional normal distribution giving the probability distribution errors. This leads to the statistical theory of regression analysis.[17] In a sense, this approach combines the ideas of Sections 4 and 5, since it in effect treats both the hypotheses in **B** as idealized and the evidence e as erroneous. In this case, the distance of the observed multinomial sample and a hypothesis in **B** is measured by the Mahalanobis distance D^2 (see (1.33)).

The important point is that the classical statistical theories of least square estimation and regression analysis have a natural motivation as methods for choosing the most truthlike elements from a given set of quantitative hypotheses.

CHAPTER 8

SINGULAR STATEMENTS

The similarity approach to the logical and the epistemic problems of truthlikeness, developed in general terms in Chapters 6 and 7, is based on the assumption that it is possible to associate to each cognitive problem **B** a distance function $\Delta: \mathbf{B} \times \mathbf{B} \to \mathbb{R}$. In the next four chapters, I shall show how function Δ can be found for various types of cognitive problems. In the present chapter, problems involving singular statements are discussed.

8.1. SIMPLE QUALITATIVE SINGULAR STATEMENTS

By a simple qualitative singular statement I mean a first-order sentence about the properties of one individual (which may be any kind of object, event or process). The following definition for the degree of truthlikeness of such statements was suggested in Niiniluoto (1977b), p. 143, and Niiniluoto (1978b), p. 300.

Let $\mathbf{Q}$ be the set of the Q-predicates of a monadic language L, and let ε be a normalized distance function on $\mathbf{Q}$. For an individual a there is one and only one Q-predicate in $\mathbf{Q}$ which a satisfies; let us denote it by Q_*. If we now claim that $Q_i(a)$, then the seriousness of our mistake depends on the distance $\varepsilon(Q_*, Q_j)$ between the Q-predicates. More generally, if $\alpha(x)$ is a pure predicate expression in L, then $\alpha(x)$ can be represented by a disjunction of Q-predicates:

$$\vdash \alpha(x) \equiv \bigvee_{i \in \mathbf{I}_a} Q_i$$

(see (2.5)). Then the distance of the claim $\alpha(a)$ from the truth $Q_*(a)$ can be defined as a function of the distances $\varepsilon(Q_*, Q_j), j \in \mathbf{I}_a$:

$$\Delta(Q_*(a), \alpha(a)) = \mathrm{red}(\{\varepsilon(Q_*, Q_j) \mid j \in \mathbf{I}_a\}).$$

(See Figure 1.) The degree of truthlikeness for $\alpha(a)$ is then

$$\mathrm{Tr}(\alpha(a), Q_*(a)) = 1 - \Delta(Q_*(a), \alpha(a)).$$

The basic idea here is very simple. If $Q_i(a)$ and $Q_j(b)$, then the

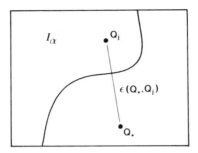

Fig. 1. Definition of $\Delta(Q_*(a), a(a))$.

distance $\varepsilon(Q_i, Q_j)$ expresses the *dissimilarity* between a and b, and $1 - \varepsilon(Q_i, Q_j)$ the *similarity* between a and b. If we have two claims $Q_i(a)$ and $Q_j(a)$ about the same individual a, then $\varepsilon(Q_i, Q_j)$ expresses the distance between these claims (cf. Chapter 6.7). Hence, $\varepsilon(Q_*, Q_i)$ is the distance of claim $Q_i(a)$ from the truth $Q_*(a)$. Further, $\mathrm{Tr}(Q_i(a), Q_*(a)) = 1 - \varepsilon(Q_*, Q_i)$ equals the degree of similarity between two individuals, one satisfying Q_i, and the other Q_*.

If the Q-predicates in $\mathbf{Q}$ are defined by the primitive predicates $M_1, \ldots, M_k$, as the conjunction $(\pm)M_1(x) \, \& \, \cdots \, \& \, (\pm)M_k(x)$ (see (2.2)), and if the distance $\varepsilon_1(Q_i, Q_j)$ is defined as the number of different signs in the corresponding conjuncts of Q_i and Q_j, divided by k (see (3.5)), then indeed $\varepsilon_1(Q_i, Q_j)$ equals the mean character difference (1.36) of Q_i-individuals and Q_j-individuals, and $1 - \varepsilon_1(Q_i, Q_j)$ equals the simple matching coefficient (1.44). Further, $\varepsilon_2(Q_i, Q_j) = \sqrt{\varepsilon_1(Q_i, Q_j)}$ equals $1/\sqrt{k}$ times the taxonomic distance (1.37). These distances between Q-predicates are thus systematically connected with those definitions of the degrees of dissimilarity and similarity which count all the matches (positive and negative) among the relevant properties.[1]

These ideas can be immediately generalized to monadic conceptual systems L with k families of predicates $\mathcal{M}_j = \{M_i^j \mid i = 1, \ldots, n_j\}$, $j = 1, \ldots, k$, and Q-predicates $Q_{j_1 \ldots j_k}(x)$, $1 \leq j_1 \leq n_1, \ldots, 1 \leq j_k \leq n_k$ (see Chapter 2.2). Let d_i be a normalized distance function on $\mathcal{M}_i$ ($i = 1, \ldots, k$), and define a distance function ε on the class $\mathbf{Q}$ of Q-predicates by the weighted Minkowski-metrics

(1) $\quad \varepsilon_p(Q_{j_1 \ldots j_k}, Q_{m_1 \ldots m_k}) = \left(\sum_{i=1}^{k} \xi_i d_i(M_{j_i}^i, M_{m_i}^i)^p \right)^{1/p},$

where $\xi_i \geq 0$ and $\xi_1 + \cdots + \xi_k = 1$ (see Chapter 3.2). If a is an individual, let us denote by $Q_{*_1 \ldots *_k}(x)$ the unique Q-predicate in $\mathbf{Q}$ that a satisfies. Then the distance of the claim $Q_{j_1 \ldots j_k}(a)$ from the truth can be defined by

(2) $\quad \Delta(Q_{*_1 \ldots *_k}(a), Q_{j_1 \ldots j_k}(a)) = \varepsilon_p(Q_{*_1 \ldots *_k}, Q_{j_1 \ldots j_k}).$

If $\alpha(x)$ is the disjunction of the Q-predicates in set $\mathbf{I}_\alpha$, then the distance of $\alpha(a)$ from the truth is defined by

(3) $\quad \Delta(Q_{*_1 \ldots *_k}(a), \alpha(a))$
$= \mathrm{red}(\{\varepsilon_p(Q_{*_1 \ldots *_k}, Q_{j_1 \ldots j_k}) \mid (j_1 \ldots j_k) \in \mathbf{I}_\alpha\})$

For example, the reduction function red on $\mathbf{Q}$ could be defined in analogy with the min-max-function Δ_{mm}^γ or the min-sum-function $\Delta_{ms}^{\gamma\gamma'}$ on cognitive problems $\mathbf{B}$ in chapter 6.4. Then the *degree of truthlikeness* of $\alpha(a)$ is defined by

(4) $\quad \mathrm{Tr}(\alpha(a), Q_{*_1 \ldots *_k}(a)) = 1 - \Delta(Q_{*_1 \ldots *_k}(a), \alpha(a)).$

As a special case of the general definition, we have the language L with one family $\mathcal{M}_1 = \{M_1^1, \ldots, M_n^1\}$, so that $k = 1$. Then the Q-predicates Q_i of L are simply the predicates M_i^1 in $\mathcal{M}_1$ ($i = 1, \ldots, n$), and (1) reduces to

$$\varepsilon_p(Q_i, Q_j) = d_1(M_i^1, M_j^1),$$

where d_1 is the distance function on $\mathcal{M}_1$. If

$$\vdash \alpha(x) \equiv \bigvee_{i \in \mathbf{I}_\alpha} M_i^1(x),$$

then

(5) $\quad \Delta(Q_*(a), \alpha(a)) = \mathrm{red}(\{d_1(M_*^1, M_i^1) \mid i \in \mathbf{I}_\alpha\}).$

EXAMPLE 1. Suppose we are interested in the number of moons of a certain planet a. Let

$$M_i^1(x) = x \text{ has } i \text{ planets},$$

take $\mathcal{M}_1 = \{M_i^1(x) \mid i = 1, \ldots, 20\}$, and define d_1 on $\mathcal{M}_1$ by

$$d_1(M_i^1, M_j^1) = |i - j|/20.$$

SINGULAR STATEMENTS

Assume $M_5^1(a)$ is true. Then

$$\sum_{i=1}^{20} d_1(M_*^1, M_i^1) = \frac{13}{2}$$

and, by using (6.40) and (6.87),

$$\Delta_{ms}^{\gamma\gamma'}\left(M_*^1(a), \bigvee_{i \in I_a} M_i^1(a)\right) = \frac{\gamma}{20} \min_{i \in I_a} |i - 5| + \frac{\gamma'}{130} \sum_{i \in I_a} |i - 5|.$$

Hence, for example, the degree of truthlikeness for the claim that a has i planets is

$$\mathrm{Tr}(M_i^1(a), M_*^1(a)) = 1 - \left(\frac{\gamma}{20} + \frac{\gamma'}{130}\right)|i - 5|,$$

and for the claim that the number of planets is in the interval $[m_1, m_2]$, where $5 \leqq m_1 < m_2 \leqq 20$,

$$\mathrm{Tr}\left(\bigvee_{i=m_1}^{m_2} M_i^1(a), M_*^1(a)\right) = 1 - \frac{\gamma}{20}(m_1 - 5) -$$

$$- \frac{\gamma'}{260}(m_2 - m_1)(m_1 + m_2).$$

Another important special case is the language L with k dichotomous families $\mathcal{M}_i = \{M_i, \sim M_i\}$, where $d_i(M_i, \sim M_i) = 1$ for $i = 1, \ldots, k$. Here $q = |\mathbf{Q}| = 2^k$. Then the most convenient choice in (1) is to take $p = 1$ and $\xi_i = 1/k$ for all $i = 1, \ldots, k$. Then the distance (1) between Q_i and Q_j reduces to the city-block metric $\varepsilon_1(Q_i, Q_j)$, discussed above. This distance function is balanced (cf. Chapter 6.3) in the sense that

$$\mathrm{av}(i, \mathbf{Q}) = \frac{1}{q} \sum_{Q_j \in \mathbf{Q}} \varepsilon_1(Q_i, Q_j) = \frac{1}{2} \quad \text{for all } i.$$

Hence, the choice of red $= \Delta_{ms}^{\gamma\gamma'}$ in (3) gives

(6) $\quad \mathrm{Tr}(\alpha(a), Q_*(a)) = 1 - \gamma \min_{j \in I_a} \varepsilon_1(Q_*, Q_j) -$

$$- \frac{2\gamma'}{q} \sum_{j \in I_a} \varepsilon_1(Q_*, Q_j)$$

and

(7) $\quad \text{Tr}(Q_i(a), Q_*(a)) = 1 - \left(\gamma + \dfrac{2\gamma'}{q}\right) \varepsilon_1(Q_*, Q_j).$

EXAMPLE 2. Let L be a monadic language with three primitive predicates

$M_1(x) = x$ is white
$M_2(x) = x$ has feathers
$M_3(x) = x$ lays eggs.

Let a be a hen, so that $Q_*(a)$ is $M_1(a)$ & $M_2(a)$ & $M_3(a)$. Then the degree of truthlikeness for the claim that a is white, has feathers, but does not lay eggs is

$$1 - \left(\gamma + \dfrac{\gamma'}{4}\right) \cdot \dfrac{1}{3}.$$

The degree of truthlikeness for the claim that a is white is

$$1 - \gamma \cdot 0 - \dfrac{\gamma'}{4}\left(0 + \dfrac{1}{3} + \dfrac{1}{3} + \dfrac{2}{3}\right) = 1 - \dfrac{\gamma'}{3}.$$

The latter degree, for a true but not very informative statement, is smaller than the former, for a false but relatively informative statement, if and only if the weight γ for the truth factor is not too high when compared to γ', i.e., iff $\gamma < 3\gamma'/4$.

Equations (1)–(4) define also a *comparative* concept of truthlikeness for simple singular statements:

(8) $\quad a_1(a)$ is more truthlike than $a_2(a)$ iff
$\Delta(Q_{*_1 \ldots *_k}(a), a_1(a)) < \Delta(Q_{*_1 \ldots *_k}(a), a_2(a)).$

Let us say that $Q_{j_1 \ldots j_k}$ is *uniformly at least as close to* $Q_{*_1 \ldots *_k}$ as $Q_{m_1 \ldots m_k}$ if

$d_i(M^i_{j_i}, M^i_{*_i}) \leq d_i(M^i_{m_i}, M^i_{*_i}) \quad$ for all $i = 1, \ldots, k.$

Then the following sufficient condition holds for any choice of $p < \infty$ and $\xi_i \geq 0$ in (1):

(9) If $Q_{j_1 \ldots j_k}$ is uniformly at least as close to $Q_{*_1 \ldots *_k}$ as $Q_{m_1 \ldots m_k}$, but not conversely, then $Q_{j_1 \ldots j_k}(a)$ is more truthlike than $Q_{m_1 \ldots m_k}(a)$.

Similarly, $a_1(x)$ is uniformly closer to $Q_{*_1 \ldots *_k}$ than $a_2(x)$ if every Q-predicate in $\mathbf{I}_{a_1}$ is uniformly closer to $Q_{*_1 \ldots *_k}$ than every Q-predicate in $\mathbf{I}_{a_2}$. This is again a sufficient condition for the claim (8). When the weights ξ_i are equal, a weaker sufficient condition can be formulated as follows: each Q-predicate in $\mathbf{I}_{a_1}$ can be correlated one-to-one to a Q-predicate in $\mathbf{I}_{a_2}$, so that the former is uniformly closer to $Q_{*_1 \ldots *_k}$ than the latter, and the possibly remaining Q-predicates in $\mathbf{I}_{a_2}$ are farther from the truth than the correlated ones. Thus,

(10) Assume $\xi_i = 1/k$ for $i = 1, \ldots, k$. If there is a bijective mapping $\eta: \mathbf{I}_{a_1} \to J$, $J \subseteq \mathbf{I}_{a_2}$, such that each $Q_{j_1 \ldots j_k}$ in $\mathbf{I}_{a_1}$ is uniformly closer to $Q_{*_1 \ldots *_k}$ than $\eta(Q_{j_1 \ldots j_k})$, and the best Q-predicate in J is uniformly closer to $Q_{*_1 \ldots *_k}$ than the Q-predicates in $\mathbf{I}_{a_2} - J$ (if $\mathbf{I}_{a_2} - J \neq \emptyset$), then $a_1(a)$ is more truthlike than $a_2(a)$.

Condition (10) entails that comparative judgments about the truthlikeness of statements expressed wholly in the vocabulary of one family $\mathcal{M}_i$ are independent of the other families $\mathcal{M}_j, j \neq i$.

(11) $M^i_j(a)$ is more truthlike in L than $M^i_m(a)$ iff
$d_i(M^i_*, M^i_j) < d_i(M^i_*, M^i_m)$.

(9) and (10) are not necessary conditions: the claim that $a_1(a)$ is more truthlike than $a_2(a)$ can hold even if $a_1(a)$ is not uniformly better than $a_2(a)$ — but in the general case this will depend on the weights ξ_i of the families $\mathcal{M}_i$ ($i = 1, \ldots, k$).

EXAMPLE 3. Suppose two anthropologists try to identify the sex and the age (at the time of death) of a particular member a of *Homo erectus*, on the basis of its skull and partial skeleton found in excavations. Let $\mathcal{M}_1 = \{M_1^1, M_2^1\}$ and $\mathcal{M}_2 = \{M_1^2, M_2^2, M_3^2\}$, where

$M_1^1(x) = x$ is female
$M_2^1(x) = x$ is male
$M_1^2(x) = x$ is a child (age between 0 and 20 years),
$M_2^2(x) = x$ is an adult (age between 20 and 40 years),
$M_3^2(x) = x$ is old (age over 40 years),

and assume

$d_1(M_1^1, M_2^1) = 1$
$d_2(M_1^2, M_2^2) = d_2(M_2^2, M_3^2) = 1/2$
$d_2(M_1^2, M_3^2) = 1$.

Let us suppose that a in fact is a female child, and let

$\alpha_1(x) = x$ is a male child
$\alpha_2(x) = x$ is an adult female.

Then, by using ε_1 in (1), the degree of truthlikeness of $\alpha_2(a)$ is greater than $\alpha_1(a)$ if and only if $\xi_1 > \xi_2/2$ (see Figure 2). Here $\alpha_2(a) = M_1^1(a)$ & $M_2^2(a)$ is not uniformly better than $\alpha_1(a) = M_2^1(a)$ & $M_1^2(a)$, since the former is correct about sex and the latter about age.

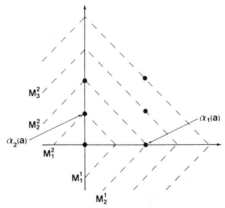

Fig. 2. ε_1-distances from M_1^1 & M_1^2 when $\xi_1 = \xi_2 = 1$.

By the results of Chapter 2.7, it is immediately clear that the above treatment of simple monadic singular statements can be generalized to polyadic cases, where a Q-predicate $Q_i(x)$ describes not only the one-place properties of x but also the relations of x to itself and to some given individuals $b_1, \ldots, b_n$.

8.2. DISTANCE BETWEEN STATE DESCRIPTIONS

In our treatment of simple qualitative statements in Section 1, the reduction function red was defined on the set $\mathbf{Q}$ of Q-predicates — not on a set of sentences. It might therefore seem that Section 1 was not strictly speaking an application of the similarity approach of Chapter 6. However, this impression is misleading, since in fact the statements $Q_i(a)$ are identical with the *state descriptions* of a monadic language L_1^k with *one individual constant a*. So what we were actually doing in Section 1 was the application of the similarity approach and functions red to the cognitive problem $\{Q_i(a) \mid Q_i \in \mathbf{Q}\}$, consisting of the state descriptions of language L_1^k (cf. Example 4.1).

Let us now generalize our treatment of singular sentences to arbitrary state descriptions $s(i_1, \ldots, i_N) \in \mathscr{Z}_N^k$, i.e., sentences of the form

$$Q_{i_1}(a_1) \& \cdots \& Q_{i_N}(a_N)$$

in the monadic language L_N^k with k families of predicates and $N < \omega$ individual constants $a_1, \ldots, a_N$ (see Chapter 2.3). Assume that L_N^k fits the universe U, and let

$$s_* = Q_{*_1}(a_1) \& \cdots \& Q_{*_N}(a_N)$$

be the true state description of U in $\mathscr{Z}_N^k$.

To introduce a metric Δ on $\mathscr{Z}_N^k$, I assume that all individuals $a_1, \ldots, a_N$ are equally important. Then the distance between $s(i_1, \ldots, i_N)$ and $s(m_1, \ldots, m_N)$ could be defined as the sum of the distances $\varepsilon(Q_{i_1}, Q_{m_1}), \ldots, \varepsilon(Q_{i_N}, Q_{m_N})$. When ε is the normalized city block metric ε_1 on $\mathbf{Q}$ (see Section 1), it is natural to normalize the metric on $\mathscr{Z}_N^k$ by dividing the above sum by the number N. Thus,

(12) $\quad \Delta(s(i_1, \ldots, i_N), s(m_1, \ldots, m_N)) = \dfrac{1}{N} \sum\limits_{j=1}^{N} \varepsilon(Q_{i_j}, Q_{m_j})$.

When $N = 1$, (12) reduces to

$$\Delta(s(i_1), s(m_1)) = \varepsilon(Q_{i_1}, Q_{m_1}),$$

just as it should.

In particular, the distance of a state description $s(i_1, \ldots, i_N)$ from the truth $s_* = s(*_1, \ldots, *_N)$ is

$$(13) \quad \Delta(s_*, s(i_1, \ldots, i_N)) = \frac{1}{N} \sum_{j=1}^{N} \varepsilon(Q_{*_j}, Q_{i_j}).$$

(Cf. Niiniluoto, 1979a.) This formula has a very natural interpretation: we play a game of placing N individuals $a_1, \ldots, a_N$ to their correct places $Q_{*_1}, \ldots, Q_{*_N}$ in a classification system $\mathbf{Q}$, and our total score is the average of the individual errors $\varepsilon(Q_{*_1}, Q_{i_1}), \ldots, \varepsilon(Q_{*_N}, Q_{i_N})$ (see Figure 3).[2]

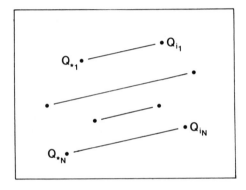

Fig. 3. Definition of $\Delta(s_*, s(i_1, \ldots, i_N))$.

For finite families of predicates, the normalized metric ε_1 on $\mathbf{Q}$ takes only a finite number of values between 0 and 1. For dichotomous families, these values are $0, 1/k, \ldots, i/k, \ldots, 1$. Let r_{iN} be the number of a_j's, $j = 1, \ldots, N$, with $\varepsilon(Q_{i_j}, Q_{m_j}) = i/k$ in (12). Then the distance (12) can be rewritten in the form

$$(14) \quad \frac{1}{k} \sum_{i=0}^{k} \frac{r_{iN}}{N} \cdot i.$$

Assume now that the number N of individuals grows without limit. Let r_i be the limit of r_{iN}/N for $N \to \infty$. Thus, r_i is the limiting proportion of individuals a_j with a mistake of size i/k. Hence, $r_0 + \cdots + r_k = 1$.

Hence, the limit of formula (14) for $N \to \infty$, i.e.,

(15) $\quad \dfrac{1}{k} \sum_{i=0}^{k} r_i \cdot i,$

expresses the distance between two *infinite* state descriptions $\bigwedge \{Q_{ij} \mid j < \omega\}$ and $\bigwedge \{Q_{mj} \mid j < \omega\}$ (cf. Niiniluoto, 1979a). In this way it is possible to define a distance Δ even in the space $\mathscr{Z}_\omega^k$ of the state descriptions of the language L_ω^k. This Δ is only a pseudometric, since r_i can be 0 even if $r_{iN} > 0$ for all N: if the number of mistakes of size i/k is bounded from above by some finite number, Δ does not count them in the limit.

As each sentence in language L_N^k has a $\mathscr{Z}$-normal form (2.19), also each singular statement α about the individuals $a_1, \ldots, a_N$ can be expressed as a disjunction of state descriptions s in the range $R(\alpha)$ of α. For such a sentence α, we may then define its degree of truthlikeness relative to the target s_*:

(16) $\quad \mathrm{Tr}(\alpha, s_*) = 1 - \Delta_{\mathrm{ms}}^{\gamma\gamma'}(s_*, R(\alpha)).$

In the special case of dichotomous families with the normalized distance ε_1, the cognitive problem $(\mathscr{Z}_N^k, \Delta)$ is balanced with $\mathrm{av}(i, \mathscr{Z}_N^k) = 1/2$ for all i. As $|\mathscr{Z}_N^k| = q^N$, (16) gives

(17) $\quad \mathrm{Tr}(\alpha, s_*) = 1 - \gamma \min_{s \in R(\alpha)} \Delta(s_*, s) - \dfrac{2\gamma'}{q^N} \sum_{s \in R(\alpha)} \Delta(s_*, s)$

and, for $s \in \mathscr{Z}_N^k$,

(18) $\quad \mathrm{Tr}(s, s_*) = 1 - \left(\gamma + \dfrac{2\gamma'}{q^N} \right) \Delta(s_*, s).$

These formulas reduce to (6) and (7), when $N = 1$.

It follows from (13) and (16) that $s(i_1, \ldots, i_N)$ is more truthlike than $s(m_1, \ldots, m_N)$ if the following condition holds

(19) $\quad \varepsilon(Q_{*_j}, Q_{i_j}) \leq \varepsilon(Q_{*_j}, Q_{m_j})$ for all $j = 1, \ldots, N$, where the strict inequality $<$ is valid at least for one $j = 1, \ldots, N$.[3]

By (9), it is sufficient for (19) that each Q_{i_j} is uniformly at least as close to Q_{*_j} as Q_{m_j}. In analogy with (10), the following sufficient condition holds for any statements a_1 and a_2 in L:

(20) If there is a bijective mapping $\eta: R(a_1) \to S$, $S \subseteq R(a_2)$, such that each $s \in R(a_1)$ is closer to the truth than $\eta(s)$, and the best element of S is closer to the truth than any of the elements of $R(a_2) - J$, then a_1 is more truthlike than a_2.

For example, (20) entails the following condition, which is also directly obtained from (16):

(21) Assume $\Delta_{\min}(s_*, a_1) = \Delta_{\min}(s_*, a_2)$. Then a_2 is more truthlike than a_1 iff

$$\sum_{s \in R(a_1)} \Delta(s_*, s) < \sum_{s \in R(a_2)} \Delta(s_*, s).$$

Condition (20) entails again (11), i.e., comparative judgments about statements within one family of predicates do not change if new families are added to the language L. Moreover, (20) implies that such comparative judgments about statements concerning one individual a_j are independent of other individuals a_i, $i \neq j$.

(22) $Q_{i_1}(a_j)$ is more truthlike in L than $Q_{i_2}(a_j)$ iff
$\varepsilon(Q_{i_1}, Q_*) < \varepsilon(Q_{i_2}, Q_*)$.

Let L_{n+1}^k be the monadic language with k one-place predicates, and assume we have observed a sample e_n^c of n individuals $a_1, \ldots, a_n$ such that $n_i \geq 0$ of them satisfy the Q-predicates Q_i ($i = 1, \ldots, q$). Then formula (2.86) for Hintikka's generalized combined system of inductive logic tells that

$$P(Q_j(a_{n+1}) \mid e_n^c) \approx \frac{n_i + 1}{n + c},$$

when n and parameter α are sufficiently large with respect to $q = 2^k$. Therefore, the estimated degree of truthlikeness for the claim $a(a_{n+1})$ on e_n^c is

(23) $\quad \text{ver}(a(a_{n+1}) \mid e_n^c) \approx \sum_{i=1}^{q} \frac{n_i + 1}{n + c} \text{Tr}(a(a_{n+1}), Q_i(a_{n+1}))$

(see (7.6)). In particular, by (7),

(24) $\operatorname{ver}(Q_j(a_{n+1}) \mid e_n^c) \approx 1 - \left(\gamma + \frac{2\gamma'}{q} \right) \sum_{i=1}^{q} \frac{n_i + 1}{n + c} \cdot \varepsilon_1(Q_i, Q_j)$

$= 1 - \frac{(\gamma + 2\gamma'/q)}{n + c} \left[\frac{q}{2} + \sum_{i=1}^{q} n_i \varepsilon_1(Q_i, Q_j) \right].$

Thus, (24) will be maximized by choosing Q_j so that

$$\sum_{i=1}^{q} n_i \varepsilon_1(Q_i, Q_j)$$

is minimized. At least when n is sufficiently large, this will happen when Q_j is that predicate for which the observed relative frequency n_j/n is the largest, i.e., when Q_j is the most probable Q-predicate on e_n^c.

For maximally informative claims about a_{n+1}, i.e., for hypotheses of the form $Q_j(a_{n+1})$, high posterior probability on e_n^c goes together with high expected verisimilitude on e_n^c. The converse does not hold, however (cf. Chapter 7.2). For example, suppose $c = 1$ and $n_i = n$. If n is large,

$P(Q_j(a_{n+1}) \mid e_n^n) \approx 1 \quad \text{iff} \quad j = i$
$\phantom{P(Q_j(a_{n+1}) \mid e_n^n)} \approx 0 \quad \text{iff} \quad j \neq i.$

Hence, by (24),

$\operatorname{ver}(Q_i(a_{n+1}) \mid e_n^n) \approx 1$

$\operatorname{ver}(Q_j(a_{n+1}) \mid e_n^n) \approx 1 - \left(\gamma + \frac{2\gamma'}{q} \right) \varepsilon_1(Q_i, Q_j).$

We thus see that the expected verisimilitude of $Q_j(a_{n+1})$ on e_n^n, when $j \neq i$, is inversely related to the distance between the predicates Q_i and Q_j. When, e.g., $k = 10$, $\gamma = 1/2$, $\gamma' = 1/4$, and $\varepsilon_1(Q_i, Q_j) = 1/k = 1/10$, then $\operatorname{ver}(Q_j(a_{n+1}) \mid e_n^n) \approx 1 - (1/2 + 2/(4 \cdot 2^{10}))/10 = .95$. This also illustrates the fact that in general probability and expected truthlikeness do not covary with each other: $P(Q_j(a_{n+1}) \mid e_n^n) \approx 0$ even if $\operatorname{ver}(Q_j(a_{n+1}) \mid e_n^n) \approx .95$; for a tautology g, $P(g \mid e_n^n) = 1$ but $\operatorname{ver}(g \mid e_n^n) = 1 - \gamma'$ (which equals .75 in the above numerical example).

The ideas of this section can be immediately generalized to the state descriptions $Ct^{(0)}(a_1, \ldots, a_N)$ of a polyadic first-order language L with individual constants $a_1, \ldots, a_N$ (see Ch. 2.7).

8.3. DISTANCE BETWEEN STRUCTURE DESCRIPTIONS

If our cognitive aim is to find out the number N_i^* of individuals in the different cells Q_i ($i = 1, \ldots, q$) of a monadic classification system L_N^k, then the target of our cognitive problem ζ_N^k is the true structure description $S(N_1^*, \ldots, N_q^*)$ (see Chapter 2.4). The elements of ζ_N^k are all the structure descriptions $S(N_1, \ldots, N_q)$, with $N_1 + \cdots + N_q = N$, of the language L with the Q-predicates $Q_1, \ldots, Q_q$ (see example 4.2).

To define a distance function Δ on ζ_N^k, note that a structure description $S(N_1, \ldots, N_q)$ corresponds one-to-one to a q-tuple of natural numbers, so that ζ_N^k is a countable subspace of $\mathbb{R}^q$. We may therefore apply here the Minkowski metrics (1.8):

$$(25) \quad \Delta(S(N_1, \ldots, N_q), S(M_1, \ldots, M_q)) = \left(\sum_{i=1}^{q} |N_i - M_i|^p \right)^{1/p}.$$

The definition (25) with $p = 1$ is in effect suggested in Niiniluoto (1978b), p. 448. In the case $p = 2$, (25) is almost identical — except normalization — with Pearson's χ^2, which is the standard statistical measure for the fit between two numerical distributions (see Example 1.11). The distance of a structure description $S(N_1, \ldots, N_q)$ from the truth, when the city-block metric is used, is then

$$\sum_{i=1}^{q} |N_i - N_i^*|.$$

As the normalized form of this distance we can use

$$(26) \quad \frac{1}{2N} \sum_{i=1}^{q} |N_i - N_i^*|.$$

In the language L_ω^k with an infinite number ω of individuals, replace a structure description $S(N_1, \ldots, N_q)$, with $0 \leq N_i \leq \omega$, $i = 1, \ldots, q$, by the q-tuple $\langle N_1 | \omega, \ldots, N_q | \omega \rangle$ of the limiting proportions of individuals in the cells $Q_1, \ldots, Q_q$. Then $N_1 | \omega + \ldots N_q | \omega = 1$, and formula (25) can be directly used for the definition of a metric Δ on ζ_ω^k.

It is possible to give similar results for the truthlikeness of structure descriptions as for state descriptions in Section 2. Instead of going to details and examples, let us observe a problem in the definition (25).

Assume, for example, that $N_1^* = N$ and $N_i^* = 0, i = 2, \ldots, q$. Let S claim that cell Q_j contains N individuals, and S' claim that cell Q_m contains N individuals. If now Q_j is very close Q_1, while Q_m is distant from Q_1, then intuitively S would seem to be closer to the truth than S' (cf. Tichý, 1978). In other words, measure (25) is independent of the distances between the Q-predicates $Q_1, \ldots, Q_q$. We shall postpone the discussion of this issue to Chapter 9.2, where the cognitive problem ζ_N^k is extended to the problem defined by the constituents of a monadic language with identity.

8.4. QUANTITATIVE SINGULAR STATEMENTS

The theory of truthlikeness for singular quantitative statements is obtained as a direct generalization of the treatment in Sections 1 and 2. It was shown in Chapter 3.2 that a quantitative state space $\mathbf{Q} \subseteq \mathbb{R}^k$, generated by k real-valued quantities $h_i: E \to \mathbb{R}, i = 1, \ldots, k$, is a limiting case of qualitative spaces of Q-predicates generated by k families of predicates. A state space $\mathbf{Q}$ might as well be an infinite dimensional Hilbert space. But in any case $\mathbf{Q}$ will have a metric ε which measures the distance $\varepsilon(Q, Q')$ between two points Q and Q' in $\mathbf{Q}$.

For the real line $\mathbb{R}$ (i.e., when $k = 1$), all the Minkowski metrics (1.8) reduce to the geometrical distance between two real numbers:

(27) $\quad \varepsilon(Q, Q') = |Q - Q'| \quad$ for $Q, Q' \in \mathbb{R}$.

Its square $(Q - Q')^2$ defines also a distance function, but it leads to the same comparative concept of truthlikeness for simple quantitative statements of the form $Q(a)$ (i.e., claims that the value $h_1(a)$ equals Q). When Q_* is the true state of individual a,

(28) $\quad Q(a)$ is more truthlike than $Q'(a)$ iff $|Q - Q_*| < |Q' - Q_*|$.

For $k \geq 2$, the most convenient choice is the Euclidean metric in $\mathbb{R}^k$: if

$Q = (Q_1, \ldots, Q_k) \in \mathbb{R}^k$ and $Q' = (Q'_1, \ldots, Q'_k) \in \mathbb{R}^k$,

$$(29) \quad \varepsilon_2(Q, Q') = \left(\sum_{i=1}^{k} (Q_i - Q'_i)^2 \right)^{\frac{1}{2}}.$$

Singular statements relative to state space **Q** correspond to its measurable subsets $\mathbf{H} \subseteq \mathbf{Q}$: $\mathbf{H}(a)$ says that the state of individual a belongs to set **H**. Thus, $\mathbf{H}(a)$ can be represented as an infinite disjunction

$$\mathbf{H}(a) \equiv \bigvee_{Q \in \mathbf{H}} Q(a),$$

so that $\mathbf{H}(a)$ is a partial answer to the continuous cognitive problem $\mathbf{B} = \{Q(a) \mid Q \in \mathbf{Q}\}$ with the target $Q_*(a)$ (cf. Example 4.10).

For reasons explained in Chapter 6.7, it is natural to assume that this continuous problem **B** is restricted in the following way: all the sharp hypotheses in $D(\mathbf{B})$ have to be connected subsets of **Q** — for example, points when $\mathbf{Q} = \mathbb{R}$, points or lines when $\mathbf{Q} = \mathbb{R}^2$, points or lines or planes when $\mathbf{Q} = \mathbb{R}^3$, etc. All the other elements of $D(\mathbf{B})$ correspond to sets $\mathbf{H} \subseteq \mathbf{Q}$ with a positive measure.

As the metric (29) is not normalized, we can define the distance between two singular statements $\mathbf{H}_1(a)$ and $\mathbf{H}_2(a)$ by $\delta^{\beta a a'}(\mathbf{H}_1, \mathbf{H}_2)$, defined in (6.132). When $\beta < 1$, we shall assume that **Q** is a bounded subspace of $\mathbb{R}^k$; when $\beta = 1$, **Q** may be $\mathbb{R}^k$ itself. Then the distance between a singular statement $\mathbf{H}(a)$ and a sharp statement $Q(a)$ is

$$(30) \quad \Delta(Q(a), \mathbf{H}(a)) = \frac{1+\beta}{2} \min_{Q' \in \mathbf{H}} \varepsilon_2(Q, Q') +$$

$$+ \frac{1-\beta}{2} \max_{Q' \in \mathbf{H}} \varepsilon_2(Q, Q') +$$

$$+ \alpha' \int_{\mathbf{H}} \varepsilon_2(Q, Q') \, dQ',$$

and the degree of truthlikeness of $\mathbf{H}(a)$ relative to $Q_*(a)$ by

$$(31) \quad Tr(\mathbf{H}(a), Q_*(a)) = 1/(1 + \Delta(Q_*(a), \mathbf{H}(a)))$$

(cf. (6.85′)). In particular, for sharp hypotheses,

(32) $\quad \Delta(Q_*(a), Q(a)) = \varepsilon_2(Q_*, Q).$

If $Q_* \in \mathbf{H}$, i.e., $\mathbf{H}(a)$ is true, then

$$\Delta(Q_*(a), \mathbf{H}(a)) = \frac{1-\beta}{2} \max_{Q' \in \mathbf{H}} \varepsilon_2(Q, Q') +$$

$$+ \alpha' \int_{\mathbf{H}} \varepsilon_2(Q_*, Q) \, dQ.$$

The comparative results (8)–(11) can then be generalized to the case of quantitative state space $\mathbf{Q}$. In analogy with Section 2, it is also possible to define state descriptions of the form $Q_{i_1}(a_1) \& \cdots \& Q_{i_n}(a_n)$, where $Q_{i_j} \in \mathbf{Q}$ for $j = 1, \ldots, n$, and a distance function between them.

EXAMPLE 4. Let $k = 1$ and $\mathbf{Q} = \mathbb{R}$, and choose $\beta = 1$ in (30). Assume $Q_*(a)$ is true. Let $\mathbf{H}$ be the union of $m \geq 1$ disjoint intervals $[Q_{11}, Q_{12}], \ldots, [Q_{m_1}, Q_{m_2}]$ in $\mathbb{R}$, where $Q_* \leq Q_{11} < Q_{12} < Q_{21} < \cdots < Q_{m_1} < Q_{m_2}$. Then, by (30),

$$\Delta(Q_*(a), \mathbf{H}(a)) = (Q_{11} - Q_*) + \alpha' \sum_{i=1}^{m} \int_{Q_{i_1}}^{Q_{i_2}} (Q - Q_*) \, dQ$$

$$= (Q_{11} - Q_*) + \alpha' \sum_{i=1}^{m} (Q_{i_2} - Q_{i_1}) \left(\frac{Q_{i_2} + Q_{i_1}}{2} - Q_* \right).$$

In particular, for $m = 1$,

$$\Delta(Q_*(a), \mathbf{H}(a)) = (Q_{11} - Q_*) + \alpha'(Q_{12} - Q_{11}) \left(\frac{Q_{12} + Q_{11}}{2} - Q_* \right).$$

Here $(Q_{12} - Q_{11})$ is the length of the interval $[Q_{11}, Q_{12}]$, and

$((Q_{12} + Q_{11})/2 - Q_*)$ is the geometrical distance of the midpoint of $[Q_{11}, Q_{12}]$ from the truth Q_*.

EXAMPLE 5. Let $k = 1$ and $\mathbf{Q} = [0, 1] \subseteq \mathbb{R}$, and choose $\beta = 1$ in (30). Assume $h_1(a) = 1/4$, so that $Q_* = 1/4$. Let $\mathbf{H}(a)$ be the tautology, i.e., $\mathbf{H} = [0, 1]$. Then

$$\Delta(Q_*(a), \mathbf{H}(a)) = a' \left[\int_0^{1/4} \left(\frac{1}{4} - Q \right) dQ + \int_{1/4}^1 \left(Q - \frac{1}{4} \right) dQ \right]$$

$$= 5a'/16.$$

This value is greater than

$$\Delta(Q_*(a), Q(a)) = \left| Q - \frac{1}{4} \right|$$

if and only if

$$\frac{1}{4}\left(1 - \frac{5a'}{4}\right) < Q < \frac{1}{4}\left(1 + \frac{5a'}{4}\right).$$

EXAMPLE 6. Let $k = 2$ and $\mathbf{Q} = \mathbb{R}^2$. Choose $\beta < 1$ in (30). Assume $h_1(a) = 0$ and $h_2(a) = 0$, so that $Q_* = (0, 0)$. Let $\mathbf{H} = \{(Q_1, Q_2) \mid c_1 \leq Q_1 \leq d_1, c_2 \leq Q_2 \leq d_2\}$, where $0 \leq c_1, 0 \leq c_2$. Then

$$\Delta(Q_*(a), \mathbf{H}(a)) = \frac{1 + \beta}{2} \sqrt{c_1^2 + c_2^2} + \frac{1 - \beta}{2} \sqrt{d_1^2 + d_2^2} +$$

$$+ a' \int_{c_1}^{d_1} \int_{c_2}^{d_2} \sqrt{x^2 + y^2} \, dx \, dy.$$

We have so far assumed that the truth about the individual a is represented by a point $Q_* \in \mathbf{Q}$. However, if truth is indefinite with respect to the conceptual framework of $\mathbf{Q}$, the target has to be represented by a connected subset $\mathbf{H}_*$ of $\mathbf{Q}$ (see Chapter 6.8). The distance of the singular statement $\mathbf{H}(a)$ from the indefinite truth $\mathbf{H}_*(a)$

can be measured by

$$(33) \quad \delta^{\beta aa'}(\mathbf{H}_*, \mathbf{H}) = \beta \Delta_{\min}(\mathbf{H}_*, \mathbf{H}) + \frac{1-\beta}{2} \max_{Q \in \mathbf{H}_*} \Delta_{\min}(Q, \mathbf{H}) +$$
$$+ \frac{1-\beta}{2} \max_{Q \in \mathbf{H}} \Delta_{\min}(Q, \mathbf{H}_*)$$
$$+ \alpha \int_{\mathbf{H}_*} \Delta_{\min}(Q, \mathbf{H}) \, dQ +$$
$$+ \alpha' \int_{\mathbf{H}} \Delta_{\min}(\mathbf{H}_*, Q) \, dQ$$

(cf. (6.132)). Hence, the distance of $Q(a)$ from the indefinite truth $\mathbf{H}_*(a)$ is

$$(34) \quad \delta^{\beta aa'}(\mathbf{H}_*, Q) = \frac{1+\beta}{2} \Delta_{\min}(Q, \mathbf{H}_*) + \frac{1-\beta}{2} \Delta_{\max}(Q, \mathbf{H}_*) +$$
$$+ \alpha \int_{\mathbf{H}_*} \Delta(Q', Q) \, dQ'.$$

EXAMPLE 7. Assume $k = 1$, $\mathbf{Q} = \mathbb{R}$, and $\beta = 1$. Let the indefinite truth be represented by the interval $\mathbf{H}_* = [Q_*, Q'_*] \subseteq \mathbb{R}$, and let $\mathbf{H}_0 = [Q_0, Q'_0]$. Assume first that $Q_* < Q'_* \leq Q_0 < Q'_0$. Then by (33)

$$\delta^{1aa'}(\mathbf{H}_*, \mathbf{H}_0) = (Q_0 - Q'_*) + \alpha \int_{Q_*}^{Q'_*} \Delta_{\min}(Q, \mathbf{H}_0) \, dQ +$$
$$+ \alpha' \int_{Q_0}^{Q'_0} \Delta_{\min}(Q, \mathbf{H}_*) \, dQ$$
$$= (Q_0 - Q'_*) + \alpha \int_{Q_*}^{Q'_*} (Q_0 - Q) \, dQ +$$
$$+ \alpha' \int_{Q_0}^{Q'_0} (Q - Q'_*) \, dQ$$
$$= (Q_0 - Q'_*) + \alpha \ell(\mathbf{H}_*)(Q_0 - m(\mathbf{H}_*)) +$$
$$+ \alpha' \ell(\mathbf{H}_0)(m(\mathbf{H}_0) - Q'_*),$$

where
$$\ell(\mathbf{H}_*) = (Q'_* - Q_*)$$
is the length of interval $\mathbf{H}_*$, and
$$m(\mathbf{H}_*) = \frac{Q_* + Q'_*}{2}$$
is the midpoint of interval $\mathbf{H}_*$. This result can also be written in the form
$$\delta^{1aa'}(\mathbf{H}_*, \mathbf{H}_0) = (1 + a\ell(\mathbf{H}_*) + a'\ell(\mathbf{H}_0))(Q_0 - Q'_*) +$$
$$+ \frac{a}{2}\ell(\mathbf{H}_*)^2 + \frac{a'}{2}\ell(\mathbf{H}_0)^2.$$

Secondly, assume that $\mathbf{H}_*$ and $\mathbf{H}_0$ partly overlap, so that $Q_* < Q_0 < Q'_* < Q'_0$. Then by (33)

$$\delta^{1aa'}(\mathbf{H}_*, \mathbf{H}_0) = a \int_{Q_*}^{Q_0} (Q_0 - Q)\,dQ + a' \int_{Q'_*}^{Q'_0} (Q - Q'_*)\,dQ$$
$$= \frac{a}{2}(Q_0 - Q_*)^2 + \frac{a'}{2}(Q'_0 - Q'_*)^2.$$

Thirdly, assume that $\mathbf{H}_0$ entails $\mathbf{H}_*$, so that $Q_* < Q_0 < Q'_0 < Q'_*$. Then

$$\delta^{1aa'}(\mathbf{H}_*, \mathbf{H}_0) = a \int_{Q_*}^{Q_0} (Q_0 - Q)\,dQ + a \int_{Q'_0}^{Q'_*} (Q - Q'_0)\,dQ$$
$$= \frac{a}{2}(Q_0 - Q_*)^2 + \frac{a}{2}(Q'_* - Q'_0)^2.$$

Hence, when $\mathbf{H}_0 = [Q_0, Q_0]$,

$$\delta^{1aa'}(\mathbf{H}_*, Q_0)$$
$$= (Q_0 - Q'_*) + a\ell(\mathbf{H}_*)(Q_0 - m(\mathbf{H}_*)), \quad \text{if } Q_* < Q'_* \leq Q_0$$
$$= \frac{a}{2}(Q_0 - Q_*)^2 + \frac{a}{2}(Q'_* - Q_0)^2, \quad \text{if } Q_* \leq Q_0 \leq Q'_*.$$

The estimation problem for the truthlikeness of quantitative singular statements $H(a)$ relative to $Q \subseteq \mathbb{R}^k$ can be solved by assuming that there is an epistemic multi-dimensional probability distribution with a continuous density function $p: Q \to \mathbb{R}$ over the space Q, and by defining the *expected distance* of $H(a)$ from the truth $Q_*(a)$ by

$$(35) \quad \mathrm{Exp}_e \, \Delta(Q_*(a), H(a)) = \int_Q p(Q \mid e) \Delta(Q(a), H(a)) \, dQ,$$

where e is the evidence statement (cf. (7.8)). Then

$$\mathrm{Exp}_e \, \Delta(Q_*(a), H_1(a)) < \mathrm{Exp}_e \, \Delta(Q_*(a), H_2(a))$$
$$\text{iff} \quad \mathrm{ver}(H_1(a) \mid e) > \mathrm{ver}(H_2(a) \mid e),$$

where the expected verisimilitude is defined by

$$(36) \quad \mathrm{ver}(H(a) \mid e) = \int_Q p(Q \mid e) \mathrm{Tr}(H(a), Q(a)) \, dQ$$

(cf. (7.7) and (31)).

For the case $k = 1$, formula (35) for $H = Q_0$ reduces to the form

$$(37) \quad \mathrm{Exp}_e \, \Delta(Q_*(a), Q_0(a)) = \int_\mathbb{R} p(Q \mid e) \Delta(Q, Q_0) \, dQ$$

$$= \int_{-\infty}^{\infty} p(Q \mid e) \mid Q - Q_0 \mid dQ.$$

It is easy to prove that the expected distance (37) is minimized by choosing Q_0 as the median of the posterior distribution $p(Q \mid e)$, i.e.,

$$\int_{-\infty}^{Q_0} p(Q \mid e) \, dQ = 1/2.$$

(Blackwell and Girshick, 1954, p. 302).

Definitions (35) and (36) are studied in more detail in Chapter 12.5, where they are related to the Bayesian theory of point and interval estimation, when the distance Δ plays the role of a cognitive loss function for a decision problem.

CHAPTER 9

MONADIC GENERALIZATIONS

This chapter shows how one can define a metric on the set of constituents of a monadic first-order language. Section 1 treats the case of monadic constituents without identity, and Section 2 monadic constituents with identity. In Section 3, the distance functions between constituents recommended here are systematically compared with, and defended against, the alternative suggestions by Tichý and Oddie. Finally, Sections 4 and 5 study the definition and the estimation of degrees of truthlikeness for existential and universal generalizations.

9.1. DISTANCE BETWEEN MONADIC CONSTITUENTS

Let Γ be the set of constituents C_i, $i = 1, \ldots, t$, for a monadic first-order language L_N^k with k families of predicates but without identity (see Chapter 2.5). Let $\mathbf{Q}$ be the set of Q-predicates Q_j, $j = 1, \ldots, q$, of L_N^k. Then constituent C_i claims that certain Q-predicates $\mathbf{CT}_i \subseteq \mathbf{Q}$ are exemplified in the universe U, and the others $-\mathbf{CT}_i$ are empty. If $\mathbf{CT}_*$ is the class of Q-predicates that are actually non-empty in U, then the true constituent C_* in Γ has the form

$$\bigwedge_{j \in \mathbf{CT}_*} (\exists x) Q_j(x) \, \& \, (x) \left[\bigvee_{j \in \mathbf{CT}_*} Q_j(x) \right].$$

The set Γ defines a finite cognitive problem with $t = 2^q$ elements and with the target C_*; $D(\Gamma)$ consists of all the consistent generalizations of L_0^k (see Example 4.3).

The following distance function was proposed in Niiniluoto (1977b). Two constituents C_i and C_j in Γ can be compared with respect to the q claims ('non-empty' or 'empty') that they make relative to the cells determined by the Q-predicates $Q_1, \ldots, Q_q$. The distances $d(C_i, C_j)$ between C_i and C_j can thus be defined as the number of *different* claims about the cells $Q_1, \ldots, Q_q$ that they make. If a constituent C_i is correlated by a binary sequence $\langle x_1^i, \ldots, x_q^i \rangle$ of length q, where for all

$m = 1, \ldots, q$

$$x_m^i = 1, \quad \text{if } Q_m \in \mathbf{CT}_i$$
$$ = 0, \quad \text{otherwise,}$$

then $d(C_i, C_j)$ is the Hamming distance between $\langle x_1^i, \ldots, x_q^i \rangle$ and $\langle x_1^j, \ldots, x_q^j \rangle$:

(1) $\qquad d(C_i, C_j) = \sum_{m=1}^{q} |x_m^i - x_m^j|.$

(Cf. Example 1.6.) This formula (1) could be further modified by giving each cell Q_m a weight $\xi_m \geq 0$ ($m = 1, \ldots, q$):

(2) $\qquad d(C_i, C_j) = \sum_{m=1}^{q} \xi_m |x_m^i - x_m^j|,$

where $\xi_1 + \cdots + \xi_q = 1$.

Formula (1) can be written in a more perpicious form by noting that it in fact equals the *cardinality of the symmetric difference* between the sets $\mathbf{CT}_i$ and $\mathbf{CT}_j$:

(3) $\qquad d(C_i, C_j) = |\mathbf{CT}_i \, \Delta \, \mathbf{CT}_j|.$

(See Niiniluoto, 1978a, b.) (Cf. Example 1.7.) $\mathbf{CT}_i \, \Delta \, \mathbf{CT}_j$ contains precisely those cells where C_i and C_j disagree; (3) tells how many such cells there exists. Thus, $d(C_*, C_i)$ is the combined size of the areas I and II in Figure 1. Hence,

$$d(C_i, C_j) = 0 \quad \text{iff} \quad \mathbf{CT}_i \, \Delta \, \mathbf{CT}_j = \emptyset \quad \text{iff} \quad C_i = C_j.$$

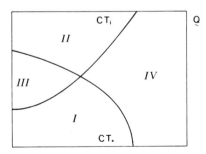

Fig. 1. Constituents C_i and C_*.

Function (3) takes values between 0 and q, and there are $\binom{q}{r}$ constituents in Γ at the distance of r from a given constituent C_i. Hence,

$$\sum_{j=1}^{t} d(C_i, C_j) = \sum_{r=0}^{q} \binom{q}{r} r = q2^{q-1}$$

for all $i = 1, \ldots, t$, and

$$d(C_i, C_j) = q \quad \text{iff} \quad \mathbf{CT}_i = -\mathbf{CT}_j.$$

An illustration for $q = 4$ is given in Figure 2, where the center corresponds to the constituent C_i with $\mathbf{CT}_i = \{Q_1, Q_3, Q_4\}$.

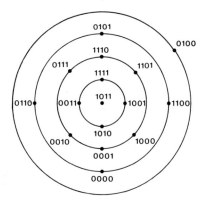

Fig. 2. Constituents of a language with 4 Q-predicates.

Let d_c be the normalized form of the distance (3):

(4) $\quad d_c(C_i, C_j) = \dfrac{1}{q} |\mathbf{CT}_i \, \Delta \, \mathbf{CT}_j|.$

Then

$$0 \leq d_c(C_i, C_j) \leq 1$$
$$d_c(C_i, C_j) = 0 \quad \text{iff} \quad i = j$$
$$d_c(C_i, C_j) = 1 \quad \text{iff} \quad \mathbf{CT}_i = -\mathbf{CT}_j$$
$$d_c(C_i, C_j) = d_c(C_j, C_i)$$

$$\frac{1}{t} \sum_{j=1}^{t} d_c(C_i, C_j) = \frac{1}{2} \quad \text{for all } i = 1, \ldots, t.$$

Hence, Γ with the distance d_c is a balanced cognitive problem.

When I discovered (1) in 1974, I did not guess that it is in fact 100 years old. The second edition of W. S. Jevons's book *The Principles of Science* (1877) contains — as I found out to my surprise in the spring of 1977 — a detailed discussion on Q-predicates generated by two or three one-place predicates, and on the monadic constituents relative to such Q-predicates (see Jevons, 1958, pp. 134—143). But this is not all: Jevons gives also a report on a paper published in *Proceedings of the Manchester Literary and Philosophical Society* (vol. xvi, February, 1877) by the famous mathematician and philosopher W. K. Clifford. In that paper, Clifford discussed the constituents generated by four one-place predicates, with the aim of giving a manageable classification of the 2^{16} or 65536 resulting "selections of combinations" (i.e., monadic constituents). The basic tool in Clifford's calculations was a measure of the 'distance' between constituents — this measure is precisely that given in (1) and (3). For the case of two monadic predicates (cf. Figure 2), the constituents at distance one from origin are called by him 'proximates'; 'mediates' are at distance two, 'ultimates' at distance three, and the 'observe' at distance four.

For these historical reasons, it is appropriate to call d_c defined by (4) the *Clifford measure* for the distance between monadic constituents.

The Clifford measure can be motivated by noting that a constituent C_i corresponds to the set $\mathbf{CT}_i$ which is as it were a 'figure' or a 'painting' in the 'table' $\mathbf{Q}$. The distance between two such 'paintings' should be defined by pairing their corresponding features and by calculating the overall difference between such pairs. We have seen in Chapter 1.3 that this is indeed a standard procedure in pattern recognition, where 'patterns' or 'figures' are represented by a set of their features, and the distances in a feature space are defined by the number of matches and mismatches. Indeed, a basic strategy in optical character recognition is to replace a figure by a 'binary image', and then to measure the fit between such images by the Hamming distance. A monadic constituent *is* in the literal sense a binary image (cf. Figure 1.16), and the Clifford measure d_c *is* precisely the (normalized) Hamming distance.

Another way of motivating the Clifford measure d_c is to note that the number of true (non-tautologous) generalizations in language L_0^k entailed by constituent C_i is a function of the set $\mathbf{CT}_i \, \Delta \, \mathbf{CT}_*$ (see Niiniluoto, 1982a). More precisely, the number of true existential statements entailed by C_i depends on $\mathbf{CT}_i \cap \mathbf{CT}_*$ (i.e., area III in Figure 1), and the number of true universal generalizations entailed by C_i depends on $-\mathbf{CT}_i \cap -\mathbf{CT}_*$ (i.e., area IV in Figure 1). For example, if C_i is C_*, it entails all the true generalizations; if $\mathbf{CT}_i \cap \mathbf{CT}_* = \emptyset$, all the existential statements entailed by C_i are false; if $-\mathbf{CT}_i = -\mathbf{CT}_*$, C_i entails all the true generalizations. Now the union of these sets III and IV is the complement of the symmetric difference between $\mathbf{CT}_i$ and $\mathbf{CT}_*$:

$$-(\mathbf{CT}_i \triangle \mathbf{CT}_*) = (\mathbf{CT}_i \cap \mathbf{CT}_*) \cup (-\mathbf{CT}_i \cap -\mathbf{CT}_*).$$

To satisfy our cognitive interest in finding truth, it seems natural to require that the closeness of C_i to C_* should be proportional to the sum

$$|\mathbf{CT}_i \cap \mathbf{CT}_*| + |-\mathbf{CT}_i \cap -\mathbf{CT}_*|,$$

and, hence, to the number of true existential and universal generalizations entailed by C_i. The complement of this quantity, i.e.,

$$q - |\mathbf{CT}_i \cap \mathbf{CT}_*| - |-\mathbf{CT}_i \cap -\mathbf{CT}_*|,$$

is precisely the measure (3).

These considerations illustrate the fact that a constituent C_i can make two different types of mistakes relative to the cells in $\mathbf{Q}$: it can claim that a non-empty cell is empty (*error of type I*) or that an empty cell is non-empty (*error of type II*). (Cf. Figure 1 again.) The value $d_c(C_i, C_*)$ of the Clifford measure is then simply the *total number of errors of type I and II*, divided by the size q of $\mathbf{Q}$. All the errors of both kinds are thus assumed to have equal weights.

In spite of its plausibility, the Clifford measure d_c is not applicable in all contexts, since it depends only on the *number* of mistakes and ignores any consideration of their *seriousness*. Tichý's (1976) general measure of truthlikeness, when applied to monadic languages, gives results which differ radically from those based on d_c, since it follows from his requirements that the distance $d(C_i, C_*)$ should be also a function of the distances between Q-predicates — measured in Tichý (1976) by the non-normalized city block distance $\varepsilon_1(Q_j, Q_m)$ (see (3.5)). This limitation of d_c can be illustrated by an example where the language L contains only one family of predicates $\mathcal{M}$ with a distance function ε.

EXAMPLE 1. Let L be a monadic language with one family $\mathcal{M} = \{M_j^1(x) \mid j = 1, \ldots, 500\}$, where

$$M_j^1(x) = x \text{ is an atom with the weight } j,$$

and

$$\varepsilon(M_j^1, M_m^1) = |j - m|/500.$$

Here the universe consists of the atoms of a certain substance which actually exist in nature. Then a constituent C_i in L gives a list of the atomic weights of all the natural isotopes of this substance. Suppose that $\mathbf{CT}_* = \{M_{250}^1\}$, $\mathbf{CT}_1 = \{M_{255}^1\}$, $\mathbf{CT}_2 = \{M_{252}^1, M_{255}^1\}$, and $\mathbf{CT}_3 = \{M_{50}^1\}$. Then definition (4) gives the result

$$d_c(C_*, C_1) = d_c(C_*, C_3) = \tfrac{1}{250} < \tfrac{3}{500} = d_c(C_*, C_2).$$

It is clear that d_c fails here to reflect the distances from the true atomic weight 250: intuitively C_1 should be better than C_3, since

$$\varepsilon(M_{250}^1, M_{255}^1) = \tfrac{1}{100} < \tfrac{40}{100} = \varepsilon(M_{250}^1, M_{50}^1).$$

It is thus desirable to find a modification of the Clifford measure d_c which is sensitive to the *cognitive seriousness* of the mistaken claims of a false constituent C_i. At the same time, such a measure should be a function of the symmetric difference $\mathbf{CT}_i \triangle \mathbf{CT}_*$ — but not merely of its cardinality. It was proposed in my paper for the Jyväskylä Conference in 1976 that d_c can be modified by assigning to each mistake of C_i a number which indicates its seriousness (see Niiniluoto, 1978a, p. 447).

Suppose we evaluate the distance between two constituents C_i and C_j, where intuitively C_i plays the role of the true constituent. If C_j makes an error of type II with respect to a cell Q_m (i.e., $Q_m \in \mathbf{CT}_j - \mathbf{CT}_i$), then this false existence claim is not very serious provided that by C_i there really are individuals in some cell close to Q_m. Similarly, if C_j makes an error of type I with respect to cell Q_m (i.e., $Q_m \in \mathbf{CT}_i - \mathbf{CT}_j$), then this false non-existence claim is the less serious the closer from Q_m one can find a cell empty by C_i. Let

$$\varepsilon_{uv} = \varepsilon_1(Q_u, Q_v)$$

be the normalized city-block measure on **Q**. Then, let us define

(5) $\quad d_J(C_i, C_j) = \sum_m \beta(Q_m),$

where

(6) $\quad \beta(Q_m) = 0, \quad\quad\quad\quad\quad$ if $m \notin \mathbf{CT}_i \triangle \mathbf{CT}_j$
$\quad\quad\quad\quad = \min_{u \in \mathbf{CT}_i} \varepsilon_{um}, \quad$ if $m \in \mathbf{CT}_j - \mathbf{CT}_i$
$\quad\quad\quad\quad = \min_{u \notin \mathbf{CT}_i} \varepsilon_{um}, \quad$ if $m \in \mathbf{CT}_i - \mathbf{CT}_j$.

If $\mathbf{CT}_i = \emptyset$, then the weight of type II error $\min \varepsilon_{um}, u \in \mathbf{CT}_i$, is defined to be 1; similarly, if $-\mathbf{CT}_i = \emptyset$, then the weight of type I error $\min \varepsilon_{um}, u \notin \mathbf{CT}_i$, is defined to be 1.[1]

The *Jyväskylä measure* d_J is a generalization of the Clifford measure d_c in the following sense: if ε in (5) is the trivial metric, so that $\varepsilon_{uv} = 1$ for all u and v, then (5) reduces to (3). Further, d_J gives intuitively satisfactory results in Example 1:

$d_J(C_*, C_1) = \frac{5}{500} + \frac{1}{500} = \frac{6}{500}$
$d_J(C_*, C_2) = \frac{5}{500} + \frac{2}{500} + \frac{1}{500} = \frac{8}{500}$
$d_J(C_*, C_3) = \frac{200}{500} + \frac{1}{500} = \frac{201}{500}.$

An interesting feature in definition (6) is that d_J is not symmetric. It has been argued by Tversky that similarity metrics in general are not symmetric (see Chapter 1.4). David Lewis has suggested that degrees of similarity between possible worlds need not be symmetric, since what is important in evaluating the distance of world U from world V may depend upon some facts or features of world U (Lewis, 1973, p. 51). In our case, the non-symmetry of d_J is rooted in the distinction between the two types of errors: the errors of type II in the evaluation of $d_J(C_i, C_j)$ turn out to be errors of type I in the evaluation of $d_J(C_j, C_i)$.

A symmetric distance function between monadic constituent can be obtained by treating the two types of errors in a symmetrical fashion: if C_j commits an error of type I with respect to Q_m, evaluate the seriousness of this mistake by the minimum distance of Q_m from $\mathbf{CT}_j$ (rather than from $-\mathbf{CT}_i$). The resulting function d_w is called the *weighted symmetric difference* (cf. Niiniluoto, 1978a, p. 448; 1978b, p. 303):

(7) $\quad d_w(C_i, C_j) = \sum_{m \in Q} \beta(Q_m),$

where

$$\begin{aligned}
\beta(Q_m) &= 0, & &\text{if } m \notin \mathbf{CT}_i \, \Delta \, \mathbf{CT}_j \\
&= \min_{u \in \mathbf{CT}_i} \varepsilon_{um}, & &\text{if } m \in \mathbf{CT}_j - \mathbf{CT}_i \\
&= \min_{u \in \mathbf{CT}_j} \varepsilon_{um}, & &\text{if } m \in \mathbf{CT}_i - \mathbf{CT}_j.
\end{aligned}$$

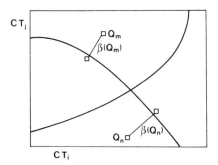

Fig. 3. Definition of d_J.

Again d_w reduces to (3), when ε is the trivial metric on $\mathbf{Q}$. In contrast with d_J, measure d_w is symmetric: $d_w(C_i, C_j) = d_w(C_j, C_i)$.

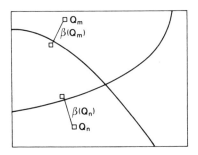

Fig. 4. Definition of d_w.

EXAMPLE 2. Let L be as in Example 1, and assume

$$\mathbf{CT_*} = \{230, 232, 233, 234, 235, 236, 237, 238, 239, 240\}$$
$$\mathbf{CT_1} = \{230, 233, 234, 235, 236, 237, 238, 239, 240\}$$
$$\mathbf{CT_2} = \{230, 232, 233, 234, 235, 237, 238, 239, 240\}.$$

(See Figure 5.) Here C_1 and C_2 make no errors of type II, but both of them make one error of type I: C_1 falsely claims that isotopes 232 do

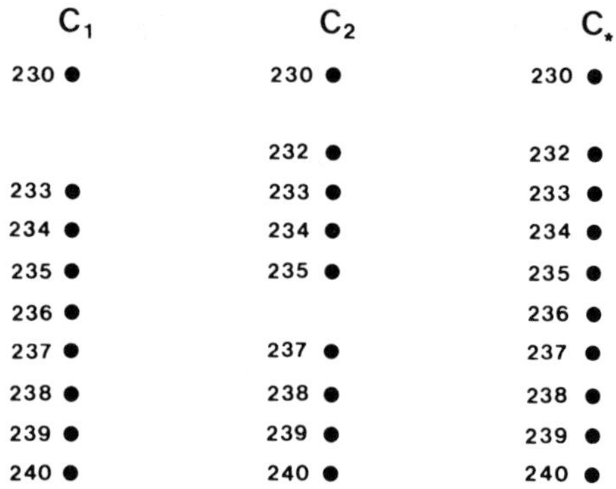

C_1	C_2	C_*
230 ●	230 ●	230 ●
	232 ●	232 ●
233 ●	233 ●	233 ●
234 ●	234 ●	234 ●
235 ●	235 ●	235 ●
236 ●		236 ●
237 ●	237 ●	237 ●
238 ●	238 ●	238 ●
239 ●	239 ●	239 ●
240 ●	240 ●	240 ●

Fig. 5. Existing isotopes according to C_1, C_2, and C_*.

not exist, while C_2 falsely claims that isotopes 236 do not exist. The former mistake is less serious than the latter, since 232 is closer to the really non-existing 231 than 236 is to 231 or 241. Measure d_J satisfies this intuition, but d_w not, since

$$d_J(C_*, C_1) = \tfrac{1}{500} < \tfrac{5}{500} = d_J(C_*, C_2)$$
$$d_w(C_*, C_1) = \tfrac{1}{500} = d_w(C_*, C_2).$$

Example 2 suggests that the weighted symmetric difference d_w does not treat adequately false non-existence claims, and that d_w is less satisfactory than d_J as a measure of distance from the truth. This observation can be made sharper, when we have considered the distance functions of Tichý and Oddie in Section 3.

The weights (6) and (7) are obviously not the only possible choices for formula (5). For example, we could replace the minimum function in (6) by the average function:

(8) $\beta(Q_m) = 0,$ if $m \notin \mathbf{CT}_i \triangle \mathbf{CT}_j$

$$= \frac{1}{|\mathbf{CT}_i|} \sum_{u \in \mathbf{CT}_i} \varepsilon_{um}, \quad \text{if } m \in \mathbf{CT}_j - \mathbf{CT}_i$$

$$= \frac{1}{|-\mathbf{CT}_i|} \sum_{u \notin \mathbf{CT}_i} \varepsilon_{um}, \quad \text{if } m \in \mathbf{CT}_i - \mathbf{CT}_j.$$

A similar modification could be made in (7). However, these measures do not have any intuitive advantages over d_J and d_w.

Roberto Festa (1982b) has made the interesting observation that a doubly weighted combination of the functions (6) and (7) leads to a family of distance measures $d_F^{\pi\rho}$, $0 \leq \pi \leq 1$, $0 \leq \rho \leq 1$, which contains d_J and d_w as special cases:

(9) $\beta(Q_m) = 0,$ if $m \notin \mathbf{CT}_i \triangle \mathbf{CT}_j$

$$= \frac{\pi\rho \min_{u \in \mathbf{CT}_i} \varepsilon_{um} + (1-\pi)(1-\rho) \min_{u \notin \mathbf{CT}_i} \varepsilon_{um}}{\pi\rho + (1-\pi)(1-\rho)} \quad \text{if } m \in \mathbf{CT}_j - \mathbf{CT}_i$$

$$= \frac{(1-\pi)\rho \min_{u \in \mathbf{CT}_i} \varepsilon_{um} + \pi(1-\rho) \min_{u \notin \mathbf{CT}_i} \varepsilon_{um}}{(1-\pi)\rho + \pi(1-\rho)} \quad \text{if } m \in \mathbf{CT}_i - \mathbf{CT}_j.$$

Hence,

$$d_F^{1\rho} = d_J$$
$$d_F^{\pi 1} = d_w.$$

Festa notes further that the special case $d_F^{\frac{1}{2}\frac{1}{2}}$ is both symmetric (like d_c and d_w, but unlike d_J) and specular (like d_c and d_J, but unlike d_w). Here a distance function d between monadic constituents is called *specular* if $d(C_i, C_j) = d({}^sC_i, {}^sC_j)$, where ${}^sC_j$ is the photographic 'negative' of C_j, i.e., every positive occurence of a Q-predicate in C_j is replaced by a negative occurence of the same predicate, and vice versa.

Specularity in Festa's sense is an appealing adequacy condition which is favourable to the measures d_c, d_J, and $d_F^{\frac{1}{2}\frac{1}{2}}$, but not to d_w or to Tichý's d_T (see the next section). $d_F^{\frac{1}{2}\frac{1}{2}}$ is the only known distance function which takes account of the distances between Q-predicates and is symmetric and specular at the same time. Still, in my view the assignment (9) of the cognitive seriousness of errors is intuitively less perspicious than (6).

Let us still mention one further kind of modification of the Clifford measure. Function d_c, defined by (4), gives an equal weight to errors of kind I and to errors of kind II. However, in some cognitive situations we might be more interested in positive matches between C_i and C_j (cf. area III in Figure 1) than in negative matches (area IV) — or vice versa. One of the measures proposed in Tuomela (1978) for the distance between monadic statements gives the following result when applied to the special case constituents:

$$(10) \quad d_{Tu}(C_i, C_j) = \gamma \cdot \frac{1}{q} |\mathbf{CT}_i \triangle \mathbf{CT}_j| + \\ + (1-\gamma) \cdot \left[\beta \cdot \frac{|\mathbf{CT}_i \triangle \mathbf{CT}_j|}{|\mathbf{CT}_i \cup \mathbf{CT}_j|} + \\ + (1-\beta) \frac{|-\mathbf{CT}_i \triangle -\mathbf{CT}_j|}{|-\mathbf{CT}_i \cup -\mathbf{CT}_j|} \right].$$

Festa (1982b) observes that (10) can be written in the form

$$(11) \quad d_{Tu}(C_i, C_j) = \gamma d_c(C_i, C_j) + \\ + (1-\gamma) \left[\beta \left(1 - \frac{|\mathbf{CT}_i \cap \mathbf{CT}_j|}{|\mathbf{CT}_i \cup \mathbf{CT}_j|} \right) + \\ + (1-\beta) \frac{|-\mathbf{CT}_i \cap -\mathbf{CT}_j|}{|-\mathbf{CT}_i \cup -\mathbf{CT}_j|} \right]$$

so that $d_{Tu}(C_i, C_j)$ is a weighted average of a 'Clifford-factor' and a 'structure factor', where the latter is the weighted combination of the degree of relative positive disagreement and the degree of relative negative disagreement between C_i and C_j. When $\beta = \frac{1}{2}$, d_{Tu} is both symmetric and specular, but it does not take into account the distances between Q-predicates. Tuomela's second, more complicated measure combines d_{Tu} with a new factor which is defined as a sum of certain average distances between Q-predicates (see (10) in Tuomela, 1978).

In the situations, where it is more important to be right about the existence claims rather than about non-existence claims, the simplest adequate measure is obtained by modifying d_J in the following way: multiply the penalty of type I errors in (6) by a constant $\rho_I > 1$ and the penalty of type II errors by a constant $\rho_{II} < 1$. Then d_J is the limiting case with $\rho_I = \rho_{II} = 1$.

9.2. MONADIC CONSTITUENTS WITH IDENTITY

Let $\Gamma_{=}^{(d)}$ be the set of depth-d constituents of the monadic language $L_N^{k=}$, and let $C_*^{(d)}$ be the true constituent of depth d. Then the elements $C_i^{(d)}$ of $\Gamma_{=}^{(d)}$ assert for each Q-predicate Q_j, $j = 1, \ldots, q$, that Q_j is occupied by 0 or 1 or $\cdots$ or $d - 1$ or *at least d* individuals (see Chapter 2.6). If $d \geq N$, the constituents in $\Gamma_{=}^{(d)}$ are equivalent to the structure descriptions of L_N^k (cf. Chapter 8.3). We shall assume here that the size N of the universe is ω. Then, for all $d \geq 1$, each constituent C_i in $\Gamma_{=}^{(d)}$ corresponds one-to-one to a q-tuple $\langle N_1^i, \ldots, N_q^i \rangle$, where $0 \leq N_j^i \leq d$ for all $j = 1, \ldots, q$.

A simple generalization of the Clifford measure is the following distance function on $\Gamma_{=}^{(d)}$:

$$(12) \quad d(C_i^{(d)}, C_m^{(d)}) = \sum_{j=1}^{q} |N_j^i - N_j^m|$$

(cf. Niiniluoto, 1978a, p. 448; see also (8.25) above).

EXAMPLE 3. Let $L_\omega^{2=}$ contain two predicates $M_1(x) =$ 'x is a prime number' and $M_2(x) =$ 'x is even'. Assume that $L_\omega^{2=}$ is given a standard interpretation on the set ω of natural numbers. Then the true depth-3 constituent $C_*^{(3)}$ corresponds to the sequence $\langle 1, 3, 3, 3 \rangle$, when the Q-predicates are defined as in Example 2.2. Then $C_i^{(3)}$ with $\langle 1, 0, 3, 3 \rangle$ is by (12) equally close to $C_*^{(3)}$ as $C_j^{(3)}$ with $\langle 1, 3, 3, 0 \rangle$ is.

The definition (12) faces the same problem as the Clifford measure, for it presupposes that all mistakes have equal weights. One natural way of modifying (12) is the following: an absolute error about the true number N_j^* is more serious when N_j^* is small than when N_j^* is large. In other words, the seriousness of the errors about cell Q_j is inversely proportional to N_j^*. This idea leads to a non-symmetric distance of $C_m^{(d)}$ from $C_i^{(d)}$:

$$\sum_{j=1}^{q} \frac{|N_j^i - N_j^m|}{N_j^i}.$$

This non-symmetric measure — which obviously is a Manhattan variant of Pearson's χ^2-statistic — is inapplicable in cases where some $N_j^i = 0$. Thus, we could replace it by

$$(13) \quad \sum_{j=1}^{q} \frac{|N_j^i - N_j^m|}{N_j^i + 1}.$$

Measure (13) is still independent of the distance between the cells Q_j. It is not easy to see how it could be modified in this respect in the same way as d_J in Section 1. A suggestion towards this direction was made in Niiniluoto (1978b), p. 303, but this idea has never been worked out. However, Tuomela's (1978) treatment of theory-distance implies two definitions which are essentially combinations of d_{Tu} and a normalization of (12). Developing this idea, we might propose the following measure — at least as a first approximation. Let C_i and C_j be the constituents in L_N^k entailed by $C_i^{(d)}$ and $C_j^{(d)}$, respectively. Then

$$(14) \quad d(C_i^{(d)}, C_j^{(d)}) = \rho d_J(C_i, C_j) + \rho' \sum_{j=1}^{q} \frac{|N_i^j - N_m^j|}{N_i^j + 1},$$

where $\rho > 0, \rho' > 0$.

Measures like (14) are relatively complicated and will not be pursued here further. Namely, it seems to me that monadic constituents in L_N^k and in $L_N^{k=}$ are fundamentally different kinds of targets — and they lead to different types of cognitive problems. An indication of this difference can be seen already in our difficulty in deciding whether monadic constituents with identity belong to the chapter on singular statements (cf. Ch. 8.3) or to the chapter on generalizations. These constituents invite us to consider cognitive problems where a certain number of individuals (with no revealed identity) have to be placed into their correct places in a classification system. Monadic constituents without identity have a different kind of relation to individuals and to singular statements, as we shall see in the next section.

9.3. TICHÝ-ODDIE DISTANCES

Pavel Tichý (1976) proposed a measure of truthlikeness which is generally applicable to the sentences of full first-order language (see also Chapter 10.1 below). In the special case of monadic languages, his definition for the distance between monadic constituents takes the following form.

A *linkage* η between sets $\mathbf{CT}_i$ and $\mathbf{CT}_j$ is a surjective mapping of the larger of the sets onto the smaller one. The cardinality card(η) of linkage η is then $\max\{|\mathbf{CT}_i|, |\mathbf{CT}_j|\}$. The *breadth* $B(\eta)$ of linkage η is defined as the average distance between the pairs of Q-predicates that η links with each other:

(15) $\quad B(\eta) = \dfrac{1}{\text{card}(\eta)} \sum_{\langle Q_u, Q_v \rangle \varepsilon \eta} \varepsilon(Q_u, Q_v),$

where ε is the normalized city-block distance on $\mathbf{Q}$. Tichý's distance between monadic constituents C_i and C_j is defined as the breadth of the narrowest linkage between $\mathbf{CT}_i$ and $\mathbf{CT}_j$:

(16) $\quad d_T(C_i, C_j) = \min\{B(\eta) \mid \eta \text{ is a linkage between } \mathbf{CT}_i \text{ and } \mathbf{CT}_j\}.$

It follows immediately that d_T is symmetric, non-specular, and satisfies the condition

$$d_T(C_i, C_j) = 0 \quad \text{iff} \quad C_i = C_j.$$

Further, d_T reaches its maximum value 1 in the following case:

(17) $\quad d_T(C_i, C_j) = 1 \quad \text{iff} \quad \mathbf{CT}_i = \{Q_u\}, \mathbf{CT}_j = \{Q_v\}, \varepsilon_{uv} = 1.$

The difference between d_T and the measures based on the symmetric difference $\mathbf{CT}_i \Delta \mathbf{CT}_j$ (such d_c, d_J, and d_w of Section 1) can be illustrated by an example (cf. Niiniluoto, 1978a, p. 446; 1978b, p. 304).

EXAMPLE 4. Let C_1, C_2, and C_* be the constituents

$\mathbf{CT}_1 = \{Q_1, Q_3\}$
$\mathbf{CT}_2 = \{Q_1, Q_2, Q_3\}$
$\mathbf{CT}_* = \{Q_3, Q_4\}$

of the monadic language L_N^k with two primitive predicates M_1 and M_2. (See Figure 6.) Here both C_1 and C_2 make an error of type II relative to cell Q_4; C_1 makes an error of type I relative to cell Q_1, and C_2 makes the additional error type I relative to cell Q_2. It seems therefore that C_1 should be closer to the truth C_* than C_2. Measures d_c, d_J, and d_w give precisely this result:

$$d_c(C_*, C_1) = \tfrac{2}{4} < \tfrac{3}{4} = d_c(c_*, C_2)$$
$$d_J(C_*, C_1) = \tfrac{1}{2} + \tfrac{1}{2} = 1 < \tfrac{3}{2} = \tfrac{1}{2} + \tfrac{1}{2} + \tfrac{1}{2} = d_J(C_*, C_2)$$
$$d_w(C_*, C_1) = 1 < \tfrac{3}{2} = d_w(C_*, C_2),$$

Fig. 6.

but d_T behaves in the opposite way:

$$d_T(C_*, C_1) = \tfrac{1}{2}(\tfrac{1}{2} + \tfrac{1}{2}) = \tfrac{1}{2}(0 + 1) = \tfrac{1}{2}$$
$$d_T(C_*, C_2) = \tfrac{1}{3}(0 + \tfrac{1}{2} + \tfrac{1}{2}) = \tfrac{1}{3}.$$

While in Example 4 I see C_1 as making *two* errors (with respect to cells Q_1 and Q_4), and C_2 as making one *additional* error (with respect to cell Q_2), Tichý defends his measure d_T by claiming the opposite: "It is C_1 which shares C_2's error of positing some M_1's and makes the additional error of positing no non-M_2's" (Tichý, 1978, p. 183). It is thus evident that Tichý and I endorse different ways of counting the mistakes that constituents make. If there is a general principle underlying Tichý's intuitive judgment of Example 4, it seems to be this: in measuring the distance between monadic constituents, attention should be paid to the *marginal distribution*, i.e., to the number of M_i's and non-M_i's, rather than to the cells Q_j. This observation is reinforced by Tichý's judgment about the following example (cf. Tichý, 1976; Niiniluoto, 1978a, p. 449; 1978b, p. 305).

EXAMPLE 5. Consider the generalizations

(B1) $(\exists^1 x) M_1(x) \,\&\, (x) \sim M_2(x)$

(B2) $(\exists^2 x) M_1(x) \,\&\, (x) \sim M_2(x)$

(B3) $(\exists^2 x) M_1(x) \,\&\, (\exists^1 x) M_2(x) \,\&\, (x)(M_1(x) \supset \sim M_2(x))$

(B4) $(\exists^2 x) M_1(x) \,\&\, (\exists^2 x) M_2(x) \,\&\, (x)(M_1(x) \supset \sim M_2(X))$,

where the truth is that the cells $Q_1, Q_2, Q_3,$ and Q_4 contain 1, 1, 1, and ω individuals, respectively. Tichý implicitly assumes that generalizations (B1)–(B4) claim the existence of at least three individuals in the cell Q_4. With this qualification, (B1)–(B4) are equivalent to the following depth-3 constituents of language $L_N^{2=}$: $\langle 0, 1, 0, 3 \rangle$, $\langle 0, 2, 0, 3 \rangle$, $\langle 0, 2, 1, 3 \rangle$, and $\langle 0, 2, 2, 3 \rangle$ (see Figure 5). Tichý argues that intuitively "(B1)–(B3) are increasingly better approximations to the truth", while (B3) and (B4) "seem equidistant from the truth". The same result is delivered by his formal definition of truthlikeness. Instead, our measure (12) for the distance between monadic constituents with identity gives the distances

$$d(C_*, B1) = 2$$
$$d(C_*, B2) = 3$$
$$d(C_*, B3) = 2$$
$$d(C_*, B4) = 3,$$

which contradicts Tichý's intuitive and theoretical results.

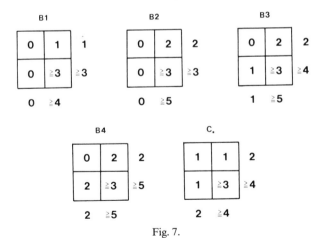

Fig. 7.

Figure 7 shows that the marginal distributions of B1—B3 are indeed increasingly better approximations of the true marginal distribution of C_*. However, it is clear that appeal to the marginal distribution, instead of the *cell distribution*, is beside the point. The use of the differences about the cells is the standard procedure in statistical measures of fit between two distributions. There is also the decisive fact that cell distributions determine marginal distributions, but not vice versa (cf. Niiniluoto, 1978b, pp. 305—306). For example, the entirely distinct 2×2 tables in Figure 8 have the same marginal values, and the constituents of Figure 9 are of course different even though they agree in positing the existence of M_1's, non-M_1's, M_2's, and non-M_2's. It is thus clear that constituents are primarily statements about the cells Q_j, only secondarily about the M_i's, so that their distances should be defined in terms of the cell distributions.

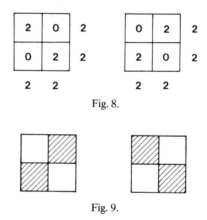

Fig. 8.

Fig. 9.

While Tichý's intuitive defense of his approach is here unsatisfactory, and does not lend support to his formal theory, we may try to find the deeper underlying motivation somewhere else. It is argued in Niiniluoto (1979a) that Tichý's theory can be at least understood — but not accepted — by considering his view about the nature of generalizations.

Consider an example, suggested by Tichý (in correspondence).

EXAMPLE 6. Let L_N^3 be a monadic language with the primitive predicates M_1, M_2, M_3, and let C_1, C_2, and C_* be the constituents in Figure 10. Then Tichý would claim that C_2 is closer to C_* than C_1,

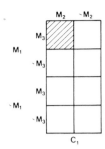

Fig. 10.

since

$$d_T(C_1, C_*) = 1$$
$$d_T(C_2, C_*) = \tfrac{1}{7}(\tfrac{0}{3} + \tfrac{1}{3} + \tfrac{1}{3} + \tfrac{1}{3} + \tfrac{2}{3} + \tfrac{2}{3} + \tfrac{2}{3}) = \tfrac{3}{7},$$

while my measures d_c and d_J give the opposite result:

$$d_c(C_1, C_*) = \tfrac{2}{8} < \tfrac{6}{8} = d_c(C_2, C_*)$$
$$d_J(C_1, C_*) = 1 + \tfrac{1}{3} = \tfrac{4}{3}$$
$$d_J(C_2, C_*) = \tfrac{1}{3} + \tfrac{1}{3} + \tfrac{1}{3} + \tfrac{2}{3} + \tfrac{2}{3} + \tfrac{2}{3} = 3.$$

It is clear that in Example 6 Tichý cannot be thinking about the *true generalizations* entailed by C_1 and C_2 (C_1 entails more of them than C_2), but rather the *distributions of individuals* that these constituents allow. Assume that C_1 is replaced by an infinite conjunction $Q_1(a_1)$ & $Q_1(a_2)$ & ... claiming that all the individuals $a_1, a_2, \ldots$ belong to cell M_1 & M_2 & M_3. Similarly, C_* claims that all individuals belong to cell $Q_8 = {\sim}M_1$ & ${\sim}M_2$ & ${\sim}M_3$. Then C_1 places each individual to the maximally distant cell from its correct place — and, as also (17) shows, the distance between C_1 and C_* is maximal. This judgment indeed follows if we apply the measure (8.15) for the distance between two infinite state descriptions: for C_1 relative to C_*, we have $r_0 = r_1 = r_2 = 0$ and $r_3 = 1$, so that (8.15) yields the value 1. By the same reasoning, C_2 can be regarded as an infinite disjunction of infinite state descriptions for which $r_0 > 0$, $r_1 > 0$, $r_2 > 0$, and $r_3 = 0$. Each of these disjuncts is closer to C_* than C_1, so that also their average (cf. (8.13)) is closer to the truth than C_1.

In brief, the motivation for Tichý's claim about Example 6 can be understood by the fact that according to C_2 all the individuals of the universe must be closer to their true places than according to C_1. This formal result also follows from our definition of the distance between state descriptions.

This point can be generalized: Tichý's distance measure d_T can be obtained as a distance between certain state descriptions (cf. Figure 11).

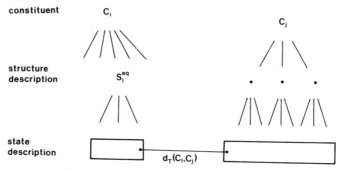

Fig. 11. The generation of Tichý's distance d_T.

Let us say that an infinite structure description $S(N_1, \ldots, N_q)$ is *equally distributed* if all the limiting non-zero numbers N_j of individuals in different cells Q_j are equal. Similarly, an infinite state description s is equally distributed if it entails an equally distributed structure description. For example, s is equally distributed if it entails $S(\frac{1}{3}, \frac{1}{3}, \frac{1}{3}, 0, \ldots, 0)$. For each constituent C_i, let S_i^{eq} be the equally distributed structure description which divides all the individuals evenly to the cells in $\mathbf{CT}_i$. Define the distance $d(s, S_i^{eq})$ of S_i^{eq} from an infinite state description s by

(18) $\quad d(s, S_i^{eq}) = \min_{s' \text{ entails } S_i^{eq}} d(s, s')$

where $d(s, s')$ is defined by (8.15). If $s' = \bigwedge_{n < \omega} Q_{i_n}(a_n)$ is a state description in the range of S_i^{eq}, then the value (18) equals

$$\min_\pi d\left(s, \bigwedge_{n < \omega} Q_{i_n}(\pi a_n)\right)$$

where π is a permutation of the individuals $\{a_n \mid n < \omega\}$.

Assume now C_i is the wider of two constituents C_i and C_j, so that $|\mathbf{CT}_i| \geq |\mathbf{CT}_j|$. Then correlate C_j with an infinite state description s_j so that $d(s_j, S_i^{eq})$ is the minimum. Define $d(C_j, C_i)$ as this minimum value:

(19) $\quad d(C_j, C_i) = \min\limits_{s \text{ entails } C_j} d(s, S_i^{eq}).$

Then $d(C_j, C_i)$ defined by (19) equals Tichý's distance $d_T(C_j, C_i)$.

Thus, in Example 6

$$S_1^{eq} = S(1, 0, 0, 0, 0, 0, 0, 0)$$
$$S_2^{eq} = S(0, \tfrac{1}{7}, \tfrac{1}{7}, \tfrac{1}{7}, \tfrac{1}{7}, \tfrac{1}{7}, \tfrac{1}{7}),$$

so that the distances of C_1 and C_2 from the true state description

$$\bigwedge_{n < \omega} Q_8(a_n)$$

are by (18) and (19)

$$\tfrac{1}{3}(0 \cdot 0 + 0 \cdot 1 + 0 \cdot 2 + 1 \cdot 3) = 1$$
$$\tfrac{1}{3}(\tfrac{1}{7} \cdot 0 + \tfrac{3}{7} \cdot 1 + \tfrac{3}{7} \cdot 2 + 0 \cdot 3) = \tfrac{3}{7},$$

respectively.

In Example 4,

$$S_1^{eq} = S(\tfrac{1}{2}, 0, \tfrac{1}{2}, 0)$$
$$S_2^{eq} = S(\tfrac{1}{3}, \tfrac{1}{3}, \tfrac{1}{3}, 0),$$

so that by (18) and (19)

$$d(C_1, C_*) = \min\limits_{s \in R(C_*)} d(s, S(\tfrac{1}{2}, 0, \tfrac{1}{2}, 0))$$
$$= \tfrac{1}{2}(\tfrac{1}{2} \cdot 0 + 0 \cdot 1 + \tfrac{1}{2} \cdot 2) = \tfrac{1}{2}$$

which is obtained by letting $s \in R(C_*)$ divide the individuals evenly to Q_3 and Q_4, and

$$d(C_2, C_*) = \min\limits_{s \in R(C_*)} d(s, S(\tfrac{1}{3}, \tfrac{1}{3}, \tfrac{1}{3}, 0))$$
$$= \tfrac{1}{2}(\tfrac{1}{3} \cdot 0 + \tfrac{2}{3} \cdot 1 + 0 \cdot 2) = \tfrac{1}{3}$$

which is obtained by letting $s \in R(C_*)$ divide the individuals to cells Q_3 and Q_4 in the proportions $\tfrac{1}{3}$ and $\tfrac{2}{3}$.

A similar treatment can be given to Example 4 involving monadic constituents with identity: as such constituents $C_i^{(d)}$ are in fact structure descriptions, their distance from a state description s can be defined again by (18), and then the distance between two such constituents by (19). Again, the result equals the value of Tichý's distance d_T.

This derivation of measure d_T shows — for me at least — why one might be inclined to define the distances between constituents in Tichý's manner. At the same time, it makes visible some of the *ad hoc* features of his definition of $d_T(C_i, C_j)$. Why should we always treat the wider constituent C_i as equally distributed, when the other constituent does not usually satisfy this condition? Why is the case that in evaluating $d_T(C_i, C_j)$ and $d_T(C_m, C_j)$, $m \neq i$, the state description s_j replacing C_j may be different in the two cases?

The point of these questions is not only the fact that it is an arbitrary matter to choose one state description S_i^{eq} to represent a monadic constituent C_i without identity. They also indicate that Tichý has not made it clear what the targets of his cognitive problems are — we have in fact seen that the state description acting as the target in (19) is variable.

Further, it appears that Tichý is interested in *put-these-individuals-to-their-right-places* problems rather than in *paint-these-cells-with-their-right-colours* problems. Both of them may be legitimate and worth studying. For the former type of problems, the natural target would be the true state description; for the latter, the true monadic constituent. In my view, the former should be treated by the methods of Chapter 8 that concern state descriptions of individuals, the latter by the distance functions given in Sections 1 and 2.

To highlight the difference between these two types of problems, let us still observe some essential features that a monadic constituent C_* without identity has as a target. Such a sentence C_* includes the cells $Q_j \in \mathbf{Q}$ as the primary units of information, and the distribution of individuals within the non-empty classes $\mathbf{CT}_i$ is regarded as irrelevant. For this reason, measures d_c and d_J are intended to be sensitive to the number and seriousness of different *kinds* of errors, not to any cumulative number of 'individual' errors. Therefore, unlike d_T, they cannot be derived from the distances of any state descriptions.

For a problem with C_i as a target, as soon as a mistake of type I relative to cell Q_j has been made once, it becomes irrelevant whether it is repeated several times. This is not satisfied by distance measures for state descriptions. It is also crucially important that two errors of the same degree of seriousness may be of different kinds and therefore both count in the evaluation of the distances — a feature that an

MONADIC GENERALIZATIONS

average function fails to take into account. These points can be illustrated by an example.

EXAMPLE 7. According to our distance measure (8.12), the sentences

$$\sim M_1(a_1) \,\&\, M_2(a_1) \,\&\, \sim M_1(a_2) \,\&\, M_2(a_2)$$

and

$$\sim M_1(a_1) \,\&\, M_2(a_1) \,\&\, M_1(a_2) \,\&\, \sim M_2(a_2)$$

are equally distant from the truth

$$\sim M_1(a_1) \,\&\, \sim M_2(a_1) \,\&\, \sim M_1(a_2) \,\&\, \sim M_2(a_2).$$

Corresponding to this fact, Tichý's measure d_T — unlike d_c and d_J — gives the result that in Figure 12 constituents C_1 and C_2 (as well as C_3) are equally close to C_*. This is the case in spite of the fact that C_3 is inconsistent with *all* the true universal generalizations, and C_2 excludes many of the true existential and universal statements entailed by C_1.

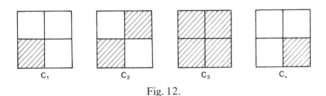

Fig. 12.

If distance function d_T is applicable to different cognitive problems from those involving d_c and d_J, i.e., to singular rather than general problems, it is no wonder that they give deviating results. However, if we nevertheless wish to regard d_T as a rival to d_c and d_J for cognitive problems with monadic constituents C_i in L_N^k, then a crucial difference between them can be seen to be the following: d_T uses *averages* of errors where d_c and d_J use *sums* of errors (cf. Example 6). It may seem a little surprising that Tichý recommends an averaging procedure both over a *disjunction* of constituents (see Chapter 6.4) and over a *conjunction* of existence and non-existence claims. But we have in fact seen that the average function in d_T is really 'inherited' from the distance function between state descriptions (where it is reasonable). The following example (cf. Niiniluoto, 1979a, p. 375) shows that this feature of d_T leads to clearly unacceptable consequences.

332   CHAPTER NINE

EXAMPLE 8. Let L be the isotope language of Example 1, and let T_1, T_2, and T_* be the constituents of Figure 13. Then the narrowest Tichý-linkage between T_1 and T_* correlates 230 with 230, and 250 and 270 with 250 so that its breadth is

$$d_T(T_*, T_1) = \tfrac{1}{3} \cdot \tfrac{20}{500} = \tfrac{1}{75}.$$

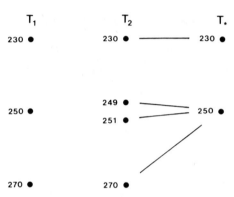

Fig. 13. Tichý-linkage between T_2 and T_*.

The narrowest linkage between T_2 and T_* correlates 230 with 230, and 249, 251, and 270 with 250, so that its breadth is

$$d_T(T_*, T_2) = \tfrac{1}{4} \left(\tfrac{1}{500} + \tfrac{1}{500} + \tfrac{20}{500} \right) = \tfrac{11}{1000} < \tfrac{1}{75}.$$

Thus, d_T gives the strange result that a physicist who first supports theory T_1 can make it better by changing the true existential claim 'There are isotopes with weight 250' to two false claims 'There are isotopes of weights 249 and 251'. By making more mistakes than T_1, theory T_2 gets a better average score than T_1 — and therefore d_T prefers T_2 to T_1.

In his response to this example, Oddie (1981) agrees with my judgment that T_1 should be more truthlike than T_2. However, he argues that the difficulty arises from Tichý's concept of a linkage: "it is thus not the *averaging* procedure which produces the result Niiniluoto dislikes: it is Tichý's unfair pairing procedure". Oddie suggests that the linkage η between **CT**$_2$ and **CT**$_*$ in Figure 13 gives an 'unfair weighting', since only one element of **CT**$_2$ is correlated with 230 and three elements of

CT$_2$ with 250. And he points out that the method of *fair* linkages proposed in Oddie (1979), p. 231, where the links are distributed as evenly as possible, produces the same result that I find desirable.

This proposal does not work, however, as the following example shows (see Niiniluoto, 1982a, p. 293).

EXAMPLE 9. Make theory T_2 of Example 8 even worse by adding two further false existential claims to it: T_3 claims that the existing isotopes are 225, 230, 235, 249, 251, and 270. Then a Tichý-linkage, which is also fair in Oddie's sense, is represented in Figure 14. It follows that Tichý and Oddie are both committed to saying that T_3 is closer to the truth T_* than theories T_1 and T_2.

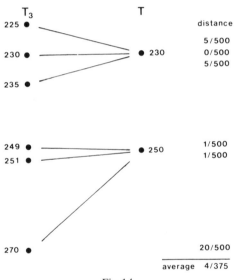

Fig. 14.

We may thus conclude that Oddie's modification of d_T is not more adequate than the original. It amounts to a different reduction of monadic constituents to representative state descriptions than Tichý's linking method, but *any* such attempt is bound to be arbitrary.

Oddie (1981) presents still one counterexample to my measures. Its essential contents can be simply represented by a further isotope example (Niiniluoto, 1982a, p. 294).

334   CHAPTER NINE

EXAMPLE 10. Let C_1, C_2, and C_* be the monadic constituents of Figure 15. Then according to d_c and d_J, C_1 is closer to the truth C_* than C_2, while d_T and d_w give the opposite result.

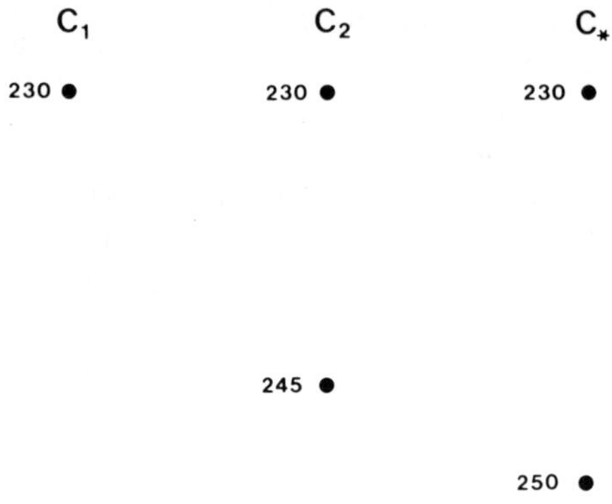

$$
\begin{array}{ccc}
C_1 & C_2 & C_* \\
230 \bullet & 230 \bullet & 230 \bullet \\
 & & \\
 & 245 \bullet & \\
 & & 250 \bullet \\
\end{array}
$$

Fig. 15.

Is C_1 really a 'hopeless theory', as Oddie thinks? I think this would be the case only if C_1, C_2, and C_* were *purely existential* statements

$$D_1 = (\exists x)Q_{230}(x)$$
$$D_2 = (\exists x)Q_{230}(x) \& (\exists x)Q_{245}(x)$$
$$D_* = (\exists x)Q_{230}(x) \& (\exists x)Q_{250}(x).$$

Then it might be reasonable to think that D_2 is closer to D_* than D_1. This is the ordering that d_w and Tichý's d_T would give to the original C's. Interestingly enough, the same result is also obtained, when D_1, D_2, and D_* are expressed as disjunctions of constituents and my d_J is combined with the min-sum distance $\Delta_{ms}^{\gamma\gamma'}$ (γ' sufficiently large). However, the replacement of C_1 with D_1 ignores the fact that constituent C_1 contains also a negative part which implies the non-existence of certain kinds of individuals. Thus, besides the true existential claim 'There are isotopes 230', C_1 makes several true non-existence claims ('There are no isotopes 231', 'There are no isotopes 232', etc.). In particular, it makes an important true claim 'There are no isotopes 245', about which constituent C_2 is mistaken. As a result, C_2 implies fewer true universal generalizations than C_1. It is not at all clear that this loss could be compensated by the fact that the false existential statement, 'There are isotopes 245', is not very far from the true statement 'There are isotopes 250'. Why should a slightly false existence claim be a heavier credit than a true non-existence claim?

It is important to distinguish *false claims of non-existence* from mere *failures to make a correct existence claim*. Constituents C_1 and C_2 commit the former type of error (i.e., error type I) with respect to isotope 250. Measure d_J accounts for this fact by assigning to them a penalty which equals the distance of 250 from the closest really non-existing cell 249. On the other hand, the purely existential claims D_1 and D_2 fail to assert that isotopes 250 exist, so that one might measure the seriousness of these errors by the distance of 250 from the nearest isotopes that do exist by these statements. This is precisely what measure d_w does.

Further evidence that d_w and d_T do not adequately account for the non-existence claim of constituents is provided by Example 2, where d_T agrees with d_w (and differs from d_J) that C_1 and C_2 are equally close to C_*. It remains as a matter of further dispute whether d_w or d_T is a better measure of the similarity of the existential parts of constituents. The following example seems to give a case where d_w is more plausible than d_T: if C_3 claims the existence of isotopes 216, 230, and 235, then d_T claims that C_3 is closer to C_* than C_1 (cf. Example 10), and d_w claims the opposite.

9.4. EXISTENTIAL AND UNIVERSAL GENERALIZATIONS

When Γ is the set of the constituents of the monadic language L_N^k ($0 \leq N \leq \omega$), its disjunctive closure $D(\Gamma)$ consists of all the consistent generalizations

$$g = \bigvee_{i \in J_g} C_i$$

expressible in L_N^k without individual constants (cf. (2.25)). The degree of truthlikeness $\mathrm{Tr}(g, C_*)$ of an arbitrary generalization $g \in D(\Gamma)$ can then be defined by $1 - \mathrm{red}(\{\Delta(C_*, C_i) \mid i \in J_g\})$ (cf. Chapter 6). We shall illustrate the results, when Δ is chosen as d_C or d_J, and red is $\Delta_{mm}^{\gamma\gamma'}$ or $\Delta_{nms}^{\gamma\gamma'}$.

The most interesting generalizations in $D(\Gamma)$ are those which make some definite existence or non-existence claims relative to the cells Q_m,

$m = 1, \ldots, q$. If g_j is such a statement, let us denote by $\mathbf{PC}_j$ its positive existence claims and by $\mathbf{NC}_j$ its negative existence claims, i.e.,

$$i \in \mathbf{J}_{g_j} \quad \text{iff} \quad \mathbf{PC}_j \subseteq \mathbf{CT}_i \subseteq \mathbf{Q} - \mathbf{NC}_j.$$

In other words, the distributive normal form of g_j contains all the monadic constituents C_i which claim that at least the cells $\mathbf{PC}_j$ are exemplified and at least the cells $\mathbf{NC}_j$ are empty. The set $\mathbf{QM}_j = \mathbf{Q} - (\mathbf{PC}_j \cup \mathbf{NC}_j)$ contains then the Q-predicates about which g_j does not say anything definite (the question mark area in Figure 16). If $\mathbf{QM}_j = \emptyset$, i.e., $\mathbf{PC}_j \cup \mathbf{NC}_j = \mathbf{Q}$, then g_j is a constituent. If $\mathbf{PC}_j = \emptyset$, g_j is a *universal* generalization; if $\mathbf{NC}_j = \emptyset$, g_j is a purely *existential* statement.

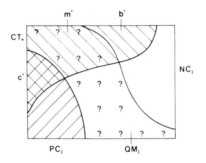

Fig. 16. Positive and negative claims made by generalization g_j.

Let us first choose the Clifford measure d_C as the distance on Γ. Let us denote

$$c = |\mathbf{PC}_j|$$
$$m = |\mathbf{QM}_j|$$
$$b = |\mathbf{NC}_j|.$$

Let C_* be the true constituent, and assume that

$$|CT_* \cap \mathbf{PC}_j| = c'$$
$$|CT_* \cap \mathbf{QM}_j| = m'$$
$$|CT_* \cap \mathbf{NC}_j| = b'.$$

Hence, $0 \leq c' \leq c$, $0 \leq m' \leq m$, $0 \leq b' \leq b$, and $c' + m' + b' =$

… $|\mathbf{CT_*}|$. Then

$$\Delta_{\min}(C_*, g_j) = (c - c' + b')/q$$
$$\Delta_{\max}(C_*, g_j) = (c - c' + b' + m)/q$$

$$\Delta_{\text{sum}}(C_*, g_j) = \frac{1}{2^{q-1}} \cdot \sum_{j=0}^{m} \binom{m}{j}(c - c' + b' + j)/q$$

$$= \frac{1}{2^{q-1}} \left[\frac{c - c' + b'}{q} \cdot 2^m + \frac{m}{q} \cdot 2^{m-1} \right]$$

$$= \frac{1}{q 2^{q-m}} [2(c - c' + b') + m]$$

$$\Delta_{\text{mm}}^{\gamma}(C_*, g_j) = (c - c' + b')/q + (1 - \gamma)m/q$$

$$\Delta_{\text{nms}}^{\gamma\gamma'}(C_*, g_j) = \gamma(c - c' + b')/q +$$
$$+ \frac{\gamma'}{q 2^{q-m}} [2(c - c' + b') + m]$$

$$= \left(\gamma + \frac{2\gamma'}{2^{q-m}} \right) \frac{(c - c' + b')}{q} + \frac{\gamma' m}{q 2^{q-m}}.$$

Hence, the degree of truthlikeness of g_j,

(20) $\quad \text{Tr}(g_j, C_*) = 1 - \Delta_{\text{nms}}^{\gamma\gamma'}(C_*, g_j)$

$$= 1 - \left(\gamma + \frac{2\gamma'}{2^{q-m}} \right) \frac{(c - c' + b')}{q} - \frac{\gamma' m}{q 2^{q-m}}$$

decreases with $c - c'$ (i.e., the number of wrong existence claims), b' (i.e., the number of wrong non-existence claims), and m (i.e., the informative weakness of g_j).

If g_j is a constituent C_j, so that $m = 0$, then (20) reduces simply to

$$1 - \left(\gamma + \frac{2\gamma'}{2^q}\right) d_C(C_*, C_j)$$

(cf. formula (6.88)).

If g_j is a universal generalization, so that $c = 0$ and $m = q - b$, then (20) implies

(21) $\quad \text{Tr}(g_j, C_*) = 1 - \left(\gamma + \frac{\gamma'}{2^{b-1}}\right) \cdot \frac{b'}{q} - \gamma' \cdot \frac{q-b}{q 2^b}.$

This value decreases with the number b' of mistakes, and increases with b (i.e., the number of excluded cells). Here b may be called the degree of *boldness* of generalization g_j, since high b is correlated with high information content.[2] In other words, for fixed b', $\text{Tr}(g_j, C_*)$ increases, when the information content of g_j, or

$$\text{cont}_u(g_j) = \frac{2^q - 2^{q-b}}{2^q - 1},$$

increases.

If g_j is a purely existential statement, so that $b = 0$ and $m = q - c$, then (20) implies

(22) $\quad \text{Tr}(g_j, C_*) = 1 - \left(\gamma + \frac{\gamma'}{2^{c-1}}\right) \cdot \frac{c - c'}{q} - \gamma' \cdot \frac{q - c}{q 2^c}.$

This value decreases with the number $c - c'$ of mistakes, and (for fixed $c - c'$) increases with c. Again, increase of c is here associated with an increase in the information content of g_j.

Similar results can be proved for the distance between two generalizations g_1 and g_2. Instead of the general case, let us illustrate the situation with universal statements. Let

$$u_1 = |NC_1 - NC_2|$$
$$u_2 = |NC_2 - NC_1|$$
$$u_3 = |NC_1 \cap NC_2|.$$

Then

(23) $\quad \delta_{nms}^{aa'}(g_1, g_2) = \dfrac{a}{2^q} \sum\limits_{i=0}^{u_2} \binom{u_2}{i} 2^{q-(u_1+u_2+u_3)} \cdot \dfrac{i}{q} +$

$\qquad\qquad\qquad + \dfrac{a'}{2^q} \sum\limits_{i=0}^{u_1} \binom{u_1}{i} 2^{q-(u_1+u_2+u_3)} \cdot \dfrac{i}{q}$

$\qquad\qquad\qquad = \dfrac{1}{q 2^{u_1+u_2+u_3}} [a u_2 2^{u_2-1} + a' u_1 2^{u_1-1}]$

which decreases with u_3 (i.e., the number of shared claims) and increases with u_1 and u_2, where $u_1 + u_2 = |\mathbf{NC}_1 \triangle \mathbf{NC}_2|$.[3]

If the Clifford measure d_C is replaced with the Jyväskylä measure d_J, similar results follow. However, it is not possible to write general formulas like (20)–(23) for this case, since now it is not only the *number* of mistakes, but also the *location* of the mistakes, that matters. For example, we may have two generalizations g_1 and g_2 with the same values of b, c, b', and c', but still g_1 is closer to the truth C_* than g_2, since

$$\sum_{v \in PC_1} \min_{u \in CT_*} \varepsilon_{uv} < \sum_{v \in PC_2} \min_{u \in CT_*} \varepsilon_{uv}$$

and

$$\sum_{v \in NC_1} \min_{u \notin CT_*} \varepsilon_{uv} < \sum_{v \in NC_2} \min_{u \notin CT_*} \varepsilon_{uv}.$$

EXAMPLE 11. Let L_0^2 be the monadic language with two families of predicates, $\mathcal{M}_1 = \{M_1, \sim M_1\}$, $\mathcal{M}_2 = \{M_1^2, M_2^2, M_3^2\}$, where

$\qquad M_1(x) =$ 'x is a raven'
$\qquad M_1^2(x) =$ 'x is black'
$\qquad M_2^2(x) =$ 'x is grey'
$\qquad M_3^2(x) =$ 'x is white',

340   CHAPTER NINE

and

$$d_1(M_1, \sim M_1) = 1$$
$$d_2(M_1^2, M_2^2) = d_2(M_2^2, M_3^2) = \tfrac{1}{2}$$
$$d_2(M_1^2, M_3^2) = 1.$$

Let g_1 be the universal generalization 'All ravens are grey', and g_2 'All ravens are white'. The truth is represented by the constituent C_* in Figure 17. Here $q = 6$, $b_1 = b_2 = 2$, $b'_1 = b'_2 = 1$, so that by (21)

$$\mathrm{Tr}(g_1, C_*) = \mathrm{Tr}(g_2, C_*) = 1 - \left(\gamma + \frac{\gamma'}{2}\right) \cdot \frac{1}{6} - \gamma' \cdot \frac{4}{6 \cdot 4}$$

$$= 1 - \frac{\gamma}{6} - \frac{\gamma'}{4}.$$

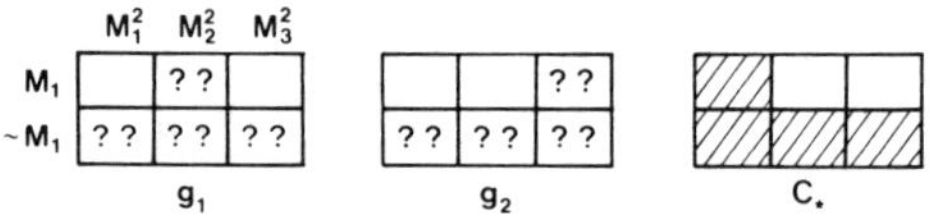

Fig. 17.

When d_C is used, g_1 and g_2 thus have the same degree of truthlikeness. However, d_J gives the intuitively correct result that

$$\mathrm{Tr}(g_1, C_*) > \mathrm{Tr}(g_2, C_*).$$

Similarly, if g_3 and g_4 are the existential statements 'There are grey ravens' and 'There are white ravens', respectively (see Figure 18), then $c = 4$ and $c' = 0$, so that by (22)

$$\mathrm{Tr}(g_3, C_*) = \mathrm{Tr}(g_4, C_*) = 1 - \left(\gamma + \frac{\gamma'}{8}\right) \cdot \frac{4}{6} - \gamma' \cdot \frac{2}{6 \cdot 16}$$

$$= 1 - \frac{2}{3}\gamma - \frac{5\gamma'}{48}.$$

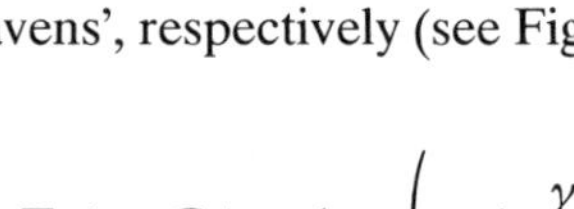

Fig. 18.

Again, the use of d_J instead of d_C implies that

$$\mathrm{Tr}(g_3, C_*) > \mathrm{Tr}(g_4, C_*).$$

9.5. ESTIMATION PROBLEM FOR GENERALIZATIONS

Assume now that the true constituent C_* is unknown, and let e_n^c be a description in L_N^k of a sample of n individuals exemplifying c cells $\mathbf{CT}_e$. Then in Hintikka's generalized combined system of inductive logic the relative probabilities $P(C_i/e_n^c)$ of constituents C_i compatible with evidence e_n^c can be expressed by

$$(24) \quad P(C_i/e_n^c) \approx \frac{(a/n)^{w_i - c}}{(1 + a/n)^{q - c}}$$

when n is sufficiently large (see (2.84)). Here $w_i = |\mathbf{CT}_i|$ is the width of constituent C_i. When c is fixed and $n \to \infty$, (24) approaches one if and only if $w_i = c$, i.e., $\mathbf{CT}_i = \mathbf{CT}_e$ and C_i is the minimally wide constituent C^c compatible with e_n^c (see (2.79)). By (7.21) and (7.22) it follows that the estimated degree of verisimilitude for $g \in D(\Gamma)$ on e_n^c approaches $\mathrm{Tr}(g, C^c)$:

$$(25) \quad \mathrm{ver}(g/e_n^c) \underset{\substack{n \to \infty \\ c \text{ fixed}}}{\to} \mathrm{Tr}(g, C^c),$$

and

$$(26) \quad \mathrm{ver}(g/e_n^c) \underset{\substack{n \to \infty \\ c \text{ fixed}}}{\to} 1 \quad \text{iff} \quad \vdash g \equiv C^c.$$

Further,

$$(27) \quad \mathrm{ver}(g/e_n^c) \underset{\substack{n \to \infty \\ c \text{ fixed}}}{\to} \mathrm{Tr}(g, C_*) \quad \text{iff} \quad C^c = C_*.$$

In general, high posterior probability is neither sufficient nor necessary for judging a generalization as highly truthlike. For tautologies, the posterior probability is of course one but the degree of estimated verisimilitude is $1 - \gamma'$ (cf. (7.11)).

In Hintikka's system, C^c is, among constituents compatible with e_n^c, the least probable initially and the most probable given e_n^c (for large n). It therefore maximizes the difference $P(g/e_n^c) - P(g)$, which thus serves among constituents as an indicator of truthlikeness.

Let us now assume for simplicity that the distance on Γ is defined by d_C. Then by (20)

$$\mathrm{Tr}(C_i, C^c) = 1 - \left(\gamma + \frac{\gamma'}{2^{q-1}}\right) d_C(C^c, C_i).$$

Hence, by (25), for large values of n,

(28) $\quad \mathrm{ver}(C_i/e_n^c) \approx 1 - \left(\gamma + \frac{\gamma'}{2^{q-1}}\right) d_C(C^c, C_i),$

i.e., $\mathrm{ver}(C_i/e_n^c)$ decreases, when the distance between C_i and C^c increases. If C_i is compatible with e_n^c, then simply

$$d_C(C^c, C_i) = (w_i - c)/q,$$

so that, by (24), the posterior probability $P(C_i/e_n^c)$ covaries with the expected degree of verisimilitude of C_i on e_n^c:

(29) Let n be sufficiently large. For constituents C_i and C_j compatible with e_n^c,

$\mathrm{ver}(C_i/e_n^c) < \mathrm{ver}(C_j/e_n^c)$ iff $P(C_i/e_n^c) < P(C_j/e_n^c)$.

(See Niiniluoto, 1977b, p. 139; 1979b, p. 252.)

On the other hand, if evidence e_n^c falsifies C_i, the posterior probability $P(C_i/e_n^c)$ is of course zero. Let $c' < c$ be the number of Q-predicates in $\mathbf{CT}_i$ which are exemplified in e_n^c. (Thus, C_i is compatible with e_n^c if and only if $c' = c$.) The distance between C_i and C^c is now

$$d_C(C^c, C_i) = (c - c')/q + (w_i - c')/q = (c + w_i - 2c')/q.$$

For fixed c and w_i, $\mathrm{ver}(C_i/e_n^c)$ therefore increases with the value c'.

Let $g \in D(\Gamma)$ be compatible with e_n^c. Then the *degree of corroboration* of g on evidence e_n^c has been defined by Hintikka (1968b) as follows:

(30) $\quad \mathrm{corr}(g/e_n^c) = \min_{i \in \mathbf{J}_g(e)} P(C_i/e_n^c),$

where $\mathbf{J}_g(e)$ includes those constituents in the normal form of g that are compatible with e_n^c.[4] This measure $\mathrm{corr}(g/e_n^c)$ has a high value only if g

does not allow constituents that are improbable relative to e_n^c. Asymptotically corr behaves like ver:

$$\text{corr}(g/e_n^c) \underset{\substack{n \to \infty \\ c \text{ fixed}}}{\to} 1 \quad \text{iff} \quad \vdash g \equiv C^c.$$

When n is sufficiently large, the minimum of $P(C_i/e_n^c)$, $i \in \mathbf{J}_g(e)$, is by (24) obtained by maximizing the difference $w_i - c$, i.e., by maximizing the distance $d_C(C^c, C_i)$. Therefore, $\text{corr}(g/e_n^c)$ increases when $\Delta_{\max}(C^c, g)$ decreases, so that $\text{corr}(g/e_n^c)$ is an indicator of the degree of information about the truth in g (cf. Chapter 5.4 and 6.1).

Let g be a universal generalization in L_N^k which claims that certain b cells are empty. If g is compatible with evidence e_n^c, then $w_i - c$ is maximized in $i \in \mathbf{J}_g(e)$ by choosing $w_i = q - b$. Hence, by (24) we have for sufficiently large n

$$(31) \quad \text{corr}(g/e_n^c) \approx \frac{(a/n)^{q-b-c}}{(1 + a/n)^{q-c}}$$

(cf. Niiniluoto and Tuomela, 1973, p. 131), while by (2.85)

$$(32) \quad P(g/e_n^c) \approx \frac{1}{(1 + a/n)^b}.$$

Thus, $P(g/e_n^c)$ decreases but $\text{corr}(g/e_n^c)$ increases with the index of boldness b. To compare these values with $\text{ver}(g/e_n^c)$, put $b' = 0$ in (21), since g is compatible with C^c, and obtain by (25)

$$(33) \quad \text{ver}(g/e_n^c) \approx \text{Tr}(g, C^c) = 1 - \gamma' \cdot \frac{q-b}{q 2^b}.$$

Thus, $\text{ver}(g/e_n^c)$ also increases with b, so that Hintikka's measure of corroboration $\text{corr}(g/e_n^c)$ is a good comparative indicator of the truthlikeness for universal generalizations compatible with the evidence:

(34) Let n be sufficiently large. For universal generalizations g_i and g_j compatible with e_n^c,

$\text{ver}(g_i/e_n^c) < \text{ver}(g_j/e_n^c)$ iff $\text{corr}(g_i/e_n^c) < \text{corr}(g_j/e_n^c)$.

(See Niiniluoto, 1977b, p. 140; 1979b, p. 252.)[5]

However, if g is a universal generalization that is incompatible with evidence e_n^c, then $\text{corr}(g/e_n^c) = 0$, while $\text{ver}(g/e_n^c)$ may be nevertheless high. If g claims that b' of the cells exemplified in e_n^c are empty, then, by (25), we have for sufficiently large n

$$\text{ver}(g/e_n^c) \approx \sum_{u=0}^{q-c-b+b'} \sum_{v=0}^{b-b'} \binom{q-c-b+b'}{u} \binom{b-b'}{v} \cdot$$

$$\cdot \frac{(a/n)^{u+v}}{\left(1+\frac{a}{n}\right)^{q-c}} \text{Tr}(g, C^{uv}),$$

where C^{uv} is a constituent of width $c+u+v$, and, by (21),

$$\text{Tr}(g, C^{uv}) = 1 - \left(\gamma + \frac{\gamma'}{2^{b-1}}\right) \cdot \frac{b'+v}{q} - \gamma' \cdot \frac{q-b}{q2^b}.$$

Combinatorial calculations give then the result

$$(35) \quad \text{ver}(g/e_n^c) \approx 1 - \gamma' \cdot \frac{q-b}{q2^b} - \left(\gamma + \frac{\gamma'}{2^{b-1}}\right) \cdot \frac{b'}{q} -$$

$$- \left(\gamma + \frac{\gamma'}{2^{b-1}}\right) \cdot \frac{b-b'}{q} \cdot \frac{1}{(1+n/a)}.$$

When g is compatible with e_n^c, so that $b' = 0$, (35) reduces to

$$(36) \quad 1 - \gamma' \cdot \frac{q-b}{q2^b} - \left(\gamma + \frac{\gamma'}{2^{b-1}}\right) \cdot \frac{b}{q} \cdot \frac{1}{(1+n/a)}.$$

When $n \to \infty$, (35) approaches the limit

$$\text{Tr}(g, C^c) = 1 - \gamma' \cdot \frac{q-b}{q2^b} - \left(\gamma + \frac{\gamma'}{2^{b-1}}\right) \cdot \frac{b'}{q},$$

and (36) approaches (33), i.e.,

$$1 - \gamma' \cdot \frac{q-b}{q2^b}.$$

In the extreme case, where $b = 0$ and g is a tautology, this value reduces to $1 - \gamma'$. These formulas indicate that asymptotically the best universal generalization is the *boldest generalization compatible with the evidence*:

(37)   When $n \to \infty$ and c is fixed, among universal generalizations the highest estimated degree of truthlikeness on evidence e_n^c is possessed by g_j such that $\mathbf{NC}_j = -\mathbf{CT}_e$.

For other universal generalizations g_i, the asymptotic value of $\text{ver}(g_i/e_n^c)$ decreases when the distance between g_i and g_j grows (cf. (23)).

The methodological situation discussed above can be complicated at least in two different ways. First, it may be assumed that the evidence is described in a language L_0 and the generalizations in a language L which is an extension of L_0. If the evidence e is in this sense 'incomplete' with respect to the language L, then there will be several constituents C_i of L which asymptotically ($n \to \infty$) receive non-zero posterior probabilities on e_n^c in L_0. These are precisely those constituents of L which entail the constituent C^c of L_0.[6] We have here a case that satisfies the conditions of (7.23).

Secondly, evidence e_n^c and generalization g may be expressed in the same language L_0, but there is available also theoretical evidence T, where T is a theory expressible in an extension L of L_0. The expected verisimilitude of g on e_n^c and T is then

$$\text{ver}(g/e_n^c \ \& \ T) = \sum_{i=1}^{t} P(C_i/e_n^c \ \& \ T)\text{Tr}(g, C_i).$$

It follows that asymptotically ($n \to \infty$) $\text{ver}(g/e_n^c \ \& \ T)$ equals the value $\text{Tr}(g, C_i)$, where C_i is the minimally wide L_0-constituent compatible with both e_n^c and T.[7]

CHAPTER 10

POLYADIC THEORIES

The treatment of monadic generalizations, developed in Chapter 9, is extended here to first-order languages L with relations. First, the distance between polyadic constituents in L is defined in Section 1. Then it is shown how this definition — together with the theory of distributive normal forms — allows us to measure distances between complete theories in L (Section 2) and between L-structures (Section 3). These distance functions finally lead to a general definition of the degree of truthlikeness of first-order theories in language L (Section 4).

10.1. DISTANCE BETWEEN POLYADIC CONSTITUENTS

Let $\Gamma_L^{(n)}$ be the set of the constituents of depth $n \geq 1$ in a polyadic first-order language L without identity (see Chapter 2.7.). The non-logical vocabulary of L is assumed to be finite and contain at least one non-monadic predicate. Then each depth-n constituent $C_i^{(n)}$ in $\Gamma_L^{(n)}$ describes in a systematic way all the different sequences of n inter-related individuals that can be found in a universe — or all the different samples of size n which can be drawn (with replacement) from a possible world. More precisely, to each constituent $C_i^{(n)}$ there corresponds a finite set of finite inverted trees all the branches of which are of length n. The structure of these trees is described by the attributive constituents $Ct_k^{(n-1)}$ of depth n occurring in $C_i^{(n)}$ (see formulas (2.44) and (2.55)). If L contains the identity $=$, the structure of constituents remains the same, but now the samples described by a constituent are drawn without replacement, so that the quantifiers are given an exclusive interpretation (see (2.47)).

It could be assumed that the predicates of L constitute families in the same sense as in the monadic case: in addition to families of one-place predicates, we might have families of mutually exclusive m-place predicates with a distance function defined on each of them. For simplicity, I shall restrict my attention in this chapter to the case, where the non-logical vocabulary of L consists of a finite list of logically independent primitive predicates.

To define a distance function $d^{(n)}$ on $\Gamma_L^{(n)}$, we have to be able to measure the distance between two trees (with a special structure). Two examples of tree metrics were given in Chapter 1: the Fu metric d_F in Example 1.9 and the Boorman-Olivier metric d_{BO} in Example 1.10. It is illustrative for our purposes to see how these metrics could be applied to the case of monadic constituents. Suppose that a monadic constituent C_i is represented by a simple labeled tree with q branches as in Figure 1, where a node i is labeled 'black' if $Q_i \in \mathbf{CT}_i$, and 'white' otherwise. Let substitution be the operation of changing a black node to white or vice versa. Then the Fu metric leads to the following definition: the distance $d_F(C_i, C_j)$ of the trees C_i and C_j is the minimum number of substitutions needed to transform C_i to C_j (cf. 1.27)). It immediately follows that the Fu metric equals the non-normalized Clifford measure d_C:

(1) $\quad d_F(C_i, C_j) = q \cdot d_C(C_i, C_j).$

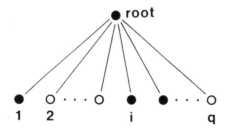

Fig. 1. A monadic constituent as a labeled tree.

Alternatively, constituent C_i could be represented by a tree where the nodes correspond to its positive existence claims (cf. Figure 2). Then to derive C_j from C_i the nodes in $\mathbf{CT}_i - \mathbf{CT}_j$ have to be deleted, and new nodes corresponding to the Q-predicates in $\mathbf{CT}_j - \mathbf{CT}_i$ have to be inserted. The number of these operations — deletions and branchings — is then the Fu distance between the trees C_i and C_j. Again, this number equals (1). Also the d_{BO}-distance between C_i and C_j is a linear function of the Clifford-distance:

$$d_{BO}(C_i, C_j) = q \cdot d_C(C_i, C_j) \cdot \max\{w_i, w_j\}.$$

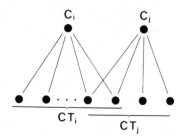

Fig. 2. Constituents with positive existence claims.

It is easy to see how these ideas can be modified so that measures d_J and d_W are obtained as special cases of a Fu metric. Instead of deleting the nodes in $\mathbf{CT}_i - \mathbf{CT}_j$, we may think that they are substituted by elements of $\mathbf{CT}_j$; the prize for this operation may be taken to be the distance of the Q-predicates in question. When the prize of the insertion of the new elements $\mathbf{CT}_j - \mathbf{CT}_i$ is defined in a suitable way, the Fu metric between C_i and C_j is a linear function of $d_J(C_i, C_j)$ or $d_W(C_i, C_j)$.

Similarly, if the tree in Figure 1 is labeled by the number N_j of individuals in each cell Q_j, $j = 1, \ldots, q$, and the prize of the substitution of N_j with N'_j is equal to $|N'_j - N_j|$, the Fu metric gives the distance (9.12) between monadic constituents with identity.

Tichý's measure d_T can also be compared with the Fu metric in an interesting way: if η is a linkage between $\mathbf{CT}_i$ and $\mathbf{CT}_j$ (where $w_i \geqslant w_j$), then the sum

$$\sum_{(Q_u, Q_v) \in \eta} \varepsilon_{uv}$$

can be viewed as the prize of the method η of transforming the tree C_i to the tree C_j. The minimum value of this *sum* would then correspond to the Fu distance between C_i and C_j. Tichý's $d_T(C_i, C_j)$ is instead defined as the *average* of the values ε_{uv} for linked Q-predicates. This observation gives us a further reason to doubt the validity of Tichý's averaging procedure in the definition of d_T (cf. Chapter 9.3).

Let now $C_i^{(n)}$ and $C_j^{(n)}$ be two unreduced depth-n constituents in L. Then $C_i^{(n)}$ is a finite tree such that each terminal node x_n is associated by a complete description $Ct^{(0)}(x_1, \ldots, x_n)$ of the properties and the relations of the elements $x_1, \ldots, x_n$ in the branch from the common root x_0 down to x_n (cf. Figure 2.4). Let $\mathbf{D}_i^{(n)}$ be the set of such descriptions associated with constituent $C_i^{(n)}$, i.e., $\mathbf{D}_i^{(n)}$ is a list of all different n-sequences of individuals that exist according to $C_i^{(n)}$. Just as a monadic constituent asserts that certain kinds of individuals exist, a depth-n constituent asserts that certain kinds of n-sequences of individuals exist.

The first proposal for a distance function on $\Gamma_L^{(n)}$ was made by Tichý (1976), who treated depth-d constituents as 'd-families' of 'd-kinds'. He suggested that the distance between two n-kinds is defined as the number of divergent answers they give to the yes-or-no questions in terms of L, i.e.,

(2) $\quad \varepsilon(Ct_1^{(0)}(x_1, \ldots, x_n), Ct_2^{(0)}(y_1, \ldots, y_n))$

is the number of atomic sentences of L which are true about $x_1, \ldots, x_n$ but false about $y_1, \ldots, y_n$. This distance is thus a direct generalization of the city block measure ε used in the monadic case (see (3.5)). Tichý (1976) then proposed that the distance between $C_i^{(n)}$ and $C_j^{(n)}$ is defined as the breadth of the narrowest linkage between the n-families $\mathbf{D}_i^{(n)}$ and $\mathbf{D}_j^{(n)}$ corresponding to $C_i^{(n)}$ and $C_j^{(n)}$. In other words,

(3) $\quad d(C_i^{(n)}, C_j^{(n)}) = d_\mathrm{T}(\mathbf{D}_i^{(n)}, \mathbf{D}_j^{(n)}),$

where d_T is defined formally by (9.16) and (2).

An alternative proposal, a direct generalization of the Clifford measure d_C when applied to the n-sequences, was made in my Jyväskylä paper:

(4) $\quad d_\mathrm{D}(C_i^{(n)}, C_j^{(n)}) = |\mathbf{D}_i^{(n)} \Delta \mathbf{D}_j^{(n)}|.$

(See Niiniluoto, 1978a, p. 451.) This measure could be normalized by dividing its value by the total number of possible n-sequences that can be described in L.

Just as d_C, measure d_D does not take into account the distances between the n-sequences. This was in effect observed by Tichý (1976) in his criticism of the concept he calls 'verisimilitude$_p$'. Therefore, it was suggested in Niiniluoto (1978a) that d_D could be modified in the same way as d_J was obtained from d_C — by using ε as the distance between the elements of the D-sets, but normalized by dividing the value (2) by the maximum distance between two n-sequences. Let us denote the resulting measure by d_DJ.

In his Section 5 ('Objections to Niiniluoto'), Tichý (1978) presents a number of arguments against the proposal d_D. Let us consider his examples.

EXAMPLE 1. Let L be the language with three two-place predicates P, Q, and R. Let $C_1^{(2)}$ claim that $P(x, y)$, $Q(x, y)$, and $R(x, y)$ for any x and y, and let $C_2^{(2)}$ and $C_*^{(2)}$ say that the relations form the patterns in Figure 3.

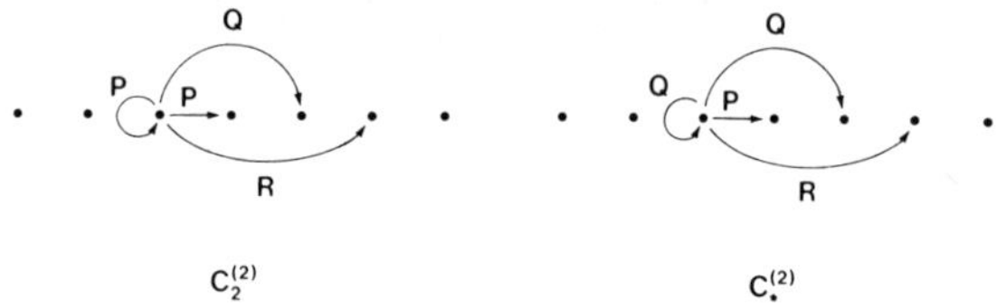

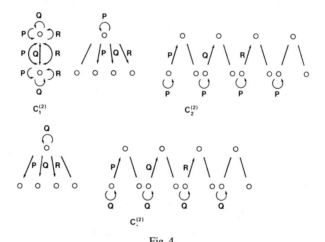

Fig. 3.

The structure of these depth-2 constituents is given in Figure 4.

Fig. 4.

Then

$$d_D(C_*^{(2)}, C_1^{(2)}) = 10$$
$$d_D(C_*^{(2)}, C_2^{(2)}) = 16,$$

even though intuitively $C_2^{(2)}$ should be closer to the truth $C_*^{(2)}$ than $C_1^{(2)}$. This example works well against definition d_D, but as a criticism against Niiniluoto (1978a) it is not quite fair, since d_{JD}, which recognizes the similarity of the elements of $D_2^{(2)}$ and $D_*^{(2)}$, gives the opposite result from d_D:

$$d_{DJ}(C_*^{(2)}, C_1^{(2)}) = \tfrac{59}{8} = 7\tfrac{3}{8}$$
$$d_{DJ}(C_*^{(2)}, C_2^{(2)}) = \tfrac{16}{8} = 2.$$

EXAMPLE 2. Let L be as in Example 1, and let $F_1^{(3)}$ and $F_*^{(3)}$ describe the two-dimensional infinite grid in Figure 5

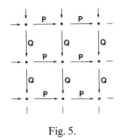

Fig. 5.

and also the following unique patterns on that grid:

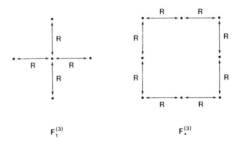

Fig. 6.

Then the constituents $F_1^{(3)}$ and $F_*^{(3)}$ include precisely the same 3-sequences, so that $d_D(F_*^{(2)}, F_1^{(2)}) = 0$, and for the same reason $d_{DJ}(F_*^{(2)}, F_1^{(2)}) = 0$. However, it is easy to see that there are 4-sequences

352   CHAPTER TEN

that distinguish the patterns of Figure 6; for example the sequence

is admitted by the expansion of $F_*^{(3)}$ to the depth 4, but not by that of $F_1^{(3)}$. Hence, d_D and d_{DJ} are able to make a difference between F_1 and F_* at depths greater than 3. This observation illustrates the general point, to be discussed later, that distances from the truth should not be expected to be invariant under increase in quantificational depth.

EXAMPLE 3. Let L be the language with one two-place predicate R, and let $G_1^{(2)}$ and $G_2^{(2)}$ be the constituents in Figure 7. Then $G_1^{(2)}$ is a true description at depth 2 of an infinite universe where R is a linear ordering with a first element and without a last element, while $G_2^{(2)}$ is a true description at depth 2 of an infinite universe where R is a linear ordering without a first and a last element. However, these two different constituents correspond to the same D-set of 2-sequences, so that their distance according to d_D (and d_{DJ}) is zero. In this case, the sets $\mathbf{D}_1^{(n)}$ and $\mathbf{D}_2^{(n)}$ remain the same for any finite depth n — the difference of the universes described by $G_1^{(2)}$ and $G_2^{(2)}$ is revealed not until at the infinite depth ω.

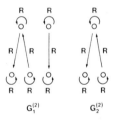

Fig. 7.

Example 3 shows convincingly that the idea of treating depth-n constituents by means of *sets* of n-sequences, rather than as *trees* with branches of length n, sometimes ignores some relevant structural features of constituents. Tichý (1978) observes that this is not only a problem for my d_D (and d_{DJ}), but also his own measure (3), proposed in Tichý (1976), is unsatisfactory, since it "neglects branching". (He attributes this criticism to Graham Oddie.)

There is one way we could try to save measures like d_D which are based on the D-sets of n-sequences. We could grant that d_{DJ} accounts for the differences in the sequences of individuals that can be drawn from the universe, and add a new factor which is able to reflect the differences in the tree structure of the constituents. So let τ_i and τ_j be the unlabeled trees corresponding to constituents $C_i^{(n)}$ and $C_j^{(n)}$, respectively, where we have enumerated the terminal elements of the branches. Then structural distance between the trees τ_1 and τ_2 could be measured by the Boorman–Olivier metric $d_{BO}(\tau_i, \tau_j)$. For example, the structural distance of $G_1^{(2)}$ and $G_2^{(2)}$ in Figure 7 is non-zero on this definition. Then the distance between constituents $C_i^{(n)}$ and $C_j^{(n)}$ could be defined as a weighted combination of d_{DJ} and d_{BO}:

$$(5) \quad d(C_i^{(n)}, C_j^{(n)}) = \rho d_{DJ}(C_i^{(n)}, C_j^{(n)}) + \rho' d_{BO}(\tau_i, \tau_j)$$
$$= \rho d_J(D_i^{(n)}, D_j^{(n)}) + \rho' d_{BO}(\tau_i, \tau_j),$$

where $\rho > 0$, $\rho' > 0$.

This measure — which is somewhat similar to the proposal (9.11) in the monadic case — seems to work nicely in a number of examples, but one cannot help feeling that it is a little artificial. Therefore, I shall proceed to consider the suggestions made in Tichý (1978), Oddie (1979), and Niiniluoto (1978b).

Instead of treating constituents $C_i^{(n)}$ as trees, where each terminal node x_n is labeled with a sentence $Ct^{(0)}(x_1, \ldots, x_n)$ describing the whole branch ending with x_n, let us consider *reduced* constituents in the sense of Chapter 2.7. This means that each node x_{n-m} in $C_i^{(n)}$ is associated with a description $A^{(m)}(x_1, \ldots, x_{n-m-1}; x_{n-m})$ of the properties of x_{n-m} and its relations to the nodes $x_1, \ldots, x_{n-m-1}$ higher up in the same branch (see Figure 8). In Tichý's terminology, this $A^{(m)}$-sentence is a conjunction of *level* $n - m$. The number of the conjuncts in $A^{(m)}$, i.e., the number of atomic formulas of L which contain the variable x_{n-m}, is denoted by $q(n - m)$. The distance between two $A^{(m)}$-sentences can be defined as the number of atomic formulas that appear unnegated in one of them and negated in the other, i.e.,

(6) $\quad \varepsilon(A_{i_1}^{(m)}(x_1, \ldots; x_{n-m}), A_{i_2}^{(m)}(y_1, \ldots; y_{n-m}))$

is the usual city block distance (cf. (2)).

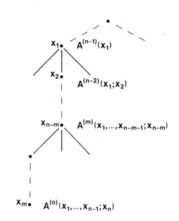

Fig. 8. A branch in a reduced constituent.

Let N_i and N_j be the sets of nodes of two depth-n constituents $C_i^{(n)}$ and $C_j^{(n)}$, respectively. Tichý (1978) defines a *linkage* between $C_i^{(n)}$ and $C_j^{(n)}$ as the smallest binary relation $\eta \subseteq N_i \times N_j$ that preserves the direct subordination relations in the associated trees: if $\langle x, y \rangle \in \eta$ and $x' \in N_i$ is directly subordinated to x, then there is a $y' \in N_j$ such that $\langle x', y' \rangle \in \eta$ and y' is directly subordinated to y (and vice versa). It follows that η links only nodes at the same level; moreover, if $\langle x, y \rangle \in \eta$, then η or η^{-1} is a surjection from the larger of the sets sub(x) and sub(y) of directly subordinated elements onto the smaller. Oddie (1979) modifies this definition by requiring that the linkage η is as *fair* as possible. Both Tichý and Oddie then define the *breadth* of linkage η by

(7) $\quad \sum_{\langle x, y \rangle \in \eta} \varepsilon(x, y) \Big/ \sum_{\langle x, y \rangle \in \eta} q(\text{the level of } x \text{ and } y),$

where $\varepsilon(x, y)$ is the distance (6) between the A-statements about x and y, respectively. The distance between constituents $C_i^{(n)}$ and $C_j^{(n)}$ is then

(8) $d_T(C_i^{(n)}, C_j^{(n)})$ = the breadth of the narrowest linkage η between $C_i^{(n)}$ and $C_j^{(n)}$.

When $n = 1$, this definition reduces to the d_T-measure (9.16) between monadic constituents.

It is clear that the Tichý-Oddie definition (8) can be criticized in the same way as in the monadic case. The following example is very similar to the counter-arguments of Chapter 9.3 (cf. Niiniluoto, 1978b, p. 319).

EXAMPLE 4. Let L be a language with identity = and four two-place predicates P, Q, R, and S. Let $G_1^{(2)}$, $G_2^{(2)}$, and $G_*^{(2)}$ be the constituents illustrated in Figure 9. Intuitively it seems to me quite obvious that $G_1^{(2)}$ should be judged to be closer to the truth $G_*^{(2)}$ than $G_2^{(2)}$ is, since the latter conjunctively adds new mistakes to those made by the former. However, d_T gives the opposite result:

$$d_T(G_*^{(2)}, G_1^{(2)}) = \frac{0 + 0 + 0 + 0 + 8}{4 + 4 + 12 + 12 + 12} = \frac{2}{11}$$

$$d_T(G_*^{(2)}, G_2^{(2)}) = \frac{0 + 0 + 0 + 0 + 0 + 8 + 0 + 4}{4 + 4 + 4 + 12 + 12 + 12 + 12 + 12}$$

$$= \frac{1}{6} < \frac{2}{11}.$$

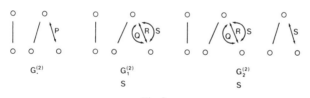

Fig. 9.

Again we see that the trouble with Tichý's measure arises from his application of an averaging procedure over conjunctively made mistakes. In fact, his own statement of the motivation for d_T appears to be different from the actual proposal (7): the distance between two constituents is defined in terms of the "most economical" linkage between their nodes, and a "linkage is the more economical the smaller the *overall number* of divergences between linked nodes" (Tichý, 1978, p. 181; my italics). This idea — which could be justified by modifying the Fu metric with 'prizes' for substitution — is destroyed by definition (7), which replaces the overall error by a special kind of *average*. The resulting measure would be radically different — and in my view much more plausible — if the normalization in (7) would be changed as follows:

$$(9) \qquad \sum_{\langle x, y \rangle \in \eta} [\varepsilon(x, y)/q(\text{the level of } x \text{ and } y)].$$

This simple but crucial modification leads to intuitively correct result in Example 4: the distance between $G_1^{(2)}$ and $G_*^{(2)}$ would be 2/3, and that between $G_2^{(2)}$ and $G_*^{(2)}$ would be 1.

In the monadic case, (9) is comparable to d_J and d_W. In fact, it would agree with $d_J(C_i, C_j)$ and $d_W(C_i, C_j)$ in those cases where $\mathbf{CT}_i \subseteq \mathbf{CT}_j$; but in other cases, it would use a different matching procedure than d_J and d_W.

These observations lead us to study the possibility of generalizing the measures d_J and d_W so that the branching structure of polyadic constituents is taken into account (see Niiniluoto, 1978b). Let us first observe that a depth-n constituent $C_j^{(n)}$ asserts the existence of certain attributive constituents $\mathbf{CT}_j^{(n-1)}$:

$$\bigwedge_{i \in \mathbf{CT}_j^{(n-1)}} (\exists x_1) Ct_i^{(n-1)}(x_1) \mathrel{\&} (x_1) \left[\bigvee_{i \in \mathbf{CT}_j^{(n-1)}} Ct_i^{(n-1)}(x_1) \right]$$

(cf. (2.44)). Thus, a straightforward generalization of the Clifford measure d_C would be the following:

$$(10) \qquad d_C^{(n)}(C_i^{(n)}, C_j^{(n)}) = |\mathbf{CT}_i^{(n-1)} \triangle \mathbf{CT}_j^{(n-1)}|.$$

For $n = 1$, $d_C^{(n)}$ reduces to d_C. Measure $d_C^{(n)}$ works clearly better than d_D: it gives the correct result in Example 3. However, just as in the monadic case, it ignores the distances between the attributive constituents $Ct^{(n-1)}$ about which $C_i^{(n)}$ and $C_j^{(n)}$ disagree. In particular, in Example 1 it gives the unintuitive result that

$$d_C^{(2)}(C_*^{(2)}, C_1^{(2)}) = 6 < 10 = d_C^{(2)}(C_*^{(2)}, C_2^{(2)}).$$

To modify the Clifford metric $d_C^{(n)}$ along the lines of d_J, a measure for

POLYADIC THEORIES 357

the distance between two attributive constituents of depth $n - 1$ is needed. Following Tuomela (1978) and Niiniluoto (1978b), this distance can be defined *recursively*: a reduced $Ct^{(n-1)}$-predicate $Ct_i^{(n-1)}(x_1)$ is equivalent to a conjunction

$$A_i^{(n-1)}(x_1) \& B_i^{(n-1)}(x_1),$$

where $B_i^{(n-1)}$ is a list of reduced $Ct^{(n-2)}$-predicates. Each member $Ct_k^{(n-2)}(x_1, x_2)$ of this list has again the form

$$A_k^{(n-2)}(x_1; x_2) \& B_k^{(n-2)}(x_1, x_2),$$

where $B_k^{(n-2)}(x_1, x_2)$ is a list of reduced $Ct^{(n-3)}$-predicates, and so on (cf. (2.46)). The distance between two $Ct^{(n-1)}$-predicates should thus be a function of the distance of their A-parts (as given by (6)) and of their B-parts. This method of evaluating the distance of $Ct^{(n-1)}$-predicates leads us to consider distances between $Ct^{(n-2)}$-predicates, $Ct^{(n-3)}$-predicates, etc. Eventually this process ends with reduced $Ct^{(0)}$-predicates which contain only the $A^{(0)}$-parts, so that their distance is given by (6). If the distance of the B-parts is defined in analogy with the symmetric difference measure d_W, the following measure is obtained (cf. Niiniluoto, 1978b, p. 318).

For $0 < m \leq n - 1$, the distance between two reduced $Ct^{(m)}$-predicates is defined by

$$(11) \quad \varepsilon^{(m)}(Ct_j^{(m)}(x_1, \ldots, x_{n-m}), Ct_k^{(m)}(y_1, \ldots, y_{n-m}))$$
$$= \varepsilon(A_j^{(m)}(x_1, \ldots; x_{n-m}), A_k^{(m)}(y_1, \ldots; y_{n-m}))/q(n-m) +$$
$$+ \sum_i \alpha^{(m)}(Ct_i^{(m-1)}),$$

where i ranges over all reduced $Ct^{(m-1)}$-predicates and

$$(12) \quad \alpha^{(m)}(Ct_i^{(m-1)}) = 0 \quad \text{if } i \notin \mathbf{CT}_j^{(m-1)} \Delta \mathbf{CT}_k^{(m-1)}$$
$$= \min_{u \in \mathbf{CT}_j^{(m-1)}} \varepsilon^{(m-1)}(Ct_u^{(m-1)}, Ct_i^{(m-1)})$$
$$\text{if } i \in \mathbf{CT}_k^{(m-1)} - \mathbf{CT}_j^{(m-1)}$$
$$= \min_{u \in \mathbf{CT}_k^{(m-1)}} \varepsilon^{(m-1)}(Ct_u^{(n-1)}, Ct_i^{(n-1)})$$
$$\text{if } i \in \mathbf{CT}_j^{(m-1)} - \mathbf{CT}_k^{(m-1)}$$

(cf. definition (9.7) of d_W). In (11), the distance of the A-parts is defined by the normalized city block measure. When $m = 0$, $Ct_i^{(0)}(x_1, \ldots, x_n)$ is of the form $A_i^{(0)}(x_1, \ldots; x_n)$, so that

$$\alpha^{(0)} = \varepsilon/q(n)$$

gives us the basis of recursion. When the distance between $Ct^{(n-1)}$-predicates has been defined in (12) recursively by function d_W, the distance between two depth-n constituents $C_i^{(n)}$ and $C_j^{(n)}$ can be obtained by applying function d_J to sets $\mathbf{CT}_i^{(n-1)}$ and $\mathbf{CT}_j^{(n-1)}$:

(13) $\quad d_J^{(n)}(C_i^{(n)}, C_j^{(n)}) = \sum_k \beta^{(n-1)}(Ct_k^{(n-1)}),$

where $\beta^{(n-1)}$ is defined as β in (9.6):

(14) $\quad \beta^{(n-1)}(Ct_k^{(n-1)}) = 0 \qquad \text{if } k \notin \mathbf{CT}_i^{(n-1)} \triangle \mathbf{CT}_j^{(n-1)}$

$$= \min_{u \in \mathbf{CT}_i^{(n-1)}} \varepsilon^{(n-1)}(Ct_u^{(n-1)}, Ct_k^{(n-1)})$$
$$\text{if } k \in \mathbf{CT}_j^{(n-1)} - \mathbf{CT}_i^{(n-1)}$$

$$= \min_{u \notin \mathbf{CT}_i^{(n-1)}} \varepsilon^{(n-1)}(Ct_u^{(n-1)}, Ct_k^{(n-1)})$$
$$\text{if } k \in \mathbf{CT}_i^{(n-1)} - \mathbf{CT}_j^{(n-1)}.$$

If $n = 1$, definition (13) reduces d_J. A similar generalization $d_W^{(n)}$ can be given to the symmetric difference function d_W.

To illustrate the behaviour of $d_J^{(n)}$, let us enumerate the 11 $Ct^{(1)}$-predicates in Figure 3 in the order they appear from the left to right, so that

$$\mathbf{CT}_1^{(1)} = \{Ct_1^{(1)}\}$$
$$\mathbf{CT}_2^{(1)} = \{Ct_2^{(1)}, Ct_3^{(1)}, Ct_4^{(1)}, Ct_5^{(1)}, Ct_6^{(1)}\}$$
$$\mathbf{CT}_*^{(1)} = \{Ct_7^{(1)}, Ct_8^{(1)}, Ct_9^{(1)}, Ct_{10}^{(1)}, Ct_{11}^{(1)}\}.$$

Then, for example,

$$\varepsilon^{(1)}(Ct_1^{(1)}, Ct_2^{(1)}) = \tfrac{2}{3} + \sum_i \alpha^{(1)}(Ct_i^{(0)})$$
$$= \tfrac{2}{3} + \tfrac{8}{9} + \tfrac{9}{9} + \tfrac{8}{9} + \tfrac{8}{9} + \tfrac{8}{9} = \tfrac{47}{9} = 5\tfrac{2}{9}$$
$$\varepsilon^{(1)}(Ct_2^{(1)}, Ct_3^{(1)}) = \tfrac{3}{3} + \tfrac{7}{9} + \tfrac{7}{9} + \tfrac{9}{9} = \tfrac{32}{9} = 3\tfrac{5}{9}$$
$$\varepsilon^{(1)}(Ct_1^{(1)}, Ct_6^{(1)}) = \tfrac{3}{3} + \tfrac{8}{9} + \tfrac{8}{9} + \tfrac{9}{9} = \tfrac{34}{9} = 3\tfrac{7}{9}$$
$$\varepsilon^{(1)}(Ct_2^{(1)}, Ct_7^{(1)}) = \tfrac{2}{3}$$
$$\varepsilon^{(1)}(Ct_3^{(1)}, Ct_8^{(1)}) = 0 + \tfrac{2}{9} + \tfrac{2}{9} = \tfrac{4}{9}.$$

Further,

$$d_J^{(2)}(C_*^{(2)}, C_1^{(2)}) = \beta^{(1)}(Ct_1^{(1)}) + \beta^{(1)}(Ct_7^{(1)}) + \beta^{(1)}(Ct_8^{(1)}) + \beta^{(1)}(Ct_9^{(1)})$$
$$+ \beta^{(1)}(Ct_{10}^{(1)}) + \beta^{(1)}(Ct_{11}^{(1)})$$
$$= \tfrac{32}{9} + 5 \cdot \tfrac{1}{9} = \tfrac{37}{9} = 4\tfrac{1}{9}$$
$$d_J^{(2)}(C_*^{(2)}, C_2^{(2)}) = \varepsilon^{(1)}(Ct_2^{(1)}, Ct_7^{(1)}) + \varepsilon^{(1)}(Ct_3^{(1)}, Ct_8^{(1)}) +$$
$$+ \varepsilon^{(1)}(Ct_4^{(1)}, Ct_9^{(1)}) + \varepsilon^{(1)}(Ct_5^{(1)}, Ct_{10}^{(1)}) +$$
$$+ \varepsilon^{(1)}(Ct_6^{(1)}, Ct_{11}^{(1)}) + 5 \cdot \tfrac{2}{9} =$$
$$= \tfrac{2}{3} + \tfrac{4}{9} + \tfrac{4}{9} + \tfrac{4}{9} + \tfrac{4}{9} + \tfrac{5}{9} = \tfrac{27}{9} = 3.$$

Similarly,

$$d_W^{(2)}(C_1^{(2)}, C_*^{(2)}) = \tfrac{32}{9} + \tfrac{47}{9} + \tfrac{32}{9} + \tfrac{32}{9} + \tfrac{32}{9} + \tfrac{34}{9} = \tfrac{209}{9} = 23\tfrac{2}{9}$$
$$d_W^{(2)}(C_2^{(2)}, C_*^{(2)}) = \tfrac{2}{3} + \tfrac{4}{9} + \tfrac{4}{9} + \tfrac{4}{9} + \tfrac{4}{9} + \tfrac{2}{3} + \tfrac{4}{9} + \tfrac{4}{9} + \tfrac{4}{9} + \tfrac{4}{9} = \tfrac{44}{9} = 4\tfrac{8}{9}.$$

In Example 4, there are four $Ct^{(1)}$-predicates so that

$$\mathbf{CT}_*^{(1)} = \{Ct_1^{(1)}, Ct_2^{(1)}\}$$
$$\mathbf{CT}_1^{(1)} = \{Ct_1^{(1)}, Ct_3^{(1)}\}$$
$$\mathbf{CT}_2^{(1)} = \{Ct_1^{(1)}, Ct_3^{(1)}, Ct_4^{(1)}\}.$$

Here

$$\varepsilon^{(1)}(Ct_1^{(1)}, Ct_2^{(1)}) = 0 + \tfrac{2}{12} = \tfrac{1}{6}$$
$$\varepsilon^{(1)}(Ct_1^{(1)}, Ct_3^{(1)}) = 0 + \tfrac{6}{12} = \tfrac{1}{2}$$
$$\varepsilon^{(1)}(Ct_2^{(1)}, Ct_3^{(1)}) = 0 + \tfrac{2}{12} + \tfrac{6}{12} = \tfrac{2}{3}$$
$$\varepsilon^{(1)}(Ct_2^{(1)}, Ct_4^{(1)}) = 0 + \tfrac{2}{12} + \tfrac{2}{12} = \tfrac{1}{3}$$
$$\varepsilon^{(1)}(Ct_3^{(1)}, Ct_4^{(1)}) = 0 + \tfrac{4}{12} + \tfrac{2}{12} = \tfrac{1}{2}.$$

Hence,

$$\begin{aligned} d_J^{(2)}(G_*^{(2)}, G_1^{(2)}) &= \beta^{(1)}(Ct_2^{(1)}) + \beta^{(1)}(Ct_3^{(1)}) \\ &= \varepsilon^{(1)}(Ct_1^{(1)}, Ct_2^{(1)}) + \varepsilon^{(1)}(Ct_1^{(1)}, Ct_3^{(1)}) \\ &= \tfrac{1}{6} + \tfrac{1}{2} = \tfrac{2}{3} \\ d_J^{(2)}(G_*^{(2)}, G_2^{(2)}) &= \beta^{(1)}(Ct_2^{(1)}) + \beta^{(1)}(Ct_3^{(1)}) + \beta^{(1)}(Ct_4^{(1)}) \\ &= \tfrac{1}{6} + \tfrac{1}{2} + \tfrac{1}{6} = \tfrac{5}{6}. \end{aligned}$$

Let us summarize the definition of $d_J^{(n)}$ in (11)–(14). A depth-n constituent $C_i^{(n)}$ asserts that certain attributive constituents $\mathbf{CT}_i^{(n-1)}$ are exemplified and others are empty. To apply our old favourite measure d_J at the depth n, we have used recursively the symmetric difference measure d_W to obtain a tree metric $\alpha^{(n-1)}$ between $Ct^{(n-1)}$-predicates (i.e., trees with branches of length n). For $n = 1$, $\alpha^{(0)}$ is simply the normalized city-block measure, and $d_J^{(1)}$ equals d_J defined in the monadic case.[1]

In the case of language L^-, there is one complication that should be mentioned. When quantifiers are given an exclusive interpretation (cf. Chapter 2.7), a depth-$(d+1)$ constituent may assert that there are precisely d individuals in the universe. But then at depths $n \geq d+1$ the expansion of this constituent cannot describe any n-sequences of different individuals, so that its attributive constituents of depths $n > d$ are still d-trees rather than n-trees. How can we measure the distance of two trees with branches of different lengths? A natural answer is suggested by the Fu metric: use the operation of stretch $n - d$ times to extend the branches of the shorter tree, and label the new nodes in the most economical way. Then add the factor $(n - d)$, perhaps with a suitable weight, to the distance that $\alpha^{(n-1)}$ gives for the n-trees. For example, if $L_N^{k=}$ is the monadic language with identity, if $C_1^{(d+1)}$ asserts that there are precisely d individuals in cell Q_i, and $C_*^{(n)}$ asserts that Q_i is occupied by at least n individuals, then at depth n the distance of $C_*^{(n)}$ and $C_1^{(d+1)}$ is a function of $(n-d)$ (cf. the definition (9.12)).

Let $d(n)$ be the maximum value of the distances $d_J^{(n)}(C_i^{(n)}, C_j^{(n)})$ for $C_i^{(n)} \in \Gamma_L^{(n)}$ and $C_j^{(n)} \in \Gamma_L^{(n)}$. Then it is clear that, for any $n < \omega$, $d(n)$ is finite, so that $d_J^{(n)}$ can be normalized by dividing its values by $d(n)$.

If g is now an arbitrary depth-n' generalization in language L, it has a distributive normal form

$$(15) \quad \vdash g \equiv \bigvee_{j \in \Delta_g^{(n)}} C_j^{(n)}$$

at any depth $n \geq n'$ (cf. (2.57)). If $C_*^{(n)}$ is the true constituent at depth n, then the *distance of g from the truth at depth n* is defined by

$$(16) \quad \Delta_{\text{nms}}^{\gamma\gamma'}(C_*^{(n)}, g) = \gamma \min_{j \in \Delta_g^{(n)}} d_J^{(n)}(C_*^{(n)}, C_j^{(n)})/d(n) +$$

$$+ \gamma' \sum_{j \in \Delta_g^{(n)}} d_J^{(n)}(C_*^{(n)}, C_j^{(n)}) \Big/ \sum_{j \in \Gamma_L^{(n)}} d_J^{(n)}(C_*^{(n)}, C_j^{(n)}),$$

and the degree of *truthlikeness* of g at depth n by

$$(17) \quad \text{Tr}(g, C_*^{(n)}) = 1 - \Delta_{\text{nms}}^{\gamma\gamma'}(C_*^{(n)}, g).$$

These definitions are *relativized to depth*: the true constituents $C_*^{(n)}$ and $C_*^{(n')}$ at different depths $n \neq n'$ are also two different targets. If $n < n'$, then the expansion of $C_*^{(n)}$ to depth n' is logically stronger than $C_*^{(n)}$, and it need not be the case that comparative distances from $C_*^{(n)}$ are preserved when the target truth becomes 'deeper'.[2] Therefore, it may happen that at depth n g_1 is closer to the truth $C_*^{(n)}$ than g_2, but the expansion of g_2 to depth $n' > n$ is closer to the truth $C_*^{(n')}$ than the expansion of g_1 to depth n' (cf. Example 2 above).

The measures (16) and (17) behave qualitatively in a similar way as their special cases in the monadic languages. Moreover, we can in principle extend the method of estimating degrees of truthlikeness to the polyadic case (cf. the remarks at the end of Chapter 2.10 and Chapter 9.4).

There is nevertheless one important difference between the monadic and polyadic cases. As we are dealing in (17) with an objective concept of truthlikeness, it is natural to assume that the normal form (15) of g does not contain any inconsistent constituents. However, it has been mentioned in Chapter 2.7 that there is no effective method for finding this normal form, where even the non-trivially inconsistent constituents have been eliminated. This consequence of the undecidability of first-order logic means that the functions Δ and Tr defined in (16) and (17) are not recursive in the general case.[3] The next section shows that nevertheless there may be effective ways of making comparative judgment of truthlikeness even for polyadic languages.

10.2 COMPLETE THEORIES

According to (2.63), a theory Σ in language L is complete if and only if Σ is axiomatizable by a monotone sequence $\langle C^{(n)} \mid n < \omega \rangle$ of constituents of L. Here $C^{(n+1)}$ is subordinate to $C^{(n)}$, so that $C^{(n+1)} \vdash C^{(n)}$ for all $n < \omega$. Let us now consider the cognitive problem $\mathscr{CT}_L$ of identifying the true complete theory $\Sigma_* = \text{Th}(\Omega_*)$ about the actual L-structure Ω_* (cf. Example 4.9).

Two cases have to be distinguished. First, assume that Σ_* is *finitely axiomatizable*. This means that the monotone basis $\langle C_*^{(n)} \mid n < \omega \rangle$ for Σ_* is likewise finite, i.e., there is a finite depth $n_0 < \omega$ such that Σ_* is axiomatizable by the true depth-n_0 constituent $C_*^{(n_0)}$. Hence, $\Sigma_* = Cn(C_*^{(n_0)})$, and $\vdash C_*^{(n_0)} \equiv C_*^{(n_0 + j)}$ for $j = 1, 2, \ldots$. Secondly, if Σ_* is *not finitely* axiomatizable, then for each $n < \omega$ there are logically stronger, 'deeper' truths $C_*^{(n+j)}$ about the world Ω_*, so that Σ_* has to be expressed by the whole infinite sequence of constituents $C_*^{(n)}$, $n = 1, 2, \ldots$.

Let Σ_1 and Σ_2 be two complete theories in L with the bases $\langle C_1^{(n)} \mid n < \omega \rangle$ and $\langle C_2^{(n)} \mid n < \omega \rangle$, respectively. To introduce a distance Δ on the space $\mathscr{CT}_L$ of complete theories in L, it is natural to consider the limit of the distances $d_J^{(n)}(C_1^{(n)}, C_2^{(n)})$ when n grows without limit (see Figure 10). In other words, we define

(18) $\quad \Delta(\Sigma_1, \Sigma_2) = \lim_{n \to \infty} d_J^{(n)}(C_1^{(n)}, C_2^{(n)}).$[4]

A computationally simpler measure is obtained by using the Clifford measure $d_c^{(n)}$ instead of $d_J^{(n)}$:

(19) $\quad \Delta_c(\Sigma_1, \Sigma_2) = \lim_{n \to \infty} d_c^{(n)}(C_1^{(n)}, C_2^{(n)}).$

Here Δ_c on $\mathscr{CT}_L$, unlike Δ, is symmetric.

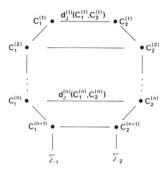

Fig. 10. Distance between complete theories.

The distance of a complete theory from the truth Σ_* is by (18) $\Delta(\Sigma_*, \Sigma)$. This distance function Δ on $\mathscr{CT}_L$ gives us the opportunity of defining the degrees of truthlikeness $\mathrm{Tr}(\Sigma, \Sigma_*)$ for complete theories in the standard way (cf. Niiniluoto, 1978a, p. 454).

The distances Δ and Δ_c are obviously only semimetrics on $\mathscr{CT}_L$, since the limit of $d^{(n)}(C_1^{(n)}, C_2^{(n)})$ may be zero for two different theories Σ_1 and Σ_2. Therefore, it also may happen that $\Delta(\Sigma_*, \Sigma_1) = \Delta(\Sigma_*, \Sigma_2)$ even if intuitively one of them is closer to the truth than the other. It is therefore desirable to find a stronger definition of the *comparative* concept of truthlikeness for complete theories. The following idea was independently suggested in 1976 by Niiniluoto (1978a) and Oddie (1979):

(20) Σ_1 is *more truthlike* than Σ_2 iff, for some $n_0 < \omega$,
$$d_J^{(n)}(C_*^{(n)}, C_1^{(n)}) < d_J^{(n)}(C_*^{(n)}, C_2^{(n)})$$
for all $n \geq n_0$.

(20) says that Σ_1 is closer to the truth Σ_* than Σ_2 if the distances of $C_1^{(n)}$ from $C_*^{(n)}$ dominate those of $C_2^{(n)}$ from some finite depth n_0 onwards. Hence,

(21) If $\Delta(\Sigma_*, \Sigma_1) < \Delta(\Sigma_*, \Sigma_2)$, then Σ_1 is more truthlike than Σ_2, but not conversely.

CHAPTER TEN

EXAMPLE 5. Let $L^=$ be a polyadic language with identity = and one two-place predicate R. Assume that $L^=$ is interpreted on an infinite domain U, where R is reflexive and symmetric. Then the true theory Σ_* is axiomatized by the monotone sequence $\langle C_*^{(n)} \mid n < \omega \rangle$, where

$$C_*^{(n)} = (\exists x_1) \ldots (\exists x_n) \bigwedge_{\substack{i=1,\ldots,n \\ j=1,\ldots,n}} R(x_i, x_j)$$

$$\& \; (x_1) \ldots (x_n) \bigwedge_{\substack{i=1,\ldots,n \\ j=1,\ldots,n}} R(x_i, x_j).$$

Since quantifiers are here given an exclusive interpretation, $C_*^{(n)}$ is equivalent to a statement with the numerical quantifier $\exists^{\geq n}$ (cf. Chapter 2.6): $C_*^{(n)}$ entails that there are *at least n* individuals in universe U, i.e.,

$$C_*^{(n)} \vdash (\exists^{\geq n} x)(x = x).$$

Let Σ_1 and Σ_2 be the complete theories in L which correctly claim that R is reflexive and symmetric, but mistakenly assert that the size of the universe U is $n_1 < \omega$ or $n_2 < \omega$, respectively, where $n_1 > n_2$. Then Σ_1 is axiomatizable by the depth-$(n_1 + 1)$ constituent

$$C_1^{(n_1+1)} = (\exists x_1) \ldots (\exists x_{n_1}) \left[\bigwedge_{\substack{i=1,\ldots,n_1 \\ j=1,\ldots,n_1}} R(x_i, x_j) \; \& \; (x) \bigvee_{i=1}^{n_1} (x = x_i) \right].$$

Similarly, Σ_2 is axiomatizable by a constituent $C_2^{(n_2+1)}$ of depth $n_2 + 1$. Thus,

$$C_1^{(n_1+1)} \vdash (\exists^{n_1} x)(x = x)$$
$$C_2^{(n_2+1)} \vdash (\exists^{n_2} x)(x = x).$$

It follows that up to depth n_2 the theories Σ_1, Σ_2, and Σ_* are equivalent; for depths $n_2 < n \leq n_1$, Σ_1 and Σ_* are equivalent while Σ_2 differs from them; for $n > n_1$, both Σ_1 and Σ_2 differ from Σ_*, but Σ_1 is closer to $C_*^{(n)}$ than Σ_2 (see Figure 11). Hence, by (20), Σ_1 is more truthlike than Σ_2. Nevertheless, in the limit $n \to \infty$, both $d_j^{(n)}(C_*^{(n)}, C_1^{(n)})$ $d_j^{(n)}(C_*^{(n)}, C_2^{(n)})$ — as functions of $|n - n_1|$ and $|n - n_2|$, respectively — grow towards infinity, so that $\Delta(\Sigma_*, \Sigma_1) = \Delta(\Sigma_*, \Sigma_2)$.

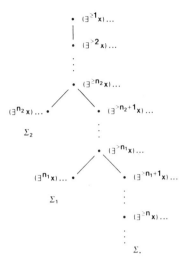

Fig. 11. Theories Σ_1, Σ_2, and Σ_*.

10.3. DISTANCE BETWEEN POSSIBLE WORLDS

Modern treatments of counterfactual conditionals have been based on the idea that it is possible to speak of the overall similarity between possible worlds (cf. also Chapter 6.1). David Lewis (1973) takes the similarity relation on the class **M** of possible worlds to be a primitive, undefined concept. The similarity approach to truthlikeness, which allows us to measure the distances between world-descriptions, yields an explicit definition for the similarity and dissimilarity between worlds — when the concept of 'possible world' is relativized to a language L (cf. Chapter 6.2).[5] The observation that the theory of truthlikeness for a first-order language L gives a distance Δ on the space $\mathbf{M}_L$ of L-structures was made in 1976 independently by Niiniluoto (1978a) and Oddie (1979).

Let Ω_1 and Ω_2 be two L-structures in the class $\mathbf{M}_L$. Let $\Sigma_1 = \mathrm{Th}(\Omega_1)$ and $\Sigma_2 = \mathrm{Th}(\Omega_2)$ be the theories of structures Ω_1 and Ω_2, respectively (cf. (2.49)). Then the distance of 'worlds' Ω_1 and Ω_2 obviously depends on the distance between the corresponding theories. As Σ_1 and Σ_2 are complete theories, their distance $\Delta(\Sigma_1, \Sigma_2)$ is given by (18) or (19). Thus, we may define

(22) $\Delta(\Omega_1, \Omega_2) = \Delta(\mathrm{Th}(\Omega_1), \mathrm{Th}(\Omega_2))$.

The distance of structure $\Omega \in \mathbf{M}_L$ from the actual L-structure $\Omega_* \in \mathbf{M}_L$ is then $\Delta(\Omega_*, \Omega)$. The function Δ defined in (22) is a semi-metric which is unique up to elementary equivalence:

(23)   If $\Omega_1 \equiv \Omega_1'$ and $\Omega_2 \equiv \Omega_2'$, then $\Delta(\Omega_1, \Omega_2) = \Delta(\Omega_1', \Omega_2')$.

Thus $\Delta_c(\Omega_1, \Omega_2)$ is a metric on the quotient space $\mathbf{M}^* = \mathbf{M}_L|\equiv$. To obtain finer distance functions on $\mathbf{M}_L$, stronger languages than L — such as infinitary or higher-order languages — should be studied (cf. Chapter 2.8).

A similarity relation on $\mathbf{M}_L$ can be defined in analogy with (20): instead of using limits, we may stipulate that

(24) L-structure Ω_1 is closer to the actual world Ω_* than L-structure Ω_2 iff the theory $\text{Th}(\Omega_1)$ is more truthlike than $\text{Th}(\Omega_2)$.

(24) says that Ω_1 is closer to Ω_* than Ω_2 if the 'finite approximation' of Ω_1 is, from some depth n onwards, closer to Ω_* than the 'finite approximation' of Ω_2.

Definitions (22) and (24) apply to a wide range of cases: from complicated ones, where the true theory $\text{Th}(\Omega_*)$ is not finitely axiomatizable, to L-structures relative to a poor first-order language L.[6] The following examples suffice to illustrate how our earlier results about the distances between linguistic entities can be reinterpreted as indicating distances between set-theoretical structures.

EXAMPLE 6. Let L be a language with identity $=$, one individual constant a, names $\mathbf{r}$ for real numbers $r \in \mathbb{R}$, and one unary function symbol $\bar{h}$. Let $\Omega_1 = \langle \{a\}, r_1 \rangle$ and $\Omega_2 = \langle \{a\}, r_2 \rangle$, where $r_1, r_2 \in \mathbb{R}$, be two L-structures. Then the complete theories $\text{Th}(\Omega_1)$ and $\text{Th}(\Omega_2)$ of these structures are axiomatized by the statements

$$(\exists!x)(x = a \ \& \ \bar{h}(x) = \mathbf{r}_1)$$
$$(\exists!x)(x = a \ \& \ \bar{h}(x) = \mathbf{r}_2),$$

respectively. These statements are essentially singular, and by Chapter 8.4 their distance is $|r_1 - r_2|$ (or its normalization). Hence, by (22),

$$\Delta(\Omega_1, \Omega_2) = |r_1 - r_2|.$$

EXAMPLE 7. Let L be a monadic language L_N^1 with one unary predicate. Let $\Omega_1 = \langle X, A_1 \rangle$ and $\Omega_2 = \langle X, A_2 \rangle$, where $A_1 \subseteq X$, $A_2 \subseteq X$, be two L-structures. Then the complete theories of Ω_1 and Ω_2 in L are state descriptions, and their distance, by (8.12), is

$$\Delta(\Omega_1, \Omega_2) = |A_1 \triangle A_2|/N.$$

EXAMPLE 8. Let L be a monadic language L_0^2 with two unary predicates. Let $\Omega_1 = \langle X, A_1, B_1 \rangle$ and $\Omega_2 = \langle X, A_2, B_2 \rangle$ be two L-structures, where $A_1 \cap B_1 \neq \emptyset$, $A_1 \cap -B_1 = \emptyset$, $-A_1 \cap B_1 \neq \emptyset$, $-A_1 \cap -B_1 \neq \emptyset$, $A_2 \cap B_2 \neq \emptyset$, $A_2 \cap -B_2 \neq \emptyset$, $-A_2 \cap B_2 = \emptyset$, and $-A_2 \cap -B_2 \neq \emptyset$. Then the complete theories of Ω_1 and Ω_2 are monadic constituents, and their Clifford distance yields $\Delta(\Omega_1, \Omega_2) = 1/2$.

EXAMPLE 9. Let L be a language with identity $=$ but without any predicates or individual constants. Let $\Omega_1 = X_1$ and $\Omega_2 = X_2$ be two sets with $|X_1| = n_1$ and $|X_2| = n_2$. Then the complete theories of Ω_1 and Ω_2 in L are the statements $(\exists^{n_1} x)(x = x)$ and $(\exists^{n_2} x)(x = x)$, respectively. It follows that the distance $\Delta(\Omega_1, \Omega_2)$ is a function of $|n_1 - n_2|$.

EXAMPLE 10. Let Ω_1, Ω_2, and Ω_3 be the three partially ordered sets of Figure 12. Let $L^=$ be the language with one two-place predicate. Then the depth-3 constituents $C_1^{(3)}$, $C_2^{(3)}$, and $C_3^{(3)}$ that are the complete theories of Ω_1, Ω_2, and Ω_3 are given in Figure 13. It follows that $\Delta(\Omega_3, \Omega_1) = \Delta(\Omega_3, \Omega_2)$.

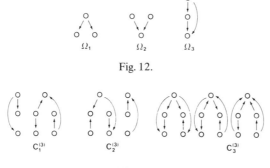

Fig. 12.

Fig. 13.

10.4. FIRST-ORDER THEORIES

The results of Sections 2 and 3 make it possible to extend the concept of truthlikeness to arbitrary first-order theories. Let Σ be a theory in language L, and let Σ_* be the true complete theory $\text{Th}(\Omega_*)$ about the actual L-structure Ω_*. Theory Σ can then be considered as a partial answer to a cognitive problem with the target Σ_* — in general Σ will be a (finite or infinite) disjunction of complete theories in $\mathscr{CT}_L$. In other words, for each theory Σ in L there is a set I_Σ of complete theories in $\mathscr{CT}_L$ such that

$$\vdash \Sigma \equiv \bigvee_{T \in I_\Sigma} T.$$

When Δ is a normalized metric on $\mathscr{CT}_L$, the distance of Σ from the truth Σ_* can be defined by $\Delta^{\gamma\gamma'}_{\text{nms}}(\Sigma_*, \Sigma)$, i.e.,

$$(25) \quad \gamma \inf_{T \in I_\Sigma} \Delta(\Sigma_*, T) + \gamma' \cdot \frac{\sum_{T \in I_\Sigma} \Delta(\Sigma_*, T)}{\sum_{T \in \mathscr{CT}_L} \Delta(\Sigma_*, T)}$$

and the degree of truthlikeness of Σ by

$$(26) \quad \text{Tr}(\Sigma, \Sigma_*) = 1 - \Delta^{\gamma\gamma'}_{\text{nms}}(\Sigma_*, \Sigma).$$

(Cf. Niiniluoto, 1978a, p. 455.)

Just as well we could have used the distance Δ defined on $\mathbf{M}_L$: the distance of theory Σ from the truth, represented by the actual L-structure Ω_*, is

$$(27) \quad \Delta^{\gamma\gamma'}_{\text{nms}}(\Omega_*, \text{Mod}(\Sigma)) = \gamma \inf_{\Omega \in \text{Mod}(\Sigma)} \Delta(\Omega_*, \Omega) +$$

$$+ \gamma' \cdot \frac{\sum_{\Omega \in \text{Mod}(\Sigma)} \Delta(\Omega_*, \Omega)}{\sum_{\Omega \in \mathbf{M}_L} \Delta(\Omega_*, \Omega)}$$

and

$$(28) \quad \text{Tr}(\Sigma, \Sigma_*) = 1 - \Delta^{\gamma\gamma'}_{\text{nms}}(\Omega_*, \text{Mod}(\Sigma)).$$

According to (26), theory Σ_1 is more truthlike than theory Σ_2 if $\Delta_{nms}^{\gamma\gamma'}(\Sigma_*, \Sigma_1) < \Delta_{nms}^{\gamma\gamma'}(\Sigma_*, \Sigma_2)$. But again this condition can be made finer: Σ_1 is more truthlike than Σ_2 if the class I_{Σ_1} of complete theories dominates the class I_{Σ_2} of complete theories in the sense of (20). A sufficient condition for comparative judgments of truthlikeness can be given as follows:

(29) Assume that there is an injective mapping $\eta: I_{\Sigma_1} \to I_{\Sigma_2}$ such that the range of η includes the element of I_{Σ_2} closest to Σ_* and, for each $T \in I_{\Sigma_1}$, T is more truthlike than $\eta(T)$. Then theory Σ_1 is more truthlike than theory Σ_2.

The condition given in (29) means that, from some depth n onwards, the 'bundle' of monotone sequences of constituents corresponding to Σ_1 is closer to the true constituent at these depths than the 'bundle' corresponding to Σ_2.

EXAMPLE 11. (Cf. Oddie, 1979.) Let Σ be a false theory, and let $\Sigma' = \Sigma \cap \Sigma_*$ be the truth content of Σ. Then Σ' is axiomatized by the disjunction of Σ_* and the complete theories in I_Σ, i.e., $I_{\Sigma'} = I_\Sigma \cup \{\Sigma_*\}$. At each finite depth n, the content of Σ' at n is axiomatized by

$$C_*^{(n)} \vee \bigvee_{i \in I_\Sigma} C_i^{(n)},$$

where $\langle C_i^{(n)} \mid n < \omega \rangle$ is the monotone basis of a complete theory T_i in I_Σ. By the properties of the min-sum measure $\Delta_{ms}^{\gamma\gamma'}$ (cf. Chapter 6), this means that from some depth n_0 onwards

$$\Delta_{ms}^{\gamma\gamma'}\left(C_*^{(n)}, C_*^{(n)} \vee \bigvee_{i \in I_\Sigma} C_i^{(n)}\right) < \Delta_{ms}^{\gamma\gamma'}\left(C_*^{(n)}, \bigvee_{i \in I_\Sigma} C_i^{(n)}\right),$$

so that Σ' dominates Σ in the sense of (29). Hence Σ' is more truthlike than Σ.

If the language L is semantically indeterminate, so that the actual world is represented by a class $\Theta_L^* \subseteq \mathbf{M}_L$ of L-structures (cf. Chapter 4.4), then the complete true theory Σ_* has to be replaced by theory $\mathrm{Th}(\Theta_L^*)$. The distance of a L-structure Ω from the actual world is then defined by $\delta(\mathrm{Th}(\Theta_L^*), \mathrm{Th}(\Omega))$, where $\delta = \delta_{nms}^{aa'}$ is given by (6.121). The degree of truthlikeness of a first-order theory Σ in L is then defined by

$$\mathrm{Tr}(\Sigma, \mathrm{Th}(\Theta_L^*)) = 1 - \delta(\mathrm{Th}(\Theta_L^*), \Sigma)$$

or

$$\text{Tr}(\Sigma, \text{Th}(\Theta_L^*)) = 1 - \delta(\Theta_L^*, \text{Mod}(\Sigma)).$$

The definitions given in this section presuppose a 'statement view' of theories, where a theory is a set of sentences Σ in language L and its truth is evaluated with respect to the whole of the actual L-structure $\Omega_* \in \mathbf{M}_L$ (cf. Chapter 2.9). In the alternative 'structuralist' conception, the domain of a theory Σ is divided into several intended applications $\mathcal{T}$, so that theory can be represented by the pair $\langle \Sigma, \mathcal{T} \rangle$. Some of the applications $\Omega \in \mathcal{T}$ may be substructures of Ω_*, some may be idealized representations of counterfactual situations. To evaluate the claim $\mathcal{T} \subseteq \text{Mod}(\Sigma)$ of the theory $\langle \Sigma, \mathcal{T} \rangle$, it is natural to evaluate its overall ability to account for the applications in $\mathcal{T}$. Then we in effect view a theory as an attempt to solve many cognitive problems at the same time. Formally, if theory Σ does not give any answer relative to an application Ω, it corresponds to the tautologous answer 'Don't know'. Therefore, it is natural to define the *cognitive value* of theory Σ relative to Ω by the degree $\text{Tr}(\Sigma, \text{Th}(\Omega))$ minus the truthlikeness of a tautology, i.e., $1 - \gamma'$. Hence, by (28) this value equals

(30) $\text{Tr}(\Sigma, \text{Th}(\Omega)) - (1 - \gamma') = \gamma' - \Delta(\Omega, \text{Mod}(\Sigma)).$

Then (30) is positive if and only if the description of structure Ω by theory Σ is not worse than a tautology. The overall cognitive value of theory $\langle \Sigma, \mathcal{T} \rangle$ is then the sum of its values relative to $\Omega \in \mathcal{T}$:

(31) $\quad \text{CV}(\Sigma, \mathcal{T}) = \sum_{\Omega \in \mathcal{T}} \xi_\Omega (\gamma' - \Delta(\Omega, \text{Mod}(\Sigma)))$

$$= \gamma' \sum_{\Omega \in \mathcal{T}} \xi_\Omega - \sum_{\Omega \in \mathcal{T}} \xi_\Omega \Delta(\Omega, \text{Mod}(\Sigma)),$$

where the weight $\xi_\Omega \geq 0$ indicates the pragmatic importance of application Ω. The value $\text{CV}(\Sigma, \mathcal{T})$ can be regarded as a 'realist' explicate of the cognitive *problem-solving ability* of theory Σ (cf. Niiniluoto, 1980a, p. 450).[7]

Let $\langle \Sigma_1, \mathcal{T}_1 \rangle$ and $\langle \Sigma_2, \mathcal{T}_2 \rangle$ be two theories. To compare their cognitive values, we have to pool their applications to one set $\mathcal{T} = \mathcal{T}_1 \cup \mathcal{T}_2$.[8] Then Σ_1 is cognitively superior to Σ_2 if and only if $CV(\Sigma_1, \mathcal{T}) > CV(\Sigma_2, \mathcal{T})$, i.e.,

(32) $\quad \sum_{\Omega \in \mathcal{T}} \xi_\Omega \Delta(\Omega, \text{Mod}(\Sigma_1)) < \sum_{\Omega \in \mathcal{T}} \xi_\Omega \Delta(\Omega, \text{Mod}(\Sigma_2))$.

In the special case, where all the weights ξ_Ω are equal, condition (32) reduces to

$$\sum_{\Omega \in \mathcal{T}} \Delta(\Omega, \text{Mod}(\Sigma_1)) < \sum_{\Omega \in \mathcal{T}} \Delta(\Omega, \text{Mod}(\Sigma_2)).$$

A sufficient condition for (32), which is independent of the weights ξ_Ω, is the uniform dominance of Σ_1 over Σ_2 relative to $\mathcal{T}$:

(33) $\quad$ If $\Delta(\Omega, \text{Mod}(\Sigma_1)) < \Delta(\Omega, \text{Mod}(\Sigma_2))$ for all $\Omega \in \mathcal{T}$, then $CV(\Sigma_1, \mathcal{T}) > CV(\Sigma_2, \mathcal{T})$.

Condition (33) is applicable also in those situations, where $\mathcal{T}$ is infinite. In the general case, without uniform dominance, the weights ξ_Ω are indispensable for comparative judgments about the cognitive value of theories.

CHAPTER 11

LEGISIMILITUDE

Chapters 9 and 10 have not paid any attention to the distinction between lawlike and accidental generalizations: the sentences and theories studied there are statements in an extensional first-order language. L. J. Cohen argued in 1980 that this is a general defect of recent theories of truthlikeness: science aims, he says, at 'legisimilitude' rather than 'verisimilitude'. In this chapter, I show how the problem of legisimilitude can be solved within the similarity approach by choosing a suitable lawlike sentence as the target. This proposal is compared with Oddie's and Kuipers's related suggestions. The approach is also extended to the definition of degrees of truthlikeness for quantitative laws — both deterministic and probabilistic laws. These ideas lead naturally to a more general theory of idealization and approximation for quantitative statements.

11.1. VERISIMILITUDE VS LEGISIMILITUDE

L. J. Cohen (1980) argues that there is an important concept which is not adequately captured by the recent theories of truthlikeness. Cohen points out that, instead of truth about the actual world, much of science pursues "physically necessary truth" or "truth about other physically possible worlds as well as about the actual one". He concludes that

> it is legisimilitude that much of science seeks, not verisimilitude. Likeness to truth would be only of inferior value unless it were also likeness to law. All genuine laws are truths, but not all truths — not even all general truths — are laws, so truthlikeness does not entail law-likeness.

Cohen's point is well made, since the earlier accounts of truthlikeness (cf. Chapters 8—10) have not paid attention to the important distinction between lawlike and accidental generalizations.

Information seeking in science is characterized by a cognitive interest in general truths about the world, not in any of its singular or accidental features. For this reason, I have earlier considered the true constituent C_*, or their monotone sequence $\langle C_*^{(n)} | n < \omega \rangle$, as the most appropriate target of cognitive problems in science (cf. Niiniluoto, 1977b, 1980a). In the same spirit, I have argued that my measures of truthlikeness for generalizations are better than Tichý's and Oddie's, since they reflect more adequately our interest in the general features of the world (cf. Niiniluoto, 1979a). However, this claim — even if partly true (cf. Chapter 9.3) — ignores the fact that the accidental features of the world include not only singular but also general facts. In other words, part of the information given by the true constituent C_* may fail to be lawlike and is therefore only accidentally true. For this reason, it is desirable to find a concept of legisimilitude which reflects our interest in the lawlike aspects of the world.

Many influential philosophers of science deny that the distinction between lawlike and accidental generalizations has semantical import (cf. Chapter 2.11). They are of course welcome to directly apply our earlier results about *veri*similitude to those extensional generalizations which they are willing to regard as important on the basis of pragmatic reasons. On the other hand, if lawlike statements are intensional rather than extensional, then the problem of legisimilitude has to be treated separately from the theory of verisimilitude. This is the assumption that is made in this chapter (cf. also Chapter 4).

A distinction similar to Cohen's is made by Theo Kuipers (1982), who argues that the earlier discussion has conflated the problems of 'descriptive' and 'theoretical' verisimilitude. He generously asserts that "Niiniluoto has given the adequate definition of descriptive verisimilitude", which explicates the idea of closeness to the strongest descriptive statement about the actual world. But the problem of theoretical verisimilitude is instead concerned with closeness to the strongest 'theoretical' statement that is 'theoretically true', i.e., true in all physically possible worlds. Thus, the point of Kuipers's 'theoretical verisimilitude' is essentially the same as that of Cohen's 'legisimilitude' (even though their proposed explicates are quite different).

The first reply to Cohen was given by Oddie (1982), who construes laws as second-order relations between predicates. He outlines an extension of Hintikka's theory of distributive normal forms to higher-order logic, and thereby extends the theory of truthlikeness to cover genuine laws.[1] Another reply to Cohen was given in my paper for the 1981 Congress in Polanica (see Niiniluoto, 1983b), where laws are treated as physically necessary generalizations (cf. Chapter 2.11). By using the general strategy of the similarity approach, the concept of

374 CHAPTER ELEVEN

legisimilitude is then explicated as closeness to a suitably chosen lawlike sentence. It turns out that this suggestion and Kuipers's (1982) independent account of theoretical verisimilitude are much closer to each other than it was first realized.

11.2. DISTANCE BETWEEN NOMIC CONSTITUENTS

Oddie (1982) represents a lawlike statement 'All X's are Y' by a second-order sentence '$N(X, Y)$'. He therefore adds a second-order *necessitation relation* N to the original first-order language L. Let us assume, for simplicity, that the extended language $L(N)$ does not allow for quantification over properties. In this case, the constituents (of depth d) of $L(N)$ can be defined by joining to the constituents (of depth d) of L a conjunctive list of the N-relations that obtain between the basic predicates. Thus, in the monadic case each constituent C_i^N of $L(N)$ is a conjunction of some constituent C_j of L with a N-part

$$\bigwedge_{\substack{i,j=1 \\ i \neq j}}^{k} (\pm) N(O_i, O_j)$$

where C_j has to be compatible with the N-part. (C_j cannot claim the existence of individuals in a Ct-predicate which is necessarily empty by the N-part.) Let C_*^N be the true constituent of $L(N)$. The definition of d_T is then extended by Oddie to measure the distance between arbitrary constituents of $L(N)$. Further, for any sentence h in $L(N)$, he defines the degree of *verisimilitude* of h by

(1) $\quad 1 - \dfrac{1}{|\mathbf{I}_h^N|} \sum\limits_{i \in \mathbf{I}_h^N} d_T(C_i^N, C_*^N),$

where

$$\vdash h \equiv \bigvee_{i \in \mathbf{I}_h^N} C_i^N$$

is the normal form of h in $L(N)$.

Let δ_T be a measure for the distance between two statements in $L(N)$. Oddie's favourite δ_T is an extension of d_T, i.e., $\delta_T(h, g) = d_T(h, g)$ if g is a constituent, and is defined formally in (9.16). Let DC_i^N

be the disjunction of all constituents of $L(N)$ which have the same N-parts as C_i^N. Define "what h says about natural necessity" by

$$(h)_N = \vee \{DC_i^N | \text{some constituent in } DC_i^N \text{ entails } h\}.$$

Then for example,

(2) $\quad (C_*^N)_N = DC_*^N.$

Oddie's two alternative definitions for *legisimilitude* can then be written as follows:

(3) $\quad \text{leg}_1(h) = 1 - \delta_T(h, DC_*^N)$
$\quad\quad \text{leg}_2(h) = 1 - \delta_T((h)_N, DC_*^N).$

Hence, $\text{leg}_2(h) = \text{leg}_1((h)_N)$.

Oddie notes that $\text{leg}_1(h)$ is maximal for $h = DC_*^N$ (which is logically weaker than C_*^N), and, by (2), $\text{leg}_2(h)$ is maximal for $h = C_*^N$. However,

$$\text{leg}_2(C_i^N) = \text{leg}_2(C_*^N)$$

for any constituent C_i^N which has the same N-part as C_*^N. On the other hand, the value (1) is maximal only for $h = C_*^N$. Therefore, Oddie concludes, only the concept of verisimilitude is able to justify our preference for "clearly the most desirable theory" C_*^N.

Oddie's comparison here presupposes that it is better to know the lawlike *and* accidental features of the world than the lawlike features alone. While in some sense this is not implausible, it should be remarked that Cohen's original contrast between verisimilitude and legisimilitude is not quite the same as Oddie's — it is a criticism of measures of truthlikeness which, unlike (1), do not make any distinction between lawlike and accidental information about the world.

One restrictive presupposition of the above treatment of the monadic case is the assumption that the necessitation relation N holds between the primitive predicates of the language. The logical form of natural laws may be more complex, however. For example, it may be necessary that all individuals which are *both* O_1 *and* O_2 are O_3 as well. To express this with the second-order relation N — i.e., by something like '$N(O_1 \& O_2, O_3)$' — we need a name for a conjunctive predicate. Such names will be available, if the language $L(N)$ is enriched so that it contains quantification over properties, for then we can express sentences which postulate the existence of a conjunction of the properties O_1 and O_2. Oddie's general treatment of higher-order languages covers such extended situations. What I find problematic, however, is that Oddie thereby has to admit in his ontology such entities as negative and conjunctive properties — creatures that at least some philosophers find suspectible.[2]

Another problem for Oddie's approach concerns the formal properties of the necessitation relation N. Intuitively it might seem plausible to assume that N is transitive: if $N(F, G)$ and $N(G, H)$, then $N(F, H)$. But then all the conjuncts in the N-part of a constituent C_i^N are not independent. Oddie's examples are formulated so that they presuppose the non-transitivity of N.

Of course the use of measure d_T and the average function in definitions (1) and (3) makes Oddie's concept of legisimilitude vulnerable to the same kind of criticism as was given in Chapters 6.6 and 9.3.

Instead of representing laws of the form 'All F's are G' by '$N(F, G)$', another approach — without complex properties — can be based upon the following formulation:

(4) $\quad \Box(x)(F(x) \supset G(x))$,

where $\Box$ is an operator of *physical* (nomic, causal) *necessity*. For example, there is no problem in formulating '$N(O_1 \& O_2, O_3)$' by '$\Box(x)((O_1(x) \& O_2(x)) \supset O_3(x))$'. The sentence (4) is true in world w if and only if the generalization '$(x(F(x) \supset G(x))$' is true in all worlds w' which are physically possible relative to w (see (2.93)). Thus, (4) asserts that the formula $F(x) \& \sim G(x)$ has an empty extension in each physically possible world. We have seen earlier that statements of the form (4) can be expressed as finite disjunctions of nomic constituents B_i, where B_i is a sentence in $L_N^k(\Box)$ of the form

(5) $\quad \bigwedge_{j \in CT_i} \Diamond (\exists x) Q_j(x) \& \Box(x) \left[\bigvee_{j \in CT_i} Q_j(x) \right]$.

(see (2.96)). Nomic constituent thus claims that certain kinds CT_i of individuals are physically possible and other kinds physically impossible.

The set $\mathcal{N}_N^k$ of nomic constituents in $L_N^k(\Box)$ defines a cognitive problem where the true nomic constituent B_* is the target (cf. Example 4.5). It is therefore natural to define the concept of legisimilitude relative to this problem $\mathcal{N}_N^k$. For any sentence h in $L_N^k(\Box)$, let $h_\Box$ be the disjunction of the nomic constituents B_i compatible with h, i.e., $h_\Box$

is the B-normal form of h. Then to define the degree of legisimilitude of h we need only a distance function Δ on $\mathcal{N}_N^k$. But since the logical structure of (5) is almost the same as that of ordinary monadic constituents, we can readily apply the Clifford measure d_C:

(6) $\quad d_C(B_i, B_j) = |\mathbf{CT}_i \Delta \mathbf{CT}_j|/q$,

or the Jyväskylä measure $d_J(B_i, B_j)$ defined by (9.5). The *degree of legisimilitude* of h can then be defined by

(7) $\quad \mathrm{leg}_3(h) = \mathrm{Tr}(h_\square, B_*) = 1 - \Delta_{ms}^{\gamma\gamma'}(B_*, h_\square)$.

(See Niiniluoto, 1983b, p. 328.) Obviously $\mathrm{leg}_3(h)$ has its maximum value if and only if h is equivalent to the true nomic constituent B_*:

(8) $\quad \mathrm{leg}_3(h) = 1 \quad \text{iff} \quad \vdash h \equiv B_*$.

Otherwise $\mathrm{leg}_3(h)$ behaves in the same way as our ordinary notion of truthlikeness.

While definition (7) of leg_3 corresponds closely to Oddie's concept of legisimilitude$_2$, a notion corresponding to his concept of verisimilitude (1) is obtained by choosing the conjunction $B_* \& C_*$ as the target sentence. The distance $\Delta(B_i \& C_j, B_k \& C_\ell)$ between two conjunctions $B_i \& C_j$ and $B_k \& C_\ell$ can be defined by

(9) $\quad \frac{1}{2}(\Delta(B_i, B_k) + \Delta(C_j, C_\ell))$.

For each sentence h in $L_N^k(\square)$, let $Dh_\square$ be the disjunction of all conjunctions $B_i \& C_j$ which are compatible with h, i.e., $Dh_\square$ is the $B \& C$-normal form of h. Then we may define

(10) $\quad \mathrm{leg}_4(h) = M(Dh_\square, B_* \& C_*)$.

(See Niiniluoto, 1983b, p. 328.) Then function $\mathrm{leg}_4(h)$ has its maximum value if and only if h is equivalent to $B_* \& C_*$:

(11) $\quad \mathrm{leg}_4(h) = 1 \quad \text{iff} \quad \vdash h \equiv (B_* \& C_*)$.

According to the viewpoint of Chapter 5.2, leg_3 and leg_4 are measures of truthlikeness for different cognitive problems: leg_3 for $\mathcal{N}_N^k$ with the target B_*, and leg_4 for the combined problem $\mathcal{N}_N^k \oplus \Gamma_N^k$ with the target $B_* \& C_*$. One of them is not 'better' than the other, but the choice between them depends on our cognitive interests at a particular situation.

As a genuine alternative to leg_3, statements in $L_N^k(\square)$ could be represented as disjunctions of statements of the form

(12) $$\bigwedge_{i=1}^{q} (\pm)\Diamond C_i$$

(see (2.100)). The true statement of this form, which is logically stronger than B_*, could be chosen as the target in a definition of legisimilitude (see Niiniluoto, 1983b, p. 328). For reasons stated already in Chapter 2.11, nomic constituents are here preferred to this alternative approach.

Just like Oddie's leg_1 and leg_2, the functions leg_3 and leg_4 differ clearly from the concept of 'inductive reliability' that Cohen offers as an explicate of legisimilitude (*op. cit.*, p. 504).[3] Cohen's concept satisfies a conjunction principle to the effect that

(13)   If $\text{leg}(h_1) > \text{leg}(h_2)$, then $\text{leg}(h_1 \,\&\, h_2) < \text{leg}(h_1)$.

To see that, e.g., leg_3 does not satisfy (13), take $h_1 = B_* \vee B_1$ and $h_2 = B_* \vee B_2$, where $0 < \Delta(B_*, B_1) < \Delta(B_*, B_2)$. Then $\text{leg}_3(h_1 \,\&\, h_2) = \text{leg}_3(B_*) = 1 > \text{leg}(h_1)$. (Cf. Oddie, 1982, p. 358.)

EXAMPLE 1. Let $L_N^3(\Box)$ be a monadic intensional language with three primitive predicates M_1, M_2, and M_3, and let

$$\begin{aligned}
Q_1(x) &= M_1(x) \,\&\, M_2(x) \,\&\, M_3(x) \\
Q_2(x) &= M_1(x) \,\&\, M_2(x) \,\&\, {\sim} M_3(x) \\
Q_3(x) &= M_1(x) \,\&\, {\sim} M_2(x) \,\&\, M_3(x) \\
Q_4(x) &= M_1(x) \,\&\, {\sim} M_2(x) \,\&\, {\sim} M_3(x) \\
Q_5(x) &= {\sim} M_1(x) \,\&\, M_2(x) \,\&\, M_3(x) \\
Q_6(x) &= {\sim} M_1(x) \,\&\, M_2(x) \,\&\, {\sim} M_3(x) \\
Q_7(x) &= {\sim} M_1(x) \,\&\, {\sim} M_2(x) \,\&\, M_3(x) \\
Q_8(x) &= {\sim} M_1(x) \,\&\, {\sim} M_2(x) \,\&\, {\sim} M_3(x).
\end{aligned}$$

Assume that the true nomic constituent B_* states that the Q-predicates $CT_* = \{Q_1, Q_5, Q_7, Q_8\}$ and only them are physically possible, and let the true constituent C_* state that the cell Q_1 is exemplified. Note that B_* entails the laws

$$\begin{aligned}
h_1 &= \Box(x)(M_1(x) \supset M_2(x)) \\
h_2 &= \Box(x)(M_2(x) \supset M_3(x)).
\end{aligned}$$

Let C_1 and C_2 be the constituents with $\mathbf{CT}_1 = \{Q_1, Q_5\}$ and $\mathbf{CT}_2 = \{Q_1, Q_6\}$, respectively. If the Clifford distance d_C is used in (7) and (10), we have

$$\text{leg}_3(h_1 \,\&\, C_1) = 1 - \gamma' \cdot \frac{(0+1+1+2+1+2+2+3+1+2+2+3+2+3+3+4)}{8} / 128$$

$$= 1 - \gamma'/32$$

$$\text{leg}_3(h_1 \,\&\, C_2) = 1 - \gamma/8 - \gamma' \cdot \frac{(1+2+2+3+2+3+3+4+2+3+3+4+3+4+4+5)}{8} / 128$$

$$= 1 - \gamma/8 - 3\gamma'/64,$$

and

$$\text{leg}_4(h_1 \,\&\, C_1) = 1 - \gamma/16 - 3\gamma'/128$$
$$\text{leg}_4(h_1 \,\&\, C_2) = 1 - \gamma/8 - \gamma'/16.$$

Here $(h_1 \,\&\, C_1)_\square$ and $(h_1 \,\&\, C_2)_\square$ are different, since C_2, unlike C_1, entails the falsity of law h_2. But as h_2 is true, $h_1 \,\&\, C_2$ should differ more from the truth B_* than $h_1 \,\&\, C_1$, just as leg_3 and leg_4 imply.

The definition for leg_3 can be generalized in three different directions. First, in the form (7) it applies to laws of coexistence. To develop the same idea for laws of succession, we have to define the distance between constituents of the form

$$(14) \quad \bigwedge_{\langle i,j \rangle \in T} \Diamond (\exists x)(Q_i^t(x) \,\&\, Q_j^{t+1}(x)) \,\&$$

$$\&\, \Box(x) \left[\bigvee_{\langle i,j \rangle \in T} (Q_i^t(x) \,\&\, Q_j^{t+1}(x)) \right]$$

(see (2.105)). If B_1^t and B_2^t are statements of the form (14) with the transformation sets T_1 and T_2, respectively, then a Clifford-type measure between them would be

$$(15) \quad d_C(B_1^t, B_2^t) = |T_1 \triangle T_2|/q^2.$$

Again (15) fails to reflect the similarity of the pairs in T_1 and T_2. To obtain a measure which accounts for the distances between the physically possible transformations, the following idea suggests itself. Set T in (14) is a one-many relation from $\{1, 2, \ldots, q\}$ to

$\{1, 2, \ldots, q\}$: For each $i \in \{1, \ldots, q\}$, T associates the set $T(i) = \{j | \langle i, j \rangle \in T\}$ of (the indices of) physically possible states after state Q_i. Then (15) equals

$$d_C(B_1^t, B_2^t) = \sum_{i=1}^{q} |T_1(i) \triangle T_2(i)|/q^2.$$

Alternatively we could measure the distance between $T_1(i)$ and $T_2(i)$ — which are sets of Q-predicates — by the Jyväskylä measure $d_J(T_1(i), T_2(i))$, based upon the normalized city block measure ε between Q-predicates. Then the overall distance of B_2^t from B_1^t is the sum over i of the d_J-distances:

(16) $\quad d_J(B_1^t, B_2^t) = \sum_{i=1}^{q} d_J(T_1(i), T_2(i)).$

In the special case, where T_1 and T_2 are functions, so that for all $i = 1, \ldots, q$, $T_1(i)$ and $T_2(i)$ are singletons, $\{Q_{i_1}\}$ and $\{Q_{i_2}\}$ say, the distance (16) reduces to

(17) $\quad \sum_{i=1}^{q} \left(\frac{1}{q} + \varepsilon(Q_{i_1}, Q_{i_2}) \right) = 1 + \sum_{i=1}^{q} \varepsilon_{i_1 i_2}.$

Secondly, the monadic language $L_N^k(\Box)$ can be defined through k families of predicates $\mathcal{M}_i$, $i = 1, \ldots, k$, so that the distances between Q-predicates depend on the distance functions d_i on $\mathcal{M}_i$.

Thirdly, the concept of a nomic constituent can be generalized to polyadic languages $L(\Box)$ (cf. Chapter 2.11), so that in principle the definitions of leg_3 and leg_4 can be extended to arbitrary nomic generalizations and theories within $L(\Box)$.

Finally, if Hintikka's system of inductive logic is applied directly to evaluate the prior and posterior probabilities $P(B_i/e_n^c)$ of nomic constituents on observational evidence e_n^c (cf. Chapter 2.10), then the results of Chapter 9.4 concerning the estimated degrees of verisimilitude $\text{ver}(g/e_n^c)$ for existential and universal generalizations g in L_N^k can be reinterpreted — as *estimated degrees of legisimilitude*.

To conclude this section, let us still compare our approach to the ideas of Kuipers (1982). A difficulty for this task arises from the fact that he presents his account of 'theoretical verisimilitude' in a propositional language, but he also suggests a general model-theoretic formulation. For each statement a, let $\|a\|$ be the set of possible worlds where a is true. Then a is 'theoretically true' if a is true in all physically possible worlds W^{ph}, i.e., $W^{ph} \subseteq \|a\|$. The strongest theoretically true statement, i.e., T such that $\|T\| = W^{ph}$, is called "the true theory" or "the theoretical truth". Kuipers's solution to the problem of theoretical verisimilitude is then

(18)   A theory B is closer to the theoretical truth T than A iff $\|B\| \Delta \|T\| \subset \|A\| \Delta \|T\|$,

where $\subset$ means strict set-theoretical inclusion. This is formally the same as Miller's definition (5.37). When T is a complete theory, (17) has the unintuitive consequence that for false theories A and B the logically stronger is always more truthlike than the weaker (see (5.33)). However, Kuipers characterizes "the theoretical context" so that the true theory T is not complete. Moreover, Kuipers tells that he started "the investigation of verisimilitude from the conviction (in the line of the structuralist approach) that empirical theories are not, and should not be, complete" (*ibid.*, p. 357).

It seems to me that this formulation is misleading. Kuipers himself urges that the terminology of actual and physically possible worlds is "misleading", since it suggests that "different physically possible worlds cannot be realized, by nature or experimentator, at the same time" (*ibid.*, p. 346). In his later papers, he has explicitly spoken of "the set of empirically realized physically possible worlds at time t" (see Kuipers, 1984a). But of course it is a hallmark of possible *worlds* that only one of them can be actual or realized at the same time. Therefore, a natural interpretation of Kuipers's framework is to replace his 'possible worlds' by 'possible kinds' of individuals or structures. Indeed, if worlds are replaced with Q-predicates, several of the latter may be realized at the same time. On this analysis, applied to monadic languages, Kuipers's T is the set of Q-predicates that the true theory claims to be physically possible; similarly, A and B are the corresponding sets for two rival theories. But then the true theory

$$\bigwedge_{i \in T} \Diamond (\exists x) Q_i(x) \ \& \ \bigwedge_{i \notin T} \sim \Diamond (\exists x) Q_i(x)$$

is nothing else than the true nomic constituent of $L_N^k(\Box)$, and the if-part of the definition (18) (not its only if-part) is a direct consequence of the Clifford distance (6) or the Jyväskylä distance between two nomic constituents.

Another possible interpretation of Kuipers's approach would be to regard the true theory T as a statement of the form (12), which lists the physically possible constituents, but then more than one of these 'possibilities' could be realized at the same time.

Kuipers's framework is of course not limited to monadic cases, but neither is mine. The above interpretation suggests that there is a remarkable convergence between two quite independently developed approaches — my treatment of legisimilitude in Niiniluoto (1983b) and Kuipers's (1982) analysis of theoretical verisimilitude.

11.3. DISTANCE BETWEEN QUANTITATIVE LAWS

Let $\mathbf{Q} \subseteq \mathbb{R}^k$ be the quantitative state space relative to quantities $h_i: E \to \mathbb{R}$, $i = 1, \ldots, k$, and let $\Delta = \varepsilon_2$ be the Euclidean metric on $\mathbf{Q}$. It was seen in Chapters 3.3 and 3.4 that laws of coexistence and laws of succession relative to $\mathbf{Q}$ are generalizations of the concept of nomic constituent in L_N^k. To solve the problem of legisimilitude for quantitative laws, all that we need is a metric on the set of laws relative to space $\mathbf{Q}$. Two alternative approaches to this question are discussed here in the case of deterministic laws; the probabilistic laws are studied separately in Section 5 below.

Assume that there is a true law of coexistence of the form

(19) $f_*(h_1(x), \ldots, h_k(x)) = 0$,

and let $\mathbf{F}_*$ be the set of points $Q \in \mathbf{Q}$ which satisfy the equation $f_*(Q) = 0$ (cf. (3.11) and (3.14)). Then the true law (19) claims that $\mathbf{F}_* \subseteq \mathbf{Q}$ is the set of physically possible states in $\mathbf{Q}$. Let us choose (19) as the target of a cognitive problem $\mathbf{B}$. Technically, we may then take the elements of $\mathbf{B}$ to be functions $f: \mathbb{R}^k \to \mathbb{R}$ which determine a set $\mathbf{F} = \{Q \in \mathbf{Q} \mid f(Q) = 0\}$ of states. For simplicity, I require that $\mathbf{B}$ includes only such sets $\mathbf{F}$ which have a positive measure and are connected (cf. Chapter 8.4).

For two elements f_1 and f_2 of $\mathbf{B}$, their distance $\Delta(f_1, f_2)$ obviously should depend on the sets $\mathbf{F}_1$ and $\mathbf{F}_2$, associated with f_1 and f_2, respectively. In Chapter 6.7, we have already seen how the distance between two subsets $\mathbf{H}_1$ and $\mathbf{H}_2$ of $\mathbf{Q}$ is measured by a function $\delta^{\gamma a a'}$.

However, in that case $\mathbf{H}_1$ and $\mathbf{H}_2$ are disjunctions of singular statements; in our present problem f_1 claims conjunctively that each state in $\mathbf{F}_1$ is physically possible. For this reason, the distance $\Delta(f_1, f_2)$ cannot be defined by $\delta^{\gamma a a'}(\mathbf{F}_1, \mathbf{F}_2)$, but rather by a function that takes its model from the treatment of nomic constituents in Section 2.

The simplest possibility is again a non-normalized Clifford measure

$$(20) \quad \Delta_c(f_1, f_2) = \int_{\mathbf{F}_1 \Delta \mathbf{F}_2} dQ.$$

If the distances between the elements of $\mathbf{Q}$ are taken into account, Δ_c can be replaced by Δ_J:

$$(21) \quad \Delta_J(f_1, f_2) = \int_{\mathbf{F}_2} \Delta_{\inf}(Q, \mathbf{F}_1) \, dQ + \int_{\mathbf{F}_1} \Delta_{\inf}(Q, \mathbf{Q} - \mathbf{F}_1) \, dQ.$$

Measures Δ_c and Δ_J have then the same general properties as the functions d_c and d_J in Section 2. If $\mathbf{Q}$ has a finite measure q, then (20) and (21) can be normalized by dividing their values by q.

EXAMPLE 2. Let $\mathbf{Q} = \mathbb{R}$, and let $\mathbf{F}_1 = [a, b]$ and $\mathbf{F}_2 = [c, d]$ be two intervals in $\mathbb{R}$. If F_1 and F_2 are disjoint, so that $a < b < c < d$, then

$$\Delta_c(f_1, f_2) = \int_a^b dx + \int_c^d dx = (b - a) + (d - c)$$

$$= \ell(\mathbf{F}_1) + \ell(\mathbf{F}_2)$$

$$\Delta_J(f_1, f_2) = \int_{c-b}^{d-c} x \, dx + \int_a^{(a+b)/2} (x - a) \, dx +$$

$$+ \int_{(a+b)/2}^b (b - x) \, dx$$

$$= (d - c)\left(\frac{d + c}{2} - b\right) + \tfrac{1}{4}(a - b)^2$$

$$= \ell(\mathbf{F}_2)(m(\mathbf{F}_2) - b) + \tfrac{1}{4}\ell(\mathbf{F}_1)^2.$$

If $\mathbf{F}_1$ and $\mathbf{F}_2$ overlap, so that $a < c < b < d$ and $c - a < d - c$, then
$$\Delta_c(f_1, f_2) = (c - a) + (d - b)$$
$$\Delta_J(f_1, f_2) = \int_0^{d-b} x \, dx + \int_a^c (x - a) \, dx$$
$$= \tfrac{1}{2}(d - b)^2 + \tfrac{1}{2}(a + c)^2.$$

If $\mathbf{F}_1$ is included in $\mathbf{F}_2$, so that $c < a < b < d$, then
$$\Delta_c(f_1, f_2) = (a - c) + (d - b)$$
$$\Delta_J(f_1, f_2) = \int_c^a (a - x) \, dx + \int_b^d (x - b) \, dx$$
$$= \tfrac{1}{2}(a - c)^2 + \tfrac{1}{2}(d - b)^2.$$

EXAMPLE 3. Let $\mathbf{Q} = \mathbb{R}^2$ and
$$\mathbf{F}_1 = \{(x, y) \mid \sqrt{x^2 + y^2} \leq r_1\}$$
$$\mathbf{F}_2 = \{(x, y) \mid r_2 \leq \sqrt{x^2 + y^2} \leq r_3\},$$
where $0 < r_1 < r_2 < r_3$ (see Figure 1). Then
$$\Delta_c(f_1, f_2) = \pi r_1^2 + \pi(r_2^2 - r_3^2)$$
$$\Delta_J(f_1, f_2) = \int_0^{2\pi} \int_{r_2}^{r_3} (r - r_1) r \, d\varphi \, dr + \int_0^{2\pi} \int_0^{r_1} (r_1 - r) r \, d\varphi \, dr$$
$$= 2\pi r_3^2 \left(\frac{r_3}{3} - \frac{r_1}{2} \right) -$$
$$- 2\pi r_2^2 \left(\frac{r_2}{3} - \frac{r_1}{2} \right) + \frac{1}{3} \pi r_1^3.$$

The proposed treatment of quantitative laws is based on the assumption that the specification of all the physically possible states is chosen as the target of a cognitive problem. In other words, the question is the following: which combinations of the values of $h_1, \ldots, h_k$ are nomically

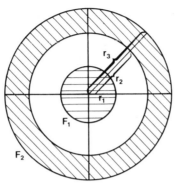

Fig. 1.

possible and which not? But there is another type of question that is common in connection with quantities: how do the values of h_k depend on the values of $h_1, \ldots, h_{k-1}$? This question presupposes that $h_1, \ldots, h_{k-1}$ are 'independent' variables and h_k is the 'dependent' variable, i.e., there is a true law of the form

(22) $\quad h_k(x) = g_*(h_1(x), \ldots, h_{k-1}(x))$,

where $g_*: \mathbf{Q}_0 \to \mathbb{R}$ may be assumed to be a continuous real-valued function and $\mathbf{Q}_0 \subseteq \mathbb{R}^{k-1}$ is the space of the physically possible values of $h_1, \ldots, h_{k-1}$.

Let us now choose (22) as the target of a cognitive problem G. Technically, the elements of this problem G can be taken to be continuous functions $g: \mathbf{Q}_0 \to \mathbb{R}$ of $k-1$ variables (cf. Example 4.12). Each element g of G thus states a possible law of coexistence about the dependence of $h_k(x)$ on $h_1(x), \ldots, h_{k-1}(x)$ (cf. (3.12)).

The distance between two elements g_1 and g_2 of G can now be defined by the Minkowski metrics

(23) $\quad \Delta(g_1, g_2) = \left(\int_{Q_0} |g_1(z) - g_2(z)|^p \, dz \right)^{1/p}.$

As special cases of (23), we have the city-block metric

(24) $\quad \Delta_1(g_1, g_2) = \int_{Q_0} |g_1(z) - g_2(z)| \, dz,$

the Euclidean metric

$$(25) \quad \Delta_2(g_1, g_2) = \left(\int_{Q_0} (g_1(z) - g_2(z))^2 \, dz \right)^{\frac{1}{2}},$$

and the Tchebycheff metric

$$(26) \quad \Delta_3(g_1, g_2) = \sup_{z \in Q_0} |g_1(z) - g_2(z)|.$$

These distance measures for quantitative laws were proposed in Niiniluoto (1982c). Here $\Delta_1(g_1, g_2)$ is in effect the volume (for $k = 2$: area) between the surfaces (for $k = 2$: curves) defined by the functions g_1 and g_2, while $\Delta_3(g_1, g_2)$ is the maximum distance between these surfaces (see Example 1.3).

The distance of law $g \in G$ from the truth g_* is measured by $\Delta(g_*, g)$. For all $i = 1, 2, 3$,

$$\Delta_i(g_*, g) = 0 \quad \text{iff} \quad g(x) = g_*(x) \quad \text{for all } x \in Q_0.$$

Comparative judgments of truthlikeness can then be defined as follows: law g_1 is *more truthlike$_i$* than law g_2 if and only if $\Delta_i(g_*, g_1) < \Delta_i(g_*, g_2)$.

If Q_0 is infinite, the values $\Delta(g_*, g_1)$ and $\Delta(g_*, g_2)$ may both be infinite, even though g_1 is uniformly closer to g_* than g_2. This problem can be handled in two different ways. First, let $B^r \subseteq Q_0$ be a finite ball of radius $r > 0$ and center in the origin of $\mathbb{R}^{k-1}$. Then we may normalize measure (22) by the following limiting procedure:

$$(27) \quad \Delta'(g_1, g_2) = \lim_{r \to \infty} \left[\left(\int_{B^r \cap Q_0} |g_1(z) - g_2(z)|^p \, dz \right)^{1/p} \Big/ \int_{B^r \cap Q_0} dz \right].$$

Then Δ' is a semimetric and $0 \leq \Delta'(g_1, g_2) \leq 1$. Secondly, we may keep the definition (23), but stipulate that g_1 is more truthlike than g_2 if and only if, for some r_0,

$$\left(\int_{B^r \cap Q_0} |g_1(z) - g_*(z)|^p \, dz\right)^{1/p} <$$
$$< \left(\int_{B^r \cap Q_0} |g_2(z) - g_*(z)|^p \, dz\right)^{1/p}$$

for all $r \geq r_0$.

EXAMPLE 4. Let $Q_0 = \mathbb{R}$, and let

$g_1(z) = a_1 z \quad$ for all $z \in \mathbb{R}$
$g_2(z) = a_2 z \quad$ for all $z \in \mathbb{R}$
$g_*(z) = az \quad$ for all $z \in \mathbb{R}$,

where $0 < a < a_1 < a_2$. Then g_1 is more truthlike than g_2.

EXAMPLE 5. Assume that the true law of refraction for a light ray passing from air to water is given by the sine law of Snell

$$\frac{\sin \alpha}{\sin \beta} = \frac{4}{3}.$$

Here α is the angle of incidence and β the angle of refraction (see Figure 2). Hence, the dependence of β from α is expressed by

$$\beta = g_*(\alpha) = \arcsin(\tfrac{3}{4} \sin \alpha).$$

Let g_1 and g_2 be the laws of Ptolemy and Grosseteste, respectively:

$g_1(\alpha) = 33\alpha/45 - \alpha^2/400$
$g_2(\alpha) = \alpha/2.$

(Cf. Niiniluoto, 1984b, p. 172.) The values of these functions are

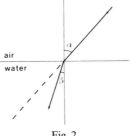

Fig. 2.

illustrated in Table I. It follows that g_1 is uniformly closer to g_* than g_2, and therefore $\Delta_i(g_*, g_1) < \Delta_i(g_*, g_2)$ for $i = 1, 2, 3$.

TABLE I

α	$\alpha/2$	$33\alpha/45 - \alpha^2/400$	arcsin($\frac{3}{4}$ sin α)
0	0	0	0
15	7.5	10.4	11.2
30	15.0	19.7	22.0
45	22.5	27.9	32.0
60	30.0	35.0	40.5
75	37.5	40.9	46.4
90	45.0	45.7	48.6

All the Minkowski metrics (23) satisfy the following theorem:

(28) If $|g_1(x) - g_*(x)| < |g_2(x) - g_*(x)|$ for all $x \in \mathbf{Q}_0$, then $\Delta(g_*, g_1) < \Delta(g_*, g_2)$ or $\Delta'(g_*, g_1) < \Delta'(g_*, g_2)$.

However, if g_1 is not uniformly better than g_2, then the metrics Δ_1, Δ_2, and Δ_3 may give different results. This is illustrated by the following example.

EXAMPLE 6. Let $\mathbf{Q}_0 = \mathbb{R}^+ = \{x \in \mathbb{R} \mid x \geq 0\}$, and let

$$g_1(x) = 2 \quad \text{for all } x \in \mathbf{Q}_0$$
$$g_2(x) = 3 \quad \text{for all } x \in \mathbf{Q}_0$$
$$g_3(x) = \frac{2x + 9}{x + 1} \quad \text{for all } x \in \mathbf{Q}_0.$$

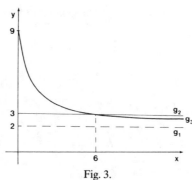

Fig. 3.

(See Figure 3.) Assume that g_3 is the true law. Then

$$\Delta_3(g_3, g_1) = 7 > 6 = \Delta_3(g_3, g_2).$$

To see that Δ_1 and Δ_2 yield the opposite result, let $r > 6$. Then

$$\int_0^r |g_1(x) - g_3(x)|\, dx = \int_0^r \left(\frac{2x+9}{x+1} - 2\right) dx$$

$$= 7 \int_0^r \frac{1}{x+1}\, dx = 7 \left|_0^r \log(x+1) = 7\log(r+1)\right.$$

$$\int_0^r |g_2(x) - g_3(x)|\, dx = \int_0^6 \left(\frac{2x+9}{x+1} - 3\right) dx +$$

$$+ \int_6^r \left(3 - \frac{2x+9}{x+1}\right) dx = \int_0^6 \frac{6-x}{1+x}\, dx + \int_6^r \frac{x-6}{x+1}\, dx$$

$$= 6 \left|_0^6 \log(1+x) - 6 \left|_6^r \log(1+x) - \int_0^6 \frac{x}{1+x}\, dx + \right.\right.$$

$$+ \int_6^r \frac{x}{x+1}\, dx = 12\log 7 - 6\log(1+r) - 6 + \log 7 +$$

$$+ (r-6) - \log(r+1) + \log 7$$
$$= r - 12 + 14\log 7 - 7\log(r+1).$$

As

$$7\log(r+1) < r - 12 + 14\log 7 - 7\log(r+1)$$

when r is sufficiently large, it follows that g_1 is more truthlike$_1$ than g_2.

Similarly,

$$\int_0^\infty \left(\frac{2x+9}{x+1} - 2\right)^2 dx = 49 \int_0^\infty \frac{1}{(x+1)^2} dx = 49$$

$$\int_0^r \left(\frac{2x+9}{x+1} - 3\right)^2 dx = \int_0^r \left(\frac{6-x}{1+x}\right)^2 dx$$

$$= 2r + 48 - \frac{r^2 + 48}{r+1} - 14\log(1+r).$$

The latter value grows without limit, when $r \to \infty$, so that g_1 is more truthlike$_2$ than g_2.

If g_1 is true, then g_2 is more truthlike$_3$ than g_3, since

$$\Delta_3(g_1, g_2) = 1 < 7 = \Delta_3(g_1, g_3).$$

g_3 is more truthlike$_1$ than g_2, since

$$\int_0^r |g_1(x) - g_3(x)| \, dx = 7\log(r+1) <$$

$$< \int_0^r |g_1(x) - g_2(x)| \, dx = \int_0^r dx = r$$

when r is sufficiently large.

When the distance $\Delta(g_1, g_2)$ between two laws g_1 and g_2 in G has been defined, we can treat disjunctions of laws in the usual way: if $h \in D(G)$ is an infinite disjunction

$$\vdash h \equiv \bigvee_{g \in I_h} g,$$

where $I_h \subseteq G$, then the distance of h from $g' \in G$ is

(29) $\quad \Delta_{\text{ms}}^{\gamma\gamma'}(g', h) = \gamma \Delta_{\inf}(g', h) + \gamma' \Delta_{\text{sum}}(g', h)$

$$= \gamma \inf_{g \in I_h} \Delta(g', g) + \gamma' \int_{I_h} \Delta(g', g) \, dg$$

(cf. (6.45)). Of two elements h_1 and h_2 of $D(G)$, h_1 is *more truthlike* than h_2 if and only if $\Delta_{\text{ms}}^{\gamma\gamma'}(g_*, h_1) < \Delta_{\text{ms}}^{\gamma\gamma'}(g_*, h_2)$.

EXAMPLE 7. Assume $\mathbf{Q}_0 = [0, 100] \subseteq \mathbb{R}$, and let

$$g_*(x) = x/2 \qquad \text{for all } x \in \mathbf{Q}_0$$
$$h(x) = x + 1 \pm 1 \qquad \text{for all } x \in \mathbf{Q}_0.$$

Thus, h is a disjunction of all the lines g^a of the form $y = x + a$, where $0 \leq a \leq 2$. (See Figure 4.) By using Δ_1, (29) gives

$$\Delta_{\text{ms}}^{\gamma\gamma'}(g_*, h) = \gamma \Delta_1(g_*, g^0) + \gamma' \int_0^2 \Delta_1(g_*, g^a) \, da$$

$$= \gamma \int_0^{100} |g_*(x) - g^0(x)| \, dx +$$

$$+ \gamma' \int_0^2 \int_0^{100} |g_*(x) - g^a(x)| \, dx \, da$$

$$= \gamma \int_0^{100} \left(x - \frac{x}{2}\right) dx +$$

$$+ \gamma' \int_0^2 \int_0^{100} \left(x + a - \frac{x}{2}\right) dx \, da$$

$$= 2500\gamma + \gamma' \int_0^2 (2500 + 100a) \, da$$

$$= 2500\gamma + 5200\gamma'.$$

This example illustrates the difference — and the similarity — between the two approaches to quantitative laws, represented by (21) and (24). If the definition (21) is formally applied to the situation of Example 7, so that $\mathbf{F}_1$ is the set of points on the line $y = x/2$ and $\mathbf{F}_2$ is the set of

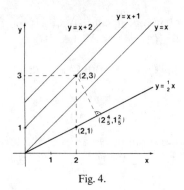

Fig. 4.

points in the bunch of lines $y = x + a$, $0 \leq a \leq 2$, then each point $(x, y) \in \mathbf{F}_2$ is correlated by measure Δ_J with the minimally distant point (x', y') in $\mathbf{F}_1$. This is illustrated in Figure 4 with the point $(2, 3)$, which is linked with point $(2\frac{4}{5}, 1\frac{2}{5})$ on line $y = x/2$. However, measure $\Delta_{ms}^{\gamma\gamma'}$, when Δ is the city-block metric Δ_1, links (x, y) in F_1 with the corresponding point (x, y') on the line $g_* -$ e.g., $(2, 3)$ with $(2, 1)$. This difference is motivated by the formulation of the underlying cognitive problem: in G it is presupposed that for each $x \in \mathbf{Q}_0$ there is a true value $g_*(x)$ of y, and it is natural to evaluate a law g in terms of the distances $|g(x) - g_*(x)|$. (In this respect, our treatment of problem G is comparable to the definition (17), and indeed with $\Delta = \Delta_1$ gives a similar measure.)

In spite of this systematic difference, the comparative results that measures of type (21) and (24) would give, when applied to the same cognitive problem, agree with each other in many cases.

The next example shows that our definitions can be applied also to the case, where function g has fewer arguments than the true function g_*.

EXAMPLE 8. Consider three alternative laws about the connection between the rest mass (m_0), relative mass (m), and velocity (v) of a physical body:

g_1: $m = m_0$
g_2: $m = m_0/(1 - v^4/c^4)^{\frac{1}{2}}$
g_*: $m = m_0/(1 - v^2/c^2)^{\frac{1}{2}}$,

where v is smaller than the velocity of light c. Then g_1 defines a plane in $\mathbb{R}^3$, while g_2 and g_* define surfaces in $\mathbb{R}^3$ which become more and more curved when v becomes larger. Both of the latter surfaces touch the plane of g_1 at $v = 0$, but elsewhere the surface of g_2 is located between the plane of g_1 and the surface of g_*. Hence, g_2 is more truthlike than the idealized law g_1.

So far we have discussed only quantitative laws of coexistence. To extend our treatment to deterministic laws of succession, recall that in the quantitative case they correspond to functions of the form $F: \mathbb{R} \times \mathbf{Q} \to \mathbf{Q}$. Here F associates with each time $t \in \mathbb{R}$ and initial state Q^{t_0} at t_0 the state Q^t at t: $Q^t = F(t, Q^{t_0})$ (see (3.25)). If Δ is the metric on state space $\mathbf{Q}$, then the distance between two laws F_1 and F_2 of this form can be measured by first taking a Minkowski-distance between the trajectories $F_1(t, Q)$ and $F_2(t, Q)$, $t \in \mathbb{R}$, i.e.,

$$\left(\int_{t_0}^{\infty} \Delta(F_1(t, Q), F_2(t, Q))^p \, dt \right)^{1/p},$$

and then summing over all the possible initial states $Q \in \mathbf{Q}$:

(30) $$d_p(F_1, F_2) = \int_Q \left(\int_{t_0}^{\infty} \Delta(F_1(t, Q), F_2(t, Q))^p \, dt \right)^{1/p} dQ$$

(cf. Niiniluoto, 1986a). In the special case, where $p = 1$ and $\Delta = \Delta_1$, (30) is obviously a generalization of the definition (17) for discrete deterministic laws of succession.

If $\mathbf{Q} \subseteq \mathbb{R}$, $F_1(t, Q)$ for fixed Q is a real-valued function of variable t, so that Δ in (30) is simply the geometrical distance in $\mathbb{R}$. In this case,

(31) $$d_1(F_1, F_2) = \int_Q \int_{t_0}^{\infty} |F_1(t, Q) - F_2(t, Q)| \, dt \, dQ$$

(32) $$d_\infty(F_1, F_2) = \int_Q \sup_{t \geq t_0} (F_1(t, Q), F_2(t, Q)) \, dQ.$$

EXAMPLE 9. Let $\mathbf{Q} = [0, 1] \subseteq \mathbb{R}$, $t_0 = 0$, and define for $x \in \mathbf{Q}$ and $t \geq 0$:

$$F_*(t, x) = xe^{-t/a}$$
$$F(t, x) = xe^{-t/b},$$

where $0 < a < b$. Then

$$d_1(F_*, F) = \int_0^1 \int_0^\infty |xe^{-t/a} - xe^{-t/b}|\, dt\, dx$$

$$= \int_0^1 x \left[\int_0^\infty e^{-t/b}\, dt - \int_0^\infty e^{-t/a}\, dt \right] dx$$

$$= (b-a) \int_0^1 x\, dx = \frac{1}{2}(b-a).$$

11.4. APPROXIMATION AND IDEALIZATION

A metric Δ on the space G of quantitative laws $g(h_1(x), \ldots, h_{k-1}(x)) = h_k(x)$ gives us the possibility to define approximate versions of many interesting methodological notions. The same concepts could be applied also to singular quantitative statements (cf. Niiniluoto, 1986a).

Two laws g_1 and g_2 in G are δ-*close* if and only if $\Delta(g_1, g_2) \leq \delta$, where $\delta > 0$ is real number. When a small treshold value $\delta > 0$ is given, g_1 and g_2 are said to be *approximate counterparts* to each other if and only if they are δ-close.

Law g_1 is *approximately deducible* from premises Σ if and only if an approximate counterpart g_2 of g_1 is deducible from Σ. When Σ is a theory, Σ *approximately explains* law g_1 if and only if Σ explains an approximate counterpart g_2 of g_1.

If law g is an approximate counterpart to the true law g_* in G, law g is *approximately true*. The choice of the Minkowski metric (23) gives us several alternative concepts of approximate truth:

(33) Law g is approximately$_i$ true within range $\mathbf{Q}_0$ (relative to degree δ) iff $\Delta_i(g_*, g) \leq \delta$ ($i = 1, 2, 3$).

(See Niiniluoto, 1982c.)[4] Thus, the city-block metric Δ_1 and the Euclidean metric Δ_2 lead to concepts of approximate truth which essentially require that the overall *sum* of the errors made by g has to be small:

(34) Law g is approximately$_1$ true in $\mathbf{Q}_0$ iff

$$\int_{Q_0} |g(x) - g_*(x)|\, dx \leq \delta.$$

(35) Law g is approximately$_2$ true in $\mathbf{Q}_0$ iff

$$\int_{Q_0} (g(x) - g_*(x))^2\, dx \leq \delta^2.$$

If g makes a large error for some values of x, it has to compensate it by being very close to truth for other values. Instead, the L_∞-metric Δ_3 leads to the notion that $g(x)$ must be close to the truth value $g_*(x)$ at *all* points x in $\mathbf{Q}_0$, i.e., g can make no large errors within its range:

(36) Law g is approximately$_3$ true in $\mathbf{Q}_0$ iff
$|g(x) - g_*(x)| \leq \delta$ for all $x \in \mathbf{Q}_0$.

Thus, in Example 6, if $\delta = 1$ and g_1 is true, g_2 is approximately$_3$ true in $[0, \infty)$, g_3 is approximately$_3$ true in $[6, \infty)$, but not in $[0, \infty)$.

A law of the form $h_k(x) = g(h_1(x), \ldots, h_{k-1}(x))$ can be used for making *exact predictions* if the precise values $h_1(a) = r_1, \ldots, h_{k-1}(a) = r_{k-1}$ of quantities $h_1, \ldots, h_{k-1}$ are known for some individual a. From these precise initial conditions and the law g, we can strictly deduce the prediction $h_k(a) = g(r_1, \ldots, r_{k-1})$ and approximately deduce statements $h_k(a) = r$ where $r \approx g(r_1, \ldots, r_{k-1})$ (i.e., $|r - g(r_1, \ldots, r_{k-1})|$ is small).

If law g is true, then all the exact predictions that g strictly entails from true precise initial conditions are likewise true. In the same situation, the exact predictions that g approximately entails are close to the truth. Further, (36) guarantees that the predictions of an approximately$_3$ true law from true initial conditions are close to the truth:

(37) If the initial condition $h_i(a) = r_i$ are true ($i = 1, \ldots, k-1$) and if the law $h_k(x) = g(h_1(x), \ldots, h_{k-1}(x))$ is approximately$_3$ true, then the prediction $h_k(a) = g(r_1, \ldots, r_{k-1})$ is close to the truth.

However, if g is approximately true in the sense of L_1- or L_2-norms, then there is no guarantee that a *single* prediction from g is close to the truth. But still the predictions from g are close to the truth *in the average*:

>   (38)   Let C be a class of predictions from the approximately$_i$ true ($i = 1, 2$) law g by using true initial conditions. If the cases in C are distributed evenly over the whole domain $\mathbf{Q}_0$ of g, then the average error of the predictions C is small.

These results (37) and (38) indicate in what sense approximate true laws are 'empirically successful' tools for prediction (cf. Niiniluoto, 1982c, 1985b).

The assumption that the precise initial conditions are known correctly is of course an idealization. But it is easy to see that errors in initial conditions can be multiplied by function g even if the law g is strictly true: the exact predictions from a true law and approximately true initial conditions need not be approximately true. For example, if $g(x) = 10^{10}x$, then small changes in the argument x induce great differenecs in the value of $g(x)$. It depends on the mathematical form of function g whether it has the characteristic known as 'robustness': g is *robust* if the value $g(x)$ of g is not too much influenced by changes in the argument x. Hence, if g is true and robust, then the predictions from g and approximately true initial conditions are close to the truth.[5]

Approximation relations between laws are symmetric but not transitive: if g_1 is an approximate counterpart to an approximately true law g_2, g_1 need not be approximately true (relative to the same degree). A sequence of laws $g_1, g_2, \ldots$ may give increasing accurate approximations of a law g, so that the sequence converges to g in the mathematical sense (cf. Chapter 1.2). A sequence of laws $g_1, g_2, \ldots$ is *convergent* if and only if for all $\varepsilon > 0$ there is a $n_0 > 0$ such that $\Delta(g_i, g_j) < \varepsilon$ for all $i, j \geq n_0$. Sequence $g_1, g_2, \ldots$ *converges to* the limit g if and only if $\Delta(g_i, g_j) \to 0$, when $i \to \infty$. When Δ is the L_∞-norm Δ_3, this concept is the same as the requirement of uniform converge of $g_1, g_2, \ldots$ to g. Finally, sequence $g_1, g_2, \ldots$ *converges to the truth* if it converges to the limit g_*.

EXAMPLE 10. For each $n \in N$, let $g_n: \mathbb{R}^2 \to \mathbb{R}$ be the function

$$g_n(x, y) = x^2(1 + y/n).$$

Then the sequence $g_1, g_2, \ldots$ does not converge uniformly to the limit function

$$g(x, y) = x^2,$$

since

$$\Delta_3(g, g_n) = \sup_{x, y \in \mathbb{R}} |x^2(1 + y/n) - x^2| = \sup_{x, y \in \mathbb{R}} |x^2 y/n| = \infty.$$

The sequence $h_1, h_2, \ldots$ defined by

$$h_n(x, y) = x^2 y + 1/n$$

converges uniformly to

$$h(x, y) = x^2 y,$$

since

$$\Delta_3(h, h_n) = \sup_{x, y \in \mathbb{R}} |1/n| = 1/n.$$

Let us next study how the approximate truth of a law is reflected on the level of its intended models (cf. Chapter 3.3). Assume that $\Omega_* = \langle D, (h_1(x))_{x \in D}, \ldots, (h_k(x))_{x \in D} \rangle$ is a model of the true law $g_*(h_1(x), \ldots, h_{k-1}(x)) = h_k(x)$. If law $g(h_1(x), \ldots, h_{k-1}(x)) = h_k(x)$ is approximately true, then it will have a model Ω which is 'close' to Ω_*. For example, if $\langle \{a\}, 5, 1/5 \rangle$ is a model of

$$h_2(x) = 1/h_1(x),$$

then $\langle \{a\}, 5, 10/51 \rangle$ is a model of

$$h_2(x) = 1/(h_1(x) + \tfrac{1}{10}).$$

To make this idea more precise, let us define the *distance between two structures* of the form

$$\Omega = \langle D, (u_1(x))_{x \in D}, \ldots, (u_k(x))_{x \in D} \rangle$$
$$\Omega' = \langle D, (v_1(x))_{x \in D}, \ldots, (v_k(x))_{x \in D} \rangle$$

by the sum over $i = 1, \ldots, k$ of the L_p-distances $L_p(u_i, v_i)$ of the corresponding functions u_i and v_i:

(39) $$\Delta(\Omega, \Omega') = \sum_{i=1}^{k} L_p(u_i, v_i),$$

where

(40) $$L_p(u_i, v_i) = \left(\sum_{x \in D} |u_i(x) - v_i(x)|^p \right)^{1/p}.$$

If D is a continuous set, the sum in (40) should be replaced by integral. If Ω and Ω' have different but intersecting domains D and D', respectively, then we may take the sum in (40) over $D \cap D'$.

If J and J' are two classes of structures, let us write $J \sim J'$ when for each $\Omega \in J$ there is $\Omega' \in J'$ such that $\Delta(\Omega, \Omega')$ is small and for each $\Omega' \in J'$ there is $\Omega \in J$ such that $\Delta(\Omega, \Omega')$ is small (cf. Moulines, 1976). When the same L_p-distance is used on the levels of laws and of structures, we have the result:

(41) $\text{Mod}(g_1) \sim \text{Mod}(g_2)$ iff g_1 and g_2 are approximate counterparts.[6]

A law g *applies approximately* to a structure Ω' if and only if g is true in a structure Ω which is close to Ω'. By (41), this is equivalent to saying that there is an approximate counterpart g' of g such that g' is true in Ω' (see Figure 5).

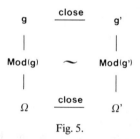

Fig. 5.

These results show that there is a close agreement between my metric approach to approximation and the method of uniformities (cf. Chapter 1.2).[7] If **M** is a space of structures of the form $\Omega = \langle D, (u_1(x))_{x \in D}, \ldots, (u_k(x))_{x \in D} \rangle$, then a *uniformity* $\mathcal{U}$ on **M** is a collection of subsets of **M** × **M** satisfying the conditions U1—U5 of Chapter 1.2. Then two structures Ω and Ω' are approximate counterparts if $\langle \Omega, \Omega' \rangle \in V$ for some (sufficiently small) entourage V in $\mathcal{U}$. We could now reformulate our definition for approximate deduction, approximate explanation, approximate truth, and convergence to the truth by replacing the metric Δ on **M** with a uniformity $\mathscr{U}$ on **M**. For example, law g is approximately true in structure Ω_* if, for some model Ω of g, $\langle \Omega, \Omega_* \rangle \in V$ holds for a (sufficiently small) V in $\mathscr{U}$.[8]

The connection between the metrics and the uniformities is in practice even closer, since in typical applications the uniformities for structures have been defined through the method illustrated in Figure 1.10, i.e., by using a suitable metric. For example, in defining the distance between two 'kinematics' $y = \langle P_y, T_y, s_y \rangle$ and $y' = \langle P_{y'}, T_{y'}, s_{y'} \rangle$, where $s: P \times T \to \mathbb{R}$ is real-valued position function, Moulines defines the base of a uniformity $\mathscr{U}$ by sets U_ε, where

$$\langle y, y' \rangle \in U_\varepsilon \quad \text{iff} \quad |s_y(p, t) - s_{y'}(p, t)| < \varepsilon$$
$$\text{for all particles } p \in P_y \cap P_{y'}$$
$$\text{and for all points of time } t \in T_y \cap T_{y'}.$$

This definition is equivalent to the use of my distance (39), when $k = 1$, $u_1 = s_y$, $v_1 = s_{y'}$, $D = P \times T$, and p is chosen as ∞ in (40). In other words, Moulines is in effect employing here a method which is equivalent to L_∞-approximation.[9]

So far we have discussed only factual or non-idealizational laws (cf. Chapter 3.3).[10] Consider now the sequence of laws

(T) $\quad \Box(x)(Cx \supset h_k(x) = g_0(h_1(x), \ldots, h_{k-1}(x)))$

(T_0) $\quad \Box(x)(Cx \ \& \ w(x) = 0 \supset h_k(x) = g_0(h_1(x), \ldots, h_{k-1}(x)))$

(T_1) $\quad \Box(x)(Cx \supset h_k(x) = g_1(h_1(x), \ldots, h_{k-1}(x), w(x)))$.

Here T is a factual law relative to the state space **Q** generated by $h_1, \ldots, h_k$, while T_0 and T_1 are defined relative to the state space $\mathbf{Q}^1$ generated by $h_1, \ldots, h_k$, and w. Further, we assume that $(x)(w(x) \neq 0)$ in the actual world, and g_1 depends on the factor $w(x)$. Then T_0 is idealizational law which contains a counterfactual assumption $w(x) = 0$. T_1 is a factual law which expresses through function g_1 how the value of h_k depends on the values of $h_1, \ldots, h_{k-1}$, and w. Law T_1 is obtained from T and T_0 by concretization.

It is not possible to state any general results about the sequence T, T_0, T_1 without further assumptions. Let us assume that the following *Principle of Correspondence* holds

(PC) $\quad \lim_{z \to 0} g_1(h_1(x), \ldots, h_{k-1}(x), z) = g_0(h_1(x), \ldots, h_{k-1}(x))$.

PC thus says that the equation of T is obtained as a limit of the equation of T_1, when the influence of the factor $w(x)$ approaches zero. Hence, if PC holds, T_1 entails T_0. Of course, T also entails T_0. But as g_0 and g_1 are different functions, T_1 and T are incompatible with each other. (For examples, see (3.24) and (3.30).)

Law T_1 describes a surface in the state space $\mathbf{Q}^1$. The idealizational law T_0 also describes a surface, but only in the subspace of $\mathbf{Q}^1$ where $w(x) = 0$. The Correspondence Principle CP means geometrically that the small surface defined by T_0 is a proper part of the large surface defined by T_1. The original law T, which defines a surface in space $\mathbf{Q}$, corresponds also to a subset of the enlarged space $\mathbf{Q}^1$; this subset coincides with the surface defined by T_1 in the subspace of $\mathbf{Q}^1$ with $w(x) = 0$, but deviates from this surface outside this subspace. Through this interpretation, we apply the measures (23) and (29) to evaluate the distance of T and T_0 from T_1.

As the choice of the function g_1 can be made in various mistaken ways, PC does not guarantee that concretization defines sequences of laws converging towards the truth. Thus, it may happen that there is a true factual law g_* relative to $\mathbf{Q}^1$, but g in $\mathbf{Q}^1$ is closer to g_* than g_1.[11] On the other hand, if T_1 is true, then T must be false, but also T_0 (as a deductive consequence of T_1 by PC) is true, so that as a more informative true statement T_1 is more truthlike than T and T_0. Further, if T_1 is approximately$_3$ true, then by (37) its consequence T_0 is also approximately$_3$ true in the subspace of $\mathbf{Q}^1$ with $w(x) = 0$. In such cases, T is typically not approximately$_3$ true.

Let us still see what happens to the intended models of T, T_0, and T_1 in the process of concretization. A typical structure Ω for $\mathbf{Q}$ satisfying T looks as follows

$$\Omega = \langle D, (h_1(x))_{x \in D}, \ldots, (h_{k-1}(x))_{x \in D},$$
$$(g_0(h_1(x), \ldots, h_{k-1}(x)))_{x \in D} \rangle.$$

This structure can be expanded to a model of T in $\mathbf{Q}^1$ in an infinite number of ways:

(42) $\langle D, (h_1(x))_{x \in D}, \ldots, (h_{k-1}(x))_{x \in D},$
$(g_0(h_1(x), \ldots, h_{k-1}(x)))_{x \in D}, (w(x))_{x \in D} \rangle,$

where $w(x) \in \mathbb{R}$ for $x \in D$. The intended models for T_0 and T_1 look as follows:

$$\Omega_0 = \langle D, (h_1(x))_{x \in D}, \ldots, (h_{k-1}(x))_{x \in D},$$
$$(g_0(h_1(x), \ldots, h_{k-1}(x)))_{x \in D}, 0 \rangle$$
$$\Omega_1 = \langle D, (h_1(x))_{x \in D}, \ldots, (h_{k-1}(x))_{x \in D},$$
$$(g_1(h_1(x), \ldots, h_{k-1}(x), w(x)))_{x \in D}, (w(x))_{x \in D} \rangle.$$

Then Ω_0 is an idealizational structure, and Ω_1 a factual structure. Structures of the form Ω_0 are included as special cases among the intended models of T_1, but the intended models of T_0 do not include factual structures.

The class of models (42) of T includes Ω_0, but by measure (39) it also includes structures which have arbitrarily great distances from Ω_0. The minimum distance of Ω_1 from the class (42) is positive, and depends on the distances

(43) $\quad |g_1(h_1(x), \ldots, h_{k-1}(x), w(x)) -$
$\quad - g_0(h_1(x), \ldots, h_{k-1}(x))|, x \in D.$

If the Principle of Correspondence PC holds between T_0 and T_1, it follows that T_0 is *approximately reducible* to T_1 in the following sense:

(44) $\quad$ For each intended model Ω_0 of T_0 there is a sequence of intended models Ω_1^m of T_1, $m = 1, 2, \ldots$, such that $\Delta(\Omega_0, \Omega_1^m) \to 0$, when $m \to \infty$.

(Cf. Niiniluoto, 1986a.) To prove (44), it is sufficient to observe that by PC the function $w(x)$ can be chosen successively so that the distances (43) approach zero.

For example, for a projectile satisfying the ballistic equations (3.28) there is a sequence of cases of projectiles with smaller and smaller resistance of air (cf. (3.30)) which indefinitely approximates the given parabolic case.

This model-theoretical analysis of approximate reduction can be compared to other similar treatments. Rantala (1979) has given an elegant analysis of the correspondence relation by explicating the closeness of models Ω_0 and Ω_1^m through to concept of 'standard approximation' from non-standard analysis.[12] Mayr (1981a, b) says that theory T_0 is approximately reducible to theory T_1, if T_0 is reducible to the completion $\hat{T}_1$ of T_1, where $\hat{T}_1$ is defined through a uniformity defined on the space of structures for T_1 (see Chapter 1.2).[13] Again we can see that there is a close connection between our result (44) and Mayr's treatment, since the former uses a metric concept of convergence, and the latter its generalized formulation by uniform spaces (cf. Chapter 1.2).

Let us conclude this section with a few remarks on the problem of appraising the comparative truthlikeness of quantitative laws on the basis of some evidence. In principle, this problem could be solved by using the general strategy of Chapter 7, i.e., by defining a probability distribution on the space of quantitative laws, and by using it in the construction of the function ver measuring expected degrees of truthlikeness. More generally, we might take the rival hypotheses to be the theories expressible in a language L with (qualitative and quantitative) theoretical terms, and the evidence e to be stated in the observational sublanguage L_0 of L. Then we could compare the values $\text{ver}(T/e)$ and $\text{ver}(T_0/e)$ for the cases, where T contains theoretical terms but theory T_0 is equivalent to an observational statement in L_0.[14]

Instead of developing these ideas here, I mention two possible ways of reformulating the estimation problem.

First, theoretical background assumptions may suggest the mathematical form of the law g up to a finite number of unknown parameters. For example, the form of $g: \mathbb{R}^{k-1} \to \mathbb{R}$ might be

$$g(x_1, \ldots, x_{k-1}) = \sum_{i-1}^{k-1} x_i^{a_i},$$

where $a_i \geq 0$. Then the problem of finding the most truthlike law of this form reduces to the problem of estimating the real-valued parameters a_i, $i = 1, \ldots, k - 1$, on the basis of some evidence (cf. the discussion in Chapter 12.5).

Secondly, the restriction of the mathematical form of the relevant function g may be a result of practical considerations — such as simplicity, economy, manageability. A scientist starts from very simple 'model', usually assuming a linear function, and proceeds to more complex polynomials only if the simpler choice leads to highly inaccurate predictions. There is some justification for this procedure: it is known that an arbitrary continuous function can be approximated by a polynomial with arbitrarily high precision (cf. Simmons, 1963).

However, the restriction to simple polynomials is usually associated with knowledge that this 'model' is idealized and strictly speaking false.

Then our estimation problem reduces to the task of finding the least false of the relevant functions. It was suggested in Chapter 7.5 that this problem can be solved by choosing the function g which has the least distance from the available data, measured by a natural Minkowski metric (7.42). If we say that a law g is *approximately valid* when its distance $D(g, E)$ from the observed evidence E is sufficiently small, then our suggestion for maximizing approximate validity amounts to the rule:

(45)   Accept law $g \in G$ on data E iff g minimizes the distance $D(g, E)$ among the elements of G.

While this rule itself can be formulated without any consideration of probabilities, an evaluation of its reliability would naturally require probabilistic assumptions about the accuracy and representativeness of the data E.[15]

11.5. PROBABILISTIC LAWS

Roger Rosenkrantz (1980) suggests that the concept of truthlikeness should be relativized to "an experiment or domain of application": the truthlikeness of a theory h is its expected support averaged over the possible outcomes x of a suitably chosen experiment. The support that x accords to h is measured by Rosenkrantz by the log-likelihood $\log P(x \mid h)$, where $P(x \mid h)$ is the average likelihood

$$\sum_{i=1}^{n} P(x \mid k_i) P(k_i) / P(h)$$

if h is a composite hypothesis $k_1 \vee \cdots \vee k_n$. Then the expected support of h over the sample space X is

(46) $\sum_{x \in X} P(x \mid h_*) \log P(x \mid h),$

where h_* is the truth. This value can be written in the form

(47) $\sum_{x \in X} P(x \mid h_*) \log P(x \mid h_*) - \sum_{x \in X} P(x \mid h_*) \log \dfrac{P(x \mid h_*)}{P(x \mid h)},$

where the latter factor equals the directed divergence $\mathrm{div}(P(x \mid h_*),$

$P(x|h)$) between the two distributions $P(x|h_*)$ and $P(x|h)$ (see (2.13)). By (47), the truthlikeness of h is greater than that of h' if

$$\text{div}(P(x|h_*), P(x|h)) < \text{div}(P(x|h_*), P(x|h')).$$

Rosenkrantz applies these ideas to cases where h and h' are probabilistic hypotheses. For example, let h and h' be binomial models which say that the probability of success in a single trial is p or q, respectively. Then the distance between h and h' relative to n trials is

$$\sum_{r=0}^{n} \binom{n}{r} p^r(1-p)^{n-r} \log[p^r(1-p)^{n-r}/q^r(1-q)^{n-r}]$$

$$= n[p\log(p/q) + (1-p)\log((1-p)/(1-q))].$$

If p^* is the true probability of success in a single trial, the distance of h from the truth is

$$p^*\log(p^*/p) + (1-p^*)\log((1-p^*)/(1-p)).$$

This value is zero if $p = p^*$.

Rosenkrantz's treatment (independently of his derivation of formulas (46) and (47)) suggests a general approach to the truthlikeness of probabilistic laws. Let h be a probabilistic law of coexistence which assigns a physical probability $P(Q_i(x)) = p_i$ to each Q-predicate Q_i in a discrete conceptual space $\mathbf{Q}$. Let the true law h_* state that these probabilities equal $P_*(Q_i(x)) = p_i^*$. Then the distance of h from h_* can be measured by the directed divergence between h_* and h:

(48) $\text{div}(h_*, h) = \sum_{Q_i \in \mathbf{Q}} p_i^* \log p_i^*/p_i.$

Hence, a probabilistic law h is more truthlike than another law h' iff $\text{div}(h_*, h) < \text{div}(h_*, h')$, i.e.,

(49) $\sum_{Q_i \in \mathbf{Q}} p_i^* \log(q_i/p_i) < 0.$

If a symmetric distance is preferred, (48) may be replaced by the divergence between h and h_*:

(50) $\text{div}(h, h_*) + \text{div}(h_*, h) = \sum_{Q_i \in \mathbf{Q}} (p_i - p_i^*)\log(p_i/p_i^*)$

(see (1.14)). Alternative definitions could be based on the distance measures mentioned in Examples 1.4 and 1.5.

If $\mathbf{Q} \subseteq \mathbb{R}^k$ is a quantitative state space, and if h and h_* define density functions f and f_*, respectively, on $\mathbf{Q}$, then the directed divergence $\text{div}(h_*, h)$ between h_* and h is defined by

$$(51) \quad \text{div}(h_*, h) = \int_Q f_*(x) \log(f_*(x)/f(x)) \, dx.$$

(See (1.15).)

Let then h be a probabilistic law of succession relative to a discrete space $\mathbf{Q} = \{Q_1, \ldots, Q_t\}$. Then h specifies a matrix of transition probabilities $p_{j/i}$ from state i to state j, where

$$\sum_{j=1}^{t} p_{j/i} = 1 \quad \text{for each } i = 1, \ldots, t.$$

(see (3.35).) If $p^*_{j/i}$ are the true transition probabilisties, then the distance of h from the truth can be measured by the sum over $i = 1, \ldots, t$ of the divergences between the distributions $p^*_{j/i}$ and $p_{j/i}$ ($j = 1, \ldots, t$):

$$(52) \quad \sum_{i=1}^{t} \text{div}(p^*_{j/i}, p_{j/i}).$$

For laws of succession relative to a continuous state space $\mathbf{Q}$, the definition of distance reduces then to the divergence between two probability measures on $\mathbf{Q}^\infty$.

CHAPTER 12

VERISIMILITUDE AS AN EPISTEMIC UTILITY

Cognitive decision theory was first developed in the early 1960s by Carl G. Hempel and Isaac Levi, who suggested that the acceptance of scientific hypotheses could be based upon the rule of maximizing 'epistemic utilities'. In contrast with various kinds of 'practical' (e.g. economic) benefits, the epistemic utilities should reflect the cognitive aims of scientific inquiry, such as truth, information, systematic power, and simplicity.

In this chapter, we shall see that the idea of using truthlikeness (as defined by our min-sum function $Tr = M_{ms}^{\gamma\gamma'}$) as an epistemic utility leads to a rational theory of scientific inference: preference between the rival answers to a cognitive problem is determined by the values of the function ver of estimated verisimilitude (Section 2). It turns out that, in a sense, this theory contains as a special case Levi's classical account of cognitive decision making: if the distance function Δ on a cognitive problem is trivial, our function Tr formally reduces to a variant of Levi's definition of epistemic utility in *Gambling with Truth* (1967) (Section 3). We shall also see that certain standard results about point estimation in Bayesian statistics can be reinterpreted in terms of cognitive decision problems where verisimilitude is the relevant utility. This observation suggests a new systematic way of approaching Bayesian interval estimation (Section 5). We shall also argue that no non-Bayesian or non-probabilistic decision rule is adequate for cognitive problems, so that the Popperian programme of verisimilitude is applicable to the problems of theoretical preference *if and only if* there is an independent solution to the traditional problem of induction (Section 4).

12.1. COGNITIVE DECISION THEORY

The classical theory of utility, as developed by Daniel Bernoulli in 1730, is based upon the principle of mathematical expectation: a rational person facing a choice between uncertain alternatives should seek to maximize the expected value of utility or 'moral worth'. Modern versions of this idea were developed by John von Neumann and Oskar Morgenstern in *Theory of Games and Economic Behaviour* (1944) and Abraham Wald in *Statistical Decision Functions* (1950).

Let $a_1, \ldots, a_n$ be alternative *acts*, and let $s_1, \ldots, s_k$ be the possible *states of nature*. One and only one of these states obtains, but its identity is unknown to the decision maker. For $1 \leq i \leq n$, $1 \leq j \leq k$, let

O_{ij} = the outcome of act a_i when s_j is the true state of nature,

and let u be a real-valued function such that

$u(O_{ij})$ = the utility of outcome O_{ij} to the decision maker.

Assume that a probability measure P is defined on the states of nature, so that $P(s_i)$ is the probability that s_i is the true state. Then the *Principle of Expected Utility* recommends the action which maximizes the expected utility relative to measure P (cf. Chapter 7.2):

(PEU) Choose the act a_i which maximizes the value of

$$\sum_{j=i}^{k} P(s_j) u(O_{ij}).$$

When P is assumed to be a subjective (personal) probability distribution, expressing the rational degree of belief of the decision maker (cf. Chapter 2.10), PEU expresses the *Bayes rule* — the central principle of Bayesian decision theory.[1]

In principle, Bayesian decision theory can be applied to acts of any kind: investing money, choosing a husband, carrying an umbrella to work etc. Hence, in particular, it is relevant to situations where we have to make a choice between some uncertain or hypothetical propositions. An interesting discussion of such decision problems was given by Bernard Bolzano in his *Wissenschaftslehre* (1837).[2]

In §317 ('Definition of the Concepts of Certainty and Probability with Respect to Thinking Beings'), Bolzano makes a distinction between proper certainty and moral certainty (sittlich, moralisch, zureichende Gewissheit). The latter is intimately related to rational action: morally certain propositions are *trustworthy* (Glaubwürdig) or *reliable* (verlässig) in the sense that it would be foolish to *act upon* the opposite assumption.

... trustworthiness has several degrees; and one and the same degree of absolute probability, which makes one proposition trustworthy for a given being, may be too low for another proposition. For it is not the low degree of probability alone which makes it foolish or illicit to think of the possible opposite of proposition *M* and to prepare for it, but certain other circumstances also play a role: Consider the damage (Gefahr) which would arise if the opposite of *M* were true and we did not prepare for it. The product of this damage (Schaden) and the degree of probability of *Neg. M* (the negation of *M*) is *the danger* to which we are subject when we do not prepare for *Neg. M*. What makes it foolish or illicit then to prepare for or consider the opposite of a given proposition *M* is that the danger in neglecting *Neg. M* is smaller than the danger in neglecting *M*. (Bolzano, 1972, p. 362; cf. Bolzano, 1930, p. 269.)

Let us try to reconstruct the logic underlying Bolzano's account of moral certainty. Let $L(M; f)$ be the "damage which would arise if the opposite of *M* were true and we did not prepare for it". If we are "not preparing for" (beachten) *Neg. M*, we obviously are acting as if *M* were true. Thus, $L(M, f)$ is the *loss* which results from *rejecting Neg. M*, i.e., *accepting M*, when *M* happens to be false. Let $r(M)$ be the "danger to which we are subject when we do not prepare for *Neg. M*", that is, $r(M)$ is the *risk* involved in accepting *M*.

According to Bolzano,

(1) $\quad r(M) = L(M, f)p(Neg.\ M/E)$,

where $p(Neg.\ M/E)$ is the 'absolute probability' of *Neg. M*, i.e., the probability of *Neg. M* on all available evidence *E*. Further, it is foolish to prepare for *Neg. M* just in case "the danger in neglecting *Neg. M* is smaller than the danger in neglecting *M*". In other words, *Neg. M* is rejectable and *M* is acceptable if and only if $r(M) < r(Neg.\ M)$. By (1), we see that

(2) $\quad r(M) < r(Neg.\ M)$ iff
$\quad\quad L(M, f)p(Neg.\ M/E) < L(Neg.\ M, f)p(M/E)$.

Hence, *M* is acceptable if and only if

(3) $\quad \dfrac{p(Neg.\ M/E)}{p(M/E)} < \dfrac{L(Neg.\ M, f)}{L(M, f)}$.

By means of Bayes's theorem, we see that *M* is acceptable if and only if

(4) $\quad \dfrac{p(E/Neg.\ M)}{p(E/M)} < \dfrac{p(M)L(Neg.\ M, f)}{p(Neg.\ M)L(M, f)}$.

where $p(M)$ is the prior probability of M and $p(Neg.\ M) = 1 - p(M)$. If acts a_0 and a_1 are defined by

a_0 = accept M (reject $Neg.\ M$)
a_1 = accept $Neg.\ M$ (reject M),

then by (4),

(5) $\quad \dfrac{a_0}{a_1} \quad \text{iff} \quad \dfrac{p(E/Neg.\ M)}{p(E/M)} \lessgtr \dfrac{p(M)L(Neg.\ M, f)}{p(Neg.\ M)L(M, f)}.$

Result (5) defines a so called *likelihood into rest* for the null hypothesis M, where the 'critical value' is determined by prior probabilities and by the loss function L. This sort of test is usually called a *Bayes-test* (see, for example, Chernoff and Moses, 1959, pp. 332–333), since it follows from the principle of maximizing expected utility (PEU) or, equivalently, minimizing expected loss. The observation that Bayes-tests for two alternative hypotheses are of the form (5) was made by L. J. Savage and D. Lindley in 1955 (see Savage, 1964, pp. 180–182). This event was in fact the starting point for the rise of the neo-Bayesian school of statistics, which treats all the basic problems of statistical inference (estimation, testing hypotheses, design of experiments) by means of the Bayes rule.

Statisticians have usually interpreted Bayesian decision theory as a method for making 'management decisions', where the relevant gains or losses are 'economic':

In proper Bayesian decision analysis, the loss function is supposed to represent a realistic economic penalty associated with the available actions. (Box and Tiao, 1973, p. 309.)

Here they follow the view that L. J. Savage (1954) has called "the behavioralistic outlook": statistics deals with "problems what to do" rather than "want to say". In other words, to 'accept' a hypothesis or to propose an estimate of a parameter is not to assert anything but rather a decision to *act as if* a certain hypothesis or estimate were true.

Similarly, some of the leading exponents of the Bayesian theory of induction — such as Rudolf Carnap, Richard Jeffrey, and Wolfgang Stegmüller — have assumed that the acceptance or rejection of hypotheses does not belong to science proper, but only to the practical application of science. Thus, Carnap argues that the theory of induction should only show how we can rationally assign inductive probabilities to rival hypotheses. In real decision situations, where a course of action has to be chosen, these probabilities then can be inserted into the Bayes rule PEU together with a practical utility function.[3]

As an alternative to the behavioralistic view, the 'cognitivists' analyse scientific inference as a process where hypotheses are tentatively accepted as true or rejected as false. Thus, critical cognitivists "reject action-analyses of belief". (Levi, 1961). Accepting a hypotheses g as true does not mean the same as acting on the basis of g (i.e., acting as if g were true). Rather, the corpus of accepted hypotheses at time t represents the tentative results of our unending quest for true information about reality. In other words, this corpus constitutes a revisable and evolving body of scientific knowledge at time t.

Hempel and Levi observed[4] — independently of each other — that decision theory can nevertheless serve as a useful tool within a cognitivist analysis of science. The acceptance or rejection of a hypotheses is an act, but the scientist *qua* scientist should measure its value in terms of the cognitive goals of his enterprise. Hempel expressed this idea in the following words:

This much is clear: the utilities should reflect the value or disvalue which the outcomes have from the point of view of pure scientific research than the practical advantages or disadvantages that might result from the application of an accepted hypotheses, according as the latter is true or false. Let me refer to the kind of utilities thus vaquely characterized as *purely scientific*, or *epistemic, utilities*. (Hempel, 1960, p. 465.)

Similarly Levi (1962) argued that the scientist's task of "replacing doubt by belief" should aim at least to true belief, and could be "tempered by other desiderata such as simplicity, explanatory power, etc.". These objectives of autonomous scientific inquiry are "quite distinct from those of economic, political, moral, etc. deliberation" (Levi, 1967a, p. 16).

Cognitive decision theory is an attempt to work out these suggestions, i.e., to analyse scientific inference as the maximization of epistemic utilities.

12.2. EPISTEMIC UTILITIES: TRUTH, INFORMATION, AND TRUTHLIKENESS

Let $\mathbf{B} = \{h_i \mid i \in I\}$ be a cognitive problem with the available evidence

e. Thus, **B** is a P-set relative to e. For a partial answer $g \in D(\mathbf{B})$, let

$u(g, t, e)$ = the epistemic utility of accepting g on the basis of e, when g is true.

$u(g, f, e)$ = the epistemic utility of accepting g on the basis of e, when g is false.

Then the *expected epistemic utility* of the acceptance of g is

$$U(g, e) = P(g|e)u(g, t, e) + P(\sim g|e)u(g, f, e).$$

The Bayes rule PEU leads now to the following principle:

(6) Accept on evidence e the answer $g \in D(\mathbf{B})$ which maximizes the value of $U(g, e)$.

If we are interested in *truth and nothing but the truth*, then a natural choice for the epistemic utilities would be

(7) $u(g, t, e) = a$
$u(g, f, e) = b$,

where a and b are constants such that $a > b$.[5]
But then

$$U(g, e) = P(g|e) \cdot a + (1 - P(g(e)) \cdot b$$
$$= P(g|e)(a - b) + b,$$

so that

$$U(g, e) = \max \quad \text{iff} \quad P(g|e) = 1.$$

If P is a regular probability measure (cf. Chapter 2.11), this implies that

$$U(g, e) = \max \quad \text{iff} \quad e \vdash g.$$

Hence, the utility assignment (7) leads to extreme conservatism: accept on e only those hypotheses g that are logically entailed by e (Levi, 1967a, 1967b).

Hempel's (1960, 1962) suggestion was to choose incremental information content as the epistemic utility:

(8) $u(g, t, e) = \text{cont}_{\text{add}}(g|e) = \text{cont}(e \supset g)$
$u(g, f, e) = -\text{cont}_{\text{add}}(g|e) = -\text{cont}(e \supset g)$

(cf. (4.60)). In fact, he restricted the choice of g to the complete answers $h_i \in B$ and to the suspension of judgment (i.e., the disjunction of all h_i's, $i \in I$), where the latter is assumed to have the epistemic utility 0.

Hence,

$$U(g, e) = P(g|e)\text{cont}(e \supset g) - (1 - P(g(e))\text{cont}(e \supset g)$$
$$= (2P(g|e) - 1)\text{cont}(e \supset g)$$
$$> 0 \quad \text{iff} \quad P(g|e) > 1|2.$$

Further,

$$U(g, e) = \max \quad \text{iff} \quad P(g|e) = 3|4.$$

As Hempel himself noted, the resulting rule,

(9) Accept $h_i \in B$ on e iff $P(h_i|e) > 1|2$,

is 'too lenient' to be satisfactory. (See also Hilpinen, 1968.)

The idea of using $\text{cont}_{\text{add}}(g|e)$ to measure the epistemic utility of g may seem plausible since the net effect of the acceptance of g on e is the addition of g to the corpus represented by e. However, it would seem natural to measure $u(g, f, e)$ by $-\text{cont}_{\text{cond}}(\sim g|e)$, since by accepting the false g we loose the potential gain of accepting the true hypothesis $\sim g$. These choices would give the undesirable result that $U(g, e) = 0$ for any $g \in D(B)$ (see Hintikka, 1968a; Hilpinen, 1968).

Similarly, the suggestion to use conditional information content as epistemic utility, i.e.,

(10) $u(g, t, e) = \text{cont}_{\text{cond}}(g|e)$
$u(g, f, e) = -\text{cont}_{\text{cond}}(\sim g|e),$

would lead to $U(g, e) = 0$. (Cf. (4.63).)

The proposal of using the absolute information content as the epistemic utility gives the choice

(11) $u(g, t, e) = \text{cont}(g)$
$u(g, f, e) = -\text{cont}(\sim g).$

Hence,

(12) $U(g, e) = P(g|e)(1 - P(g)) - (1 - P(g|e))P(g)$
$= P(g|e) - P(g).$

The same result is obtained by using transmitted information:

(13) $\quad u(g, t, e) = \text{transcont}_{\text{add}}(g\,|\,e)$
$\quad\quad u(g, f, e) = -\text{transcont}_{\text{add}}(\sim g\,|\,e).$

or

(14) $\quad u(g, t, e) = \text{transcont}_{\text{cond}}(g\,|\,e)$
$\quad\quad u(g, f, e) = -\text{transcont}_{\text{cond}}(\sim g\,|\,e).$

(Cf. (4.67).) To maximize (12), or $P(g\,|\,e) + \text{cont}(g) - 1$, g should be a hypotheses which has both a high information content and a high posterior probability on e.[6]

Definition (11) was first discussed by Levi (1963) and then defended by Hintikka and Pietarinen (1966) (see also Niiniluoto and Tuomela, 1973). Levi (1967a) rejects the utility assignment (11), since he thinks that g_1 and g_2 should have the same utility on e, if $e \vdash g_1 \equiv g_2$ but $\nvdash g_1 \equiv g_2$ (see also Levi, 1979). This is hardly conclusive, since we are interested in the 'global' virtues of the rival elements of $D(B)$, not in their content relative to the evidence e that we happen to have.[7] Moreover, the use of the conditional cont-measure would lead to unsatisfactory consequences (see (10)).

Levi's own suggestion in *Gambling With Truth* (1967a) was to measure information content or 'relief from agnosticism' by a function which is pragmatically relativized to the 'ultimate partition' $\mathbf{B} = \{h_i\,|\,i \in \mathbf{I}\}$. Assume that $\mathbf{B}$ is a finite P-set relative to evidence e. Hence, $|\mathbf{I}| < \omega$.

If

$$\vdash g \equiv \bigvee_{i \in I_g} h_i,$$

then the content of g is defined by

$$\text{cont}_u(g\,|\,e) = 1 - \frac{|\mathbf{I}_g|}{|\mathbf{I}|},$$

(See (4.50).) In particular, each complete answer $h_i \in \mathbf{B}$ has the same content

$$\text{cont}_u(h_i\,|\,e) = 1 - \frac{1}{|\mathbf{I}|} = \frac{|\mathbf{I}| - 1}{|\mathbf{I}|}.$$

Levi derives his definition of epistemic utility from four conditions, where $g \in D(\mathbf{B})$ and $h \in D(\mathbf{B})$:

(L1) $u(g, t, e) > u(h, f, e)$
(L2) $u(g, t, e) \gtreqless u(h, t, e)$ iff $\text{cont}_u(g|e) \gtreqless \text{cont}_u(h|e)$
(L3) $u(g, f, e) \gtreqless u(h, f, e)$ iff $\text{cont}_u(g|e) \gtreqless \text{cont}_u(h|e)$
(L4) $u(g, t, e) - u(h, t, e) = u(g, f, e) - u(h, f, e)$.

Here L1 says that correct answers are always epistemically prefered to errors. L2 and L3 express the idea that of two answers with the same truth value the more informative is preferable. Levi then shows that all the utility assignments satisfying (L1)–(L4) are linear transformations of the definition

(15) $u(g, t, e) = 1 - q\,\text{cont}_u(\sim g | e)$
         $u(g, f, e) = -q\,\text{cont}_u(\sim g | e)$,

where $0 \leq q \leq 1$. As $|\mathbf{I}_{\sim g}| = |\mathbf{I}| - |\mathbf{I}_g|$, (15) reduces to

(16) $u(g, t, e) = 1 - q|\mathbf{I}_g|/|\mathbf{I}|$
         $u(g, f, e) = -q|\mathbf{I}_g|/|\mathbf{I}|$.

If we let tv(g) be the truth value of g, i.e.,

         tv(g) = 1, if g is true
                 = 0, if g is false,

the epistemic utility (16) of g can be written simply as

(17)   tv(g) $- q|\mathbf{I}_g|/|\mathbf{I}|$.

By adding a constant q to (17), it gives an equivalent definition

(18)   tv(g) $+ q\,\text{cont}_u(g|e)$,

which shows that Levi's epistemic utility is essentially a weighted combination of a truth factor tv(g) and an information factor $\text{cont}_u(g|e)$. The case where $q = 0$ has to be excluded, since it reduces to (17). The parameter q can thus be interpreted as an 'index of boldness' which tells how willing the scientist is to risk error in his attempt to relieve from agnosticism.

The expected utility of accepting g on e is, by (16),

(19) $U(g, e) = P(g|e) - q|\mathbf{I}_g|/|\mathbf{I}|$.

VERISIMILITUDE AS AN EPISTEMIC UTILITY 415

On the basis of (19) and a 'rule for ties', Levi suggests the following rule of acceptance:

(20)   Reject all elements $h_i \in \mathbf{B}$ with $P(h_i|e) < q/|\mathbf{I}|$, and accept the disjunction of all unrejected elements of $\mathbf{B}$ as the strongest on the basis of e.

Risto Hilpinen (1968) has proposed that (15) is modified by replacing $\text{cont}_u(\sim g|e)$ with $\text{cont}(\sim g)$. Thus,

(21) $u(g, t, e) = 1 - q\,P(g)$
       $u(g, f, e) = -q\,P(g)$.

This utility assignment is a linear transformation of

(22) $\text{tv}(g) + q\,\text{cont}(g)$.

It satisfies conditions similar to Levi's L1—L4, and leads to the expected utility

(23) $U(g, e) = P(g|e) - q \cdot P(g)$

and the acceptance rule

(24)   Reject all elements $h_i \in \mathbf{B}$ with $P(h_i|e) < q\,P(h_i)$ and accept the disjunction of all unrejected element of $\mathbf{B}$ as the strongest on the basis of e.

If we follow Popper in regarding verisimilitude as a combination of the ideas of truth and information, then Levi's (17) and Hilpinen's (21) could be regarded as putative candidates for a measure of truthlikeness. What happens if we replace them with our favourite definition of Tr, i.e., with the min-sum function $M_{ms}^{\gamma\gamma'}$?

To apply this measure to a finite cognitive problem $\mathbf{B} = \{h_i | i \in \mathbf{I}\}$, the elements h_i of $\mathbf{B}$ have to be chosen as the 'states of nature', and the epistemic utility of accepting $g \in D(\mathbf{B})$ on evidence e, when h_i is the true state, is defined by

(25) $M_{ms}^{\gamma\gamma'}(g, h_i) = 1 - \Delta_{ms}^{\gamma\gamma'}(h_i, g)$
       $= 1 - \gamma\Delta_{\min}(h_i, g) - \gamma'\Delta_{\text{sum}}(h_i, g)$.
       $= 1 - \gamma\Delta_{\min}(h_i, g) - \gamma' \cdot \sum_{j \in I_g} \Delta_{ij} \Big/ \sum_{i \in I} \Delta_{ij}$.

(See (6.85).) The *expected utility* of accepting g on evidence e is then

(26) $U(g, e) = \sum_{i \in I} P(h_i | e) M_{ms}^{\gamma\gamma'}(g, h_i) = \text{ver}(g|e)$.

Hence, $U(g, e)$ is equal to the estimated degree of verisimilitude $\text{ver}(g|e)$ of g on e (see Chapter 7). The Bayes rule PEU leads thus to the rule of acceptance:

(27)   Accept on evidence e that answer $g \in D(\mathbf{B})$ which maximizes the value of $\text{ver}(g|e)$.

In Chapter 7, we studied $\text{ver}(g|e)$ as a measure for the rational estimate of the truthlikeness of g on e. The results concerning the behaviour of ver can now be applied to the study of the acceptance rule (27). For example, in the two extreme cases of complete ignorance and complete certainty, we have the following theorems:

(28)   Assume that $\mathbf{B}$ is a balanced cognitive problem, and $P(h_i|e) = 1/|\mathbf{I}|$ for all $i \in \mathbf{I}$. If $\gamma > 2\gamma'$, then it is rational to suspend judgment on evidence e.

(29)   Assume that $P(h_j/e) \approx 1$ for some $h_j \in B$. Then it is rational to accept h_j on evidence e.

Note that (29) applies only to complete answers $h_j \in B$: the high probability of a partial answer $g \in D(B)$ is not a sufficient condition for its acceptance. Further results for continuous cognitive problems are discussed in Section 5.

12.3 COMPARISON WITH LEVI'S THEORY

Isaac Levi's *Gambling with Truth* (1967a) — with further revisions and refinements in later works[8] — is the most sophisticated account of cognitive decision theory. In this section, my definition (25) is compared with Levi's theory. It turns out that their connection is closer and more exciting than I initially expected.

We have already in Chapter 6 explained our reason for thinking that functions like Levi's (17) or Hilpinen's (21) are not adequate explications of truthlikeness. Indeed, they violate several of our adequacy conditions M1—M13 in Chapter 6.6. In particular, while Levi's L2 coincides with our M4, his L1 is the opposite of our M10. Moreover, Levi's L3 is the negation of our M5 — and the same as (5.33), which we regarded as the stumbling block of Miller's definition of truthlikeness.

Appeal to conflicting conditions of adequacy may seem question begging: Levi and Hilpinen do not present their utility functions as explicates of verisimilitude. These comparisons are not pointless, however, since we are dealing here with rival ways of solving cognitive problems. So is it possible to find some deeper motivation for the approval or disapproval of conditions like L1 and L3?

It seems to me that the answer to this question can be found in Levi's slogan "a miss is as good as a mile" (Levi, 1986b). Levi's epistemic utility function (16) shares an important feature with Popper's attempted definitions of verisimilitude: they do not contain any element of 'likeness', 'similarity' or 'distance' (cf. Chapter 5). Of course Levi does not deny that some cognitive problems **B** may involve sets of complete answers which are in fact metric spaces. For example, if **B** is $\mathbb{R}$, it is possible to speak about the distance $|x - y|$ between two points $x \in \mathbf{B}$ and $y \in \mathbf{B}$. But while I think that it is crucially important to build such a metric into the definition epistemic utility (see also Festa, 1986), Levi denies this: his adequacy condition L3 implies that *all false complete answers have the same epistemic utility*. Indeed, by definition (16),

$$u(h_i, f, e) = -q/|\mathbf{I}| \quad \text{for all } h_i \in \mathbf{B}.$$

In my account, the epistemic value of a false complete answer h_i depends on its distance Δ_{*i} from the true complete answer h_*.

These observations suggest that, from the perspective of our theory, we might view Levi's approach as a theory for cognitive problems **B** where the distance function $\Delta: \mathbf{B} \times \mathbf{B} \to \mathbb{R}$ is *trivial*:

$$\Delta_{ij} = 1, \text{ if } i \neq j$$
$$= 0, \text{ if } i = j.$$

This means that *all complete false answers are equally far from the truth*. In this case it is possible to motivate Levi's conditions L1 and L3. First, as all complete errors are cognitively equally bad, it is not possible to find highly informative answers which make so *small* errors that they can be preferred to uninformative truths. Hence, our M10 collapses, and L1 is justified. Secondly, a false partial answer g is a disjunction of complete errors located on the same circle, and therefore by making g logically stronger we eliminate some of its 'bad' disjuncts. Hence, L3 seems to be justified in this case. (See Figure 1.)

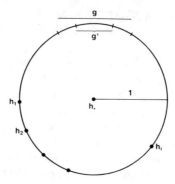

Fig. 1. Trivial metric: g' is better than g.

It is now interesting to see what happens, if the trivial metric Δ is substituted into our definition (25) of epistemic utility. In this case, the mini-sum function $M_{\text{ms}}^{\gamma\gamma'}$ of truthlikeness reduces to

$$\text{(30)} \quad 1 - \gamma' \cdot \frac{|I_g| - 1}{|I| - 1} \quad \text{if } g \text{ is true}$$

$$1 - \gamma - \gamma' \cdot \frac{|I_g|}{|I| - 1} \quad \text{if } g \text{ is false}$$

By defining

$$\alpha = \frac{|I|}{|I| - 1} \cdot \gamma',$$

(30) can be rewritten as

$$\text{(31)} \quad 1 - \alpha \cdot \frac{|I_g|}{|I|} + \frac{\alpha}{|I|} \quad \text{if } g \text{ is true}$$

$$-\alpha \cdot \frac{|I_g|}{|I|} + 2 - \gamma \quad \text{if } g \text{ is false}$$

Here $\gamma' > 0$ implies that $\alpha > 0$.

By choosing

$$\gamma = 1 - \frac{\alpha}{|I|} = 1 - \frac{\gamma'}{|I| - 1},$$

(31) reduces to

$$1 - \alpha \frac{|I_g|}{|I|} + \frac{\alpha}{|I|} \quad \text{if } g \text{ is true}$$

$$-\frac{|I_g|}{|I|} + \frac{\alpha}{|I|} \quad \text{if } g \text{ is false}$$

which is a linear transformation of Levi's measure (16). In this sense, Levi's formula for epistemic utility is obtained as a special case of our $M_{ms}^{\gamma\gamma'}(g, h_*)$ with the trivial distance function Δ. This observation was first made in my PSA 1984 paper. (Niiniluoto, 1986b.)

It should be noted that Levi's restriction $\alpha \leq 1$ cannot be derived from our constraints on γ and γ'. Indeed, through the assumed connection between these parameter, condition $\alpha \leq 1$ would require that

$$\gamma' \leq \frac{|I| - 1}{|I|} \leq \gamma,$$

which is sufficient to guarantee that every false complete answer is misleading in the sense of (6.91) (with $\Delta_{*j} = 1$), so that our M10 is violated. This observation highlights the difference between situations with trivial and non-trivial distance functions.

Similarly, the min-cont-measure

(32) $\quad \gamma(1 - \Delta_{\min}(h_*, g)) + (1 - \gamma)\text{cont}_u(g),$

(see (6.77)) reduces directly to the variant (18) of Levi's measure, with $q = (1 - \gamma)/\gamma$, when the distance function Δ is trivial. Note, however, that the restriction $0 < q \leq 1$ is equivalent to the constraint $1/2 \leq \gamma < 1$, which is stronger than our condition $0 < \gamma < 1$ in Chapter 6.5.

If $\text{cont}_u(g)$ is replaced by the cont-measure in (32), it reduces to the variant (23) of Hilpinen's measure of epistemic utility, when the distance function is trivial.

In his interesting comments on the truthlikeness approach, Levi (1986b) observes that a function like (25) makes both the 'truth factor' and the 'information factor' dependent of truth, and he suggests that I could have defined epistemic utility "as a weighted average of some measure of verisimilitude and a measure of information value", where the latter is independent of where the truth is. A measure of this form,

(33) $\quad \gamma M(g, h_*) + (1 - \gamma)C(g),$

would reduce to the min-cont function (6.77), if M is chosen to be the closeness $M_{\min}(g, h_*)$ of g to h_*. My reasons for preferring the min-sum-measure to the min-cont-measure were explained already in Chapter 6.5. On the other hand, if M is the min-max measure $M^\gamma_{\min}$, then (33) is an interesting variant of the min-max-cont-measure (6.79).

Finally, if M is chosen to be the min-sum-measure $M^{\gamma\gamma'}_{ms}$, then (33) is a new proposal which might have interesting methodological applications. However, from my viewpoint, the information factor in $M^{\gamma\gamma'}_{ms}$ expresses "information about the truth" (cf. Chapter 6.4) in a manner which is sensitive to changes in the absolute information content. For trivial distances, the use of this factor reduces to the $cont_u$-measure. As $M^{\gamma\gamma'}_{ms}$ already is a way of extending the Levian theory to situations with non-trivial distances, I am not inclined to join absolute content to it as an additional independent factor.

When we move from the original epistemic utility assignment to *expected* epistemic utilities, the difference between my similarity approach and Levi's a-miss-is-as-good-as-a-mile-doctrine need not be great. (See also Levi, 1986b). The reason for this is the following: in many applications, the likelihood function $P(e|h_i)$ decreases when the 'distance' of observation e from h_i increases (cf. Chapter 7.5). When evidence e is large enough, the posterior probabilities $P(h_i|e)$, obtained by Bayes' Theorem, have this same property. It follows that the expected utility $U(h_j, e)$, calculated from (19) by using the probabilities $P(h_i|e)$, $i \in I$, often covaries with the distance between h_j and e — so that evidence e will favour those hypotheses that are 'close' to it.

This partial agreement in the results of the two approaches does not diminish the importance of the conceptual difference in their foundations.

12.4. THEORETICAL AND PRAGMATIC PREFERENCE

The results of Section 2 show how the concept of truthlikeness can be used for expressing preferences in the context of *theory choice*. Popper (1972) formulates two different problems of preference:

(TP) *Theoretical Preference*: Given the theoretical aim of science which of the rival theories should we prefer?

(PP) *Practical Preference*: Which of the rival theories should we rationally prefer for rational action?

The rule (27) for maximizing expected verisimilitude, i.e., ver(g/e), solves the TP-problem in the Popperian spirit: it advices us to prefer the theory which tentatively can be claimed to have the highest degree of truthlikeness on e, and thus serves best the "theoretical aim of science".

The concept of "theoretical preference" can be understood in stronger or weaker sense. Popper's fallibilism usually claims that theories are 'accepted' only for the purpose of criticism and further testing (Popper, 1959, p. 415). Sometimes he is nevertheless willing to speak of the best-tested but inconclusive theories as "the science of the day" (Popper, 1974a, p. 68). Levi's (1980a) combination of corrigibilism and infallibilism takes accepted hypotheses to be "standards of serious possibility" which one regards to be 'infallibly true' — at least as long as the corpus of knowledge is not revised through 'contraction' and 'replacement'.

Levi (1980a) rejects Peirce's and Popper's fallibilism, which takes the ultimate aim of inquiry to be convergence on the true complete story of the world. He suggests that concern for truth and avoidance of error should be 'myopic', since otherwise we cannot explain the revisability of the body of knowledge accepted as infallibly true at some time t (see also Levi, 1986a).[9]

Levi's argument presupposes his principle L1 which says that all truths are preferred to all errors. This begs the question against Popper who clearly rejects L1: if L1 is denied, then it may be epistemically rational to prefer a false theory g' to another theory g even if g is true and regarded as infallibly true. We have seen in Chapter 7 that this is a real possibility (see also Niiniluoto, 1979b, p. 253; 1984b, p. 171). If the accepted theory g is a complete answer in **B**, then there cannot be any element g' in $D(\mathbf{B})$ better than g, but the situation may change if the cognitive problem is extended to a finer problem **B'** where $g \in D(\mathbf{B'})$ but $g \notin \mathbf{B'}$.

The fallibilism defended in this book agrees with Levi that hypotheses are often tested and accepted in order that they can be used as background knowledge "in subsequent inquiries in science" (Levi, 1980, p. 24). I also agree with Levi, contra Popper, that this expansion is a result of *inductive* inference. Furthermore, even if my epistemic utility assignment does not satisfy Levi's L1, avoidance of error is a serious concern of my measure of truthlikeness.

However, it is important to acknowledge also that in many inquiries it is not rational to accept *any* hypotheses *as true* in Levi's strong sense, i.e., as a standard of serious possibility. These are the inquiries where the cognitive problem is defined relative to a false presupposition (see Chapters 6.9 and 7.5). For many theoretical and practical purposes, the best we can do is to study some problem with counterfactual idealizing assumptions. In such situations, avoidance of error cannot be an absolute demand: it will be replaced by the desire to find the least false of the available answers. Inquiries of this sort are even more 'myopic' than Levi recommends: their results are not intended to be preserved in the corpus of knowledge in the long run since they can be replaced in the short run by more 'concrete' results relative to more realistic background assumptions.

Levi complains that to insist on a "double standard for serious probability" — one for theoretical, one for practical purposes — "implies an untenable dualism between theory and practice" (Levi, 1980, p. 72). The existence of such a dualism was an important message of *Gambling With Truth*: given a set of hypotheses **B** and evidence *e*, what is acceptable on *e* for belief and for action are two different things, since the former depends on epistemic, the latter on practical utilities. Sometimes it may be too risky to act as if our epistemically preferred theory were true (e.g. a hypotheses about the side effects of a medical cure). Sometimes we are ready to act upon hypotheses which are known to be false (e.g. the application of Newton's theory in technology). Hence, the problems of theoretical preference and for practical preference have the same solution only in the special case, where the epistemic utilities and the practical utilities agree with each other.[10]

This observation constitutes the main argument against Popper's (1972) solution to the pragmatic problem of preference (cf. Niiniluoto and Tuomela, 1973; Niiniluoto, 1982b).[11] Popper certainly does not support a dualism between theory and practice, since he claims that the best-tested theory gives the most rational solution both to TP- and PP-problems (Popper, 1972, pp. 15, 22): we should never 'rely' on any theory, but if we must act it is rational to act upon the theory with the highest degree of corroboration — which is also the theory that can be claimed to be closest to the truth. In this sense, Popper's solution to the problem of pragmatic preference is essentially the same as the original use of the concept of *verisimilitude* by Carneades — at least in Augustine's interpretation (see Chapter 5.2).

However, if rule (27) for maximizing expected verisimilitude ver($g|e$) gives a rational answer to a TP-problem **B**, a PP-problem with **B** may require a different solution. Of course, there may be special cases where the practical utility of acting upon $g \in D(\mathbf{B})$ is a linear function of the truthlikeness Tr(g, h_*) of g, but clearly this is not generally true. For example, suppose I am left on an island in winter, and my task is to estimate how thick the ice on the lake will need to be so that I can walk safely to the shore. Assume that h_1 underestimates the correct value of θ by the same amount as h_2 overestimates it. From a theoretical viewpoint, h_1 and h_2 are equally distant from the truth. From a practical viewpoint, the loss associated with h_1 is much greater than that with h_2, since the outcome of acting upon h_1 is cold swim and death (cf. Niiniluoto, 1982b, p. 192).

This criticism of Popper may seem irrelevant, since it in effect is limited to a contrast between two applications — cognitive and practical — of the *Bayesian inductive* solution to the problems of theoretical and pragmatic preference, since the calculation of expected utilities presupposes epistemic probabilities. So we could still ask whether the concept of verisimilitude is able to give a *non-inductive* solution to the problems of theory choice.

It was already hinted in Chapter 7.5 that this question has to be answered negatively. A conclusive proof for this claim is not possible, but I shall show that none of the standard non-probabilistic decision criteria are able to give us reason to accept any interesting hypotheses as better than their rivals — in the way our function ver does with non-uniform probability distributions. (See Niiniluoto, 1982b.).

Let us reformulate our decision problem in terms of losses rather than utilities: the states of natures are represented by $h_i \in \mathbf{B}$, the alternative acts by the acceptance of the hypotheses $g \in D(\mathbf{B})$, and loss of accepting g relative to the state h_i by the distance $\Delta(h_i, g)$ of g from h_i (see Table I). Minimizing the expected loss in this decision problem is equivalent to maximizing ver($g|e$), when Δ is the min-sum distance $\Delta_{ms}^{\gamma\gamma'}$.

Tichý (1978, pp. 188—191) has proposed a possible way of connecting verisimilitude and practical action. He suggests that if we 'put our trust', e.g., to the disjunction of two constituents h_1 and h_2, we have to make a 'blind choice' between h_1 and h_2. In other words, the choice $h_1 \vee h_2$ is what decision theorists call a *randomized act with symmetric probabilities*. In this interpretation, to accept the disjunction

TABLE I
The matrix of losses

	states		
	...	h_i	...
acts h_i	...	0	...
g	...	$\Delta(h_i, g)$	...

$h_{i1} \vee \cdots \vee h_{in}$ for practical action is equivalent to a lottery where we may 'win' each h_{ij} with the probability $1/n$. In other words, the choice of hypotheses g is equivalent to choosing one of the disjuncts h_i, $i \in \mathbf{I}_g$, in the normal form of g with probability $1/|\mathbf{I}_g|$. As the utility of a randomized act is defined — following von Neumann and Morgenstern — as the expected value of the basic act included in it, it follows that the loss associated with the acceptance of g becomes equal to

$$(34) \quad \sum_{i \in \mathbf{I}_g} \frac{1}{|\mathbf{I}_g|} \Delta(h_*, h_i),$$

which is equal to

$$\Delta_{av}(h_*, g) = \frac{1}{|\mathbf{I}_g|} \sum_{i \in \mathbf{I}_g} \Delta_{*i}.$$

Tichý does not tell how (34) is supposed to be used in decision making. But he takes the derivation of (34) as an argument for regarding his average measure $1 - \Delta_{av}$ as a reasonable explication of truthlikeness. I find this idea to be *ad hoc*. First, if g is accepted in a TP-problem, there is no reason to assume that its disjuncts have equal probabilities. Secondly, accepting g in a PP-problem means that we try to prepare for all the possibilities allowed by g, not that we select at random some disjunct h_j of g and act upon h_j.[12]

Let us next go through the standard decision criteria which do not involve probabilities (cf. Luce and Raiffa, 1957). Following Watkins' (1977) ideas on non-Bayesian theory we assume that, when evidence e has falsified some elements of **B**, these criteria are applied to the unrefuted hypotheses in **B** as the reduced set of the states of nature. We shall also assume that **B** is finite and the underlying distance function Δ on **B** is balanced.

(a) *Minimax*. The value of

$$\max_{i \in I} \Delta_{ms}^{\gamma\gamma'}(h_i, g)$$

is $(\gamma + 2\gamma'/|I|)$ if g is an element of **B**, and γ' if g is a tautology. Thus, this criticism makes no difference between the unrefuted complete answer $h_i \in \mathbf{B}$. If $\gamma|\gamma' > (|I| - 2)/|I|$, then the tautology dominates in preference over all the other hypotheses. Similarly,

$$\max_{i \in I} \Delta_{av}(h_i, g)$$

is 1 if $g \in \mathbf{B}$, and $1|2$ if g is a tautology.

(b) *Minimax risk*. As the minimum loss for each column is zero, this criterion reduces to minimax.

(c) *Pessimism-optimism index*. The value of

$$\alpha \min_{i \in I} \Delta_{ms}^{\gamma\gamma'}(h_i, g) + (1 - \alpha) \max_{i \in I} \Delta_{ms}^{\gamma\gamma'}(h_i, g)$$

(where $0 < \alpha < 1$) is $(1 - \alpha)(\gamma + 2\gamma'/|I|)$ if $g \in \mathbf{B}$, and γ' if g is a tautology. In this case, a complete answer will be better than the tautology, if the index α of optimism is sufficiently high. On the other hand, there is no basis for preferring one complete answer against another, no matter how α, γ and γ' are chosen. Similarly,

$$\alpha \min_{i \in I} \Delta_{av}(h_i, g) + (1 - \alpha) \max_{i \in I} \Delta_{av}(h_i, g)$$

is $1 - \alpha$ for $g \in \mathbf{B}$, and $1|2$ for the tautology.

(d) *Insufficient reason*. The value of

$$\sum_{i \in I} \frac{1}{|I|} \Delta_{ms}^{\gamma\gamma'}(h_i, g)$$

is $\gamma/2 + \gamma'/|I|$ if $g \in \mathbf{B}$, and γ' if g is the tautology. Furthermore,

$$\sum_{i \in I} \frac{1}{|I|} \Delta_{av}(h_i, g) = \frac{1}{|I|} \sum_{i \in I} \frac{1}{|\mathbf{I}_g|} \sum_{j \in \mathbf{I}_g} \Delta_{ij} = \frac{1}{|\mathbf{I}_g|} \sum_{j \in \mathbf{I}_g} \frac{1}{|I|} \sum_{i \in I} \Delta_{ij}$$

$$= \frac{1}{|\mathbf{I}_g|} \sum_{j \in \mathbf{I}_g} \frac{1}{|I|} \sum_{i \in I} \Delta_{ji} = \frac{1}{|\mathbf{I}_g|} \sum_{j \in \mathbf{I}_g} \frac{1}{2} = \frac{1}{2}$$

for *all* $g \in D(\mathbf{B})$.

All these results are disappointing (cf. also Chapter 7.4). They never enable us to distinguish between unrefuted complete answers. In many cases, the tautology (or the disjunction of all unrefuted elements of **B**) is preferred over other hypotheses. If instead the complete answers $h_i \in \mathbf{B}$ are the best, we are forced to treat them equally — and accept again their disjunction. In any case, preference is systematically biased towards logically weak answers — in contrast with the Popperian insistence on boldness. Our function ver instead behaves in a different way, and allows us to accept informative hypotheses.

We may therefore conclude: if anyone wishes to claim that the concept of verisimilitude can give a non-inductive solution to the problem of preference between rival hypotheses, the onus of proof lies with him.

12.5. BAYESIAN ESTIMATION

In this section, we shall see that certain standard results concerning Bayesian point estimation can be interpreted in terms of cognitive decision problems where truthlikeness is the relevant epistemic utility: given a suitable loss function, accepting a Bayesian point estimate is equivalent to maximizing expected verisimilitude (cf. Niiniluoto, 1982b). This perspective suggests a new systematic way of approaching interval estimation in decision-theoretic terms (see Niiniluoto, 1982c). Bayesian statisticians have done surprisingly little work in this direction — apparently for the reason that the choice of the appropriate loss function has seemed to be quite arbitrary. However, the requirement that the loss function should express *the distance of an interval estimate from the truth* gives an interesting restriction to permissible choices. It also turns out that the form of this distance function determines the question whether the Bayesian decision maker has a general preference to point estimates over interval estimates or *vice versa*.[13]

For simplicity, I restrict my attention here to cases, where θ is a real-valued parameter. The range $\Theta \subseteq \mathbb{R}$ of the possible values of θ is called the *parameter space*. Let $f(x/\theta)$ be the density function of a probability distribution for the values x of a random variable **x**. The range **X** of the values of **x** is called the *sample space*. For a fixed sample x, $f(x/\theta)$ as a function of the parameter θ is known as the *likelihood function*.

The Bayesian strategy for making inferences about the known value of θ on the basis of sample x presupposes that there is a *prior probability distribution* for θ. Let $g(\theta)$ be the density function of this distribution.[14] Then, for each measurable set $A \in \Theta$, the value

$$(35) \quad g(A) = \int_A g(\theta)\,d\theta$$

expresses the personal probability of the statistician for the claim that the value of θ belongs to A. According to Bayes' formula, the *posterior probability* density $g(\theta/x)$ for θ given x is obtained by multiplying prior probabilities with likelihoods:

$$(36) \quad g(\theta/x) = \frac{g(\theta)f(x/\theta)}{f(x)},$$

where

$$f(x) = \int_\Theta g(\theta)f(x/\theta)\,d\theta.$$

Hence, for $A \subseteq \Theta$,

$$(37) \quad g(A/x) = \int_A g(\theta/x)\,d\theta$$

expresses the conditional personal probability that $\theta \in A$ given x. (cf. Chapter 7.4).

A Bayesian theory of interval estimation can be based directly on the posterior distribution. Let $0 \leq \alpha \leq 1$. Then an interval $I = [c, d] \subseteq \Theta$ is a $100\alpha\%$ Bayesian *confidence interval* for θ if $g(I/x) = \alpha$. For example, if $\alpha = .99$, the statistician is 99% certain that the confidence interval covers the true value of θ. x is called the *confidence level* of I

(to distinguish it from the 'significance level' of the Neyman–Pearson theory). Given x and α, the choice of I can be made in many cases unique by requiring that the values of $g(\theta/x)$ outside I are smaller than the values inside I (see Lindley, 1965; Box and Tiao, 1973).

A different Bayesian approach to estimation has been based upon decision theory (cf. Wald, 1950; Raiffa and Schlaifer, 1961; Ferguson, 1967). An *estimator* $d: \mathbf{X} \to \Theta$ is a real-valued function which assigns to each sample $x \in \mathbf{X}$ a *point estimate* $d(x) \in \Theta$ of the parameter θ. When x is observed, estimator d recommends us to choose $d(x)$ as the 'best' estimate of the unknown value of θ. Let $L: \Theta \times \Theta \to \mathbb{R}$ be a real-valued *loss function* which tells that $L(\theta', \theta)$ is the loss resulting from the adoption of the estimate θ' when θ is the true value of the parameter. Then the Bayes rule PEU in decision making, i.e., the demand to *minimize expected loss*, leads to the following principle:

(38) For each $x \in \mathbf{X}$, choose estimate $d(x) \in \Theta$ so that the posterior expected loss

$$\int_\Theta L(d(x), \theta) g(\theta/x) \, d\theta$$

is the minimum.

The value of $d(x)$ defined by (38) is usually called the *Bayes-estimate* of θ given x. Some standard results concerning Bayesian point estimation are given in Table II.

TABLE II
Bayesian point estimation

Loss function $L(\theta', \theta)$	Bayes-estimate
$\lvert \theta' - \theta \rvert$	the median of $g(\theta/x)$
$(\theta' - \theta)^2$	the mean of $g(\theta/x)$
$\begin{cases} a_1 \lvert \theta' - \theta \rvert & \text{if } \theta' < \theta \\ a_2 \lvert \theta' - \theta \rvert & \text{if } \theta' \geqslant \theta \end{cases}$	any fractile of $g(\theta/x)$
$\begin{cases} 1 & \text{if } \lvert \theta' - \theta \rvert > \varepsilon \\ 0 & \text{if } \lvert \theta' - \theta \rvert \leqslant \varepsilon \end{cases}$ ($\varepsilon > 0$ arbitrarily small)	the mode of $g(\theta/x)$

The Bayesian school of statistics is divided with respect to attitudes towards point estimation. While those theorists who have worked in the decision-theoretical framework have mainly concentrated on point estimation, some authors like Lindley (1965) do not see any point in the whole idea of point estimates.

Box and Tiao (1973), who favour a theory of "highest posterior density intervals", give a critical discussion of point estimation in their Appendix A5.6. They reasonably argue that all types of posterior distributions cannot be described or summarized by a single number. But they also make it clear that they are thinking about the instrumental value of estimates: "The first question a Bayesian should ask is whether or not point estimates provide a useful tool for making inferences." They admit the possibility of a decision-theoretic analysis of estimation, where the relevant losses are interpreted in practical or 'economic' terms, but conclude that the choice of the loss function is 'arbitrary', since "it seems very dubious that a general loss function could be chosen in advance to meet all situations".

This criticism does not exclude the possibility that for some types of estimation problems the most reasonable loss function may be uniquely (or almost uniquely) determined. We shall argue below that this is the case with *cognitive* estimation problems. Let us first illustrate this in connection with point estimation.

Assume that the decision maker uses the loss function $L(\theta', \theta) = |\theta' - \theta|$ in evaluating his point estimates θ' of θ (see Table II). Then, by the Bayes Rule (38), he should choose θ' so that it minimizes the value of

$$(39) \qquad \int_\Theta |\theta' - \theta| g(\theta/x) \, d\theta.$$

As $|\theta' - \theta|$ expresses the *distance of θ' from the truth θ*, it can also be said that the minimization of (39) is equivalent to the *maximization of the expected verisimilitude* of the estimate θ'. In other words, the Bayes rule leads here to the recommendation to accept that hypothetical value θ' which has the greatest estimated truthlikeness given evidence x (see Chapter 7).

A decision maker who follows the rule (39) is thus in effect using truthlikeness as his 'epistemic utility'. Such an interest in truthlikeness — in contrast to various kinds of practical utilities — is justified in those situations where we wish to find the 'theoretically best' estimate of θ, i.e., the estimate which is the best from the cognitive viewpoint in the light of available evidence.[15]

Let us now extend this idea so that the relevant estimates include intervals. The first Bayesian account of estimation along these lines was given by Levi (1967a, 1976, 1980).

First we should note that point estimates and interval estimates can be regarded as rival answers to the same cognitive problem, expressed by the question 'Which is the true value of parameter θ?'. Point estimates $\theta' \in \mathbb{R}$ are potential complete answers, and interval estimates partial answers to this problem. Indeed, points can be regarded as *degenerate intervals*, and interval estimates as *infinite disjunctions of point estimates*.

Intervals are usually very cautious answers to estimation problems. In contrast, if a point estimate θ' of θ is regarded as a degenerate interval $[\theta', \theta']$, then it is in fact a 0% Bayesian confidence interval for θ. Is it not curious that some statisticians recommend for us 99% confidence intervals, and some 0% intervals! Why don't we ever see 50% intervals? The choice between these alternatives seems to be an entirely arbitrary matter, which is not solved by any systematic principle in the Bayesian theory of estimation — in spite of Savage's (1962) claim that one of the main virtues of the Bayesian approach is to build the subjective ingredients of statistical inference into the theory.

A decision-theoretic account of estimation helps us to handle this issue. Let **Int** be the class of intervals in θ, including the degenerate ones.

Int $= \{[a, b] \mid -\infty \leq a \leq +\infty, -\infty \leq b \leq +\infty, a, b \in \Theta, a \leq b\}$.

Note that if Θ is the whole real line $\mathbb{R}$, we include $\mathbb{R}$ itself in **Int**. An *interval estimator* $D: \mathbf{X} \to \mathbf{Int}$ is then a function D which assigns to each sample $x \in \mathbf{X}$ an interval $D(x) \in \mathbf{Int}$. The element Θ in **Int** is essentially the disjunction of all possible estimates, and therefore it corresponds to the suspension of judgment (cf. Levi, 1967a).

Let $L: \mathbf{Int} \times \Theta \to \mathbb{R}$ be a real-valued loss function indicating the loss $L(I, \theta)$ of adopting the estimate $I \in \mathbf{Int}$ when θ is the true value of the parameter. Define the *posterior loss* of estimate I given x by

(40) $\quad EL(I, x) = \int_\Theta L(I, \theta) g(\theta/x) \, d\theta.$

Then the general Bayes strategy, when limited to choices within **Int**, leads to the following rule:

(41)   For each $x \in \mathbf{X}$, choose estimate $D(x) \in \mathbf{Int}$ so that the posterior expected loss $EL(D(x), x)$ is the minimum.

(Cf. Rule (39)). The rule (41) could in principle be extended to situations, where $I \subseteq \Theta$ is any Lebesgue-measurable set. I shall consider below only cases, where I is a point, an interval or a finite union of non-degenerate intervals.

Historically the first proposal for a loss function $L(I, \theta)$, made but not studied by Abraham Wald, is

(42) $L_1(I, \theta) = 1$,   if $\theta \notin I$
        $ = 0$,   if $\theta \in I$.

(See Wald, 1950, pp. 22–23.). In other words, if K_I is the indicator function of set I (see Example 1.7), then

$$L_1(I, \theta) = 1 - K_I(\theta).$$

Here $K_I(\theta)$ is in effect the truth value (1 or 0) of the claim $\theta \in I$, so that the choice of L_1 is equivalent to the definition (7) of epistemic utility. Therefore, it leads to the result that

$$EL_1(I, x) = 1 - g(I \mid x).$$

This is unsatisfactory, since this expected loss will be minimized by choosing I to be the trivial estimate Θ.

Levi (1967a, 1967b, 1976, 1980) has applied the epistemic utility function (16) to the problem of estimation in the following way. Assume that $\Theta \subseteq \mathbb{R}$ is finite. Let $M: \Theta \to [0, 1]$ be a normalized function which expresses the pragmatic information $M(\theta)$ of the complete answer $\theta \in \Theta$. Then the epistemic utility of accepting $I \subseteq \Theta$ is

(43) $1 - q \int_I M(\theta) \, d\theta$ if $\theta \in I$

$ -q \int_I M(\theta) \, d\theta$ if $\theta \notin I$.

The rule (20) leads then to the recommendation that those elements

θ of Θ are rejected on x which satisfy $g(\theta/x) < qM(\theta)$, and the disjunction of the unrejected elements is accepted as strongest. If the posterior distribution $g(\theta/x)$ is unimodal, the accepted interval estimate I will cover the mode of $g(\theta/x)$, and the length of I depends on the index of boldness q (and on the variance of $g(\theta/x)$). If $g(\theta/x)$ is bimodal, the acceptable estimate may be a union of two intervals.

Levi also suggests that the process of exclusion can be repeated — through 'bookkeeping' — and a point estimate, identical to the mode of $g(\theta/x)$ (if that is unique), is reached by "throwing caution to the winds", i.e., by choosing $q = 1$.

The *length* $l(I)$ of interval $I = [c, d]$ is defined by

$$l(I) = d - c.$$

If I is a finite union of disjoint intervals $[c_1, d_1] \cup \ldots \cup [c_n, d_n]$, then the length of I is the sum of the length of these intervals:

$$l(I) = \sum_{i=1}^{n} (d_i - c_i).$$

If the parameter space Θ is a finite interval, and the information measure M in (43) is uniform on Θ, then Levi's utility function can be written in the following form:

(44) $\quad \begin{array}{ll} 1 - ql(I)/l(\Theta), & \text{if } \theta \in I \\ -ql(I)/l(\Theta) - 1, & \text{if } \theta \notin I. \end{array}$

The negative of (44) thus defines a loss function

(45) $\quad \begin{array}{ll} ql(I)/l\Theta), & \text{if } \theta \notin I \\ ql(I)/l(\Theta) - 1, & \text{if } \theta \in I. \end{array}$

In the case $\Theta = [0, 1]$, this reduces to

$\quad \begin{array}{ll} ql(I), & \text{if } \theta \notin I \\ ql(I) - 1, & \text{if } \theta \in I. \end{array}$

This agrees with the proposal of T. Ferguson:

(46) $\quad L_2(I, \theta) = ql(I) - K_I(\theta),$

except that he allows q to take any positive values (see Ferguson, 1967, p. 184).

Function L_2 satisfies Levi's principle that "a miss is as good as a mile", since

$$L_2([a, a], \theta) = 0, \quad \text{if } a \neq \theta$$
$$= -1, \quad \text{if } a = \theta.$$

Thus, all mistaken point estimates are equally bad. Moreover, if $q < 1$, each of them is epistemically worse than the trivial true answer Θ, i.e., suspension of judgement, no matter how small the distance $|a - \theta|$ is.

For these reasons, it seems natural to suggest that the cognitive virtues of an estimate $I = [c, d] \in \mathbf{Int}$ should depend on its length $l(I) = d - c$ and on its location with respect to the true value of θ. This location depends on the *left endpoint c* and the *right endpoint d* of I; these points together determine the *midpoint m(I)* of I:

$$m(I) = c + \frac{d-c}{2} = \frac{c+d}{2}.$$

In particular, we should expect that the loss $L([c, d], \theta)$ increases, when $l(I)$ is fixed and $|m(I) - \theta|$ increases. This means that

(47) $L([c, c], \theta)$ is an increasing function of $|c - \theta|$.

This condition is not satisfied by Levi's loss function.

Levi (1986b) suggests that there is a sense in which it is possible to speak of "approximately true" point estimates without "deploying a conception of nearness to the truth": to declare a point estimate to be "approximately true" means that the true value "falls in an interval spanning" the estimate. Accepting such a point estimate can then be construed as the acceptance of a true interval estimate which is, by (44), better than the trivial estimate Θ, since it is shorter.

However, the length of the interval which spans the point estimate θ' and covers the true value θ depends on $|\theta' - \theta|$, i.e., on the distance of θ' from the truth. The estimate θ' is 'approximately true' only if this distance is small. If θ' is sufficiently far from θ, I think θ' will be epistemically worse than the trivial answer Θ, even though here also the interval covering θ' and θ would be shorter than Θ.

Levi (1986b) also raises the interesting question whether point estimation plays any significant cognitive role within a 'Peirce-Popper vision' of inquiry. Perhaps point estimates are needed only in practical decision problems? However, I am still inclined in thinking that the estimation of some universal constant of nature, the value of some quantity, or the coefficient in a quantitative law, may serve theoretical purposes — and may be needed, e.g., as a background assumption in further inquiry. Therefore, I shall study below the general problem of estimation by exploring loss functions $L(I, \theta)$ which express the *distance of $I \in$ Int from the truth θ* so that condition (47) is satisfied. (For a more detailed account, see Niiniluoto, 1986c.).

Examples of functions which reflect the distance between an interval $I = [c, d]$ and a point θ include the following:

$$L_3(I, \theta) = \min_{t \in I} |t - \theta|$$

$$L_4(I, \theta) = \max_{t \in I} |t - \theta|$$

$$L_5(I, \theta) = \gamma L_3(I, \theta) + (1 - \gamma) L_4(I, \theta) \qquad (0 < \gamma < 1)$$

$$L_6(I, \theta) = \frac{1}{l(I)} \int_I |t - \theta| \, dt$$

$$L_7(I, \theta) = \beta l(I) + L_3(I, \theta) \qquad (0 < \beta < 1)$$

$$L_8(I, \theta) = \int_I |t - \theta| \, dt$$

$$L_9(I, \theta) = \beta L_8(I, \theta) + L_3(I, \theta) \qquad (\beta > 0).$$

L_4 has been used by R. E. Moore (1966) in the approximation theory for intervals. L_5 and L_6 were discussed in Niiniluoto (1982b).

The only measure that has been discussed by Bayesian statisticians is L_7 (see Aitchison and Dunsmore, 1968; Cox and Hinkley, 1974).

In fact, L_3–L_9 are special cases of distance measures that are familiar to us from Chapter 6. With the qualification that the distances have not been normalized, we have the following correspondences:

$$L_3 - \Delta_{min}$$
$$L_4 - \Delta_{max}$$
$$L_5 - \Delta_{mm}^\gamma$$
$$L_6 - \Delta_{av}$$
$$L_7 - \Delta_{ms}^\gamma$$
$$L_8 - \Delta_{sum}$$
$$L_9 - \Delta_{ms}^{\gamma\gamma'}.$$

More precisely, we have

(48) $L_5([c, d], \theta) = \theta - \gamma d - (1 - \gamma)c,$ if $c \leqslant d \leqslant \theta$
$= \gamma c + (1 - \gamma)d - \theta,$ if $\theta \leqslant c \leqslant d$
$= (1 - \gamma)(d - \theta),$ if $c \leqslant \theta \leqslant m(I) \leqslant d$

(49) $L_6(I, \theta) = |m(I) - \theta|,$ if $\theta \notin I$
$= \dfrac{1}{l(I)} \left(\dfrac{1}{4} l(I)^2 + (m(I) - \theta)^2 \right),$ if $\theta \in I$

(50) $L_7(I, \theta) = \beta l(I) + (\theta - d),$ if $c \leqslant d \leqslant \theta$
$= \beta l(I) + (c - \theta),$ if $\theta \leqslant c \leqslant d$
$= \beta l(I),$ if $c \leqslant \theta \leqslant d$

(51) $L_9(I, \theta) = |m(I) - \theta| \cdot (1 - \beta l(I)) - \frac{1}{4} l(I),$ if $\theta \notin I$
$= \beta(m(I) - \theta)^2 + (\frac{3}{4} l(I))^2,$ if $\theta \in I$

Further, L_5 with $\gamma = 1/2$, and L_7 with $\beta = 1/2$, agree with L_6, when $\theta \notin I$. L_7 with $\beta = 1$ reduces to L_4, when $\theta \notin I$.

As a simple variant of L_9, we may define

$$L_{10}(I, \theta) = \beta L_8(I, \theta) + L_3(I, \theta)^2$$

(cf. (6.94)). Hence,

(52) $L_{10}(I, \theta) = \frac{1}{4} l(I)^2 + (m(I) - \theta)^2 +$
$+ (\beta - 1)l(I) \cdot |m(I) - \theta|,$ if $\theta \notin I$
$= \dfrac{\beta}{4} l(I)^2 + \beta(m(I) - \theta)^2,$ if $\theta \in I.$

In particular,

(53) $L_{10}([a, a], \theta) = (a - \theta)^2.$

Niiniluoto (1986c) concludes that L_{10} has all the intuitively satisfactory desiderata for a distance measure between intervals and points. In comparison to L_7, function L_{10} has the additional virtue that it can be directly generalized to situations where I is not a single interval but a finite union of n intervals I_i, $i = 1, \ldots, n$.[16] In such a case, L_7 fails to impose any restriction on the maximum distance of I from the truth: $I = [0, 2]$ and $I' = [0, 1] \cup [99, 100]$ would be equally distant from the truth 0 by L_7, since $l(I) = l(I')$. Instead, L_{10} yields the general result:

$$L_{10}(I_1 \cup \ldots \cup I_n, \theta) = \beta(\tfrac{1}{4} l(I_i)^2 + (m(I_i) - \theta)^2)$$

$$+ \beta \sum_{\substack{j=1 \\ j \neq i}}^{n} l(I_j) |m(I_j) - \theta|, \quad \text{if } \theta \in I_i$$

$$= \Delta_{\min}(\theta, I_1 \cup \ldots \cup I_n)^2 +$$

$$+ \beta \sum_{j=1}^{n} l(I_j) |m(I_j) - \theta| \quad \text{if } \theta \notin I_1 \cup \ldots \cup I_n.$$

It is interesting to observe that L_7, L_9 and L_{10} are all generalizations of Ferguson's function L_2: if the geometric distance $L(\theta', \theta)$ is replaced by the trivial distance

$$L(\theta', \theta) = 1 \quad \text{if } \theta' \neq \theta$$
$$= 0 \quad \text{if } \theta' = \theta,$$

then $L_3(I, \theta)$ reduces to $1 - K_I(\theta)$ and $L_8(I, \theta)$ to $l(I)$, so that L_7 and L_9 are both replaced by

$$\beta l(I) + 1 - K_I(\theta) = 1 + L_2(I, \theta).$$

In other words, L_2 is a special case of L_7, L_9 and L_{10}, where each mistake in I is counted as equally serious — and distances from the truth are ignored. (Cf. the comments on Levi in Section 3.)

Festa (1986) proposes a set of eight axioms for the loss function $L(I, \theta)$ which leads to the following two-dimensional continuum of functions:

(54) $\quad L_F^{\sigma\alpha}(I, \theta) = \sigma l(I) + 2(\alpha - \sigma) |m(I) - \theta|, \quad \text{if } \theta \in I$
$\qquad\qquad\quad\; = (\alpha - 1/2) l(I) + |m(I) - \theta|, \quad \text{if } \theta \notin I$
$\qquad$ where $0 < \alpha < 1/2$ and $0 < \sigma \leq \alpha$.

Hence,

$L_F^{\sigma\alpha} = L_7$ (with $0 < \beta < 1/2$), $\quad$ if $\sigma = \alpha = \beta$

$L_F^{\sigma\alpha} = L_5$ (with $1/2 < \gamma < 1$), $\quad$ if $\sigma = \dfrac{1-\gamma}{2}$ and $\alpha = 1 - \gamma$.

L_9 and L_{10} are also functions of the factors $l(I)$ and $|m(I) - \theta|$, but mathematically more complex than those included in Festa's σ-α-continuum.

Let us now study the behaviour of these loss functions for the decision problem (41). First, it is easy to see that $EL_3(I, x)$ has its minimum value 0, if $g(I/x) = 1$. Thus, L_3 is equally unsatisfactory as L_1, since it always recommends us to accept the trivial interval Θ.

For L_4,

$$EL_4(I, x) = \int_{-\infty}^{m(I)} (d - t)g(t/x)\, dt + \int_{m(I)}^{+\infty} (t - c)g(t/x)\, dt.$$

If $c \leqslant a \leqslant d$, then

$$L_4([a, a], \theta) \leqslant L_4([c, d], \theta) \quad \text{for all } \theta \in \mathbb{R}.$$

If $c < a < d$, the above inequality $\leqslant$ can be replaced by $<$. Hence,

(55) $\quad EL_4([a, a], \theta) \leqslant EL_4(I, \theta) \quad \text{if } a \in I.$

This means that the loss function L_4 systematically favours point estimates over interval estimates. In fact, L_4 always recommends us to choose from **Int** the median of the posterior distribution (cf. Table II).

The behaviour of EL_5 depends crucially on the parameter $\gamma > 0$ which expresses the relative weight of our desire to find a true estimate, while $1 - \gamma$ is the weight for our interest of excluding bad mistakes. If $\gamma \leqslant \frac{1}{2}$, then each interval estimate is dominated by some point estimates: when $\gamma d + (1 - \gamma)c \leqslant a \leqslant \gamma c + (1 - \gamma)d$, then

$$L_5([a, a], \theta) \leqslant L_5([c, d], \theta) \quad \text{for all } \theta \in \mathbb{R}.$$

In this case, $EL_5(I, \theta)$ is minimized by the median of the posterior distribution $g(\theta/x)$. If $\gamma > \frac{1}{2}$, then interval estimates are not dominated by any point estimates, or *vice versa*, and the best interval in **Int** may be degenerate or non-degenerate.

For L_6, we have the result that the midpoint $m(I)$ of I, where $l(I) > 0$, always dominates I as an estimate:

$$L_6([m(I), m(I)], \theta) \leqslant L_6(I, \theta) \quad \text{for all } \theta \in \mathbb{R}.$$

When $\theta \in I$, the above inequality $\leqslant$ can be replaced by $<$. It follows that L_6 favours point estimates to interval estimates. In fact, L_6 gives the same recommendation is L_3: accept from **Int** the median of $g(\theta/x)$.

The definition of L_7 depends on the parameter β, $0 < \beta < 1$, which indicates the weight of the information factor $l(I)$. If $\beta \geqslant \frac{1}{2}$, each interval estimate is dominated by some point estimate: for $\beta c + (1-\beta)d < a < \beta d + (1-\beta)c$,

$$L_7([a, a], \theta) \leq L_7([c, d], \theta) \quad \text{for all } \theta \in \mathbb{R}.$$

In this case, L_7 recommends the median of $g(\theta/x)$ as the best estimate in **Int**. However, if $\beta < \frac{1}{2}$, point estimates do not dominate interval estimates (see Figure 2).

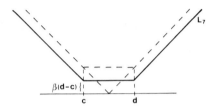

Fig. 2. $L_7([c, d], \theta)$ when $\beta < \frac{1}{2}$.

EXAMPLE 1. Assume that $g(\theta/x)$ is $N(0, \sigma^2)$, so that it is defined by

$$\varphi(t) = \frac{1}{\sigma\sqrt{2\pi}} e^{-t^2/2\sigma^2}.$$

Let $c > 0$. Then

$$(56) \quad EL_7([-c, c], x) = \int_{-\infty}^{-c} (2\beta c - c - t)\varphi(t)\,dt + \int_{-c}^{c} 2\beta c \varphi(t)\,dt +$$

$$+ \int_{c}^{\infty} (t - 2\beta c - c)\varphi(t)\,dt$$

$$= 2\beta c \int_{-\infty}^{\infty} \varphi(t)\,dt - 2c \int_{-\infty}^{-c} \varphi(t)\,dt -$$

$$- \int_{-\infty}^{-c} t\varphi(t)\,dt + \int_{c}^{+\infty} t\varphi(t)\,dt$$

$$= 2c\left(\beta - \int_{-\infty}^{-c} \varphi(t)\,dt\right) + \frac{2\sigma}{\sqrt{2\pi}} e^{-c^2/2\sigma^2}.$$

By (56),

$$EL_7([0, 0], x) = \frac{2\sigma}{\sqrt{2\pi}}$$
$$EL_7([-c, c], x) \to \infty, \quad \text{if } c \to \infty.$$

To study the behaviour of (56), take its derivative with respect to c:

(57) $\quad 2\left(\beta - \int_{-\infty}^{-c} \varphi(t) \, dt\right).$

If $c = 0$, (57) reduces to $2(\beta - \frac{1}{2}) = 2\beta - 1$. Hence, (56) increases at the point 0 (and all the other points $c \geq 0$ as well), if $\beta > \frac{1}{2}$. This agrees with our earlier observation that L_7 with $\beta > \frac{1}{2}$ favours point estimates — in this case the median 0 of the posterior distribution — over interval estimates. Assume then that $\beta < \frac{1}{2}$. Then the expected loss (56) will have a local minimum, if (57) equals 0, i.e.,

$$2\left(\beta - \int_{-\infty}^{-c} \varphi(t) \, dt\right) = 2\left(\beta - \left(1 - \int_{-\infty}^{c} \varphi(t) \, dt\right)\right) = 0.$$

This is the case, when c is chosen so that

$$\int_{-\infty}^{c} \varphi(t) \, dt = 1 - \beta,$$

that is,

(58) $\quad \int_{-c}^{c} \varphi(t) \, dt = 1 - 2\beta.$

In other words, the expected loss is minimized by a $100(1 - 2\beta)\%$ confidence interval with the midpoint 0. The choice of the index of boldness $\beta \leq \frac{1}{2}$ thus determines uniquely the confidence level of the acceptable interval estimate. When β decreases from $\frac{1}{2}$ to 0, the acceptable intervals grow from $[0, 0]$ to $\mathbb{R}$, as illustrated by Table III.

TABLE III
The confidence level as a function of parameter β

β	$100(1 - 2\beta)\%$
$\geq 1/2$	0 %
3/8	25 %
1/4	50 %
1/10	80 %
1/20	90 %
1/40	95 %
1/200	99 %
0	100 %

Function L_8 gives the same loss 0 to all point estimates. Therefore, $EL_8([a, a], x) = 0$ for all $a \in \theta$, so that L_8 cannot use any evidence to make any difference between the alternatives in θ.

The behaviour of L_9 is similar to that of L_7. Again point estimates dominate interval estimates, if β sufficiently great. These results can be given in a more straightforward way for the variant L_{10}. The formula (52) entails directly that

(59) $\quad L_{10}([m(I), m(I)], \theta) < L_{10}(I, \theta) \quad$ for all $\theta \in \mathbb{R}$.

if and only if $\beta \geq 1$. In other words, condition $\beta \geq 1$ guarantees that each interval I is dominated by its midpoint $m(I)$, so that $EL_{10}(I, x)$ is minimized by the mean of the posterior distribution $g(\theta/x)$ (cf. (53)). If $\beta = 0$, the trivial interval will be the best (cf. L_3). If $0 < \beta < 1$, EL_{10} recommends the acceptance of an $100\alpha\%$ interval estimate I, where the confidence level α increases, when the penalty of mistakes β decreases.

Some of the results obtained above are summarized in Table IV. They show which or what kinds of elements I in **Int** — point or nondegenerate interval estimates — the studied loss functions favour.[17] These consequences are in principle quite general, independent of the interpretation given to the loss functions, but for the cognitive purpose of maximizing truthlikeness we found reasons to regard L_{10} as better than the others.

TABLE IV
Bayesian interval estimation

Loss function		Bayesian estimate in **Int**
L_1		Θ
L_3		Θ
L_4		the median of $g(\theta/x)$
L_5,	$0 \leq \gamma \leq \frac{1}{2}$	the median of $g(\theta/x)$
	$\frac{1}{2} < \gamma < 1$	a point or a non-trivial interval estimate
L_6		the median of $g(\theta/x)$
L_7,	$\beta \geq \frac{1}{2}$	the median of $g(\theta/x)$
	$0 < \beta < \frac{1}{2}$	a point or a non-trivial interval estimate
L_{10},	$\beta \geq 1$	the mean of $g(\theta/x)$
	$0 < \beta < 1$	a point or a non-trivial interval estimate

We have also seen that the choice of the confidence level of a Bayesian interval estimate can be conceptualized in decision-theoretic terms by an "index of boldness" ($1 - \gamma$ in L_5, β in L_7, β in L_{10}) which has a natural interpretation in terms of the cognitive interests of the investigator. For example, β in L_{10} indicates the relative weight that is given to the seriousness of the mistakes in the interval estimate, while at the same time we wish to hit as close to the truth as possible. Thus, low values of β go together with high confidence levels.

These results can be read in two directions. Suppose your loss function is L_{10}, where the parameter β is fixed. Then you have already determined the confidence level of the interval estimates that are acceptable to you — and it is then in each case up to the evidence *via* the posterior distribution to decide what this interval is and how long it is. Conversely, if β has not yet been fixed, then your choice of the appropriate confidence level (between 0 and 1) determines the relative weight β that you assign to the cognitive aim of avoiding guesses that differ from the truth.

CHAPTER 13

OBJECTIONS ANSWERED

In this final chapter, I attempt to meet some possible objections to the similarity approach to truthlikeness. The changing attitudes towards the explication of verisimilitude are reviewed in Section 1. Sections 2 and 3 discuss queries arising from the fact that the definition of truthlikeness is relative to a language: Is there too much linguistic variance in our similarity approach, as David Miller has argued? Finally, Section 4 explores the possibilities of developing the theory of truthlikeness on a conceptual basis that is weaker than the treatment followed in this work.

13.1. VERISIMILITUDE AS A PROGRAMME

In spite of the long prehistory of fallibilism, serious work on defining a concept of truthlikeness for scientific theories was started as late as in 1960 (see Chapter 5). Popper's definition did not at first provoke much discussion in the sixties. The main reason, I guess, was that the Popper—Carnap controversy over induction was generally held to be more topical than the issue about truthlikeness. Many philosophers outside the Popperian camp felt that the problem of verisimilitude is irrelevant: they regarded it merely as an idiosyncratic problem for the critical rationalists, which becomes interesting only if one wholly accepts Popper's questionable criticism of induction. It also seemed that the emerging new approaches within the Bayesian theory of induction — Hintikka's inductive logic for generalizations and Levi's cognitive decision theory — could incorporate some (but not all) important elements of Popperian methodology into their frameworks, without paying attention to Sir Karl's definition of truthlikeness.[1]

The situation changed radically in the early seventies (cf. Niiniluoto, 1986b). New stimulus for solving the logical problem of truthlikeness came in 1973 from the news that David Miller and Pavel Tichý had refuted Popper's comparative concept of verisimilitude by showing that it cannot be used for comparisons between false theories (see Chapter 5.4). As Popper's definition had strong intuitive appeal, this negative result, and its generalizations, gave support to a sceptical attitude: perhaps there does not exist any solution to the problem of truthlikeness? (cf. Miller, 1975a).

Miller has recently made the sarcastic remark that "it almost looks as though verisimilitude was not seen to be of any importance for falsificationism until it was thought to pose a major embarrassment to it" (Miller, 1982a, p. 93). For a philosopher, who for some reason tends to think that the whole idea of approach to the truth is 'absurd',[2] 'paradoxical'[3] or represents dubious 'teleological metaphysics',[4] it is indeed tempting to regard verisimilitude as a *cul-de-sac* for Popperian falsificationism. If the concept of approximate truth or verisimilitude is indispensable for a realist conception of scientific progress, then the Miller—Tichý theorem could be thought to decide the Kuhn—Popper controversy (cf. Lakatos and Musgrave, 1970) in Kuhn's favour.

The critical rationalists and their associates have responded to this situation in various ways. Miller and Tichý themselves took their results as a challenge to develop a new formal theory of truthlikeness. Popper himself pointed out that his failure to define verisimilitude in an adequate way does not prove the problem to be insoluble. In his Appendix to the revised edition of *Objective Knowledge*, Popper declared: "I am optimistic concerning verisimilitude." (Popper, 1979, p. 371.)

A different reaction is represented by the anti-formalists who have adopted an extreme version of Popper's opposition against the "quest of precision":[5] Paul Feyerabend makes scornful remarks about "the development of epicycles to such intellectualist monsters as verisimilitude and content increase" (Feyerabend, 1979, p. 93), and Gerard Radnitzky complains about the "monster-producing and monster-barring" character of the recent logical studies in verisimilitude (Radnitzky and Andersson, 1978, p. 17).

In attempting to find a middle position, Gunnar Andersson acknowledged the importance of the task of defining verisimilitude, but at the same time argued that the absence of a "formal definition" should not prevent us from using this concept "in an intuitive way".[6] Similarly, Noretta Koertge claimed that the Miller-Tichý "criticisms do not descredit the cogency of the *intuitive* idea of truth-likeness" (Koertge, 1978, p. 276). This position is, in my view, untenable. Popper's definition of truthlikeness had a very clear motivation — and it is precisely this background intuition which collapsed under Miller's canoon. The later debates about the explication of verisimilitude have shown that perhaps no one has got an unobjectionable well-educated intuition about this concept. It is true, as Andersson points out, that sometimes the scientists have successfully used concepts intuitively without precise definitions — the early history of differential calculus is a good example. However, this is by no means a justification for the suggestion that the philosophers of science could intentionally do the same.

The problems with the definition of verisimilitude have led some members of the Popperian camp to doubt the indispensability of this concept for the theory of scientific change. John Watkins — who has personally also given strong support to the logical study of truthlikeness — has reminded us that "Popper got along well enough without the idea of verisimilitude for a quarter of a century after 1934" (Watkins, 1978, p. 365). It is indeed remarkable how slight role verisimilitude plays in Popper's 'Autobiography' (Popper, 1974a). However, it is doubtful that the comparisons of the information contents of theories — this is what Watkins suggests as the key element of the LSE philosophy of science[7] — could serve the same variety of methodological purposes as the concept of verisimilitude. Moreover, in defining the concept of scientific progress, there is no reason to assume that among false theories logical strength and content always covary with the closeness to the truth.[8]

Koertge (1978) urged strongly that "ever-increasing verisimilitude" should be made the aim of science and that "a good theory of the estimate of the degree of verisimilitude of a theory would be of great value to philosophers of science of all persuasions". But soon afterwards she became sceptical. After complaining that the possible measures of truthlikeness seem inevitably contain weights which are relative to 'circumstances', she concludes: "perhaps a notion of verisimilitude is not needed after all in order to describe the progress of science" (Koertge, 1979, p. 238). Peter Urbach (1983) has also recently expressed his disappointment that there does not seem to be any 'interesting' notion of verisimilitude — for him, this means a notion which is 'objective' in the same sense as truth, just as Popper originally intended.

If Popper's falsificationism is regarded as a methodological research programme in Lakatos's sense, one can still debate whether the concept of truthlikeness belongs to its 'hard core'. However, verisimilitude of course is not only a family quarrel among the critical rationalists. The same spectrum of opinions — ranging from enthusiasm to indifference, from optimism to outright rejection — can be found in the larger philosophical community.

For example, it is still easy to find papers, in the leading journals of the philosophy of science, which use concepts like 'approximate truth' in a presystematic sense. On the other hand, already in the year 1974, a new programme for explicating truthlikeness was initiated by Pavel Tichý, Risto Hilpinen, David Miller, and Ilkka Niiniluoto — and was soon joined by Raimo Tuomela and Graham Oddie (cf. Chapter 6). This approach takes seriously the idea that the degree of truthlikeness of a statement depends on its similarity with the true state of affairs — and thereby employs an effective 'positive heuristics' which Popper's programme lacked.

The supporters of the similarity approach include several philosophers who do not share the Popperian prejudice against induction: Hilpinen, Niiniluoto, and Tuomela — and later Theo Kuipers and Roberto Festa — have all worked in the formal theory of inductive inference. Thus, "the interest in truthlikeness is not incompatible with simultaneous interest in Sir Karl's *bête noir*, inductive logic" (Niiniluoto, 1978b, p. 282). In this spirit, I have argued, since Niiniluoto (1977b), that the similarity approach has to be extended from the logical to the epistemic problem of truthlikeness.

At the same time, especially after 1977, the problems of truth and realism have become perhaps the hottest issues in the contemporary philosophy of science. Joseph Sneed's and Wolfgang Stegmüller's 'structuralist' reconstruction of Thomas Kuhn, Nicholas Rescher's 'methodological pragmatism', Hilary Putnam's rejection of 'metaphysical realism', Larry Laudan's problem-solving approach to scientific progress, and Bas van Fraassen's 'constructive empiricism' are all variations of a forceful neo-pragmatist or neo-instrumentalist trend which now poses the major challenge to scientific realism.[9] Perhaps the strongest criticism has been voiced by Laudan, who concludes his recent *Science and Values* with the following harsh judgement:

The realist offers us a set of aims for science with these features: (1) we do not know how to achieve them (since there is no methodology for warranting the truthlikeness of universal claims); (2) we could not recognize ourselves as having achieved those aims even if, mysteriously, we had managed to achieve them (since the realist offers no epistemic, as opposed to semantic, tokens of truthlikeness); (3) we cannot even tell whether we are moving closer to achieving them (since we generally cannot tell for any two theories which one is closer to the truth); and (4) many of the most successful theories in the history of science (e.g., aether theories) have failed to exemplify them. In my view, any one of these failings would be sufficient to raise grave doubts about the realist's proposed axiology and methodology of science. Taken together, they seem to constitute as damning an indictment of a set of cognitive values as we find anywhere in the historical record. Major epistemologies of the past (e.g., classical empiricism, inductivism, instrumentalism, pragmatism, infallibilism, positivism) have been abandoned on grounds far flimsier than these. (Laudan, 1984, p. 137).

As a summary of the results of the similarity approach to truthlikeness, this book — together with Niiniluoto (1984b) — tries to show that Laudan's "confutation of convergent realism" is premature. Degrees of truthlikeness for virtually all kinds of scientific statements can be defined, and these degrees can be rationally estimated on evidence.

But before concluding, we still have to consider some important questions which have been raised as objections against the whole similarity approach to truthlikeness.

13.2. THE PROBLEM OF LINGUISTIC VARIANCE

Perhaps the most common objection to the similarity approach is that it appears to make truthlikeness too relative to the choice of the language. Miller (1974b) presented this criticism against Tichý's measure in propositional logic, but later he extended it to the case of singular quantitative statements (Miller, 1975) and constituents (Miller, 1978).

Before going to Miller's argument, let us first see what principles of *linguistic invariance* our definition of truthlikeness satisfies. Let L be the language of the cognitive problem **B** with the target h_*. Then the degrees of truthlikeness for $g \in D(\mathbf{B})$ are *preserved under logical equivalence* in the following sense:

(1) If $g_1 \in D(\mathbf{B})$, $g_2 \in D(\mathbf{B})$, and $\vdash g_1 \equiv g_2$,
then $\mathrm{Tr}(g_1, h_*) = \mathrm{Tr}(g_2, h_*)$.

The same conclusion holds for estimated degrees of verisimilitude, if P is a probability measure for language L in a Hintikka-type inductive logic (cf. 5°—6° in Chapter 2.10):

(2) If $\vdash e_1 \equiv e_2$ and $\vdash g_1 \equiv g_2$, then
$\mathrm{ver}(g_1 | e_1) = \mathrm{ver}(g_2 | e_2)$.

Principles (1) and (2) quarantee that degrees of truthlikeness — and, hence, comparative claims of verisimilitude — are invariant under logically equivalent reformulations of the relevant statements.

Principles (1) and (2) involve only a single language L. The problem of linguistic variance concerns the behavior of Tr and ver in *language-shifts* from one conceptual framework L to another L'. To what extent should the notion of truthlikeness be insensitive to such shifts?

Let us consider first the simple case, where language L is a sublanguage of language L', so that $\lambda \subseteq \lambda'$ holds for the non-logical predicates λ and λ' of L and L', respectively.[10] If each new term in $\lambda' - \lambda$ is explicitly definable in terms of the old terms λ, then the shift from L to L' is *conservative*: L', together with the explicit definitions as the meaning postulates $\mathcal{MP}'$, is essentially the same language as L. This means that every sentence g' in L' is equivalent (relative to $\mathcal{MP}'$) to a sentence g in L. Further, if P and P' are inductive probability measures for L and L', respectively, with the same parameter values, then the following holds:

(3) $\quad P(g|e) = P'(g|e \,\&\, \mathcal{MP}')\quad$ for all e and g in L.[11]

(See Niiniluoto and Tuomela, 1973, p. 58). A similar result holds for truthlikeness:

(4) Let L' be a conservative extension of L, where the terms in $\lambda' - \lambda$ are explicitly definable in terms of λ. Let $\mathbf{B}' = \{h'_i | i \in I\}$ be a cognitive problem in L', and let $\mathbf{B} = \{h_i | i \in I\}$ be the corresponding cognitive problem in L with $\mathcal{MP}' \vdash h_i \equiv h'_i$ for all $i \in I$. Further, let $\mathcal{MP}' \vdash e \equiv e'$ and $\mathcal{MP}' \vdash g \equiv g'$, where $g \in D(\mathbf{B})$ and $g' \in D(\mathbf{B}')$. Then $\text{Tr}(g, h_*) = \text{Tr}(g', h'_*)$ and $\text{ver}(g|e) = \text{ver}(g'|e')$.

Result (4) shows in which precise sense truthlikeness is invariant under conservative extensions of the language.

An essential feature of conservative language-shifts is the fact that the target does not change (cf. Chapter 5.2): when the logical depth is fixed, the most informative truth h_* in L is the same as the most informative truth h'_* in L', i.e., $\mathcal{MP}' \vdash h_* \equiv h'_*$. This situation changes radically, when L' is a non-conservative extension L, since then L' has a greater expressive power than L. In such cases, the relevant cognitive problem $\mathbf{B}'$ in L' is finer than the corresponding problem $\mathbf{B}$ in L, i.e., $\mathbf{B} \leq \mathbf{B}'$ and $\mathbf{B} \neq \mathbf{B}'$ (see Chapter 4.2). Here *truth* is still preserved in the sense that $h'_* \vdash h_*$ but $h_* \nvdash h'_*$ (see (4.24)). However, a similar requirement cannot be directed to *truthlikeness*, since the *target has changed* from h_* to h'_*, and the cognitive problems **B** and **B'** are not the same any more.

This reasoning can be illustrated by taking L to be a monadic first-order language with two Q-predicates Q_1 and Q_2, and four constituents

$$\begin{align} C_1 &= \sim(\exists x)Q_1(x) \,\&\, \sim(\exists x)Q_2(x) \\ C_2 &= (\exists x)Q_1(x) \,\&\, \sim(\exists x)Q_2(x) \\ C_3 &= \sim(\exists x)Q_1(x) \,\&\, (\exists x)Q_2(x) \\ C_4 &= (\exists x)Q_1(x) \,\&\, (\exists x)Q_2(x). \end{align}$$

If L' is obtained by adding to L a new monadic predicate M, then each Q-predicate Q_i of L ($i = 1, 2$) splits into two subcells, defined by $Q_i(x) \,\&\, M(x)$ and $Q_i(x) \,\&\, \sim M(x)$. If Q_i is empty, then both of its subcells in L' remain empty. But several things may happen to the subcells of a non-empty Q-predicate — as long as at least one of them remains empty. Thus, in Figure 1, constituent C_1 is equivalent to a single constituent in L', while C_2 and C_3 both split into three L'-constituents, and C_4 into nine L'-constituents. The same phenomenon repeats itself, if still further new predicates are added to language L'.

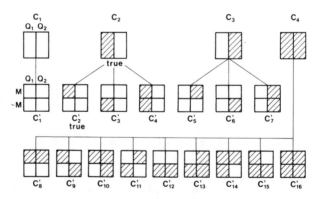

Fig. 1. Conceptual enrichment.

Assuming that the (absolute) information content of a generalization depends on the number of constituents in its normal form, it follows that non-existence claims in L remain informative in non-conservative extensions of L to richer languages L', but existence claims in L

become less and less informative. By the same token, if information about the truth in $g \in D(\mathbf{B})$ is a function of the maximum distance $\Delta_{max}(h_*, g)$ (see Chapter 6.4), then this quantity is not invariant under conceptual enrichment. For example, when the distance of constituents is measured, e.g., by the Clifford-function Δ_c, if $h_* = C_2$ in L and $h'_* = C'_2$ in L', then

$$\Delta_{max}(h_*, C_1) = \tfrac{1}{2} = \Delta_{max}(h_*, C_4)$$
$$\Delta_{max}(h'_*, C_1) = \tfrac{1}{4} < \tfrac{3}{4} = \Delta_{max}(h'_*, C_4).$$

This is in effect the argument that is given in Niiniluoto (1977b), Section 3, for the view that truthlikeness need not be generally preserved when "greater depths of analysis" are considered: as truthlikeness is a combination of a truth-factor and an information-factor, it cannot be expected to be "analogous to truth in all important respects". The argument can be repeated, if Δ_{max} is replaced by our favourite information-factor Δ_{sum}. Thus, the following holds:

(5) Let L' be a non-conservative extension of L. Let $g_1 \in D(\mathbf{B})$ and $g_2 \in D(\mathbf{B})$ be statements in L (and thereby in L'). Then it may happen that truthlikeness ordering is changed in conceptual enrichment: $\text{Tr}(g_1, h_*) > \text{Tr}(g_2, h_*)$ in L, but $\text{Tr}(g_1, h'_*) < \text{Tr}(g_2, h'_*)$ in L'.

The argument for (5) can be strengthed by noting that also the truth-factor Δ_{min} can be sensitive to conceptual change. For example, suppose that, in Figure 1, C_4 is true in L, and C'_{14} is true in L'. Then

$$\Delta_{min}(C_4, C_2) = \tfrac{1}{2} = \Delta_{min}(C_4, C_3)$$
$$\Delta_{min}(C'_{14}, C_2) = \tfrac{1}{4} < \tfrac{2}{4} = \Delta_{min}(C'_{14}, C_3).$$

Our general attitude towards truthlikeness, as explained in Chapter 5.2, is in fact a generalization of result (5): the truthlikeness of a statement g is not a measure of its distance from the 'whole truth', but from a chosen target, i.e., from the most informative true statement at some given depth of analysis. Hence, comparative claims of truthlikeness are not generally invariant with respect to conceptual enrichment, if this language-shift implies a change of the target of our cognitive problem.[12]

Miller admits that Niiniluoto (1977b) "has argued, perhaps convincingly, that distances from the truth may quite reasonably vary as we go from one language to a richer, more expressive one" (Miller, 1978b, p. 199). His objection to the similarity approach concerns instead two 'intertranslatable' or 'equally expressive' languages: "truthlikeness, like truth itself, has no appeal if not invariant under translation" (Miller, 1975, pp. 214–215). In the same spirit, Popper (1976), too, endorses the requirement that a measure of verisimilitude should be "invariant with respect to translations into other languages".

Let h, r, and w be the propositions

$h = $ 'It is hot'
$r = $ 'It is rainy'
$w = $ 'It is windy'.

If all of these sentences are true, Tichý's measure for the distance between two propositional constituents (see (6.17)) tells that $g = \sim h \,\&\, r \,\&\, w$ is closer to truth than $g' = \sim h \,\&\, \sim r \,\&\, \sim w$. Miller (1974b) points out, however, that the $h-r-w$-language is intertranslatable with the $h-m-a$-language with

$m = $ 'It is a Minnesotan weather'
$a = $ 'It is an Arizonan weather',

where 'Minnesotan' means 'hot and wet or cold and dry' and 'Arizonan' means 'hot and windy or cold and still'. In this case, we can define m and a in terms of h, r, and w:

(6) $\quad m =_{df} (h \equiv r)$
$\quad\quad a =_{df} (h \equiv w)$,

and conversely r and w by h, m, and a:

(7) $\quad r =_{df} (h \equiv m)$
$\quad\quad w =_{df} (h \equiv a)$.

In the new language, truth is represented by $h \,\&\, r \,\&\, w$, g by $\sim h \,\&\, \sim m \,\&\, \sim a$, and g' by $\sim h \,\&\, m \,\&\, a$. Thus, by Tichý's criterion, g' is here closer to the truth than g.

In his reply, Tichý (1976) denied the equivalence of the truths $h \,\&\, r \,\&\, w$ and $h \,\&\, m \,\&\, a$. These statements do not "come from the same logical space", and their Carnapian ranges (cf. (2.16)), far from being identical, have "nothing in common". In my view, this defence is insufficient (cf. Niiniluoto, 1977b, Note 19; 1978a, Note 10), since the definition of the 'logical space' relative to propositional language seems to me too arbitrary: truthlikeness may be relative to the *concepts* that we choose for our description of reality, not to a system of unanalysed primitive *propositions* (cf. Niiniluoto, 1978b, p. 710).

In his later, more interesting reply, Tichý (1978) argues that Miller has confused the object language and metalanguage readings of the intertranslatability judgments (6) and (7). If (6) and (7) are only metalinguistic definitions, then $\sim h \,\&\, r \,\&\, w$ and $\sim h \,\&\, \sim m \,\&\, \sim a$ express different states of affairs — and may be assigned different degrees of truthlikeness. But if (6) introduces new symbols to an extended $h-r-w-m-a$-language, then $\sim h \,\&\, r \,\&\, w$ and $\sim h \,\&\, \sim m \,\&\, \sim a$ indeed become 'logically equivalent' and will have the same degree of truthlikeness.

Tichý's response to Miller is here in fact the same that I have earlier used as a defence of inductive logic against the change of 'interlinguistic contradictions' (cf. Salmon, 1963; Michalos, 1971). Let us illustrate this with an example of Michalos (see Niiniluoto (1972); Niiniluoto and Tuomela (1973), pp. 165—178). Let L be a monadic language with primitive predicates R and B, and L' another monadic language with primitive predicates R, U, and M. Let g be the generalization $(x)(R(x) \supset B(x))$ in L, and g' the generalization $(x)(R(x) \supset (U(x) \,\&\, M(x)))$ in L'. Suppose that e describes in L a sample of five individuals exemplifying the cells $R \,\&\, B$, $\sim R \,\&\, B$, and $\sim R \,\&\, \sim B$, and e' describes in L' a sample of five individuals exemplifying the cells $R \,\&\, U \,\&\, M$, $\sim R \,\&\, U \,\&\, M$, $\sim R \,\&\, U \,\&\, \sim M$, $\sim R \,\&\, \sim U \,\&\, M$, and $\sim R \,\&\, \sim U \,\&\, \sim M$. In Hintikka's combined system $(\lambda(w) = w)$ with $\alpha = 0$, we have

$$P(g|e) = .727 \text{ (in } L)$$
$$P(g'|e') = .287 \text{ (in } L').$$

But if the predicates in L and L' are interpreted by the semantical rules

(8) 'Rx' means that x is Roman
     'Bx' means that x is a bachelor
     'Ux' means that x is unmarried
     'Mx' means that x is male,

then e and e' can be assumed to describe the same sample of five individuals. Further, generalization g ('All Romans are bachelors') and g' ('All Romans are unmarried men') differ from each other 'purely linguistically', and the fact that they receive different probabilities in L and L' amounts to an 'interlinguistic contradiction'.

The proper answer to this argument is to note that the metalinguistic rules (8) cannot quarantee the equivalence of g and g'. The alleged contradiction disappears, if the connection between the predicates of L and L' is expressed by a meaning postulate

(9) $(x)(B(x) \equiv (U(x) \& M(x)))$.

The explicit definition (9) (let us call it h) can be expressed neither in L nor in L'. For that purpose we need a language L'' with R, B, U, and M as primitive predicates. Even in L'' the definition h is not *logically* true, but it can be adopted in L'' as an analytically true *meaning postulate* (cf. Chapter 2.1). Hence, in L'', g and g' are really equivalent relative to h, i.e., $h \vdash g \equiv g'$, and Hintikka's system in L'' gives consistently the result

$$P(g|e \& h) = P(g'|e' \& h) = .287.$$

Even though this defence of inductive logic seems to me entirely correct, the proposed way of resolving 'interlinguistic contradictions' cannot be the whole story in connection with truthlikeness. This is shown by Miller (1978b), who — through an ingenious and important construction — extends his criticism from propositional to monadic (and *a fortiori* to polyadic) first-order languages.

Miller's first argument is directed against those definitions of truthlikeness for generalizations which rely on a distance function ε between the Q-predicates of a monadic language (see Miller, 1978b, pp. 202, 210). My Jyväskylä measure d_J, Tichý's d_T, and Tuomela's distance function are examples of such definitions (see Chapter 9). Even more directly, Miller's argument concerns my proposals for defining the truthlikeness of singular sentences by means of a metric on the class of Q-predicates (see Chapter 8).

Assume that language L contains two primitive monadic predicates M_1 and M_2 which are logically independent from each other. Let L' be another monadic language with the primitive predicates M_1 and M_3. If M_3 is definable by M_1 and M_2 through

(10) $(x)(M_3(x) \equiv (M_1(x) \equiv M_2(x)))$,

then M_2 is definable by M_1 and M_3 through

(11) $(x)(M_2(x) \equiv (M_1(x) \equiv M_3(x)))$.

Hence, L and L' are intertranslatable via the explicit definitions (10) and (11). Moreover, relative to these definitions (call them h and h'), the classes of Q-predicates $\mathbf{Q}$ and $\mathbf{Q}'$ of L and L', respectively, are the same:

(12) $\quad h \vdash (M_1(x) \ \& \ M_2(x)) \equiv (M_1(x) \ \& \ M_3(x))$
$\quad\quad\quad h \vdash (M_1(x) \ \& \sim M_2(x)) \equiv (M_1(x) \ \& \sim M_3(x))$
$\quad\quad\quad h \vdash (\sim M_1(x) \ \& \ M_2(x)) \equiv (\sim M_1(x) \ \& \sim M_3(x))$
$\quad\quad\quad h \vdash (\sim M_1(x) \ \& \sim M_2(x)) \equiv (\sim M_1(x) \ \& \ M_3(x))$.

Therefore, as Miller observes, each constituents C_i of L can be translated "conjunct by conjunct" to a constituent C'_i of L', and the Clifford measure d_c between constituents is preserved: $d_c(C_i, C_j) = d_c(C'_i, C'_j)$. However, the distances between the corresponding Q-predicates are not invariant: if ε is the Manhattan distance (3.5), then

(13) $\quad \varepsilon(M_1(x) \ \& \ M_2(x), \sim M_1(x) \ \& \ M_2(x)) = 1 \neq 2 =$
$\quad\quad\quad = \varepsilon(M_1(x) \ \& \ M_3(x), \sim M_1(x) \ \& \sim M_3(x))$.

The same result holds of course for any Minkowski metric of the form (3.3). It follows that the relative d_J-distances between constituents need not be preserved in the language-shift from L to L': it is possible that

$$d_J(C_*, C_i) < d_J(C_*, C_j)$$
$$d_J(C'_*, C'_i) > d_J(C'_*, C'_j).$$

Similar conclusions follow for singular statements.

Again, if L and L' are extended to a third language L'' with M_1, M_2, and M_3 as the primitive predicates, then relative to the meaning postulate h there are only four Q-predicates in L'':

$\quad\quad M_1(x) \ \& \ M_2(x) \ \& \ M_3(x)$
$\quad\quad M_1(x) \ \& \sim M_2(x) \ \& \sim M_3(x)$
$\quad\quad \sim M_1(x) \ \& \ M_2(x) \ \& \sim M_3(x)$
$\quad\quad \sim M_1(x) \ \& \sim M_2(x) \ \& \ M_3(x)$,

and no 'contradiction' of the form (13) can be established in L''.

There is more to be said about this situation, however. (See Niiniluoto, 1978b, Note 16a). The translation from L to L' via (10) has the effect that the primitive predicates M_1 and M_3 in L' are not logically independent in the same sense as M_1 and M_2 in L. Hence, if the metric structure of the class $\mathbf{Q}$ is, e.g., Euclidean, *then the class $\mathbf{Q}'$ cannot be* *assumed to be Euclidean, too.* This can be illustrated by an example, where the class **Q** of qualitative *Q*-predicates is replaced by a quantitative state space $\mathbf{Q} = \mathbb{R}^2$. Define a bijective mapping $t: \mathbb{R}^2 \to \mathbb{R}^2$ of **Q** onto itself by

$$t(x, y) = \langle x, x + y \rangle.$$

Then, using the Euclidean distance, point $\langle x, 0 \rangle$ is closer to $\langle 0, 0 \rangle$ than point $\langle x, -x \rangle$, but the ordering of the translations is reversed: $t(x, 0) = \langle x, x \rangle$ is farther from $t(0, 0) = \langle 0, 0 \rangle$ than $t(x, -x) = \langle x, 0 \rangle$ (see Figure 2).

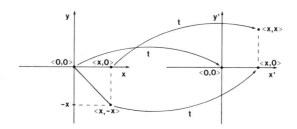

Fig. 2. Translation of R^2 onto R^2.

This example is sufficient to show that there are many one-to-one transformations from a metric space onto itself which distort the original comparative distances of points. In Figure 2, this is a result of the fact that the new $x' - y'$-axes are not independent of each other. The allowance of such transformations would have disastrous consequences to our treatment of truthlikeness and approximate truth relative to quantitative state spaces. Miller's example shows that the same holds for qualitative languages. (Cf. also Miller, 1975.)

Hence, we may conclude that *Miller's concept of intertranslatability is too weak*, that is, too liberal. As Miller's translations need not preserve metric structure, *we cannot expect that comparative truthlikeness is invariant under all translations in Miller's sense.*

Miller's (1978b) second argument differs from the first in that it is directed against the Clifford measure d_c between monadic constituents. (See Niiniluoto, 1978b, pp. 308—313). Given a monadic language L with logically independent primitive predicates $M_1, \ldots, M_k$, Miller constructs another monadic language L' with the primitive predicates

$P_1, \ldots, P_k$ such that

(i) L and L' are intertranslatable, in the sense that each M_i ($i = 1, \ldots, k$) is explicitly definable in terms of $P_1, \ldots, P_k$ and each P_i ($i = 1, \ldots, k$) is explicitly definable in terms of $M_1, \ldots, M_k$.

(ii) if the true constituent C_* in L is not the maximally wide constituent of L which claims that all Q-predicates are instantiated, then there are constituents C_i and C_j in L such that in L

$$d_c(C_*, C_i) < d_c(C_*, C_j)$$

but, for the corresponding translations in L',

$$d_c(C'_*, C'_i) > d_c(C'_*, C'_j).$$

To see how this result can be proved, assume that $k = 2$ and that C_*, C_1, C_2 are represented in Figure 3. Define $P_1 = M_1$ and

(14) $P_2(y) =_{df} (M_2(y) \equiv (\exists x) M_1(x))$.

Then C'_*, C'_1, C'_2 in L' are represented in Figure 4. Definition (14) can be written in the form

(14)' $P_2(y) =_{df} M_2(y)$, if $(\exists x) M_1(x)$ is true
         $\sim M_2(y)$, if $(\exists x) M_1(x)$ is false.

This entails that M_2 can be defined in terms of P_1 and P_2:

(15) $M_2(y) =_{df} (P_2(y) \equiv (\exists x) P_1(x))$.

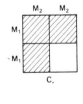

Fig. 3.

Fig. 4.

456   CHAPTER THIRTEEN

Condition (i) is thus satisfied. Moreover, $d_c(C_*, C_1) = \frac{1}{2} < 1 = d_c(C_*, C_2)$ whereas $d_c(C'_*, C'_1) = 1 > \frac{1}{2} = d_c(C'_*, C'_2)$.

This result of Miller's resembles, in an interesting way, Goodman's famous paradox concerning 'grue' emeralds. Goodman (1955) shows that there are intertranslatable monadic languages L and L' such that, contrary to our intuitive expectations, the true descriptions of a sample of emeralds in L and L', respectively, seem to confirm equally well certain non-equivalent hypotheses h and h' in L and L', respectively. The problem in the case of Goodman's paradox is to show why the hypotheses h and h', which seem to have a symmetric position with respect to the given evidence, nevertheless have different degrees of confirmation. Miller's example, as he presents it, is a sort of mirror image of this problem: one should be able to show that the intertranslatable constituents C_1 and C'_1, which in the light of d_c seem to have a non-symmetric position with respect to the truth, nevertheless have the same degrees of truthlikeness. While Goodman's riddle seems to refute all essentially syntactic approaches to induction, Miller's result seems to have the same effect against all 'linguistic' definitions of truthlikeness.

Again, it can be pointed out that constituents C_1 and C'_1 are not intertranslatable within either of languages L and L', so that from the viewpoint of L and L' they stand for different propositions. On the other hand, one can state the definition (14) within language L'' by the meaning postulate

(16) $(y)[P_2(y) \equiv (O_2(y) \equiv (\exists x)O_1(x))]$

where the primitive predicates of L'' are O_1, O_2, and P_2. Relative to the meaning postulate (16), C_1 and C'_1 are equivalent with each other, so that within L'' they should have the same degree of truthlikeness.

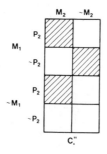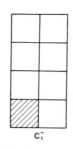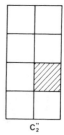

Fig. 5.

This defence of the Clifford measure d_c is not without its problems, however. To see this, note that the constituents C_*, C_1, and C_2 will be represented in language L'' by C''_*, C''_1, and C''_2 which are illustrated in Figure 5. As a result, we obtain $d_c(C''_*, C''_1) = d_c(C''_*, C''_2) = 3$, which disagrees with both of the results concerning L and L'.

This observation concerning the difference between the intertranslatable languages L and L'', where L moreover is a sublanguage of L'', gives us reason to suspect that there is something really peculiar in the character of Miller's predicate P_2. Indeed, in my view, Miller's argument serves to show how radically different two intertranslatable languages can be. Therefore, one need not expect that truthlikeness is invariant in this radical language-shift from L to L', since the predicates in L and L' seem to designate entirely different sorts of concepts.

To explain this thesis, let us first note that a distinction can be made between *monadic* and *relational properties*. It can be said that, e.g., the property of being a man is essentially monadic, while the property of being a father is essentially relational: person x is a father if and only if there exists another person y such that x is the father of y. One cannot determine whether x is a father by studying x alone; instead, one has to take into account the relations which x has to other persons. But even if the property of being a father is essentially relational, it can be designated by a monadic predicate 'is a father'. That this predicate 'is a father' designates an essentially relational property can also be seen from the fact that it can be explicitly defined by using the relational term 'x is a father of y', but not conversely.

Miller's predicate P_2 defined by (14) or (14)', has a very strange character: it does not designate a monadic or a relational property. The question whether or not P_2 applies to an individual y cannot be decided on the basis of the genuinely monadic properties of y and the relations which y bears to other individuals. Rather, it depends upon the question whether there exists individuals satisfying the predicate O_1, even though this question is quite 'accidental' so far as y is concerned, as it is not required that y has any relationship with these other individuals. Let us say that this sort of predicates are *odd predicates*.

There is one possible objection to the notion of odd predicate. As M_2 is, by (15), definable in terms of P_1 and P_2, why cannot we claim that M_2, too, is odd? Of course it might happen that both M_2 and P_2 are odd predicates. However, the main point of the distinction is the fact that if the property of being an M_2 is monadic then the predicate P_2 is odd, and conversely if the property of being a P_2 is monadic then the predicate M_2 is odd. In both cases, there is an essential difference between languages L and L': in one language the predicates designate monadic properties while in the other language they (or some of them) are odd.

It would seem natural to say that odd predicates do not designate any properties at all. As there are good reasons to doubt already the existence of negative and disjunctive properties (see Armstrong, 1978b), predicates like 'is a lion if there are black things and a non-lion if there are no black things' do not name any real properties at all. If our theory of truthlikeness is restricted to such languages L, where the primitive predicates designate genuine properties, Miller's example can be exluded.

Miller might object that the appeal to real properties commits us a questionable form of 'essentialism' or 'metaphysical realism'. If we want to avoid such metaphysical presuppositions in the theory of truthlikeness, Miller's predicate P_2 could be excluded on methodological grounds: P_2 is not a "good scientific predicate" (Tuomela, 1978, Note 4), and no one would take such an artificial construct seriously in a cognitive problem situation, where the aim is to find true information about reality.

To these remarks one may also add our earlier remarks about translations that do not preserve metric structure. Again, the definition (14) has the effect that in L' predicates P_1 and P_2 are not logically independent of each other. Therefore, it is not surprising that a metric concept like d_c is not invariant under the translation (14) between L and L'.

This general conclusion is supported by Pearce (1983), who formulates the problem of truthlikeness as the task of finding a suitable metric d on the Stone space $S(\mathscr{A}_L)$ of the Lindenbaum algebra $\mathscr{A}_L$ of a first-order language L (see Chapter 2.9). Equivalently, we wish to find a metric d on the space $\mathbf{M}_* = \mathbf{M}/\equiv$ of L-structures modulo elementary equivalence (see Chapter 4.1). Let L and L be two first-order languages (with the sets $\mathbf{M}^L$ and $\mathbf{M}^{L'}$ of structures) such that the predicates λ' are explicitly definable by the predicates λ through the meaning postulates $\mathscr{MP}$. This means that there is a translation function t from the formulas of L' to the formulas L such that for g in L', $t(g)$ is the translation of g in L. Function t induces a semantic mapping s

from the L-structures to corresponding L'-structures, since the extensions of the predicates in λ' can be calculated by $\mathscr{MP}$ from the extensions of the predicates λ. Languages L and L' are *intertranslatable* by $\mathscr{MP}$, if the inverse function t^{-1} of t is also a translation from L to L'. In this case, s^{-1} is a semantic mapping from the L'-structures to the L-structures, and the pair (t, s) is called an *equivalence*. Miller's definitions (10) and (14) define equivalence in this sense.

If $\langle t, s \rangle$ is any translation from L' to L, define the function s^*: $\mathbf{M}_*^L \to \mathbf{M}_*^{L'}$ by

$$s^*(\overline{\Omega}) = \overline{s(\Omega)} \quad \text{for } \Omega \in \mathbf{M}^L,$$

where $\overline{\Omega} = \{\Omega' \in \mathbf{M} | \Omega' \equiv \Omega\}$ is in $\mathbf{M}_*^L$. Then $\langle t, s \rangle$ is an equivalence if and only if s^* is a homeomorphism of $\mathbf{M}_*^L$ onto $\mathbf{M}_*^{L'}$. Miller's requirement about invariance with respect to intertranslatability can now be formulated as follows:

(17) Let d and d' be metrics that define the concept of truthlikeness relative to first-order languages L and L', respectively. If $\langle t, s \rangle$ is an equivalence between L and L', then
$d(\overline{\Omega}_1, \overline{\Omega}_2) = d'(s^*(\overline{\Omega}_1), s^*(\overline{\Omega}_2))$
for all Ω_1 and Ω_2 in $\mathbf{M}^L$.

However, as Pearce (1983) observes, condition (17) is clearly too strong. Homeomorphisms are 'topological mappings' which preserve *topological* structure but not generally *metric* structure.

These results leave us with an important problem: what subclass of homeomorphisms between $\mathbf{M}_*^L$ and $\mathbf{M}_*^{L'}$ should be required to preserve degrees of truthlikeness — or at least truthlikeness orderings? This question cannot be solved in this work. It is reasonable to expect that further logical study of translation will bring forward semantic conditions that restrict the choice of the class of functions s^*.[13] But it is also in the spirit of our earlier discussion to admit that the choice of a proper language is at least partly a function of our cognitive interests which in turn depend on pragmatic conditions. Thus, truthlikeness should be preserved in a translation from L to L', *if the cognitive problem does not change* within this language shift. It is by no means clear that a *purely* semantic criterion can be given for this condition. This interplay of semantic and pragmatic factors in the theory of truthlikeness will be discussed further in the two concluding sections.

13.3. PROGRESS AND INCOMMENSURABILITY

According to the *realist* view of scientific progress, science makes cognitive progress so far as it succeeds in gaining increasingly truthlike information about reality: step from theory T_1 to theory T_2 is *progressive* if T_2 is more truthlike than T_1, and this step *seems progressive* on evidence e if T_2 appears on e to be more truthlike than T_1.[14]

Let $T_1, T_2, \ldots$ be theories in language L with the true target h_*. Then

(18) $\quad T_1, T_2, \ldots, T_n, \ldots$

is a *progressive theory sequence* in L if each T_{n+1} is more truthlike than its predecessor T_n, i.e., $\text{Tr}(T_{n+1}, h_*) > \text{Tr}(T_n, h_*)$ for all $n = 1, 2, \ldots$. This condition does not require that (18) is also *convergent*; that will hold if there is a limit theory T in L such that the distance $\delta(T_n, T) \to 0$, $n \to \infty$. The sequence (18) *approaches* or *converges to the truth* in L if the limit theory T is true. Further, the sequence (18) approaches the whole truth in L if $\delta(T_n, h_*) \to 0$, $n \to \infty$. This condition entails that $\text{Tr}(T_n, h_*) \to 1$, $n \to \infty$. Examples of such convergent sequences include increasingly fitting classificatory statements, sequences of point estimates approaching to the true value of a parameter, and sequences of laws approaching to a limit function.

This reconstruction of 'convergent realism' does not yet account for the fact that rival theories in science may be associated with different conceptual systems. If T_1 is a theory in language L_1, and T_2 a theory in language L_2, can we compare T_1 and T_2 for truthlikeness? Let us consider first the case of conceptual enrichment, touched already in Section 2.

So let L_1 be sublanguage of L_2, i.e., $\lambda_1 \subseteq \lambda_2$. Then the relevant P-set $\mathbf{B}_1$ in L_1 is a subproblem of the P-set $\mathbf{B}_2$ in L_2, i.e., $\mathbf{B}_1 \leqslant \mathbf{B}_2$. If T_1 is in L_1 and T_2 in L_2, then we can compare T_1 (or its homophonic translation in L_2) and T_2 for truthlikeness in L_2. More generally, let $T_1, T_2, \ldots$ be a sequence of theories in a monotonely increasing sequence of languages $L_1, L_2, \ldots$ such that $\lambda_1 \subseteq \lambda_2 \subseteq \ldots$, then the corresponding problems $\mathbf{B}_i$ in L_i become finer:

$$\ldots \leqslant \mathbf{B}_i \leqslant \mathbf{B}_{i+1} \leqslant \ldots,$$

and the targets h_*^i in L_i become logically stronger:

$$\ldots h_*^{i+1} \vdash h_*^i \vdash \ldots$$

($i = 1, 2, \ldots$). Then we say that $T_1, T_2, \ldots$ is a progressive theory sequence if each subsequence $T_1, \ldots, T_n$ is progressive relative to its deepest target h_*^n:

$$\text{Tr}(T_1, h_*^n) < \cdots < \text{Tr}(T_n, h_*^n) \quad \text{for all } n < \omega.$$

As a language can be enriched by new concepts in many different ways, and several alternative orders, the concept of progressive theory sequence allows for 'progress branching' in Stegmüller's (1979) sense. We may even have 'progress loops' of the following type:

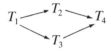

This can be illustrated by examples of quantitative theories that could stand in the relation of correspondence with each other (cf. Chapter 12.5).

EXAMPLE 1. Define languages L_1-L_4 so that they contain the following function symbols:

$$\lambda_1 = \{\bar{h}_1, \bar{h}_2\}$$
$$\lambda_2 = \{\bar{h}_1, \bar{h}_2, \bar{h}_3\}$$
$$\lambda_3 = \{\bar{h}_1, \bar{h}_3, \bar{h}_4\}$$
$$\lambda_4 = \{\bar{h}_1, \bar{h}_2, \bar{h}_3, \bar{h}_4\}.$$

Let T_1-T_4 be quantitative laws of the form

(T_1) $h_1(x) = g_1(h_2(x))$
(T_2) $h_1(x) = g_2(h_2(x), h_3(x))$
(T_3) $h_1(x) = g_3(h_2(x), h_4(x))$
(T_4) $h_1(x) = g_4(h_2(x), h_3(x), h_4(x))$.

Then we may have a progress loop: T_2 and T_3 are both more truthlike than T_1, and T_4 is more truthlike than T_2 and T_3.

A progressive theory sequence that is also convergent is illustrated in the next example.

EXAMPLE 2. Let L_i, $i = 1, 2, \ldots$ be languages with

$$\lambda_1 = \{\bar{h}_1, \bar{h}_2\}$$
$$\vdots$$
$$\lambda_i = \{\bar{h}_1, \bar{h}_2, \ldots, \bar{h}_{i+1}\}$$
$$\vdots$$

and let T_i, $i = 1, 2, \ldots$ be laws of the form

$$(T_1) \quad h_1(x) = g_1(h_2(x))$$
$$\vdots \qquad \vdots$$
$$(T_i) \quad h_1(x) = g_i(h_2(x), \ldots, h_{i+1}(x))$$
$$\vdots \qquad \vdots$$

If the functions h_2, h_3, ... have been ordered according to their 'significance' for h_1 (cf. Nowak, 1980), then the corrections that each T_{i+1} makes to the law stated by its predecessor T_i become smaller and smaller. It follows that the sequence $T_1, \ldots, T_i, \ldots$ is convergent.

Let us next consider the case of two languages L_1 and L_2 such that neither $\lambda_1 \subseteq \lambda_2$ nor $\lambda_2 \subseteq \lambda_1$. The theory of truthlikeness should be able to say something of these cases — at least if it wants to face Kuhn's and Feyerabend's challenge concerning 'incommensurable' theories.[15]

Popper did not find this problem as a serious one, since he thought that it is always possible to find a common extension of two languages L_1 and L_2: while at any moment we are "prisoners caught in the framework" of our language, a "critical discussion and comparison of the various frameworks is always possible" (Popper, 1970, p. 56). Similarly, Miller (1975b) suggested that it is always possible to combine the language of two theories into a language adequate for them both.

Niiniluoto (1977b) argued that to compare theories T_1 in L_1 and T_2 in L_2 for truthlikeness "we first have to *make them comparable* by translating them into a suitably chosen new language". Let L be the *common extension* of L_1 and L_2 with $\lambda = \lambda_1 \cup \lambda_2$. Then T_1 and T_2 have homophonic translations T'_1 and T'_2, respectively, in L, and it is possible to compare T'_1 and T'_2 for truthlikeness in L. This is different from comparing the original T_1 and T_2 directly, since the steps from L_1 and L_2 to L "may involve radical conceptual change". By using the terminology of Section 2, the target h_* relative to L is different from the targets h_*^1 in L_1 and h_*^2 in L_2, so that the cognitive problem has changed.

In fact, the introduction of the common extension L to L_1 and L_2 means that the original cognitive problems $\mathbf{B}_1$ and $\mathbf{B}_2$ relative to L_1 and L_2 are replaced by the finer problem $\mathbf{B}_1 \oplus \mathbf{B}_2$, i.e., the *combination* of $\mathbf{B}_1$ and $\mathbf{B}_2$ (see (4.16)).

Instead of ascending to the common extension of L_1 and L_2, we could also descend to their common sublanguage L_0 with $\lambda_0 = \lambda_1 \cap \lambda_2$. Hence, cognitive problems $\mathbf{B}_1$ and $\mathbf{B}_2$ are replaced by their *meet* $\mathbf{B}_1 \times \mathbf{B}_2$; this problem is non-trivial if and only if $\lambda_0 \neq \emptyset$. Here theories T_1 and T_2 fail to be elements of $\mathbf{B}_1 \times \mathbf{B}_2$, if they are not expressible in L_0. What we can compare for truthlikeness in L_0, however, are the strongest deductive consequences of T_1 and T_2 in L_0.

If $\lambda_1 \cap \lambda_2 = \emptyset$, then $\mathbf{B}_1$ and $\mathbf{B}_2$ are isolated, and the meet $\mathbf{B}_1 \times \mathbf{B}_2$ is trivial $\mathbf{0}$. In such cases, the combination $\mathbf{B}_1 \oplus \mathbf{B}_2$ is essentially a composition with two separate factors, and the target of $\mathbf{B}_1 \oplus \mathbf{B}_2$ is a conjunction $h_*^1 \& h_*^2$ of two logically independent statements. Here it seems that there is no interesting sense in which theories T_1 and T_2 in L_1 and L_2 could be comparable. By the same token, they cannot be regarded as rival to each other. For example, Quantum Mechanics and the theory of evolution are not rivals, and there is no point in trying to say that one of them is more truthlike than the other.

These ideas can be formulated as a principle:

(19) It is meaningful to compare T_1 and T_2 for truthlikeness if and only if T_1 and T_2 can be regarded as answers to the same cognitive problem.

Hence,

(20) If $\lambda_1 \cap \lambda_2 \neq \emptyset$, theories T_1 in L_1 and T_2 in L_2 (or rather their translations) are comparable in the common sublanguage L_0 of L_1 and L_2 (i.e., in problem $\mathbf{B}_1 \times \mathbf{B}_2$), in the common extension L of L_1 and L_2 (i.e., in problem $\mathbf{B}_1 \oplus \mathbf{B}_2$), and in all further extensions L' of L (i.e., in problems finer than $\mathbf{B}_1 \oplus \mathbf{B}_2$).

As Niiniluoto (1979b), p. 256, emphasizes, comparisons of T_1 and T_2 in the extended language L and in its deeper extensions L' may give different truthlikeness orderings (see (5) above).[16] So there is no *unique* way of evaluating the distance of T_1 and T_2 from 'the truth', but the results depend on the depth in which we have so far been able to explore the world. In this dynamical sense, scientific progress involves not only progressive theory-change within the boundaries of a given conceptual system, but also the development of new, more powerful frameworks that allow us to penetrate deeper into reality and to formulate theories with a great unifying power (*ibid.*, p. 257).[17]

There is one very important qualification to be made to the above analysis: the given treatment of conceptual enrichment does not pay attention to the problem of *meaning variance*. The same linguistic symbols may have different meaning in different conceptual systems, and also different predicates may have conceptual connections with each other. To study this important problem in its full generality, a well-developed theory of meaning would be needed. Here I have to limit my attention only to issues that are directly relevant to the comparative problem of truthlikeness.[18]

First, it should be noted that the strategy of renaming a predicate should be used in those cases, where two languages use the same linguistic expression to designate different properties. If M occurs as a primitive predicate in languages L_1 and L_2, but these languages are interpreted so that the property F_M in L_1 is different from F_M in L_2, then M should not be included in $\lambda_1 \cap \lambda_2$, but it should be replaced by a new symbol in one of the languages.

Secondly, if there are conceptual connections between the predicates of L_1, they should be expressed by L_1-statements as *meaning postulates*. Let $\mathcal{MP}_1$ and $\mathcal{MP}_2$ be the sets of the meaning postulates of L_1 and L_2, respectively. Then it is possible to have conceptual shifts from L_1 to L_2 with $\lambda_1 \subseteq \lambda_2$ and $\mathcal{MP}_1 \subseteq \mathcal{MP}_2$, where $\mathcal{MP}_2$ expresses analytical connections between the new terms in $\lambda_2 - \lambda_1$ and the old terms λ_1. If these connections are explicit definitions, we have a case of a conservative language-shift (see Section 2); if they imply new meaning relations between the terms in λ_1, the shift is creative. Similarly, we may have a case of two languages with compatible meaning postulates: $\lambda_1 \neq \lambda_2$ and $\mathcal{MP}_1 \neq \mathcal{MP}_2$, where $\mathcal{MP}_1$ and $\mathcal{MP}_2$ are mutually consistent. Then it would be natural to choose $\mathcal{MP} = \mathcal{MP}_1 \cup \mathcal{MP}_2$ as the set of the meaning postulates in the extension L of L_1 and L_2 with $\lambda = \lambda_1 \cup \lambda_2$. However, it is also possible to express new meaning postulates in L between the terms in λ_1 and λ_2. Even in cases, where originally λ_1 and λ_2 have no common elements, the common extension L may create an interlinguistic 'tie' between some of them. Then the target in L would not any more be simply a conjunction of two independent targets h^1_* and h^2_*. In all of these types of cases, comparisons of truthlikeness can be made as suggested earlier — but relative to the meaning postulates.

EXAMPLE 3. Let L_1 be a monadic language with one family of predicates $\mathscr{F}_1 = \{M_1^1, \ldots, M_k^1\}$, and L_2 a monadic language with one family of predicates $\mathscr{F}_2 = \{M_1^2, \ldots, M_m^2\}$. Suppose that all of these predicates are distinct, so that the cognitive problems $\mathbf{B}_1$ and $\mathbf{B}_2$ associated with L_1 and L_2, respectively, are isolated. However, $\mathscr{F}_1$ and $\mathscr{F}_2$ may nevertheless be partitions of the same attribute space (see Figure 6). For example, $\mathscr{F}_1$ and $\mathscr{F}_2$ might be two different systems of

Fig. 6. Two partitions of the same attribute space.

colour predicates which partition the space of colours into different regions (cf. Chapter 1.3). Then the meaning postulates in the common extension L of L_1 and L_2 should exlude those combinations M_i^1 & M_j^1 which combine two incompatible colours.

Example 3 suggests a method for combining two conceptual systems which carve up the same underlying reality into different pieces at different joints. The families $\mathscr{F}_1$ and $\mathscr{F}_2$ are as it were two alternative jigsaw puzzles for constructing eventually the same picture. If these puzzles have to be combined into one, then we must cut the old pieces into smaller ones, defined by the overlapping parts M_i^1 & M_j^2 ($i = 1, \ldots, k; j = 1, \ldots, m$).

Colour systems of natural languages are sometimes mentioned as examples of *incommensurable* linguistic frameworks. We have now seen that this is not quite true, since there is a 'common measure' for them — the points of the colour space. To find genuine examples of conceptual incommensurability, we need cases, where the meaning postulates $\mathscr{MP}_1$ and $\mathscr{MP}_2$ contradict each other. Then we cannot form the common extension of L_1 and L_2 any more, since $\mathscr{MP}_1 \cup \mathscr{MP}_2$ is an inconsistent set.[19]

The inconsistency of $\mathscr{MP}_1 \cup \mathscr{MP}_2$ necessarily presupposes that $\lambda_0 = \lambda_1 \cap \lambda_2 \neq \varnothing$. Hence, there must be a part of the actual world, the interpretation W_{L_0} of the common sublanguage L_0, which $\mathscr{MP}_1$ and $\mathscr{MP}_2$ describe in conflicting ways. Interestingly enough, this means that the statements that are taken to be *analytically true* in L_1 and in L_2 cannot all be *true* in W_{L_0}. In this sense, the meaning postulates $\mathscr{MP}_1$ and $\mathscr{MP}_2$ are rival *theories* — and should be treated as such in comparisons of truthlikeness.

Elaborating this idea, the following procedure for comparing incommensurable theories was proposed in Niiniluoto (1980a), p. 445. (Cf. Niiniluoto, 1984b, p. 93.) Let T_1 and T_2 be theories in L_1 and L_2, where $\mathcal{MP}_1 \cup \mathcal{MP}_2$ is inconsistent. Let $\mathcal{MP} = \mathcal{MP}_1 \cap \mathcal{MP}_2$ (this set may be empty or non-empty), possibly enlarged with bridge principles connecting λ_1 and λ_2, $\mathcal{MP}_1^* = \mathcal{MP}_1 - \mathcal{MP}_2$, and $\mathcal{MP}_2^* = \mathcal{MP}_2 - \mathcal{MP}_1$. Let L be the common extension of L_1 and L_2 with $\lambda = \lambda_1 \cup \lambda_2$ and with the meaning postulates $\mathcal{MP}$. Then the actual world W_L relative to L is an expansion of W_{L_0}. The comparative truthlikeness of T_1 and T_2 depends now on whether T_1 together with $\mathcal{MP}_1^*$ fits W_L better than T_2 together with $\mathcal{MP}_2^*$. In other words,

(21) T_1 is more truthlike than T_2 in L iff T_1 & $\mathcal{MP}_1^*$ is more truthlike than T_2 & $\mathcal{MP}_2^*$ relative to the target domain W_L.

A similar proposal applies to estimated truthlikeness: evidence statements e_1 in L_1 and e_2 in L_2 will have translations in the extended language L, and the relative truthlikeness of T_1 and T_2 in L is evaluated by the condition

(22) $\text{ver}(T_1 \& \mathcal{MP}_1^* / e_1 \& e_2) > \text{ver}(T_2 \& \mathcal{MP}_2^* / e_1 \& e_2)$.

It does not matter in (22), if e_2 happens to be inconsistent with $\mathcal{MP}_1^*$ or e_1 with $\mathcal{MP}_2^*$, since in general $\text{ver}(h/e)$ may be high even if e contradicts h. But, of course, (22) will not be applicable, if e_1 and e_2 syntactically contradict each other in L. To evaluate rival theories on logically inconsistent evidence is impossible, and this kind of situation can be resolved only by a re-examination of the evidential conditions where e_1 and e_2 were initially obtained.

EXAMPLE 4. Let L_1 be a monadic language with three families of predicates

$$\mathcal{F}_1 = \{M_1, M_2, M_3\}$$
$$\mathcal{F}_2 = \{P, \sim P\}$$
$$\mathcal{F}_3 = \{Q, \sim Q\}$$

with the usual meaning postulates for $\mathcal{F}_1$. Assume further that $\mathcal{MP}_1$ contains the meaning postulate

$$h_1 = (x)(P(x) \equiv (M_2(x) \vee M_3(x))).$$

Let L_2 contain the families $\mathscr{F}_1$, $\mathscr{F}_2$, and

$$\mathscr{F}_4 = \{R, \sim R\},$$

with the additional meaning postulate

$$h_2 = (x)(P(x) \equiv M_3(x))$$

in $\mathscr{MP}_2$. Then $\mathscr{MP}_1 \vdash (x)(M_2(x) \supset P(x))$ and $\mathscr{MP}_2 \vdash (x)(M_2(x) \supset \sim P(x))$. Relative to the assumption $\exists x M_2(x)$, the meaning postulates $\mathscr{MP}_1$ and $\mathscr{MP}_2$ contradict each other. As an interpretation of this formalism, we might have

$M_1(x) = $ 'x is a stone'
$M_2(x) = $ 'x is a plant'
$M_3(x) = $ 'x is an animal'
$P(x) \;\;\;= $ 'x is animate'
$Q(x) \;\;\;= $ 'x is a germ'
$R(x) \;\;\;= $ 'x is a spore'

Suppose then that the user of L_1 identifies certain individuals $a_1, \ldots, a_n$ as germs and plants, and concludes that all germs are plants:

$(T_1) \quad (x)(Q(x) \supset M_2(x))$.

The user of L_2 identifies certain individuals $b_1, \ldots, b_n$ as spores and plants, and concludes that all spores are plants:

$(T_2) \quad (x)(R(x) \supset M_2(x))$.

In L_1 theory T_1 entails that all germs are animate, i.e., $(x)(Q(x) \supset P(x))$, but in L_2 theory T_2 entails that all spores are inanimate, i.e., $(x)(R(x) \supset \sim P(x))$.

To compare T_1 and T_2 for truthlikeness, let L be the extension of L_1 and L_2 with the families $\mathscr{F}_1$, $\mathscr{F}_2$, $\mathscr{F}_3$, and $\mathscr{F}_4$. Let us assume that the intended reference of the terms 'germ' and 'spore' is the same, so that in L we accept as a meaning postulate

$$L_3 = (x)(Q(x) \equiv R(x)).$$

Assume that the truth is represented by the constituent C_* of Figure 7, i.e., all germs are animate plants, but there are other animate and inanimate plants. Then, by (21), T_1 is more truthlike than T_2 in L, since of the two false generalizations $T_1 \& h_1$ and $T_2 \& h_2$ the former is more truthlike than the latter.

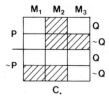

Fig. 7.

Pearce (1986a) has made the suggestion that languages L_1 and L_2 with contradictory meaning postulates need not be incommensurable, since there might be a meaning-preserving (not necessarily homophonic or literal) translation t of L_1 to L_2. Then, instead of trying to ascend to the common extension L of L_1 and L_2, as proposed in (21) and (22), T_1 and T_2 could be compared in L_2 by comparing the M- and ver-values of $t(T_1)$ and T_2. This proposal is a generalization of the approach that has been recommended above for conceptual enrichment, but Pearce does not assume that L_1 is a sublanguage of L_2. He is also able to show the existence of translations in a wide range of examples of theory change.[20] If no appropriate type of translation exists between L_1 and L_2, then T_1 and T_2 are genuinely incommensurable — and, according to principle (19), no need to compare T_1 and T_2 for truthlikeness arises.

I find Pearce's proposal highly interesting. Its relative merits to my approach deserve further study, and have to be left open here. Perhaps it is fair to remark that they need not conflict with each other, since — as Pearce himself admits — in situations where both L_1 and L_2 are clearly descriptively incomplete (cf. Example 1) it is natural to seek a common extension of L_1 and L_2. Furthermore, one might point out that even the method of translation may be 'relativistic' in similar ways as my language-relative treatment. If there happens to be two different translations t and t' from L_1 to L_2, which one should be chosen? If there is, besides translation t from L_1 to L_2, another translation from L_2 to L_1 (not identical with t^{-1}), should we compare theories T_1 and T_2 in L_1 or in L_2?

13.4. TRUTHLIKENESS AND LOGICAL PRAGMATICS

The theory of truthlikeness could follow many alternative research strategies. I don't claim that my approach is the only possible: other directions could and should be explored. Some critics have argued that the concept of similarity gives a too 'shaky basis' for verisimilitude, while some others have recommended weaker mathematical assumptions than quantitative metrics or distance functions. Brief comments on these issues are in order in this concluding section.

Peter Urbach (1983) has recently criticized Tichý and Oddie for believing that the theory of truthlikeness can be based upon an *objective concept of similarity*, which expresses the 'real' likeness between the structure described by a statement and the true structure of the world. Urbach convincingly points out that our 'intuitions' about the similarity of two structures (e.g. human faces) always presupposes that some special significance is attached to certain properties of the structures, and such weights are "strongly influenced by our constitution and our language and culture".

The point that the concept of similarity always depends on a choice of the relevant features and on the weights attached to them was made already in Niiniluoto (1977b) — which Urbach does not mention. In that article, I emphasized that degrees of truthlikeness are "semantically unambiguous but pragmatically ambiguous": measures of truthlikeness "presuppose the specification of a distance measure which reflects the relevant respects of comparison and balances them against each other" (*ibid.*, p. 129). This point has been elaborated in Chapter 1.4. Thus, while identity or complete similarity is a logical concept which can be treated in syntactical terms, incomplete similarity is partly a pragmatic notion. In particular, while truth is a logical concept, truthlikeness is to some extent pragmatically ambiguous.

This lack of 'objectivity' does not show that verisimilitude is too 'shaky' to be 'significant' or 'interesting', as Urbach claims. Even Popper has recently acknowledged that perhaps the problem of truthlikeness "cannot be solved by purely logical means", but "only by a relativization to relevant problems or even by bringing in the historical problem situation" (Popper, 1979, p. 372).[21] It is true that judgments about similarity depend on some contextual features, but our 'intuitions' may become 'clear-cut' and 'reliable' when the relevant boundary conditions have been fixed. Urbach fails here to appreciate the possibility of the kind of study that has been called 'logical pragmatics'.

For example, Urbach's observation that worlds with 9, 10, and 11 planets can be compared in many different ways is correct but not interesting: the statement 'There are 10 planets' is closer to the truth than 'There are 11 planets' *if* they are considered as answers to the cognitive problem 'What is the number of planets in our solar system?'. Other problem situations might naturally invite different kinds of similarity considerations.

Urbach's appeal to Popper's remark that verisimilitude is "not an epistemological or an epistemic idea" is likewise beside the point here: in spite of its pragmatic ambiguity, measure Tr is not *epistemic*, or relative to knowledge, in any sense. Instead, measure ver is distinguished from Tr precisely by the fact that its values are relative to evidence and therefore serve epistemological purposes. Again Urbach fails to refer to Niiniluoto (1977b) where this distinction was made.

Noretta Koertge (1979) notes that the definition of verisimilitude presupposes some weights for the relative seriousness of 'agnosticism' and 'false predictions'. This idea is built in the definition of our measures Δ_{mm}^{γ} and $\Delta_{ms}^{\gamma\gamma'}$, but it was explicitly used already in the 'index of caution' of Levi's (1967a) theory of inductive acceptance. If truthlikeness is, as Popper suggested, a combination of a truth-factor and an information-factor, then it is inevitable that these two factors have to be balanced against each other in some way. Of course, the situation does not change, if you introduce still additional factors into the picture (such as explanatory power, simplicity, etc.). Unless your system for appraising scientific statements is one-dimensional, you will need weights indicating the importance of the various factors.

Again, this kind of relativity to many-dimensional cognitive interests does not make the idea of scientific progress uninteresting. In the philosophy of science, it helps us to conceptualize the differences between various methodological theories. In the actual context of scientific inquiry, the values of the various 'pragmatic' boundary conditions are not fixed purely subjectively, but they are characteristics of the long-term scientific research traditions, so that there may be a broad consensus about them within the scientific community (cf. Kuhn, 1977, Ch. 13).

At the same time, it should be emphasized that many comparative judgments about truthlikeness are relatively robust with respect to changes in the pragmatic conditions. Two investigators who to some extent disagree on the values of the parameters γ and γ', and on the weights to be given to some of the relevant properties, may still agree on the relative cognitive merits of rival statements. In the limit, some comparative judgments are 'completely objective' in the sense that they are entirely independent of all parameter values.[22]

If one wishes to remain restricted to these non-problematic cases, where one theory is uniformly better than another, then a weak theory of truthlikeness is obtained[23] — most pairs of theories turn out to be uncomparable in it. This restricted theory is obtained as a special case of our more powerful, but also pragmatically more relative, approach.

It might be objected that our theory is too strong, since it is not necessary to introduce *quantitative* degrees of verisimilitude in order to obtain a comparative concept of truthlikeness — and the latter is our real goal, as I have repeatedly emphasized. One practical reply is that so far the quantitative approach has been the only method for explicating an interesting comparative notion of truthlikeness — i.e., a notion which goes beyond the trivial cases where one theory is uniformly closer to the true one than another. A similar method in defining a quantitative concept of similarity has been employed in many different disciplines, such as statistics, computer science, pattern recognition, linguistics, biological taxonomy, and psychology. So why should philosophers and logicians be afraid of doing the same? Moreover, perhaps the strongest argument for the quantitative approach comes for the observation that virtually in all applications the relevant P-sets **B** already are metric spaces or else they are known to be metrizable topological spaces.

Our philosophical aim in using the quantitative approach is conceptual clarity. While it certainly would be desirable to do more work in applying our framework to concrete case studies in the history of science, it is also clear that primarily we are not trying to build up for working scientists algorithms that could be effectively used for calculating the real or estimated degrees of truthlikeness of their favorite theories. Rather, our main goal is conceptual analysis, logical and epistemological clarification of a concept that is crucially important for developing a realist view of science. For the purpose of disclosing the structure of a reasonable notion of truthlikeness, the use of the quantitative tools seems equally justified as the choice to develop probability theory on a quantitative basis. Real-valued probability measures were studied almost for 300 years, before attempts at direct analyses of comparative probability relations were initiated in our century.[24]

Quantitative degrees of truthlikeness can also be used for solving the epistemic problem of truthlikeness. So far no one has presented alternatives to our function ver, which essentially relies on the numerical values of the Tr-function. Whether the epistemic problem can be treated — outside the trivial cases — with a purely comparative notion of verisimilitude is not known to me.

The quantitative approach would be unnecessary, if truthlikeness were only a topological concept. It was seen in Section 2 above that this would also help us to solve Miller's problem of linguistic invariance, since purely topological notions would be preserved within translations (homeomorphisms). However, truth-*likeness* involves essentially the idea of similarity, and the explication of that concept requires something more than a mere topology.

For us this 'something more' has been a metric or a distance function defined on the set **B** of complete answers to a cognitive problem. But perhaps we could work out our theory by employing the concept of *uniformity* which is known to be weaker than the concept of metric (see Chapter 1.2). Indeed, following G. Ludwig, the structuralist school in the philosophy of science has recently developed highly interesting account of such concepts as 'approximate explanation', 'approximate reduction', and 'approximate application' by using the concept of uniformity. Most of this work could easily be 'transferred' to the study of approximate truth (cf. Chapter 12.5) — as soon as the structuralists would be ready to single out some model or models as representing the actual world, or some theory as the true theory.

My preference for the concept of a metric, in spite of its being stronger and in this sense less general than the concept of uniformity, is partly practical. The theory of metric spaces is better known and easier to apply than Bourbaki's theory of uniform spaces. Another practical remark is that all the applications in the philosophy of science known to me, where a uniformity has been actually defined, have in fact been examples of a metric space. And, as noted already above, the spaces associated with the kind of cognitive problems studied in our book are at least metrizable spaces.

If anyone still thinks that the metric approach is too strong, I should like to encourage all attempts to develop the theory of truthlikeness on a weaker basis. But, as we have noted above, this need not relieve us from the problem of linguistic variance, since Miller-type arguments against truthlikeness can be applied also to other ways of analysing approximation (e.g., uniformities).

To those readers who are worried that the similarity account of truthlikeness does not make verisimilitude completely 'objective' or purely 'logical' I wish to cite my conclusion from Niiniluoto (1986b). If measures of verisimilitude always have a pragmatic dimension, by being dependent on our cognitive interests, this only shows that it has the same character as most of the interesting concepts in the philosophy of science — such as theory, explanation, and confirmation.[25] Science is a fallible and progressive enterprise which is run by historically developing scientific communities. Our tools for analysing science and its change should be flexible enough to take into account this richness of scientific practices. So let us not try to make our theory of science more objective — or less objective — than science in fact is.

Perhaps the most serious charge against the concept of verisimilitude has been its failure to *exist* — in spite of the many sweeping applications that it has had in the philosophy of science. This book has tried to find a cure for this strange limitation. But it should be remembered that the creature introduced in this work is still relatively young and not yet complete: the task of finding a reasonable concept of truthlikeness is a research programme for logically minded philosophers of science, and further progress in this programme is needed in many directions.

NOTES

CHAPTER 1

[1] According to John Locke, "probability is likeness to be true" (*An Essay Concerning Human Understanding*, 1689, Book 4, Ch. XV). In many modern languages, the concept of probability is derived from the Latin *verisimilitudo* by combining the words 'true' (Latin *verum*) and 'similitude' (cf. Hartigan, 1971), but this should not lead us to confuse these concepts with each other. For the connection between truthlikeness and probability, see Chapter 5.2 below. For an analysis of the first component in the notion of truthlikeness — i.e., the truth — see Chapter 4.3 below.

[2] The standard abbreviation 'iff' for 'if and only if' is used in this book.

[3] A function $h: \mathbb{R} \to \mathbb{R}$ is an order-preserving transformation if

$$x < y \text{ iff } h(x) < h(y)$$

for all x and y in $\mathbb{R}$.

[4] Note that (8) for $0 < p < 1$ is not a metric, since it does not satisfy the triangle inequality (D4). The unit sphere for L_p with $p < 1$ would look as follows:

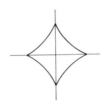

[5] log denotes here the natural logarithm.

[6] Boorman and Olivier (1973) note — without giving any details — that the definition (28) can be generalized to cases where $\mathscr{A}_1$ and $\mathscr{A}_2$ are associated with different label sets V_1 and V_2 and where all the modes in $\mathscr{A}_1$ and $\mathscr{A}_2$ are labeled.

[7] For the basic concepts and results of topology, see Bourbaki (1966) and Simmons (1963).

[8] Y is *compact* if every class of closed subsets of Y with empty intersection has a finite subclass with empty intersection.

[9] F. Bacon's (1620) method of eliminative induction can be viewed as a systematic way of exploring analogies, as Keynes (1921) observes.

[10] See the summary in Klug (1966). For the treatment of analogy in legal reasoning, see also Niiniluoto (1983b).

[11] Note that 'M' may be a singular term (like 'Socrates') or a general term (like 'man'). In the latter case, the premise 'M is P' means 'All M's are P', so that Erdmann's

NOTES 475

formulation could be changed to

> All M's are P
> All S's are N
> M is similar to N
> ―――――――――
> All S's are P.

[12] See, for example, Heywood (1976), Ruse (1973), Sneath and Sokal (1973). A nice illustration of taxonomical methods is given in the exhibition *Dinosaurs and their living relatives*, opened at the Natural History Museum, London, in 1979. It examines the question: Which living animals are most closely related to dinosaurs? On the basis of evidence about 'homologies' (i.e., "characteristics that are similar because they are inherited from a common ancestor"), it comes to the conclusion: "Birds are the closest living relatives of the dinosaurs. But some scientists think that birds and crocodiles are more closely related to each other than either of them is to dinosaurs". For example, fossil birds, fossil crocodiles, dinosaurs, pretosaurs, and thecodontians share uniquely among them the homology that they have a hole in the skull in the front of the eye socket. (See Niiniluoto, 1980b.)

[13] *Dinosaurs and Their Living Relatives*, p. 17. See also Sneath and Sokal (1973), Ch. 6.

[14] If x_i^{max} and x_i^{min} are the largest and the smallest values of x_{ij} ($j = 1, \ldots, m$), then x_{ij} may be replaced by x_{ij}/x_i^{max} or by $(x_{ij} - x_i^{min})/(x_i^{max} - x_i^{min})$. The values x_{ij} may also be 'standardized' by replacing x_{ij} with $(x_{ij} - \bar{x}_i)/s_i$. See Sneath and Sokal (1973), pp. 153–157.

[15] Multistate qualitative characters are treated in the same way — without assuming that the states of O_i can be ordered (see Sneath and Sokal, 1973, p. 115). In J. C. Gower's (1971) general similarity coefficient

$$\frac{\sum_{i=1}^{u} W_{ijk} S_{ijk}}{\sum_{i=1}^{n} W_{ijk}}$$

the value of $0 \leq S_{ijk} \leq 1$ expresses the similarity between the states x_{ij} and x_{ik} and W_{ijk} is the weight for character O_i. If $S_{ijk} = |x_{ij} - x_{ik}|$ for quantitative states, Gower's coefficient is a complement of a weighted mean character difference (36). In Chapter 3.2 below we generalize this idea to cases where Q_i may be multistate qualitative characters. For a treatment of similarity relation with vague properties, see Bugajski (1983).

[16] For pattern recognition, see Fu (1977, 1982), Chen (1973).

[17] For cluster analysis, see Chen (1973), Sneath and Sokal (1973), van Ryzin (1977), Rao (1977), Zadeh (1977).

[18] See Rosenfeld (1969, 1976, 1979).

[19] See also Carnap (1971b, 1980), Hilpinen (1973), Goodman (1972), and Eberle (1975). For a non-psychological, physicalistic interpretation of qualities, see Armstrong (1978a, b).

[20] See Tversky (1977), Beals, Krantz, and Tversky (1968), Krantz and Tversky (1975), Sattath and Tversky (1977), Schwarz and Tversky (1980).
[21] Condition (47) need not be interpreted to imply that all metrics d can be used in defining the concept of similarity.
[22] D. Lewis (1973), who uses as a primitive concept the notion of overall similarity between possible worlds, also argues that similarity is a directional relation.
[23] For criticism of nominalism, see Armstrong (1978a, b), Loux (1978), and Bunge (1977).
[24] The atoms correspond to Carnap's Q-predicates (see Chapter 2.2).
[25] For a criticism of negative and disjunctive properties, see Armstrong (1978b).
[26] While I agree with Armstrong's (1978a, b) criticism of nominalism and platonism, I am inclined to accept (against his view) a 'Stoutian' realism, where each particular has its own individual qualities or 'tropes' — and universals exist only as abstractions in the man-made World 3. However, I shall not develop these ideas here further, since nothing in this book depends on a solution to the classical problem of universals.
[27] For these terms, see Stalnaker (1968).

CHAPTER 2

[1] For a detailed discussion of monadic languages L_N^k, see Carnap (1962). For a standard treatment of predicate logic, see Monk (1976).
[2] The notion of truth is discussed also in Chapter 4.3.
[3] See Montague (1974), Hintikka (1975) and Lewis (1986). This generic notion of property should be distinguished from individual property-instances associated with particulars (cf. Note 26 to Ch. 1). In any case, we can say that a *has property* F_M *in* Ω iff $\Omega \models M(a)$.
[4] For the notion of analytical truth, see Stenius (1972) and Hintikka (1973).
[5] This idea was developed in W. E. Johnson's (1921) theory of determinables and determinates. Cf. also Armstrong (1978b) and Hautamäki (1986).
[6] Carnap (1971b) includes in monadic conceptual systems also a set of individuals Ind = $\{u_1, u_2, \ldots\}$. This is slightly unnatural, since individuals are not conceptual entities (such as e.g. individual concepts are). Moreover, it should be possible to apply a conceptual system to different domains of individuals.
[7] $^{\inf}L_\omega^k$ is a special case of the infinitary language $L_{\omega_1\omega}$ (cf. Monk, 1976).
[8] Distributive normal forms for monadic predicate logic are, in a sense, generalizations of the disjunctive normal forms of propositional logic. They were first studied in the 19th century algebraic logic, and rediscovered by G. H. von Wright in 1948 (see von Wright, 1957). See also Chapter 9.1.
[9] This concept was introduced by Carnap (1968).
[10] $\exists^{\geqslant \omega}$ is a so-called 'generalized quantifier'.
[11] Constituents of L_N^{k-} are discussed in Hilpinen (1966).
[12] Cf. Hintikka (1970b), Hintikka (1973), p. 142, n. 33, and Hintikka and Niiniluoto (1973).
[13] See Hintikka (1970b, 1973), Rantala (1986).
[14] See Rantala (1979), Karttunen (1979, 1985). Constituents for higher-order logic have been discussed by Oddie (1979, 1982).

[15] For the proof of these results, see Bell and Slomson (1969).
[16] Cf. the discussion in Tuomela (1973), Niiniluoto and Tuomela (1973).
[17] For discussions of this idea, see Suppes (1967), Sneed (1971), Wojcicki (1973), Stegmüller (1976, 1979, 1986), Niiniluoto (1984b), Ch. 6, Pearce and Rantala (1983). The 'structuralist' programme (Sneed and Stegmüller) attempts to analyse the structure of theories within a set-theoretical metalanguage without assuming the object language L at all. See also Chapter 3.3.
[18] For the 'personalist' interpretation of probability, see Kyburg and Smokler (1964).
[19] For a general survey of inductive logic, see Niiniluoto (1983d). Classical works in the field include Carnap (1950, 1952, 1962), Hintikka and Suppes (1966), Hilpinen (1968), Carnap and Jeffrey (1971), Niiniluoto and Tuomela (1973), Kuipers (1978a, b), Cohen and Hesse (1979), Jeffrey (1980). The most famous critic of inductive logic is Karl Popper (1959, 1963, 1972).
[20] Discussions of personal probabilities for an agent who is not logically omniscient are given in Garber (1983) and Niiniluoto (1983c).
[21] See Carnap (1980), Niiniluoto (1981a), Spohn (1981), Kuipers (1984b).
[22] This result is due to John Kemeny.
[23] The q-dimensional continuum was developed in Hintikka and Niiniluoto (1976) and Niiniluoto (1977a). See also Kuipers (1978a, b).
[24] My strategy in this book is to develop a theory of truthlikeness both for extensional and intensional statements. If the Humeans are right, then the application of my theory may be restricted to the extensional languages (cf. Chapters 8, 9, 10). But if they are wrong, as I believe, then the intensional case becomes important as well (cf. Chapter 11). For recent discussions about the problem of lawlikeness, see Armstrong (1978b), Levi (1980), Niiniluoto (1978c), and Fetzer (1981). Probabilistic laws are discussed in Chapter 3.5.
[25] Instead of (90), it has been suggested that laws should be analysed in terms of some new intensional connective $\Box\!\!\rightarrow$ ('if-then'), i.e., by $(x)(F(x) \; \Box\!\!\rightarrow \; G(x))$. See Lewis (1973), Nute (1980, 1981), Harper, Pearce, and Stalnaker (1980), and the special issue of *Journal of Philosophical Logic* **10**:2 (1981). Armstrong (1978b) and Dretske have proposed that laws are treated as second-order relations between properties. It is certain that (90) cannot be the whole story in the analysis of natural laws: while

$$\vdash \Box(x)(F(x) \supset G(x)) \equiv \Box(x)(\sim G(x) \supset \sim F(x)),$$

it is clear that 'All ravens are black' is lawlike, but 'All non-black things are non-ravens' is not. The trouble with the latter statement is that negative predicates do not express genuine properties (see Chapter 1.3). Therefore, we should add to the analysis (90) the material adequacy condition that the predicates 'F' and 'G' express properties.
[26] See Hintikka (1975), Lewis (1973, 1986).
[27] For systems of modal logic, see Chellas (1980).
[28] For the history of the Principle of Plenitude, see Knuuttila (1981).
[29] Hacking (1984) gives an interesting account of the role of experimentation in science.
[30] See von Wright (1971), pp. 71–73.

CHAPTER 3

[1] This is one of the questions discussed in Benacerraf and Putnam (1964).
[2] See Duhem (1954).
[3] I am not suggesting that a scientific realist should be a reductionist. Instead, I think that a reasonable ontology should accept emergent materialism with the Popperian Worlds 1, 2, and 3 (see Niiniluoto, 1984b, Ch. 9). However, the theory of truthlikeness in this book does not depend on these ontological issues.
[4] See Scott and Suppes (1958), Krantz et al. (1971).
[5] This argument from the Carnapian discrete space Q of Q-predicates to the quantitative state space was presented in 1983 in Niiniluoto (1986a).
[6] See also Suppe (1974, 1976).
[7] Our account here is simplified, since the distinction between T-theoretical and non-T-theoretical terms is not considered. Cf. Niiniluoto (1984b), Ch. 5.
[8] The problem of idealization has been fruitfully discussed within Poznan school: see Nowak (1980), Krajewski (1977). See also Niiniluoto (1986a).
[9] The propensity interpretation of probability was first developed by C. S. Peirce and Karl Popper as an alternative to the frequency interpretation (von Mises, Reichenbach). For the single-case interpretation, see Fetzer (1981) and Niiniluoto (1982d).
[10] Cf. the discussion in Levi (1980a).
[11] See also van Fraassen (1972). von Plato (1982) argues convincingly that one can in a sense associate objective physical probabilities also to deterministic systems (e.g., classical games of chance). However, single-case propensities (if they exist) are applicable only to genuinely indeterministic situations. Again, my strategy here is not to solve the problem of determinism and indeterminism, but to develop the theory of truthlikeness both for deterministic and probabilistic laws.
[12] For the theory of stochastic processes, see Parzen (1962), Cramer and Leadbetter (1967).

CHAPTER 4

[1] Laudan (1977) has argued that science is a 'problem-solving activity' rather than a 'truth-seeking activity'. In my view, this constrast is misplaced: in solving cognitive problems, science is a truth-seeking enterprise (see Niiniluoto, 1984b, Ch. 7, 11). Hintikka (1982), Sintonen (1984), and Kleiner (1985) use erotetic logic to develop the interrogative model of scientific inquiry.
[2] This characterization of closed and open questions is due to Jerzy Giedymin. Cf. Niiniluoto (1976a). The notion of 'an effective method' should be understood here in a liberal sense (cf. Levi, 1979): the problem of finding the true value of a parameter $\theta \in \mathbb{R}$ is closed, even though all the real numbers are not 'constructive'.
[3] These contexts are known to philosophers of science as 'paradigms' or 'disciplinary matrices' (Kuhn, 1962), 'research programmes' (Lakatos, 1970), or 'research traditions' (Laudan, 1977).
[4] The account of P-set in this section follows Niiniluoto (1976a). P-sets correspond to Levi's (1967a) 'ultimate partitions'.
[5] Cf. Levi (1967a).

[6] This section follows partly Niiniluoto (1985a). See also Niiniluoto (1984a). For a recent survey of theories of truth, see Haack (1978).

[7] See Chapter 5.3 for comments on Bradley.

[8] Cf. Popper (1963) and Niiniluoto (1984b), Chs. 3, 5.

[9] See also Niiniluoto (1984b), Chs. 7, 9.

[10] Davidson (1984) finds this correlation between languages and world in Tarski's satisfaction relation, where the terms (individual constants and variables) are valuated through sequences of objects (see Chapter 2.1). However, this valuation depends partly on the interpretation function V_a — which also establishes a correlation between predicates and their extensions. Cf. also LePore (1983).

[11] For a defence of this view, see Niiniluoto (1981b).

[12] Against this view, it seems to me that the Tarskian correspondence theory of truth can be applied also to Worlds 2 and 3 which are not reducible to World 1 in the physicalist way. A statement can be in correspondence relation also to (concrete and abstract) man-made reality. (See Niiniluoto, 1981b, c, 1984b, Ch. 7.)

[13] See Putnam (1978). In a more elaborate version of the argument, Putnam (1980) requires that the ideal theory T satisfies all operational constraints — thus its restriction to the observational language would be true (cf. van Fraassen's (1980) concept of 'empirical adequacy', and the concept of 'pragmatic truth' by Mikenberg, Da Costa, and Chuaqui, 1986). Again, there is no guarantee that the structure Ω_w coincides with the non-observational part of Ω_L^*. For discussion about Putnam's argument, see Merrill (1980), Hacking (1983), Pearce and Rantala (1982), Tuomela (1985).

[14] Assumption (36) would in fact make our task easier, since for us there would be no problem of linguistic variance: a metaphysical realist could simply apply our definition of truthlikeness relative to L_{id}. Cf. Chapter 12.2.

[15] See Carnap (1962) for a description of explication.

[16] For vagueness, see Fine (1975) and Sanford (1976).

[17] For fuzzy sets, see Zadeh (1965, 1975), Dubois and Prade (1980).

[18] Strategy (i) has been discussed by Prezelecki (1976).

[19] Fuzzy logic is developed in Zadeh (1975). Cf. also Katz (1982).

[20] Similarly, we may define α to be *false* if and only if α is false in each structure Ω in Θ_L^*. This characterization of truth and falsity agrees with a definition for 'absolute truth' and 'absolute falsity' in Przelecki (1969, 1976). Here the Principle of Bivalence (or the metalogical Excluded Middle) has to be given up. Alternatively, one could use the concept of falsity in a broader sense: α is false in Θ_L^* if and only if α is not true in Θ_L^*. Then those sentences which are true only in some (but not all) structures Ω in Θ_L^* would be false, but they would nevertheless be 'partly true' and 'truthlike' (cf. Chapter 6.9). However, in this case both α and $\sim \alpha$ could be false.

[21] See Popper (1959, 1963, 1972). For an evaluation of the Carnap-Popper controversy, see Niiniluoto (1973). Several authors have pointed out that Popper's insistence on high information content, or low prior probability, does not imply that also the posterior probability of a hypothesis on evidence should be low. Thus, it is possible to reconstruct some Popper's main ideas within a probabilistic theory of induction (see Levi, 1963; Hintikka and Pietarinen, 1966; Niiniluoto and Tuomela, 1973; Jeffrey, 1975; Good, 1975). For this purpose, we need a system of inductive logic (like Hintikka's) where genuine universal generalizations over infinite domains may have non-zero probabilities. The existence of such systems disproves Popper's

claim that universal laws — as a sort of infinite conjunctions — must have the probability zero. On the other hand, Popper may be right that scientists often operate with idealizational hypotheses which have the probability zero relative to the total evidence (cf. Chapter 6.9).

CHAPTER 5

[1] For this distinction, see Niiniluoto (1984b), Ch. 3.
[2] See Chapter 12.
[3] See Laudan (1981), Ch. 14.
[4] Such sentences have been called 'partially true' (pacceka sacca) in the Buddhist tradition. To mistake a part for a whole is an example where a statement corresponds only "to some extent with facts": a blind man reports that the whole elephant is like the part (ears, forehead, etc.) that he touches (see Jayatilleke, 1971, p. 57).
[5] Peirce's theory of induction has been interpreted in both ways: successive inductions give a sequence of false point estimates which converge towards the truth (cf. Niiniluoto, 1984b, Ch. 3), or a sequence of more and more narrow interval estimates (see Levi, 1980b).
[6] For a logician, it is tempting to try to give a recursive definition of the concept $\Omega \models_r h$ (h is true to the degree r in structure Ω) by imitating and modifying the clauses of Tarski's definition (2.1). For attempts toward this direction, see Weston (1977, 1981), Chuaqui and Bertossi (1985). However, as we shall see also in Chapter 6, this strategy fails, since degrees of truth (approximate truth, truthlikeness) are truth-functional only in a very limited sense. For example, it may happen that both h_1 and h_2 are approximately true in Ω, but still $h_1 \& h_2$ is a logical contradiction.
[7] The definition of the classes T and F have to be relativized to a language L: the class of all truths in all possible languages hardly makes sense. Popper is not worried about this relativity, as we can judge from his comments on the "myth of the framework" (Popper, 1970, p. 56). Problems of linguistic relativity and conceptual incommensurability are discussed below in Chapter 13.2.
[8] Hattiangadi (1975) informs that C. G. Hempel had proved essentially the same negative result already in 1970; the given proof is valid for finitely axiomatizable theories.
[9] For a comment on Agassi's (1981) attempt to save Popper's definition, see Notes 3 to Chapter 8. Cf. also Perry (1982). Newton-Smith's (1981) definition of verisimilitude is convincingly criticized by Oddie (1986).
[10] Note that $Vs(h)$ cannot be defined in terms of the cardinalities of the truth content and the falsity content of h (cf. Robinson, 1971), since each sentence h in a first-order language L has an infinite number of true and false consequences in L (see Miller, 1972). For comments on Popper's measure Vs, see also Keuth (1976) and Grünbaum (1976a, b).

CHAPTER 6

[1] A flow chart indicating the early history of the similarity approach is given in

Niiniluoto (1978b), p. 283. A line from 'Niiniluoto (1978a)' to 'Tichý (1978)' should be added to it.

[2] This remark is made also in Pearce (1983). Cf. also the discussion about metrics (distance functions) and uniformities in Chapter 13.2.

[3] The idea of replacing possible worlds by linguistic descriptions of possible worlds (such as constituents or model sets) has been a characteristic feature of Hintikka's programme in philosophical logic. See Hintikka (1973).

[4] A similar condition has been used by Kuipers (1984b) in the theory inductive analogy.

[5] Levi (1986b) proposes a general method for guaranteeing that every element $h_i \in \mathbf{B}$ has a Δ-complement $h_k \in \mathbf{B}$ at the distance $\Delta_{ik} = 1$:

$$\Delta''_{ij} = \Delta_{ij} / \max_{m \in I} \Delta_{im}.$$

For example, the normalized distances from $x \in [0, 1]$ to the point 0.25 are

$$\frac{|x - 0.25|}{0.75},$$

which takes its maximum 1, when $x = 1$. However, Levi's proposal does guarantee that Δ'' is Δ-complemented (the normalized distances from 0 and 1 to 0.5 are both equal to 1) or that Δ'' is structurally symmetric (av$(i, \mathbf{B})$ may depend on i).

[6] Cf. the discussion on point and interval estimation in Chapter 12.5.

[7] $\Delta_{sum}(h_i, g)$ always takes the extreme values 0 and 1. To guarantee that $\Delta_{min}(h_i, g)$ also takes the value 1, we could use the normalization described in Note 5.

[8] This term is taken from Rawls (1971).

[9] See Chapter 12.4 for a more detailed discussion.

[10] Tuomela (1978) has studied distances between theories, and applied them to the definition of truthlikeness via (106). His approach relies on the structure of first-order constituents (see Chapter 9.1 for some comments), so that it cannot be generalized to our abstract framework in this chapter.

[11] Przelecki (1976) suggests that a statement g is "closer to the truth" than another statement g' if g is true at least in the same structures $\Omega \in \Theta^*_L$ as g'. This condition follows from our definition (134), if g and g' agree outside the class Θ^*_L (e.g., g and g' are false in all $\Omega \notin \Theta^*_L$).

[12] For a similar criticism of the use of the Hamming distance by Reisinger (1981) in the theory of analogical reasoning in jurisprudence, see Niiniluoto (1983b).

[13] It might happen that there are b-worlds arbitrarily close to Ω^*_L, so that there are no minimally close b-worlds in Θ^b_*. (This was pointed out to me by Graham Oddie.) In that case, we include in the target those statements $h_i \in \mathbf{B}_b$ which are true in every neighbourhood of Ω^*_L.

CHAPTER 7

[1] This argument is presented, e.g., in Ayer (1974). See also Fine (1984).

[2] Quoted in Ewing (1934), p. 242.

[3] Some scientific realists believe in a metaphysical law of progress: science *must*, by some kind of natural necessity, approach to the truth. For a critical evaluation of such views, see Niiniluoto (1984c), Chapters 5, 7.

[4] It is not a trivial matter to make this method explicit, but it can be done by using the tools of pattern recognition, described in Chapter 1.3. See also Rosenfeld and Pfaltz (1968).

[5] Laudan's criticism of my solution seems to presuppose that I should give an infallible method for estimating verisimilitude (see Laudan, 1981, p. 31; 1984, p. 119). Leplin (1985) suggests similarly that 'the major difficulty' of my account of scientific progress is that "Niiniluoto's comparative measures of verisimilitude must be relativized to some theoretical knowledge that is unproblematic in context", so that my "appraisals of theoretical progress are therefore defeasible". For progress defined by my non-epistemic measure Tr, Leplin's claim is false. For the epistemic concept of verisimilitude, it is trivially true that $ver(g|e)$ is defined relative to some (non-theoretical and/or theoretical) evidence e, but it is one of my adequacy conditions — rather than a 'major difficulty' — that the appraisals by ver are fallible and revisable.

[6] Swinburne (1973) suggests that Popper's concept of truthlikeness could be defined by the expected difference between the truth content and the falsity content, but he did not work out this proposal. (It would not work, for reasons stated in Chapter 5.6.) Expected truthlikeness is also discussed by Rosenkrantz (1980). Maher (1984) recommends, too, that Laudan's scepticism can be answered by calculating the expected distance of a proposition from the truth. He does not mention that this suggestion is made already in Niiniluoto (1977b) (which he refers to in his rejection of the linguistic similarity approach to truthlikeness).

[7] It is of course possible to study variants of our definition (6), where P is an interval valued indeterminate probability distribution (cf. Levi, 1980a), or some generalization on single Bayesian conditionalization (cf. Jeffrey, 1965; Williams, 1980; van Fraassen, 1980) is used. I shall not try to study these generalizations of ver in this book.

[8] In his defense of 'causal epistemic realism', Tuomela (1985a, b) defines 'epistemic truth' by means of my function ver: statement g in a first-order language L is epistemically true if and only if $ver(g|e_n) \to 1$ when $n \to \infty$ (Tuomela, 1985b, p. 192). Further, he defines the degree of truthlikeness of g as the difference $ver(g^*|e_n) - ver(g|e_n)$, where g^* is the epistemically true constituent in L. In contrast, my theory maintains a conceptual distinction between $Tr(g, h_*)$ and (the limit of) $ver(g|e_n)$ — or, more generally, between truth and its epistemic surrogates. Cf. Chapter 4.2.

[9] See Niiniluoto (1982b). This reduction of the estimation problem for verisimilitude to the problem of induction does not mean that we have to give up fallibilism (cf. Chapter 5). The rationality of inductive inference may have 'presuppositions', but they are not absolute in a sense that would conflict with fallibilism, but only contextual (cf. Niiniluoto and Tuomela, 1973). It may be that for each cognitive problem B there are some factual assumptions taken for granted in B, but it does not follow that some factual presupposition would be valid in all cognitive problems B.

[10] Koertge (1978), p. 276, suggests that, even if the probability for a universal theory to be true were zero, it might be the case that its probability of having a positive degree of verisimilitude is non-zero. This idea holds for $PTr_{1-\varepsilon}$, and also for $PA_{1-\varepsilon}$.

[11] Of course, it also has to be assumed that the probability measure P, used in calculating the values of ver, is rational, i.e., $P(h|e)$ expresses the *rational* degree of

NOTES 483

belief in L on e. The existence of rational epistemic probabilities is a version of the classical problem of induction.
[12] The early history of this theory is discussed in Todhunter (1865).
[13] See Lindley (1965), p. 3.
[14] Note that, if b is known to be false in the evidential situation e_0, then $P(b|e_0) = 0$ and b cannot be a part of e_0.
[15] The idea of Figure 2 appeared in Niiniluoto (1980c), p. 222.
[16] See Chapter 11.
[17] See, for example, Wilks (1962).

CHAPTER 8

[1] These ideas can be applied to measure the distance between two 'homological' species of animals (see Niiniluoto, 1980b), and between 'analogical' cases in jurisprudence (see Niiniluoto, 1983b).
[2] It is essential that N is here fixed. Cf. the criticism of the use of the average function in Chapter 6.6.
[3] Agassi (1981) makes an attempt 'to save verisimilitude' by suggesting that we confine Popper's qualitative definition (5.21) to 'empirical content'. There are several problems with this proposal. First, the non-epistemic concept of truthlikeness cannot be defined in terms of *empirical* content: two different theories may have the same 'potential falsifiers', but still one of them may be closer to the truth than the other, since their theoretical parts may differ in truthlikeness. Secondly, if the language L in Popper's definition (5.21) is restricted to empirical statements, then the Miller—Tichý trivialization theorem would still be valid. Thirdly, if the class T and its complement F are taken to consist only of simple singular empirical statements and their negations ('basic statements' in Popper's sense), then T would in fact correspond to the true state description s_* in an empirical language L_N^1 with one dichotomous family. Here Agassi's definition would work, and could be generalized to theoretical languages of the form L_N^k as well, but it would coincide with our sufficient condition (19). Thus, the valid core of Agassi's proposal is a (relatively trivial) special case of the definition proposed already in Niiniluoto (1977b, 1978b).

CHAPTER 9

[1] Rosenkrantz (1980) evaluates d_1 — finding it to be intuitively more appealing than Tichý's distance measure.
[2] Cf. the terminology used in Niiniluoto and Tuomela (1973).
[3] For an alternative treatment of the distance between generalizations in L_N^k, see Tuomela (1978).
[4] See Niiniluoto and Tuomela (1973), Niiniluoto (1973).
[5] Perhaps my first intuitive idea about verisimilitude in 1973 was the vague expectation that Hintikka's measure corr might have something to do with it, since corr$(g|e)$ imposes some limits to the worst possibility allowed by g. A nice thing about the definition of ver for expected verisimilitude is that it becomes possible to *prove*

theorems like (34), which show precisely how corr and ver are related in Hintikka's system of induction. Leplin (1985) misunderstands the rules of such a game of explication when he claims that my argument is "not self-contained", since "it *assumes that empirical measures of corroboration covary with a measure of estimated verisimilitude*" (my italics).
[6] See Niiniluoto (1976b).
[7] See Niiniluoto and Tuomela (1973).

CHAPTER 10

[1] Definition 11 could be modified by giving different weights to the A-part and B-part of $Ct^{(m)}$-predicates. In the present form, (11) takes every node of a constituent to be equally important. (The same is true of Tichý's (7).) However, one could adopt a system where the weight $w(n)$ of agreement at depth n decreases with n. If $w(n)$ decreases sufficiently rapidly, our recursive method of defining the distance $d(C_i^{(n)}, C_j^{(n)})$ would lead to an ordering which agrees with the topological structure of Figure 2.9. Let us say that constituents $C_i^{(n)}$ and $C_j^{(n)}$ *agree up to depth* $d = a(i, j)$, if they are subordinate to the same depth-d constituent ($d < n$). Thus, $C_i^{(n)}$ and $C_j^{(n)}$ both entail $C_k^{(a(i,j))}$. Then in Figure 2.9 $C_i^{(n)}$ agrees *more* with $C_*^{(n)}$ than $C_j^{(n)}$, if $a(i, *) > a(j, *)$, i.e., $C_i^{(n)}$ agrees with $C_k^{(n)}$ up to a higher depth than $C_j^{(n)}$. Our definition (11–14) of distance does not coincide with this agreement relation: it is possible to find examples, where $C_1^{(2)}$ agrees with $C_*^{(2)}$ up to depth 1 but makes very radical mistakes at depth 2, while $C_2^{(2)}$ makes small errors at both depths 1 and 2, so that $d_j^{(2)}(C_*^{(2)}, C_1^{(2)}) > d_j^{(2)}(C_*^{(2)}, C_2^{(2)})$.
[2] For a similar argument, see Chapter 13.4.
[3] Oddie (1979) makes the same point about his measure.
[4] Note that there is no general quarantee for all L that the limit (18) is finite.
[5] This view about possible worlds corresponds to 'moderate modal realism' (cf. Lewis, 1986).
[6] This means that it is by no means necessary in all cases to rely on the limit definition (18) in order to obtain a measure for the distance between structures. In Chapter 11.4, we shall also see how the distance between quantitative structures (with real-valued functions) can be measured.
[7] $CV(\Sigma, \mathcal{T})$ is thus a realist variant of Laudan's concept of problem-solving effectiveness for a theory, which he tries to define without using the concept of truth. Cf. Laudan (1977), Niiniluoto (1984c), and Sintonen (1984). Note that if $\mathcal{T}$ contains only one intended application Ω, then $CV(\mathcal{T}, \{\Omega\})$ is a linear function of the truthlikeness of theory relative to structure Ω. In this sense, the problem of evaluating the problem-solving ability of theories contains as a special case the problem of truthlikeness (cf. Niiniluoto, 1984b, Ch. 4).
[8] This suggestion presupposes a charitable attidute from the supporters of the two theories: if Σ_1 has no relevance to a problem Ω in $\mathcal{T}_2 - \mathcal{T}_1$, then Σ_1 gets no credit or discredit for Ω. If Σ_2 gets credit for Ω, but the supporters of Σ_1 regard Ω as a 'pseudoproblem', this may seem unfair. Another approach for comparing Σ_1 and Σ_2, suggested by Pearce (1986b), would be to try to establish a bijective translation τ between the problems $\mathcal{T}_1$ and $\mathcal{T}_2$. Then we could apply our measure CV and say that

Σ_1 is cognitively superior to Σ_2 if and only if

$$\sum_{\Omega \in \mathcal{T}_1} \zeta_\Omega \Delta(\Omega, \mathrm{Mod}(\Sigma_1)) < \sum_{\Omega \in \mathcal{T}_1} \zeta_{\tau(\Omega)} \Delta(\tau(\Omega), \mathrm{Mod}(\Sigma_2)).$$

CHAPTER 11

[1] Oddie discussed these issues first in his unpublished Ph.D. Thesis *The Comparability of Theories by Verisimilitude and Content* (LSE, London, 1979). He develops his theory of higher-order constituents and legisimilitude in more detail in his book *Likeness to Truth* (Reidel, forthcoming 1986). This book is still unknown to me, and I shall not discuss here the higher-order approach at all. Even if I find the modal treatment of laws preferable to the use of higher-order logic, there may of course be other independent reasons to extend the theory of truthlikeness to languages richer than first-order logic.

[2] Armstrong (1980a, b), who also treats laws as second-order relations between universals, accepts conjunctive properties, but rejects negative and disjunctive ones.

[3] See also Cohen (1973, 1977). Cohen's concept of inductive support behaves in some respects in a similar way as my notion of estimated truthlikeness: if evidence e refutes a hypothesis h, its degree of support in Cohen's sense will not become zero (as the posterior probability of h on e would). This has lead some writers to guess that Cohen's explicatum would be verisimilitude rather than probability (cf. Swinburne, 1973, p. 213). Cohen's framework with a hierarchy of 'variables' is also similar to L. Nowak's (1980) treatment of idealization and concretization (cf. Section 4).

[4] Krajewski (1977, 1978) and Patryas (1977) use the Tchebycheff metric $\Delta_3(g_*, g)$ to define "the degree of inadequacy" of a law g (relative to truth g_*).

[5] These results also give a reply to Laudan's (1981, 1984) challenge to the realists: they show to what extent approximate truth is preserved in the 'downward' inference from a law to prediction. The condition that truthlikeness should be preserved in deduction (cf. Laudan, 1981; Newton–Smith, 1982) is inadequate, since, by principle M5 of Chapter 6.6, truthlikeness does not covary with logical strength among false statements. In fact, even true and highly truthlike statements (e.g., h_* itself) may have true but much less truthlike deductive consequences (e.g., h_* entails $h_* \vee h_j$, where $\Delta(h_*, h_j)$ is high), since deduction does not preserve the information factor Δ_{sum} of Tr. On the other hand, *deduction preserves degrees of truth* in the following sense: if $h \vdash g$, then $\Delta_{\min}(h_*, h) \geq \Delta_{\min}(h_*, g)$ (cf. Niiniluoto, 1982b). More complicated situations, where T is a theory in a language L with theoretical terms and g is its observational consequence, are discussed in Niiniluoto (1984b), Ch. 7. Even there the above result is valid: if h_* is the target in the language L of theory T, and g is a deductive consequence of T in the observational sublanguage L_0 of L, then

$$\Delta_{\min}(h_*, T) \geq \Delta_{\min}(h_*, g).$$

In this sense it is quaranteed that the observational consequences g of a theory T must be close to the truth, if T is close to the truth. This holds also for the case, where T contains quantitative theoretical terms (cf. also the discussion of idealization below).

[6] Just as in Chapter 10.3, this shows that the distance between two structures with real-valued functions can be obtained through a distance function defined between the linguistic descriptions of these structures.

[7] The use of uniformities in the theory of approximation for scientific theories has been advocated by Ludwig (1978), Moulines (1976, 1980), and Mayr (1980, 1981a, b, 1984). See also the comparison to Moulines in Niiniluoto (1986a).

[8] The structuralists like Stegmüller (1976, 1986) usually wish to avoid talking about truthlikeness and approximate truth, but they could apply the approach of uniformities to the explication of these concepts as soon as they would be willing to single out one structure Ω_* (or a class of structures) as representing the actual world.

[9] Similar remarks apply to examples used by Mayr (1981a, b, 1984).

[10] The following analysis of idealizational laws is based on Niiniluoto (1986a). Note, however, that here I formulate laws T, T_0, and T_1 with the physical necessity operator $\Box$, not with the counterfactual $\Box\!\!\rightarrow$ if-then-connective. For syntactical treatment of idealization and concretization, see Krajewski (1977), Nowak (1972, 1980).

[11] Note also that in some cases T, as a more informative statement than T_0, may be closer to T_1 than T_0. This may still be true if T_0 is replaced by its approximate version AT_0 (cf. Nowak, 1980):

(AT_0) $\Box(x)((x \,\&\, w(x) \leq a \supset h_k(x) \approx g_0(h_1(x), \ldots, h_{k-1}(x)))$.

Here AT_0 is less restricted but also less precise than T_0.

[12] See also Pearce and Rantala (1983, 1984).

[13] Mayr uses here a structuralist concept of strict reduction between theories. See Sneed (1971), Mayr (1980), Stegmüller (1979, 1986), and Pearce (1986b).

[14] Cf. the discussion, and results for qualitative cases, in Niiniluoto and Tuomela (1973), Niiniluoto (1976b). Obviously a great variety of results are possible. First, some theories may postulate a non-existent ontology (e.g., aether, flogiston), but still have empirical success to some extent (cf. Niiniluoto, 1984b, Ch. 7). Secondly, some theories may postulate idealized entities (e.g., ideal gas) which closely resemble real entities. Thirdly, some theories may give a literally true description of non-observable reality. Against semantical anti-realism, in all of these cases it is meaningful to speak of the truth value of a theory T — and of its degree of truthlikeness $Tr(T, h_*)$. Against methodological instrumentalism (cf. Laudan, 1977; van Fraassen, 1980), it is cognitively important to appraise theories with respect to their truthlikeness (cf. Niiniluoto, 1986a) — perhaps even to try to establish examples where $ver(T/e) > ver(T_0/e)$, where T is a genuine theory and T^0 an observational theory. Furthermore, against the position of Cartwright (1983), who accepts theoretical entities but not theoretical laws, the falsity of 'fundamental' theoretical laws does not force us to adopt an instrumentalist attitude towards them: even if theoretical explanations in science use laws that are "deemed false even in the context of use", these may nevertheless have a relatively high degree of (real and estimated) truthlikeness.

[15] For an interesting account of the testing of scientific theories, see also Laymon (1980, 1982). He suggests that a theory is confirmed if it can be shown that using more realistic initial conditions will lead to correspondingly more accurate predictions. In our terms, this means that an idealized theory T' is confirmed if we find a concretization T'' of T' which has greater approximate validity than T'.

CHAPTER 12

[1] See Savage (1954), Chernoff and Moses (1959), Ferguson (1967), Jeffrey (1965).
[2] This account Bolzano was first presented in Niiniluoto (1977c).
[3] See Carnap (1971a), Jeffrey (1965), Stegmüller (1973). For criticism, see Levi (1967a), Hilpinen (1968), Niiniluoto and Tuomela (1973).
[4] See the discussions by Rudner (1953), Jeffrey (1956), Hempel (1960, 1962), Levi (1960, 1962, 1963, 1967a).
[5] The solution of a decision problem does not change, if the utility function u is replaced by its positive linear transformation $u' = \alpha u + \beta$, $\alpha > 0$. Thus, (7) is equivalent to the choice $a = 1$ and $b = 0$, where the value of u for g equals the truth value of g.
[6] In Hintikka's system of inductive logic, among the constituents for a monadic language, the minimally wide of the constituents unrefuted by evidence e has both the highest absolute content and the highest posterior probability on e (see Chapter 2.1). Thus, this constituent maximizes (12) on e.
[7] Cf. Hintikka's (1968b) distinction between 'local' and 'global' theorizing.
[8] See Levi (1967b, 1976, 1980a, 1986a). For discussion on Levi, see the papers in Bogdan (1976, 1982).
[9] For comments on Levi's 'myopic realism' (Levi, 1986a), see Niiniluoto (1984b), Ch. 5.
[10] See also Backman (1983) and Maher (1984).
[11] See also Miller (1980) and Salmon (1981).
[12] Cf. Niiniluoto (1982b), p. 193.
[13] This section is largely based on Niiniluoto (1986c). The main results were first presented in the autumn of 1982 at the Department of Statistics, University of Helsinki, and the Department of Philosophy, Columbia University, New York. Since 1982, independent work in this area has been done by Festa (1984, 1986). In particular, the observation about the connection between Levi's and Ferguson's utility functions (see (46)) is due to Festa.
[14] I assume here that the distribution $g(\theta)$ is continuous on Θ. As an alternative, Shimony (1970) allows that a singular point in Θ may receive non-zero prior probability. Cf. Niiniluoto (1983d).
[15] Cf. Hacking's (1965) 'belief-estimates' which aim at truth regardless of the practical consequences.
[16] Levi (1986b) correctly notes that the treatment of Niiniluoto (1986c) is not directly applicable to such a generalization.
[17] Note that, more generally, non-trivial interval estimates for L_{10} mentioned in Table IV could be unions of distinct intervals if L_{10} is extended in the way suggested above. This will be the case when the posterior distribution is not unimodal.

CHAPTER 13

[1] The theory of semantic information, developed by Carnap and Bar-Hillel (cf. Chapter 4.5), was already a step in this direction. See also Levi (1963, 1967a), Hintikka

(1968a, b), Niiniluoto and Tuomela (1973), Niiniluoto (1973), Jeffrey (1975), Good (1975), Grünbaum (1976a, 1978).

[2] Cf. Hübner (1978b), p. 396.

[3] See Hübner (1978a), p. 286.

[4] See Stegmüller (1978), p. 59.

[5] "One should never try to be more precise than the problem situation demands." "The quest for precision is analogous to the quest for certainty, and both should be abandoned." (Popper, 1974a, p. 17.) I find it strange that in philosophy Popper recommends a timid method. Just as a good scientific theory should be bold and solve problems other than those it was originally designed to solve, a good philosophical theory should do the same. This is often possible by being "more precise than the problem situation demands".

[6] See Andersson (1978), p. 308.

[7] See Watkins (1978, 1984).

[8] See condition M5 in Chapter 6. Content comparison is a good method of appraisal, if we have reason to believe that we are dealing with *true* theories — not generally in the case of false theories.

[9] For evaluations of these positions, see Niiniluoto (1984b, 1986a), Leplin (1984).

[10] We generally assume here that languages L and L' have the same individual constants, but may differ in their predicates.

[11] A similar result holds for 'piecewise definable' new concepts (Niiniluoto and Tuomela, 1973).

[12] Cf. also the comments on Mott's (1978) example in Niiniluoto (1978b).

[13] It should be noted that the Miller-type arguments against truthlikeness are equally forceful against the uniformity approach to approximation. The problem of translation remains urgent for this approach as well, since homeomorphisms between $\mathbf{M}_*^L$ and $\mathbf{M}_*^{L'}$ need not preserve the concept of closeness defined by uniformities. The best general discussion on translation is given by Pearce (1986b).

[14] This is the account of scientific progress developed in Niiniluoto (1979b), (1980a), (1984b). It should be noted that, while this analysis takes truthlikeness be the primary epistemic goal of science, a more extensive treatment could also consider other relevant but secondary utilities (cf. Niiniluoto, 1984b, Ch. 7; Sintonen, 1984). Wachbroit (1986) challenges my view that a reasonable philosopher, who accepts a realist interpretation of theoretical terms, should also accept a realist theory of progress in my sense. A recent formulation to the position that there are no differences in the progress of science and art is given by Feyerabend (1984b).

[15] See Kuhn (1962, 1983), Feyerabend (1975).

[16] In contrast it is argued in Niiniluoto (1976a), against Levi (1967a), that a step from a problem to its subproblems should not be too relative to the choice of the 'ultimate partition': in particular, if g is acceptable in $\mathbf{B}$ and $\mathbf{B}_0 \leq \mathbf{B}$, then $\sim g$ should not be acceptable in $\mathbf{B}_0$ (cf. Hilpinen, 1968).

[17] Recall what we said in Chapter 4.3 about the impossibility of finding an ideal language L_{id}, applicable to each cognitive problem B. In Niiniluoto (1979b), I concluded that the theory scientific progress needs, in addition to the concept of truthlikeness, an account of the dynamic choice of conceptual frameworks. Whewell's Aphorism VIII concerning the language of science is a good starting point for such an

account: "Terms must be constructed and appropriated so as to be fitted to enunciate simply and clearly true general propositions". This criterion for the choice of language is "objective" in an obvious sense, but it is not *a priori*, since the ability of a language to express interesting and informative truths depends upon what the world is like.

[18] For model-theoretic treatments of the semantics of scientific terms, see Przelecki (1969), Tuomela (1973), Pearce (1984).

[19] Such cases of languages which have no common (meaning-preserving) extension were discussed in the 'radical conventionalism' of Ajdukiewiez (see Giedymin, 1978). Cf. also Giedymin (1971).

[20] We could thus study theory sequences $T_1, T_2, \ldots$ which satisfy the condition that, for each $i = 1, 2, \ldots$, theory T_i is translatable by a translation τ_i to the next theory T_{i+1}. Then our concept of progressive theory sequence could be extended to such situations. Boos (1983) has studied a special case of such sequences, where each T_i is 'interpretable' in T_{i+1}.

[21] In fact, Popper made already in 1959 the following remark on similarity: "These diagrams show that things may be similar in *different respects*, and that any two things which are from one point of view similar may be dissimilar from another point of view. Generally, similarity, and with it repetition, always presuppose the adoption of *a point of view*: some similarities or repetitions will strike us if we are interested in one problem, and others if we are interested in other problem." (Popper, 1959, p. 422.) Popper seems to have forgotten this insight in his early work on verisimilitude.

[22] Maher (1984) argues that the similarity approach to truthlikeness is 'fundamentally flawed', since the distance measures proposed by Tichý and Niiniluoto are, in Miller's sense, relative to language. Such a language-relative conception, he thinks, "will not do the job for which verisimilitude was introduced in the first place", i.e., to allow us to say that science is making progress. He proceeds to an account where "the distance of a theory from the truth is not an objective but a *subjective* property of the theory" — this is done by first constructing a utility function by means of general representation theorems for utility (cf. Savage, 1954) and then defining distance from truth as a function of utilities. (This reverses the idea of our Chapter 12.) What I find puzzling here is the assumption made by Maher that it is better for the analysis of scientific progress to relativize truthlikeness to subjective persons rather than to such an intersubjective and even objectively appraisable factor as language.

[23] This is the position supported, at least as the first task for a theory of truthlikeness, by Kuipers.

[24] See Fine (1973).

[25] Cf. the discussion of the pragmatics of explanation in Sintonen (1984).

BIBLIOGRAPHY

Agassi, J., 'Verisimilitude: Comment on David Miller', *Synthese* **30** (1975), 199—204.
Agassi, J., 'To Save Verisimilitude', *Mind* **90** (1981), 576—579.
Aitchison, J. and Dunsmore, I. R., 'Linear-Loss Interval Estimation of Location and Scale Parameters', *Biometrika* **55** (1968), 141—148.
Almeder, R., 'Scientific Progress and Peircean Utopian Realism', *Erkenntnis* **20** (1983), 253—280.
Andersson, G., 'The Problem of Verisimilitude', in Radnitzky and Andersson (1978), pp. 291—310.
Armstrong, D. M., *Nominalism and Realism*, Cambridge University Press, Cambridge, 1978a.
Armstrong, D. M., *A Theory of Universals*, Cambridge University Press, Cambridge, 1978b.
Augustine, St., *Against the Academics*, The Newman Press, Westminster, 1950.
Ayer, A. J., 'Truth, Verification and Verisimilitude', in Schilpp (1974), pp. 684—692.
Backman, W., 'Practical and Scientific Rationality: A Difficulty for Levi's Epistemology', *Synthese* **57** (1983), 269—276.
Bacon, F., *The New Organon*, Bobbs-Merrill, Indianapolis, 1960.
Balzer, W., 'Sneed's Theory Concept and Vagueness', in Hartkämper and Schmidt (1981), pp. 147—163.
Balzer, W., 'Theory and Measurement', *Erkenntnis* **19** (1983), 3—25.
Balzer, W., Pearce, D. A., and Schmidt, H.-J. (eds.), *Reduction in Science: Structure, Examples, Philosophical Problems*, D. Reidel, Dordrecht, 1984.
Bar-Hillel, Y., 'Semantic Information and Its Measures', *Transactions of the Tenth Conference on Cybernetics*, Josiah Macy Jr. Foundation, New York, 1952, pp. 33—48. (Reprinted in Bar-Hillel, 1964, Ch. 17.)
Bar-Hillel, Y., *Language and Information*, Addison-Wesley and the Jerusalem Academic Press, Reading, Mass. and Jerusalem, 1964.
Bar-Hillel, Y. and Carnap, R., 'Semantic Information', *The British Journal for the Philosophy of Science* **4** (1953), 147—157.
Battacharya, A., 'On a Measure of Divergence Between Two Multinomial Populations', *Sankhya* **7** (1946), 401.
Beals, R., Krantz, D. H., and Tversky, A., 'Foundations of Multidimensional Scaling', *Psychological Review* **75** (1968), 127—142.
Bell, J. L. and Slomson, A., *Models and Ultraproducts*, North-Holland, Amsterdam, 1969.
Benacerraf, P. and Putnam, H. (eds.), *Philosophy of Mathematics: Selected Readings*, Blackwell, Oxford, 1964.
Bigelow, J., 'Possible Worlds Foundations for Probability', *Journal of Philosophical Logic* **5** (1976), 299—320.

BIBLIOGRAPHY

Bigelow, J., 'Semantics of Probability', *Synthese* **36** (1977), 459—472.
Blackwell, D. and Girshick, M. A., *Theory of Games and Statistical Decisions*, John Wiley and Sons, New York, 1954.
Bochenski, I. M., *A History of Formal Logic*, University of Notre Dame Press, Notre Dame, Indiana, 1961. (Chelsea Publ. Co., New York, 1971.)
Bogdan, R. J. (ed.), *Local Induction*, D. Reidel, Dordrecht, 1976.
Bogdan, R. J. (ed.), *Henry E. Kyburg, Jr. & Isaac Levi*, Profiles 3, D. Reidel, Dordrecht, 1982.
Bolzano, B., *Theory of Science* (ed. by R. George), Blackwell, Oxford, 1972.
Bolzano, B., *Wissenschaftslehre* III, Felix Meiner, Leipzig, 1930.
Boorman, S. D. and Olivier, D. C., 'Metrics on Spaces of Finite Trees', *Journal of Mathematical Psychology* **10** (1973), 26—59.
Boos, W., 'Limits of Inquiry', *Erkenntnis* **20** (1983), 157—194.
Bourbaki, N., *Elements of Mathematics: General Topology*, Part 1, Addison-Wesley, Reading, Mass., 1966.
Box, G. and Tiao, G., *Bayesian Inference in Statistical Analysis*, Addison-Wesley, Reading, Mass., 1973.
Bradley, F. H., *Appearance and Reality*, George Allen, London, 1893. (Ninth ed., 1930.)
Bugajski, S., 'Languages of Similarity', *Journal of Philosophical Logic* **12** (1983), 1—18.
Bunge, M., *The Myth of Simplicity*, Prentice-Hall, Englewood Cliffs, N.J., 1963.
Bunge, M., *Treatise on Basic Philosophy vol. 2, Semantics II: Interpretation and Truth*, D. Reidel, Dordrecht, 1974.
Bunge, M., *Treatise on Basic Philosophy, vol. 3: Ontology I, The Furniture of the World*, D. Reidel, Dordrecht, 1977.
Byrne, E., *Probability and Opinion: A Study of Medieval Presuppositions of Post-Medieval Theories of Probability*, M. Nijhoff, The Hague, 1968.
Cain, A. J. and Harrison, G. A., 'An Analysis of the Taxonomist's Judgment of Affinity', *Proc. Zool. Soc. Lond.* **131** (1958), 85—98.
Carnap, R., *Der logische Aufbau der Welt*, Weltkreis-Verlag, Berlin, 1928. (Transl. in *The Logical Structure of the World and Pseudoproblems in Philosophy*, The University of California Press, Berkeley and Los Angeles, 1969.)
Carnap, R., *The Logical Foundations of Probability*, The University of Chicago Press, Chicago, 1950. (2nd enlarged ed. 1962.)
Carnap, R., *The Continuum of Inductive Methods*, The University of Chicago Press, Chicago, 1952.
Carnap, R., 'The Aim of Inductive Logic', in Nagel *et al.* (1962), pp. 303—318.
Carnap, R., 'Probability and Content Measure', in P. K. Feyerabend and G. Maxwell (eds.), *Mind, Matter and Method*, University of Minnesota Press, Minneapolis, 1966, pp. 248—260.
Carnap, R., 'The Concept of Constituent-Structure', in Lakatos (1968), pp. 218—220.
Carnap, R., 'Inductive Logic and Rational Decisions', in Carnap and Jeffrey (1971a), pp. 5—31.
Carnap, R., 'A Basic System of Inductive Logic, Part I', in Carnap and Jeffrey (1971b), pp. 33—165.
Carnap, R., 'A Basic System of Inductive Logic, Part II, in Jeffrey (1980), pp. 7—155.

Carnap, R. and Bar-Hillel, Y., 'An Outline of a Theory of Semantic Information', *Technical Report No. 247*, MIT Research Laboratory in Electronics, 1952. (Reprinted in Bar-Hillel, 1964.)

Carnap, R. and Jeffrey, R. C. (eds.), *Studies in Inductive Logic and Probability*, vol. I, University of California Press, Berkeley, 1971.

Cartwright, N., 'The Truth Doesn't Explain Much', *American Philosophical Quarterly* **17** (1980), 159—163.

Cartwright, N., *How the Laws of Physics Lie?*, Oxford University Press, Oxford, 1983.

Chellas, B., *Modal Logic: An Introduction*, Cambridge University Press, Cambridge, 1980.

Chen, C. H., *Statistical Pattern Recognition*, Hayden, Rochelle Park, 1973.

Chernoff, H. and Moses, L. E., *Elementary Decision Theory*, John Wiley and Sons, New York, 1959.

Chuaqui, R. and Bertossi, L., 'Approximation to Truth and the Theory of Errors', in C. A. di Prisco (ed.), *Methods in Mathematical Logic*, Lecture Notes in Mathematics, vol. 1130, Springer-Verlag, Berlin, 1985, pp. 13—31.

Cohen, L. J., 'The Paradox of Anomaly', in R. J. Bogdan and I. Niiniluoto (eds.), *Logic, Language, and Probability*, D. Reidel, Dordrecht and Boston, 1973, pp. 78—82.

Cohen, L. J., *The Probable and the Provable*, Oxford University Press, Oxford, 1977.

Cohen, L. J., 'What has Science to Do with Truth?', *Synthese* **45** (1980), 489—510.

Cohen, L. J. and Hesse, M. (eds.), *Applications of Inductive Logic*, Oxford University Press, Oxford, 1980.

Cox, D. R. and Hinkley, D. V., *Theoretical Statistics*, Chapman and Hull, London, 1974.

Cramér, H., *Mathematical Methods of Statistics*, Princeton University Press, Princeton, 1946.

Cramér, H. and Leadbetter, M. R., *Stationary and Related Stochastic Processes*, J. Wiley, New York, 1967.

Cusanus, N. (Nicholas of Cusa), *Of Learned Ignorance*, Routledge and Kegan Paul, London, 1954.

Czekanowski, J., 'Der Differentialdiagnose der Neandertalgruppe', *Korrespondenzblatt Deutsch. Ges. Anthropol. Ethnol. Urgesch.* **40** (1909), 44—47.

Davidson, D., *Inquiries into Truth and Interpretation*, Oxford University Press, Oxford, 1984.

Dinosaurs and their Living Relatives, The British Museum (Natural History), London, 1979.

Dubois, D. and Prade, H., *Fuzzy Sets and Systems: Theory and Applications*, Academic Press, New York, 1980.

Duhem, P., *The Aim and Structure of Physical Theory*, Princeton University Press, Princeton, 1954.

Dummett, M., *Truth and Other Enigmas*, Duckworth, London, 1978.

Eberle, R., 'A Construction of Quality Classes, Improved upon the *Aufbau*', in Hintikka (1975b), pp. 55—74.

Engels, F., *Ludwig Feuerbach and the End of Classical German Philosophy*, Progress Publishers, Moscow, 1946.

Engels, L., *Dialectics of Nature*, Progress Publishers Moscow, 1934. (7th printing 1976.)

Ewing, A. C., *Idealism: A Critical Survey*, Methuen, Strand, 1934. (3rd ed. 1961.)
Ferguson, T. S., *Mathematical Statistics: A Decision-Theoretic Approach*, Academic Press, New York, 1967.
Festa, R., 'Problemi logici ed epistemologici nella nozione di verisimilitudine', *Lingua e stile* **17** (1982a), 233—268.
Festa, R., manuscript, 1982b.
Festa, R., 'Epistemic Utilities, Verisimilitude, and Inductive Acceptance of Interval Hypotheses', *7th International Congress of Logic, Methodology and Philosophy of Science, Abstracts of Sections 1, 2, 3, 4 and 7*, J. Huttegger Ohg, Salzburg, 1983, pp. 212—215.
Festa, R., 'A Measure for the Distance Between an Interval Hypothesis and the Truth', *Synthese* **67** (1986), 273—320.
Fetzer, J., *Scientific Knowledge: Causation, Explanation, and Corroboration*, D. Reidel, Dordrecht, 1981.
Feyerabend, P., 'Dialogue on Method', in Radnitzky and Andersson (1979), pp. 253—278.
Feyerabend, P., 'Xenophanes: A Forerunner of Critical Rationalism?', in G. Andersson (ed.), *Rationality in Science and Politics*, D. Reidel, Dordrecht, 1984a, pp. 95—109.
Feyerabend, P., *Wissenschaft als Kunst*, Suhrkamp, Frankfurt am Main, 1984b.
Field, H., 'Tarski's Theory of Truth', *The Journal of Philosophy* **69** (1972), 347—375.
Findlay, J. N., *The Philosophy of Hegel: An Introduction and Re-Examination*, Collier Books, New York, 2nd printing, 1966.
Fine, A., 'The Natural Ontological Attitude', in Leplin (1984), pp. 83—107.
Fine, K., 'Vagueness, Truth, and Logic', *Synthese* **30** (1975), 265—300.
Fine, T., *Theories of Probability*, Academic Press, New York, 1973.
van Fraassen, B., 'On the Extension of Beth's Semantics of Physical Theories', *Philosophy of Science* **37** (1970), pp. 325—339.
van Fraassen, B., 'A Formal Approach to the Philosophy of Science', in R. Colodny (ed.), *Paradigms and Paradoxes: The Philosophical Challenge of the Quantum Domain*, University of Pittsburgh Press, Pittsburgh, 1972, pp. 303—366.
van Fraassen, B., *The Scientific Image*, Clarendon Press, Oxford, 1980.
van Fraassen, B., 'Rational Belief and Probability Kinematics', *Philosophy of Science* **47** (1980), 165—187.
Freeman, J. B. and Daniels, C. B., 'Maximal Propositions and the Coherence Theory of Truth', *Dialogue* **17** (1978), 56—71.
Fu, K. S. (ed.), *Syntactic Pattern Recognition, Applications*, Springer-Verlag, Berlin, 1977.
Fu, K. S., *Syntactic Pattern Recognition and Applications*, Prentice-Hall, Englewood Cliffs, 1982.
Garber, D., 'Old Evidence and Logical Omniscience in Bayesian Confirmation Theory', in J. Earman (ed.), *Testing Scientific Theories* (Minnesota Studies in the Philosophy of Science X), University of Minnesota Press, Minneapolis, 1983, pp. 99—131.
Giedymin, J., 'The Paradox of Meaning Variance', *The British Journal for the Philosophy of Science* **22** (1971), 30—48.
Giedymin, J. (ed.), *Kazimier Ajdukiewicz: The Scientific World-Perspective and Other Essays, 1931—1963*, D. Reidel, Dordrecht, 1978.

Gnedenko, B. V. and Kolmogorov, A. N., *Limit Distributions for Sums of Independent Random Variables*, Addison-Wesley, Cambridge, Mass., 1954.
Good, I. J., 'Explicativity, Corroboration, and the Relative Odds of Hypotheses', *Synthese* **30** (1975), 39—73.
Goodman, N., *The Structure of Appearance*, Harvard University Press, Cambridge, Mass., 1951. (3rd ed., D. Reidel, Dordrecht, 1977.)
Goodman, N., *Fact, Fiction, and Forecast*, Harvard University Press, Cambridge, Mass., 1955. (2nd ed., Bobbs-Merrill, Indianapolis, 1965.)
Goodman, N., *Problems and Projects*, The Bobbs-Merrill Company, Indianapolis, 1972.
Gower, J. C., 'A General Coefficient of Similarity and Some of Its Properties', *Biometrics* **27** (1971), 857—871.
Grünbaum, A., 'Is the Method of Bold Conjectures and Attempted Refutations Justifiably the Method of Science?', *The British Journal for the Philosophy of Science* **27** (1976a), 105—136.
Grünbaum, A., 'Can a Theory Answer More Questions Than One of Its Rivals?', *British Journal for the Philosophy of Science* **27** (1976b), 1—22.
Grünbaum, A., 'Popper vs Inductivism', in Radnitzky and Andersson (1978), pp. 117—142.
Gudder, S. P., 'Convexity and Mixtures', *SIAM Review* **19** (1977), 221—240.
Gurwitsch, A., 'Galilean Physics in the Light of Husserl's Phenomenology', in E. McMullin (ed.), *Galileo, Man of Science*, Basic Books, New York, 1967, pp. 388—401.
Haack, S., 'Is Truth Flat or Bumpy?', in D. H. Mellor (ed.), *Prospects for Pragmatism: Essays in Memory of F. P. Ramsey*, Cambridge University Press, Cambridge, 1980, pp. 1—20.
Habermas, J., 'Reply to My Critics', in J. B. Thompson and D. Held (eds.), *Habermas: Critical Debates*, The Macmillan Press, London, 1982, pp. 214—283.
Hacking, I., *The Logic of Statistical Inference*, Cambridge University Press, Cambridge, 1965.
Hacking, I., *Representing and Intervening*, Cambridge University Press, Cambridge, 1983.
Hamming, R. W., 'Error Detecting and Error Correcting Codes', *Bell System Tech. Journal* **29** (1950), 147—160.
Haralick, R. M., 'Automatic Remote Sensor Image Processing', in Rosenfeld (1976), pp. 5—66.
Hardin, C. L. and Rosenberg, A., 'In Defense of Convergent Realism', *Philosophy of Science* **49** (1982), 604—615.
Harmon, L. D., Khan, M. K., Lasch, R., and Ramig, P. F., 'Machine Identification of Human Faces', *Pattern Recognition* **13** (1981), 97—110.
Harper, W., Stalnaker, R. and Pearce, G. (eds.), *Ifs: Conditionals, Belief, Decision, Chance, and Time*, D. Reidel, Dordrecht, 1980.
Harris, J. H., 'Popper's Definition of "Verisimilitude"', *The British Journal for the Philosophy of Science* **25** (1974), 160—166.
Harris, J. H., 'On Comparing Theories', *Synthese* **32** (1976), 29—76.
Hartigan, J. A., 'Similarity and Probability', in V. P. Godambe and D. A. Sprott (eds.), *Foundations of Statistical Inference*, Holt, Rinehart and Winston, Toronto, 1971, pp. 305—313.

Hartkämper, A. and Schmidt, H.-J. (eds.), *Structure and Approximation in Physical Theories*, Plenum Press, New York, 1981.
Hattiangadi, J. N., 'After Verisimilitude', in *Contributed Papers, 5th International Congress of Logic, Methodology and Philosophy of Science*, London, Ontario, 1975, pp. V-49—50.
Hattiangadi, J. N., 'To Save Fallibilism', *Mind* **92** (1983), 407—409.
Hautamäki, A., *Points of View and Their Logical Analysis* (Acta Philosophica Fennica 41), Helsinki, 1986.
Hegel, G. W. F., *Hegel's Logic* (transl. by W. Wallace 1873), Oxford University Press, Oxford, 3rd ed., 1975.
Hempel, C. G., 'Inductive Inconsistencies', *Synthese* **12** (1960), 439—469. (Reprinted in Hempel (1965), pp. 53—79.)
Hempel, C. G., 'Deductive-Nomological vs Statistical Explanation', in H. Feigl and G. Maxwell (eds.), *Scientific Explanation, Space, and Time*, University of Minnesota Press, Minneapolis, 1962, pp. 98—169.
Hempel, C. G., *Aspects of Scientific Explanation*, The Free Press, New York, 1965.
Hempel, C. G., 'Turns in the Evolution of the Problem of Induction', *Synthese* **46** (1981), 389—404.
Hesse, M., *Models and Analogies in Science*, University of Notre Dame Press, Notre Dame, Indiana, 1966.
Heywood, V. H., *Plant Taxonomy*, 2nd ed., Edward Arnold Ltd, London, 1976.
Hilpinen, R., 'On Inductive Generalization in Monadic First-Order Logic with Identity', in Hintikka and Suppes (1966), pp. 133—154.
Hilpinen, R., *Rules of Acceptance and Inductive Logic* (Acta Philosophica Fennica 22), North-Holland, Amsterdam, 1968.
Hilpinen, R., 'Relational Hypotheses and Inductive Inference', *Synthese* **23** (1971), 266—286.
Hilpinen, R., 'Carnap's New System of Inductive Logic', *Synthese* **25** (1973), 307—333.
Hilpinen, R., 'Approximate Truth and Truthlikeness', in Przełecki *et al.* (1976), pp. 19—42.
Hintikka, J., 'Towards a Theory of Inductive Generalization', in Y. Bar-Hillel (ed.), *Proceedings of the 1964 International Congress for Logic, Methodology, and Philosophy of Science*, North-Holland, Amsterdam, 1965, pp. 274—288.
Hintikka, J., 'A Two-Dimensional Continuum of Inductive Methods', in Hintikka and Suppes (1966), pp. 113—132.
Hintikka, J., 'The Varieties of Information and Scientific Explanation', in B. van Rootselaar and J. F. Staal (eds.), *Logic, Methodology, and Philosophy of Science III: Proceedings of the 1967 International Congress*, North-Holland, Amsterdam, 1968a, pp. 151—171.
Hintikka, J., 'Induction by Enumeration and Induction by Elimination', in Lakatos (1968b), pp. 191—216.
Hintikka, J., 'On Semantic Information', in Hintikka and Suppes (1970a), pp. 3—27.
Hintikka, J., 'Surface Information and Depth Information', in Hintikka and Suppes (1970b), pp. 263—297.
Hintikka, J., *Logic, Language-Games and Information*, Oxford University Press, Oxford, 1973.
Hintikka, J., *The Intensions of Intensionality*, D. Reidel, Dordrecht, 1975a.
Hintikka, J. (ed.), *Rudolf Carnap, Logical Empiricist*, D. Reidel, Dordrecht, 1975b.

Hintikka, J., *The Semantic of Questions and the Questions of Semantics* (Acta Philosophica Fennica 28:4), North-Holland, Amsterdam, 1976.
Hintikka, J., 'On the Logic of an Interrogative Model of Scientific Inquiry', *Synthese* **47** (1981), 69—81.
Hintikka, J. and Niiniluoto, I., 'On the Surface Semantics of Quantificational Proof Procedures', *Ajatus* **35** (1973), 197—215.
Hintikka, J. and Niiniluoto, I., 'An Axiomatic Foundation of Inductive Generalization', in Przełecki *et al.* (1976), pp. 57—81. (Also in Jeffrey, 1980, pp. 157—181.)
Hintikka, J., Niiniluoto, I. and Saarinen, E. (eds.), *Essays in Mathematical and Philosophical Logic*, D. Reidel, Dordrecht, 1979.
Hintikka, J. and Pietarinen, J., 'Semantic Information and Inductive Logic', in Hintikka and Suppes (1966), pp. 96—112.
Hintikka, J. and Suppes, P. (eds.), *Aspects of Inductive Logic*, North-Holland, Amsterdam, 1966.
Hintikka, J. and Suppes, P. (eds.), *Information and Inference*, D. Reidel, Dordrecht, 1970.
Hull, D. L., 'The Effect of Essentialism on Taxonomy — Two Thousand Years of Stasis', *The British Journal for the Philosophy of Science* **15** (1965), 314—326, and **16** (1965), 1—18.
Hübner, K., 'Some Critical Comments on Current Popperianism on the Basis of a Theory of System Sets', in Radnitzky and Andersson (1978a), pp. 279—289.
Hübner, K., 'Reply to Watkins', in Radnitzky and Andersson (1978b), pp. 393—396.
James, W., *The Will to Believe*, Dover, New York, 1956.
Jayatilleke, K. N., *Facets of Buddhist Thought*, Buddhist Publication Society, Kandy, Ceylon, 1971.
Jeffrey, R., 'Valuation and Acceptance of Scientific Hypotheses', *Philosophy of Science* **23** (1956), 237—246.
Jeffrey, R. C., *The Logic of Decision*, McGraw-Hill, New York, 1965. (2nd ed. University of Chicago Press, 1983.)
Jeffrey, R. C., 'Probability and Falsification: Critique of the Popper Program', *Synthese* **30** (1975), 95—117.
Jeffrey, R. (ed.), *Studies in Inductive Logic and Probability*, vol. II, University of California Press, Berkeley, 1980.
Jeffrey, R., 'De Finetti's Probabilism', *Synthese* **60** (1984), 73—90.
Jeffreys, H., *Theory of Probability*, 2nd ed., Clarendon Press, Oxford, 1948.
Jevons, W. S., *The Principles of Science*, Dover Publications, New York, 1958.
Johnson, W. E., *Logic*, Part I, Cambridge University Press, Cambridge, 1921. (Dover, New York, 1964.)
Jones, K., 'Verisimilitude versus Probable Verisimilitude', *The British Journal for the Philosophy of Science* **24** (1973), 174—176.
Kant, I., *Logic*, Bobbs-Merrill, Indianapolis, 1974.
Karttunen, M., 'Infinitary Languages $N_{\infty\lambda}$ and Generalized Partial Isomorphism', in Hintikka *et al.* (1979), pp. 153—168.
Karttunen, M., *Model Theory for Infinitely Deep Languages*, Annales Academiae Scientiarum Fennicae, Series A, I. Mathematica, Dissertationes 50, Helsinki, 1984.
Kasher, A., 'Verisimilitude is a Surface Concept', *The Southwestern Journal of Philosophy* **3** (1972), 21—27.

BIBLIOGRAPHY

Katz, M., 'The Logic of Approximation in Quantum Theory', *Journal of Philosophical Logic* **11** (1982), 215–228.
Keuth, H., 'Verisimilitude or the Approach to the Whole Truth', *Philosophy of Science* **43** (1976), 311–336.
Keynes, J. M., *A Treatise on Probability*, Macmillan, London, 1921. (New Edition: Harper & Row, New York, 1962.)
Kleiner, S. A., 'Interrogatives, Problems and Scientific Inquiry', *Synthese* **62** (1985), 365–428.
Klug, U., *Juristische Logik*, Dritte Auflage, Springer-Verlag, Berlin, 1966.
Knuuttila, S. (ed.), *Reforging the Great Chain of Being*, D. Reidel, Dordrecht, 1981.
Koertge, N., 'Towards a New Theory of Scientific Inquiry', in Radnitzky and Andersson (1978), pp. 253–278.
Koertge, N., 'The Problem of Appraising Scientific Theories', in P. D. Asquith and H. E. Kyburg Jr. (eds.), *Current Research in Philosophy of Science*, Philosophy of Science Association, East Lansing, 1979, pp. 228–251.
Krajewski, W., *Correspondence Principle and the Growth of Knowledge*, Reidel, Dordrecht, 1977.
Krajewski, W., 'Approximative Truth of Fact-Statements, Laws and Theories', *Synthese* **38** (1978), 275–279.
Krantz, D. H., Luce, R. D., Suppes, P., and Tversky, A., *Foundations of Measurement*, vol. I, Academic Press, New York, 1971.
Krantz, D. H. and Tversky, A., 'Similarity of Rectangles: An Analysis of Subjective Dimensions', *Journal of Mathematical Psychology* **12** (1975), 4–34.
Kuhn, T., *The Structure of Scientific Revolutions*, University of Chicago Press, Chicago, 1962. (2nd ed. 1970.)
Kuhn, T. S., *The Essential Tension*, The University of Chicago Press, Chicago, 1977.
Kuhn, T., 'Commensurability, Comparability, Communicability', in P. D. Asquith and T. Nickles (eds.), *PSA 1982*, vol. 2, Philosophy of Science Association, East Lansing, 1983, pp. 669–688.
Kuipers, T., *Studies in Inductive Probability and Rational Expectation*, D. Reidel, Dordrecht, 1978a.
Kuipers, T., 'On the Generalization of the Continuum of Inductive Methods to Universal Hypotheses', *Synthese* **37** (1978b), 255–284.
Kuipers, T. A. F., 'Approaching Descriptive and Theoretical Truth', *Erkenntnis* **18** (1982), 343–378.
Kuipers, T. A. F., 'Approaching the Truth with a Rule of Success', in P. Weingartner and C. Pühringer (eds.), *Philosophy of Science — History of Science* (7th LMPS, Salzburg 1983), *Philosophia Naturalis* **21** (1984a), pp. 244–253.
Kuipers, T., 'Two Types of Inductive Analogy by Similarity', *Erkenntnis* **21** (1984b), 63–87.
Kullback, S., *Information Theory and Statistics*, J. Wiley, New York, 1959. (Dover, 1968.)
Kuratowski, K. and Mostowski, A., *Set Theory*, North-Holland, Amsterdam, 1970.
Kyburg, H. E. Jr. and Smokler, H. (eds.), *Studies in Subjective Probability*, J. Wiley, New York, 1964.
Lakatos, I., 'Changes in the Problem of Inductive Logic', in Lakatos (1968), pp. 315–417.

Lakatos, I. (ed.), *The Problem of Inductive Logic*, North-Holland, Amsterdam, 1968.
Lakatos, I., 'Falsification and the Methodology of Scientific Research Programmes', in Lakatos and Musgrave (1970), pp. 91—195.
Lakatos, I., 'Popper on Demarcation and Induction', in Schilpp (1974), pp. 241—273.
Lakatos, I. and Musgrave, A. (eds.), *Criticism and the Growth of Knowledge*, Cambridge University Press, Cambridge, 1970.
Lakoff, G., 'Hedges: A Study in Meaning Criteria and the Logic of Fuzzy Concepts', *Journal of Philosophical Logic* **2** (1973), 458—508.
Laudan, L., 'Peirce and the Trivialization of the Self-Correcting Thesis', in R. N. Giere and R. S. Westfall (eds.), *Foundations of Scientific Method: The Nineteenth Century*, Indiana University Press, Bloomington, 1973, pp. 275—306.
Laudan, L., *Progress and Its Problems*, Routledge and Kegan Paul, London, 1977.
Laudan, L., 'A Confutation of Convergent Realism', *Philosophy of Science* **48** (1981), 19—49.
Laudan, L., *Science and Values: The Aims of Science and Their Role in Scientific Debate*, University of California Press, Berkeley, 1984.
Laymon, R., 'Idealization, Explanation, and Confirmation', in P. D. Asquith and R. N. Giere (eds.), *PSA 1980*, vol. 1, Philosophy of Science Association, East Lansing, 1980, pp. 336—350.
Laymon, R., 'Scientific Realism and the Hierarchical Counterfactual Path from Data to Theory', in P. D. Asquith and T. Nickles (eds.), *PSA 1982*, vol. 1, Philosophy of Science Association, East Lansing, 1982, pp. 107—121.
Lenin, V. I., *Materialism aad Empirio-Criticism*, International Publishers, New York, 1927.
Leplin, J. (ed.), *Scientific Realism*, University of California Press, Berkeley, 1984.
Leplin, J., Review of Niiniluoto (1984c), *Philosophy of Science* **52** (1985), 646—648.
LePore, E., 'What Model Theoretic Semantics Cannot Do?', *Synthese* **54** (1983), 167—188.
Levenshtein, V. I., 'Binary Codes Capable of Correcting Relations, Insertions and Reversals', *Sov. Phys. Dokl.* **10** (1966), 707—710.
Levi, I., 'Must the Scientist Make Value Judgments', *Journal of Philosophy* **57** (1960), 345—357.
Levi, I., 'Decision Theory and Confirmation', *Journal of Philosophy* **58** (1961), 614—625.
Levi, I., 'On the Seriousness of Mistakes', *Philosophy of Science* **29** (1962), 47—65.
Levi, I., 'Corroboration and Rules of Acceptance', *British Journal for the Philosophy of Science* **13** (1963), 307—313.
Levi, I., *Gambling with Truth*, Alfred A. Knopf, New York, 1967a.
Levi, I., 'Information and Inference', *Synthese* **17** (1967b), 369—391.
Levi, I., 'Acceptance Revisited', in Bogdan (1976), pp. 1—71.
Levi, I., 'Abduction and Demands of Information', in Niiniluoto and Tuomela (1979), pp. 405—429.
Levi, I., 'Incognizables', *Synthese* **45** (1980a), 413—427.
Levi, I., *The Enterprise of Knowledge: An Essay on Knowledge, Credal Probability, and Chance*, The MIT Press, Cambridge, Mass. and London, 1980b.
Levi, I., 'Messianic vs Myopic Realism', in P. Asquith and P. Kitcher (eds.), *PSA 1984*, vol. 2, Philosophy of Science Association, East Lansing, 1986a.

BIBLIOGRAPHY

Levi, I., 'Estimation and Error Free Information', *Synthese* **67** (1986b), 347—360.
Lewis, D., *Counterfactuals*, Blackwell, Oxford, 1973.
Lewis, D., *On the Plurality of Worlds*, Blackwell, Oxford, 1986.
Lindley, D., *Introduction to Probability and Statistics from a Bayesian Viewpoint 1—2*, Cambridge University Press, Cambridge, 1965.
Loux, M., *Substance and Attribute*, D. Reidel, Dordrecht, 1978.
Luce, R. D. and Raiffa, H., *Games and Decisions*, J. Wiley, New York, 1957.
Ludwig, D., *Die Grundstrukturen einer physikalischen Theorie*, Springer-Verlag, Berlin-Heidelberg-New York, 1978.
Ludwig, G., 'Imprecision in Physics', in Hartkämper and Schmidt (1981), pp. 7—19.
Mach, E., *Knowledge and Error*, D. Reidel, Dordrecht, 1976.
Mahalanobis, P. C., 'On the Generalized Distance in Statistics', *Proc. Nat. Inst. Sci. India* **2** (1936), 49—55.
Maher, P. L., *Rationality and Belief*, Doctoral Dissertation, University of Pittsburgh, 1984.
Mayr, D., 'Investigations of the Concept of Reduction I', *Erkenntnis* **10** (1976), 275—294.
Mayr, D., 'Investigations of the Concept of Reduction II', *Erkenntnis* **16** (1981a), 109—129.
Mayr, D., 'Approximative Reduction by Completion of Empirical Uniformities', in Hartkämper and Schmidt (1981b), pp. 55—70.
Mayr, D., 'Contact Structures, Predifferentiability and Approximation', in Balzer *et al.* (1984), 187—198.
Michalos, A., *The Popper-Carnap Controversy*, M. Nijhoff, The Hague, 1971.
Merrill, G. H., 'The Model-Theoretic Argument Against Realism', *Philosophy of Science* **47** (1980), 69—81.
Mikenberg, I., Da Costa, N. C. A., and Chuaqui, R., 'Pragmatic Truth and Approximation to Truth', *The Journal of Symbolic Logic* **51** (1986), 201—221.
Mill, J. S., *A System of Logic*, London, 1843.
Miller, D., 'The Truth-Likeness of Truthlikeness', *Analysis* **33** (1972), 50—55.
Miller, D., 'Popper's Qualitative Theory of Verisimilitude', *The British Journal for the Philosophy of Science* **25** (1974a), 166—177.
Miller, D., 'On the Comparison of False Theories by Their Bases', *The British Journal for the Philosophy of Science* **25** (1974b), 178—188.
Miller, D., 'The Accuracy of Predictions', *Synthese* **30** (1975), 159—191, 207—219.
Miller, D., 'Verisimilitude Redeflated', *The British Journal for the Philosophy of Science* **27** (1976), 363—380.
Miller, D., 'Bunge's Theory of Partial Truth is No Such Thing', *Philosophical Studies* **31** (1977), 147—150.
Miller, D., 'On Distance from the Truth as a True Distance', in Hintikka *et al.* (1978a), pp. 415—435.
Miller, D., 'The Distance Between Constituents', *Synthese* **38** (1978b), 197—212.
Miller, D., Critical Notice of R. E. Butts and J. Hintikka (eds.), Proceedings of the Fifth International Congress of Logic, Methodology and Philosophy of Science, London, Ontario, 1975, *Synthese* **43** (1980), 381—410.
Miller, D., 'Truth, Truthlikeness, Approximate Truth', *Fundamenta Scientiae* **3** (1982a), 93—101.

Miller, D., 'Impartial Truth', *Stochastica* **6** (1982b), 169—186.
Monk, J. D., *Mathematical Logic*, Springer-Verlag, New York, Heidelberg, Berlin, 1976.
Montague, R., *Formal Philosophy* (ed. by R. H. Thomason), Yale University Press, New Haven and London, 1974.
Moore, R. E., *Interval Analysis*, Prentice-Hall, Englewood Cliffs, N.J., 1966.
Mortensen, C., 'A Theorem on Verisimilitude', *Bulletin of the Section of Logic* **7** (1978), 34—43.
Mortensen, C., 'Relevance and Verisimilitude', *Synthese* **55** (1983), 353—364.
Mott, P., 'Verisimilitude by Means of Short Theorems', *Synthese* **38** (1978), 247—273.
Moulines, C. U., 'Approximative Application of Empirical Theories: A General Explication', *Erkenntnis* **10** (1976), 210—227.
Moulines, C. U., 'Intertheoretic Approximation: The Kepler—Newton Case', *Synthese* **45** (1980), 387—412.
Moulines, C. U., 'A General Scheme for Intertheoretic Approximation', in Hartkämper and Schmidt (1981), pp. 123—146.
Nagel, E., P., Suppes, and A Tarski (eds.) *Logic, Methodology and Philosophy of Science: Proceedings of the 1960 International Congress*, Stanford University Press, Standford, 1962.
von Neumann, J. and Morgenstern, O., *Theory of Games and Economic Behaviour*, 2nd ed., Princeton University Press, Princeton, 1947.
Newton-Smith, W. H., *The Rationality of Science*, Routledge and Kegan Paul, Boston, 1981.
Niiniluoto, I., Review of A. Michalos, *The Popper-Carnap Controversy*, *Synthese* **25** (1973), 417—436.
Niiniluoto, I., 'Inquiries, Problems, and Questions: Remarks on Local Induction', in Bogdan (1976a), pp. 263—296.
Niiniluoto, I., 'Inductive Logic and Theoretical Concepts', in Przełecki *et al.* (1976b), pp. 93—112.
Niiniluoto, I., 'On a K-Dimensional System of Inductive Logic', in F. Suppe and P. D. Asquith (eds.), *PSA 1976*, vol. 2, Philosophy of Science Association, East Lansing, 1977a, pp. 425—447.
Niiniluoto, I., 'On the Truthlikeness of Generalizations', in R. E. Butts and K. J. Hintikka (eds.), *Basic Problems in Methodology and Linguistics. Part Three of the Proceedings of the Fifth International Congress of Logic, Methodology and Philosophy of Science, London, Ontario, 1975*, D. Reidel, Dordrecht, 1977b, pp. 121—147.
Niiniluoto, I., 'Bolzano and Bayes-Tests', in I. Niiniluoto, J. von Plato, and E. Saarinen (eds.), *Studia Excellentia*, Reports from the Department of Philosophy 3/1977, University of Helsinki, Helsinki, 1977c, pp. 30—36.
Niiniluoto, I., 'Truthlikeness in First-Order Languages', in Hintikka *et al.* (1978a), pp. 437—458.
Niiniluoto, I., 'Truthlikeness: Comments on Recent Discussion', *Synthese* **38** (1978b), 281—329.
Niiniluoto, I., 'Dretske on Laws of Nature', *Philosophy of Science* **45** (1978c), 431—439.
Niiniluoto, I., 'Degrees of Truthlikeness: From Singular Sentences to Generalizations', *The British Journal for the Philosophy of Science* **30** (1979a), 371—376.

Niiniluoto, I., 'Verisimilitude, Theory-Change, and Scientific Progress', in Niiniluoto and Tuomela (1979b), pp. 243—264.
Niiniluoto, I., 'Scientific Progress', *Synthese* **45** (1980a), 427—462.
Niiniluoto, I., 'Dinosaurs and Verisimilitude', in I. Patoluoto *et al.* (eds.), *Semi-Ramistic Studies*, Reports from the Department of Philosophy 5/80, University of Helsinki, Helsinki, 1980b, pp. 38—44.
Niiniluoto, I., 'Analogy, Transitivity, and the Confirmation of Theories', in Cohen and Hesse (1980c), pp. 218—234.
Niiniluoto, I., 'Analogy and Inductive Logic', *Erkenntnis* **16** (1981a), 1—34.
Niiniluoto, I., 'Language, Norms, and Truth', in I. Pörn (ed.), *Essays in Philosophical Analysis* (Acta Philosophica Fennica 32), Helsinki, 1981b, pp. 168—189.
Niiniluoto, I., 'On the Truth of Norm Propositions', in I. Tammelo and A. Aarnio (eds.), *Zum Fortschritt von Theorie und Technik in Recht und Ethik, Rechtstheorie: Beiheft 3*, Duncker und Humblot, Berlin, 1981c, pp. 171—180.
Niiniluoto, I., 'Statistical Explanation Reconsidered', *Synthese* **48** (1981d), 437—472.
Niiniluoto, I., 'On Explicating Verisimilitude: A Reply to Oddie', *The British Journal for the Philosophy of Science* **33** (1982a), 290—296.
Niiniluoto, I., 'What Shall We Do with Verisimilitude?', *Philosophy of Science* **49** (1982b), 181—197.
Niiniluoto, I., 'Truthlikeness for Quantitative Statements', in P. D. Asquith and T. Nickles (eds.), *PSA 1982*, vol. 1, Philosophy of Science Association, East Lansing, 1982c, pp. 208—216.
Niiniluoto, I., 'Verisimilitude vs Legisimilitude', *Studia Logica* **17** (1983a), 315—329.
Niiniluoto, I., 'Analogy and Legal Reasoning', in U. Kangas (ed.), *Essays in Legal Theory in Honor of Kaarle Makkonen*, Oikeustiede — Jurisprudentia, The Yearbook of the Finnish Lawyer Society XVI, Vammala, 1983b, pp. 178—187.
Niiniluoto, I., 'Novel Facts and Bayesianism', *The British Journal for the Philosophy of Science* **34** (1983c), 375—379.
Niiniluoto, I., 'Inductive Logic as a Methodological Research Programme', *Scientia: Logic in the 20th Century*, Milano, 1983d, pp. 77—100.
Niiniluoto, I., 'Wissenschaft auf der Suche nach Wahrheit?', in J. Manninen and H. J. Sandkühler (eds.), *Realismus und Dialektik oder was können wir wissen?*, Dialektik 8, Pahl-Rugenstein, Köln, 1984a, pp. 9—19.
Niiniluoto, I., *Is Science Progressive?*, D. Reidel, Dordrecht, 1984b.
Niiniluoto, I., 'Truth and Legal Norms', in N. MacCormick, S. Panov, and L. L. Vallauri (eds.), *Conditions of Validity and Cognition in Modern Legal Thought*, ARSP, Beiheft Nr. 25, Franz Steiner Verlag, Wiesbaden, 1985a, pp. 168—190.
Niiniluoto, I., 'Truthlikeness, Realism, and Progressive Theory-Change', in J. Pitt (ed.), *Change and Progress in Modern Science* (Proceedings of the Fourth International Conference on History and Philosophy of Science, Blacksburg, 1982), D. Reidel, Dordrecht, 1985b, pp. 235—265.
Niiniluoto, I., 'Theories, Approximations, Idealizations', in R. Barcan Marcus, G. J. W. Dorn, and P. Weingartner (eds.), *Logic, Methodology, and Philosophy of Science VII*, North-Holland, Amsterdam, 1986a, pp. 255—289.
Niiniluoto, I., 'The Significance of Verisimilitude', in P. D. Asquith and P. Kitcher (eds.), *PSA 1984*, vol. 2, Philosophy of Science Association, East Lansing, 1986b.
Niiniluoto, I., 'Truthlikeness and Bayesian Estimation', *Synthese* **67** (1986c), 321—346.

Niiniluoto, I. and R. Tuomela, *Theoretical Concepts and Hypothetico-Inductive Inference*, D. Reidel, Dordrecht, 1973.
Niiniluoto, I. and R. Tuomela (eds.), *The Logic and Epistemology of Scientific Change*, Proceedings of a Philosophical Colloquium, Helsinki, December, 12—14, 1977. (Acta Philosophica Fennica 30), North-Holland, Amsterdam, 1979.
Nowak, L., 'Laws of Science, Theory, Measurement', *Philosophy of Science* **39** (1972), 533—548.
Nowak, L., 'Relative Truth, the Correspondence Principle and Absolute Truth', *Philosophy of Science* **42** (1975), 187—202.
Nowak, L., *The Structure of Idealization: Towards a Systematic Interpretation of the Marxian Idea of Science*, Reidel, Dordrecht, 1980.
Nute, D., *Topics in Conditional Logic*, D. Reidel, Dordrecht, 1980.
Nute, D., 'Introduction', *Journal of Philosophical Logic* **10** (1981), 127—147.
Oddie, G., 'Verisimilitude and Distance in Logical Space', in Niiniluoto and Tuomela (1979), pp. 243—264.
Oddie, G., 'Verisimilitude Reviewed', *The British Journal for the Philosophy of Science* **32** (1981), 237—265.
Oddie, G., 'Cohen on Verisimilitude and Natural Necessity', *Synthese* **51** (1982), 355—379.
Oddie, G., 'The Poverty of the Popperian Programme for Truthlikeness', *Philosophy of Science*, forthcoming 1986.
Parthasarathy, K. R., *Probability Measures on Metric Spaces*, Academic Press, New York, 1967.
Parzen, E., *Stochastic Processes*, Holden-Day, San Francisco, 1962.
Patryas, W., 'Idealization and Approximation', *Poznan Studies in the Philosophy of the Sciences and the Humanities* **3** (1977), 180—198.
Pearce, D., 'Is There Any Theoretical Justification for a Non-Statement View of Theories?', *Synthese* **46** (1981), 1—40.
Pearce, D., 'Logical Properties of the Structuralist Concept of Reduction', *Erkenntnis* **18** (1982), 307—333.
Pearce, D., 'Truthlikeness and Translation: A Comment on Oddie', *The British Journal for the Philosophy of Science* **34** (1983), 380—385.
Pearce, D., 'Research Traditions, Incommensurability and Scientific Progress', *Zeitschrift für allgemeine Wissenschaftstheorie* **15** (1984), 261—271.
Pearce, D., 'Critical Realism in Progress: Reflections on Ilkka Niiniluoto's Philosophy of Science', forthcoming 1986a.
Pearce, D., *Roads to Commensurability*, D. Reidel, forthcoming 1986b.
Pearce, D. and Rantala, V., 'Realism and Reference: Some Comments on Putnam', *Synthese* **52** (1982), 439—448.
Pearce, D. and Rantala, V., 'New Foundations for Metascience', *Synthese* **56** (1983), 1—26.
Pearce, D. and Rantala, V., 'A Logical Study of the Correspondence Relation', *Journal of Philosophical Logic* **13** (1984), 47—84.
Pearson, K., 'On a Criterion that a System of Deviations from the Probable in the Case of a Correlated System of Variables is Such that It can be Reasonably Supposed to have Arisen in Random Sampling', *Phil. Mag.* **50** (1900), 157—175.
Pearson, K., 'On the Coefficient of Racial Likeness', *Biometrika* **18** (1926), 105—117.

BIBLIOGRAPHY 503

Peirce, C. S., 'A Theory of Probable Inference' (1883), in *Collected Papers*, 2.694–754.
Peirce, C. S., *Collected Papers*, vols. 1–6 (ed. by C. Hartshorne and P. Weiss), Harvard University Press, Cambridge, Mass., 1931–35.
Perry, C. B., 'Verisimilitude and Shared Tests', *Noûs* **16** (1982), 607–612.
Pitt, J. C., *Pictures, Images and Conceptual Change: An Analysis of Wilfrid Sellars' Philosophy of Science*, D. Reidel, Dordrecht, 1981.
Plantinga, A., *The Nature of Necessity*, Oxford University Press, Oxford, 1974.
von Plato, J., 'Probability and Determinism', *Philosophy of Science* **49** (1982), 51–66.
Popkin, R. H., *The History of Scepticism from Erasmus to Descartes*, Van Gorcum & Co, Assen, 1960.
Popper, K. R., *Logik der Forschung*, Julius Springer, Wien, 1934.
Popper, K. R., 'Degree of Confirmation', *The British Journal for the Philosophy of Science* **5** (1954), 143–149. (Reprinted in Popper, 1959.)
Popper, K. R., *The Logic of Scientific Discovery*, Hutchinson, London, 1959.
Popper, K. R., 'Some Comments on Truth and the Growth of Knowledge', in Nagel, Suppes, and Tarski (1962), pp. 285–292.
Popper, K. R., *Conjectures and Refutations: The Growth of Scientific Knowledge*, Routledge and Kegan Paul, London, 1963.
Popper, K. R., 'A Theorem on Truth-Content', in P. Feyerabend and G. Maxwell (eds.), *Mind, Matter, and Method*, University of Minnesota Press, Minneapolis, 1966.
Popper, K. R., 'Normal Science and Its Dangers', in Lakatos and Musgrave (1970), pp. 51–58.
Popper, K. R., *Objective Knowledge*, Oxford University Press, Oxford, 1972. (2nd ed. 1979.)
Popper, K. R., 'Autobiography', in Schilpp (1974a), pp. 1–181.
Popper, K. R., 'Replies to My Critics', in Schilpp (1974b), pp. 961–1197.
Popper, K. R., 'A Note on Verisimilitude', *The British Journal for the Philosophy of Science* **27** (1976), 147–159.
Przełecki, M., 'Fuzziness as Multiplicity', *Erkenntnis* **10** (1976), 371–380.
Przełecki, M., Szaniawski, K. and Wójcicki, R. (eds.), *Formal Methods in the Methodology of Empirical Sciences*, Ossolineum, Wroclaw, and D. Reidel, Dordrecht, 1976.
Putnam, H., *Meaning and the Moral Sciences*, Routledge and Kegan Paul, London, 1978.
Putnam, H., 'Models and Reality', *Journal of Symbolic Logic* **45** (1980), 464–482.
Putnam, H., *Reason, Truth, and History*, Cambridge University Press, Cambridge, 1981.
Quine, W. V., *Word and Object*, The M.I.T. Press, Cambridge, Mass., 1960.
Radnitzky, G. and Andersson, G., 'Objective Criteria of Scientific Progress? Inductivism, Falsificationism, and Relativism', in Radnitzky and Andersson (1978), pp. 3–19.
Radnitzky, G. and Andersson, G. (eds.), *Progress and Rationality in Science*, D. Reidel, Dordrecht, 1978.
Radnitzky, G. and Andersson, G. (eds.), *The Structure and Development of Science*, D. Reidel, Dordrecht, 1979.
Raiffa, H. and Schlaifer, R., *Applied Statistical Decision Theory*, The M.I.T. Press, Cambridge, Mass., 1961.

Rantala, V., *Aspects of Definability* (Acta Philosophica Fennica 29:2—3), North-Holland, Amsterdam, 1977.
Rantala, V., 'Game-Theoretical Semantics and Back-and-Forth', in Hintikka *et al.* (1979), pp. 119—152.
Rantala, V., 'Correspondence and Non-Standard Models: A Case Study', in Niiniluoto and Tuomela (1979), pp. 366—378.
Rantala, V., 'Constituents', forthcoming in R. J. Bogdan (ed.), *Jaakko Hintikka*, Profiles 8, D. Reidel, Dordrecht, 1987.
Rao, C. R., 'Cluster-Analysis Applied to a Study of Race Mixture in Human Populations', in van Ryzin (1977), pp. 175—197.
Rawls, J., *A Theory of Justice*, Harvard University Press, Cambridge, Mass., 1971.
Reichenbach, H., *Experience and Prediction*, The University of Chicago Press, Chicago, 1938.
Reichenbach, H., *The Theory of Probability*, University of California Press, Berkeley and Los Angeles, 1949.
Reisinger, L., 'A Mathematical Model of Reasoning by Analogy', in *Pre-Proceedings of the International Study Congress on Logica, Informatica, Diritto*, Florence, 6—10 April, 1981, pp. 635—656.
Rescher, N., *The Coherence Theory of Truth*, Oxford University Press, Oxford, 1973.
Rescher, N., *Methodological Pragmatism*, Blackwell, Oxford, 1977.
Rice, J. R., *The Approximation of Functions, Vol. I: Linear Theory*, Addison-Wesley, Reading, Mass., 1964.
Rice, J. R., *The Approximation of Functions, Vol. 2: Nonlinear and Multivariate Theory*, Addison-Wesley, Reading, Mass., 1969.
Robinson, G. S., 'Popper's Verisimilitude', *Analysis* **31** (1971), 194—196.
Rorty, R., *Philosophy and the Mirror of Nature*, Princeton University Press, Princeton, 1980.
Rosenfeld, A., *Picture Processing by Computer*, Academic Press, New York, 1969.
Rosenfeld, A. (ed.), *Digital Picture Analysis*, Springer-Verlag, Berlin, 1976.
Rosenfeld, A., *Picture Languages: Formal Models for Picture Recognition*, Academic Press, London, 1979.
Rosenfeld, A. and Pfaltz, J. L., 'Distance Functions on Digital Pictures', *Pattern Recognition* **1**, (1968), 33—61.
Rosenkrantz, R., 'Measuring Truthlikeness', *Synthese* **45**(1980), 463—488.
Rudner, R., 'The Scientist *Qua* Scientist Makes Value Judgements', *Philosophy of Science* **20** (1953), 1—6.
Ruse, M., *The Philosophy of Biology*, Hutchinson University Library, London, 1973.
Ryzin, J. van (ed.), *Classification and Clustering*, Academic Press, New York, 1977.
Salmon, W., 'Vindication of Induction', in H. Feigl and G. Maxwell (eds.), *Current Issues in the Philosophy of Science*, Holt, Rinehart, and Winston, New York, 1961, pp. 245—257.
Salmon, W., 'Rational Prediction', *The British Journal for the Philosophy of Science* **32** (1981), 115—125.
Sanford, D., 'Competing Semantics of Vagueness: Many Values versus Super-Truth', *Synthese* **33** (1976), 195—210.
Sattath, S. and Tversky, A., 'Additive Similarity Trees', *Journal of Mathematical Psychology* **42** (1977), 319—345.

Savage, L. J., 'Subjective Probability and Statistical Practice', in L. J. Savage *et al.*, *The Foundations of Statistical Inference*, Methuen, London, 1962, pp. 9—35.
Savage, L. J., *The Foundations of Statistics*, J. Wiley and Sons, New York, 1954. (2nd ed., Dover, New York, 1972.)
Scheibe, E., 'The Approximative Explanation and the Development of Physics', in P. Suppes, L. Henkin, A. Joja, and Gr. C. Moisil (eds.), *Logic, Methodology and Philosophy of Science* IV, North-Holland, Amsterdam, 1973, pp. 931—942.
Schilpp. P. A. (ed.), *The Philosophy of Karl Popper* (The Library of Living Philosophers, vol. XIV, Books I—II), Open Court, La Salle, 1974.
Schwarz, G. and Tversky, A., 'On the Reciprocity of Proximity Relations', *Journal of Mathematical Psychology* **22** (1980), 157—175.
Scott, D., 'A Note on Distributive Normal Forms', in E. Saarinen *et al.* (eds.), *Essays in Honour of Jaakko Hintikka*, D. Reidel, 1979, pp. 75—90.
Scott, D. and Suppes, P., 'Foundational Aspects of Theories of Measurement', *Journal of Symbolic Logic* **28** (1958), 113—128.
Shimony, A., 'Scientific Inference', in R. C. Colodny (ed.), *The Nature and Function of Scientific Theories*, The University of Pittsburgh Press, Pittsburgh, 1970, pp. 79—172.
Simmons, G. F., *Introduction to Topology and Modern Analysis*, McGraw-Hill, New York, 1963.
Sintonen, M., *The Pragmatics of Scientific Explanation* (Acta Philosophica Fennica 37), The Philosophical Society of Finland, Helsinki, 1984.
Sneath, P. H. A. and Sokal, R. R., *Numerical Taxonomy: The Principles and Practice of Numerical Classification*, W. H. Freeman, San Francisco, 1973.
Sneed, J. D., *The Logical Structure of Mathematical Physics*, Reidel, Dordrecht, 1971. (2nd ed. 1979.)
Sokal, R. R., 'Distance as a Measure of Taxonomic Similarity', *Systematic Zoology* **10** (1961), 70—79.
Spinoza, B., *The Works of Spinoza*, Dover, New York, 1955.
Spohn, W., 'Analogy and Inductive Logic: A Note on Niiniluoto', *Erkenntnis* **16** (1981), 35—52.
Stalnaker, R. C., 'A Theory of Conditionals', in N. Rescher (ed.), *Studies in Logical Theory* (APQ Monograph No. 2), Blackwell, Oxford, 1968.
Stegmüller, W., 'Carnap's Normative Theory of Inductive Probability', in P. Suppes *et al.* (eds.), *Logic, Methodology and Philosophy of Science* IV, North-Holland, Amsterdam, 1973a, pp. 501—513.
Stegmüller, W., *Personelle und statistische Wahrscheinlichkeit. Probleme und Resultate der Wissenschaftstheorie und analytischen Philosophie, Band IV, Erster Halbband*, Springer-Verlag, Berlin, 1973b.
Stegmüller, W., *The Structure and Dynamics of Theories*, Springer-Verlag, New York, Heidelberg, Berlin, 1976.
Stegmüller, W., 'A Combined Approach to the Dynamics of Theories', *Theory and Decision* **9** (1978), 39—75.
Stegmüller, W., *The Structuralist View of Theories: A Possible Analogue of the Bourbaki Programme in Physical Science*, Springer-Verlag, Berlin, 1979.
Stegmüller, W., *Theorie und Erfahrung, Dritter Teilband*, Springer-Verlag, Berlin, Heidelberg, New York, Tokio, 1986.

Stenius, E., *Critical Essays* (Acta Philosophica Fennica 25), North-Holland, Amsterdam, 1972.
Suppe, F., 'The Search for Philosophic Understanding of Scientific Theories', in F. Suppe (ed.), *The Structure of Scientific Theories*, University of Illinois Press, Urbana, 1974, pp. 1—241.
Suppe, F., 'Theoretical Laws', in Przełecki *et al.* (1976), pp. 247—267.
Suppes, P., 'Models of Data', in Nagel, Suppes and Tarski (1962), pp. 252—261.
Suppes, P., 'What is a Scientific Theory?', in S. Morgenbesser (ed.), *Philosophy of Science Today*, Basic Books, New York, 1967, pp. 55—67.
Suppes, P. and Zinnes, J. L., 'Basic Measurement Theory', in R. D., Luce *et al.* (eds.), *Handbook of Mathematical Psychology*, vol. 1, Wiley, New York, 1963, pp. 1—76.
Swinburne, R., *An Introduction to Confirmation Theory*, Methuen, London, 1973.
Tarski, A., 'The Semantic Conception of Truth and the Foundations of Semantics', *Philosophy and Phenomenological Research* **4** (1944), 341—376. (Also in H. Feigl and W. Sellars, *Readings in Philosophical Analysis*, Appleton-Century-Crofts, New York, 1949, pp. 52—94.)
Tarski, A., *Logic, Semantics, and Metamathematics*, Oxford University Press, Oxford, 1956.
Tichý, P., 'On Popper's Definition of Verisimilitude', *The British Journal for the Philosophy of Science* **25** (1974), 155—160.
Tichý, P., 'Verisimilitude Redefined', *The British Journal for the Philosophy of Science* **27** (1976), 25—42.
Tichý, P., 'Verisimilitude Revisited', *Synthese* **38** (1978), 175—196.
Todhunter, I., *A History of the Mathematical Theory of Probability from the Time of Pascal to that of Laplace*, Cambridge, 1865. (Chelsea Publ. Co., Bronx, New York, 1949.)
Törnebohm, H., 'On Piecemeal Knowledge-Formation', in Bogdan (1976), pp. 297—318.
Tuomela, R., *Theoretical Concepts*, Springer-Verlag, Wien and New York, 1973.
Tuomela, R., 'Verisimilitude and Theory-Distance', *Synthese* **38** (1978), 213—246.
Tuomela, R., 'Scientific Change and Approximation', in Niiniluoto and Tuomela (1979), pp. 265—297.
Tuomela, R., 'Analogy and Distance', *Zeitschrift für allgemeine Wissenschaftstheorie* **11** (1980), 276—291.
Tuomela, R., 'Truth and Best Explanation', *Erkenntnis* **22** (1985a), 271—300.
Tuomela, R., *Science, Action, and Reality*, D. Reidel, Dordrecht, 1985b.
Tversky, A., 'Features of Similarity', *Psychological Review* **84** (1977), 327—352.
Uchii, S., 'Inductive Logic with Causal Modalities: A Probabilistic Approach', *Philosophy of Science* **39** (1972), 162—178.
Uchii, S., 'Inductive Logic with Causal Modalities: A Deterministic Approach', *Synthese* **26** (1973), 264—303.
Uchii, S., 'Induction and Causality in Cellular Space', in F. Suppe and P. D. Asquith (eds.), *PSA 1976*, vol. 2, Philosophy of Science Association, East Lansing, 1977, pp. 448—461.
Ullmann, J. R., 'Picture Analysis in Character Recognition', in Rosenfeld (1976), pp. 295—344.

Urbach, P., 'Intimations of Similarity: The Shaky Basis of Verisimilitude', *The British Journal for the Philosophy of Science* **34** (1983), 266—275.
Vetter, H., 'A New Concept of Verisimilitude', *Theory and Decision* **8** (1977), 369—375.
Wachbroit, R., 'Progress: Metaphysical and Otherwise', *Philosophy of Science*, forthcoming 1986.
Wald, A., *Statistical Decision Functions*, John Wiley and Sons, New York, 1950.
Watanabe, S., *Knowing and Guessing*, J. Wiley, New York, 1969.
Watanabe, S., 'Pattern Recognition as a Quest for Minimum Entropy', *Pattern Recognition* **13** (1981), 381—387.
Watkins, J., 'Towards a Unified Decision Theory: A Non-Bayesian Approach', in R. E. Butts and J. Hintikka (eds.), *Foundational Problems in the Special Sciences*, D. Reidel, Dordrecht, 1977, pp. 345—379.
Watkins, J., 'Corroboration and the Problem of Content-Comparison', in Radnitzky and Andersson (1978), pp. 339—378.
Watkins, J., *Science and Scepticism*, Princeton University Press, Princeton, 1984.
Weston, T. S., 'Approximate Truth' (abstract), *The Journal of Symbolic Logic* **42** (1977), 157—158.
Weston, T. S., 'Approximate Truth and Valid Operators' (abstract), *The Journal of Symbolic Logic* **46** (1981), 688.
Wilks, S. S., *Mathematical Statistics*, J. Wiley, New York, 1962.
Williams, P. M., 'Bayesian Conditionalization and the Principle of Minimum Information', *The British Journal for the Philosophy of Science* **31** (1980), 131—144.
Wójcicki, R., 'Basic Concepts of Formal Methodology of Empirical Sciences', *Ajatus* **25** (1973), 168—196.
Wójcicki, R., 'Set Theoretic Representations of Empirical Phenomena', *Journal of Philosophical Logic* **3** (1974), 337—343.
Woozley, A. D., *Theory of Knowledge*, Hutchinson, London, 1949.
von Wright, G. H., *Logical Studies*, Routledge and Kegan Paul, London, 1957.
von Wright, G. H., *Explanation and Understanding*, Cornell University Press, Ithaca, New York, 1971.
Zadeh, L., 'Fuzzy Sets', *Information and Control* **8** (1965), 338—353.
Zadeh, L., 'Fuzzy Logic and Approximate Reasoning', *Synthese* **30** (1975), 407—428.
Zadeh, L. A., 'Fuzzy Sets and Their Applications in Pattern Classification and Clustering Analysis', in van Ryzin (1977), pp. 251—299.
Ziehen, T., *Lehrbuch der Logik auf positivistischer Grundlage mit Berücksichtigung der Geschichte der Logik*, Bonn, 1920.

INDEX OF NAMES

Aarnio, A. 501
Adanson, M. 24
Agassi, J. 480, 483, 490
Aitchison, J. 434, 490
Ajdukiewics, K. 489, 493
Almeder, R. 490
Andersson, G. 443—4, 488, 490, 493—4, 496—7, 503, 507
Aquinas, T. 23, 158, 161—2, 164
Archimedes 165
Aristotle 23, 38, 135, 138, 157, 160, 266
Armstrong, D. M. 458, 475—7, 485, 490
Asquith, P. D. 497—8, 500—1, 506
Augustine, St. xv, 157—8, 161, 265—8, 422, 490
Ayer, A. J. 481, 490

Backman, W. 487, 490
Bacon, F. 166, 474, 490
Balzer, W. 490, 499
Bar-Hillel, Y. 148, 151—3, 487, 490, 492, 495
Battacharya, A. 8, 490
Bayes, T. ix, xiv, xvi, 86, 159—61, 270, 283, 285, 309, 406—11, 416, 420, 423, 425—31, 434, 441—2, 482, 491, 493, 499—501, 507
Beals, R. 476, 490
Bell, J. L. 41, 477, 490
Benacerraf, P. 478, 490
Berkeley, G. 104
Bernoulli, D. 163, 406
Bertossi, L. 480, 492
Bigelow, J. 490—1
Blackwell, D. 309, 491
Bochenski, I. M. 23, 491
Bogdan, R. J. 487, 491—2, 498, 500, 504, 506

Bogdanov, A. 174
Bolzano, B. 83, 407—8, 487, 491, 500
Bolzmann, L. 109
Boole, G. 10—1, 36—7, 41, 78—9, 83, 190—1, 197, 233, 243
Boorman, S. D. 14, 347, 353, 474, 491
Boos, W. 489, 491
Bourbaki, N. 20, 472, 474, 491, 505
Box, G. 409, 428—9, 491
Boyle, R. 111, 113—4, 166
Bradley, F. H. xv, 135, 166, 169—70, 172—4, 479, 491
Brentano, F. 134
Brouwer, L. 191
Bugajski, S. 475, 491
Bunge, M. 180—2, 476, 491, 499
Bush, R. R. 34
Butler, J. 23
Butts, R. E. 499—500, 507
Byrne, E. 158, 161—2, 491

Cain, A. J. 491
Cantor, G. 19
Carnap, R. xi, xiii—iv, xvi, 29, 32, 39, 42—4, 46, 51, 83—6, 89—90, 103, 107, 109, 138, 144, 148, 151—3, 159—60, 269, 409, 442, 450, 475—9, 487, 490—2, 495, 499—500, 505
Carneades xv, 161, 421
Cartwright, N. 486, 492
Cauchy, A. 19, 21—2
Chellas, B. 477, 492
Chen, C. H. 475, 492
Chernoff, H. 409, 487, 492
Chuaqui, R. 479—80, 492, 499
Church, A. 61
Cicero 157, 160—1
Clark, P. J. 26

INDEX OF NAMES

Clifford, W. K. 158, 313—6, 320—1, 336, 339, 347, 349, 356, 362, 367, 377, 379, 382—3, 449, 453—4, 456
Cohen, L. J. 372—3, 375, 378, 477, 485, 492, 501—2
Colodny, R. C. 493, 505
Coombs, C. H. 32
da Costa, N. C. A. 479, 499
Cox, D. R. 434, 492
Cramér, H. 15, 478, 492
Cusanus, N. xv, 165, 168—9, 171, 175, 492
Czekanowski, J. 25, 492

Daniels, C. B. 135, 493
Davidson, D. 140, 479, 492
Descartes, R. 104, 109, 134, 158, 503
Dewey, J. 136
Dice, L. R. 27—8, 34
Dorn, G. J. W. 501
Dretske, F. 477, 500
Drobisch, M. W. 23
Dubois, D. 10, 479, 492
Duhem, P. 104, 478, 492
Dummett, M. 136, 492
Dunsmore, I. R. 434, 490

Earman, J. 493
Eberle, R. 475, 492
Ehrenfeucht, A. 73
Einstein, A. 264, 266
Eisler, H. 34
Ekman, G. 34
Engels, F. xv, 164, 166—8, 172, 174, 492
Erdmann, B. 23, 474
Euclides 1, 3—4, 26, 47, 107—10, 254, 303, 382, 386, 394, 453—4
Eudoxos 165
Ewing, A. C. xv, 173—5, 177, 481, 493

Feigl, H. 495, 504, 506
Ferguson, T. S. 428, 432, 436—7, 487, 493
Festa, R. xvii, 198, 319—20, 417, 436, 445, 487, 493
Fetzer, J. 477—8, 492

Feyerabend, P. 160, 443, 462, 488, 491, 493, 503
Field, H. 140, 493
Findlay, J. N. 173, 493
Fine, A. 481, 493
Fine, K. 479, 493
Fine, T. 489, 493
de Finetti, B. 161, 496
van Fraassen, B. 109—10, 115, 445, 478—9, 482, 486, 493
Fraisse, R. 73
Fréchet, M. 1, 11
Freeman, J. B. 135, 493
Fu, K. S. 12—3, 29, 347—8, 356, 360, 475, 493

Galilei, G. 103—4, 116, 163, 494
Garber, D. 477, 493
Gauss, C. F. 164, 281—2
Giedymin, J. 478, 481, 493
Giere, R. N. 498
Girshick, M. A. 309, 491
Gnedenko, B. V. 8, 494
Godambe, V. D. 494
Gödel, K. 185
Good, I. J. 479, 488, 494
Goodman, N. 32, 35—7, 456, 475, 494
Gower, J. C. 475, 494
Gregson, R. A. M. 34
Grosseteste, R. 387
Grünbaum, A. 480, 488, 494
Gudder, S. P. 31, 494
Guild, J. 31
Gurwitsch, A. 104, 494

Haack, S. 138, 143—4, 173, 175, 178, 479, 494
Habermas, J. 136, 494
Hacking, I. 477, 479, 487, 494
Halonen, I. xvii
Haltenorth, T. 26
Hamming, R. W. 8—10, 29, 108, 258, 311, 313, 481, 494
Haralick, R. M. 494
Harding, C. L. 494
Harmon, L. D. 494
Harper, W. 477, 494

INDEX OF NAMES

Harris, J. H. 188—90, 494
Harrison, G. A. 491
Hartigan, J. A. 474, 494
Hartkämper, A. 490, 495, 499—500
Hartley, D. 166
Hattiangadi, J. A. 480, 495
Hausdorff, F. 11, 19, 22, 79, 204, 245
Hautamäki, A. 476, 495
Hegel, G. W. F. xv, 23, 135, 164, 166—7, 169, 171—3, 178, 493, 495
Heincke, F. 25
Held, D. 494
Hellinger, E. 8
von Helmholtz, H. 29
Hempel, C. G. 406, 410—12, 480, 487, 495
Henkin, L. 505
Heraclitus 172
Herbart, J. F. 29
Hesse, M. 23, 477, 492, 495, 501
Heywood, V. H. 475, 495
Hilbert, D. 5, 109, 120, 303
Hilpinen, R. xii, xvii, 89—90, 176, 181—2, 198—202, 204—5, 219—20, 233, 268, 412, 415—7, 419, 445, 475—7, 487—8, 495
Hinkley, D. V. 434, 492
Hintikka, J. xiii—xv, xvii, 39, 61, 71—2, 77—8, 82, 85, 90, 93, 99, 102, 122, 153—4, 160, 204, 206, 306, 341—3, 373, 380, 412—3, 422, 446, 451—2, 476—9, 481, 483—4, 487, 492, 495—6, 499—500, 504—5, 507
Hooke, R. 166
Hübner, K. 488, 496
Hull, D. L. 496
Hume, D. 23, 91—2, 112, 118, 158—9, 161, 477
Husserl, E. 29, 104, 494
Hyland, M. 189

Jaccard, P. 27—8, 34
James, W. 496
Jayatilleke, K. N. 480, 496
Jeffrey, R. C. 161, 270, 409, 477, 479, 482, 487—8, 491—2, 496
Jeffreys, H. 8, 496

Jevons, W. S. 23, 313, 496
Johnson, W. E. 476, 496
Joja, A. 505
Jones, K. 496

Kaipainen, A. xvii
Kangas, U. 501
Kanger, S. 93
Kant, I. 23, 162, 168, 183, 496
Karttunen, M. 74, 476, 496
Kasher, A. 496
Katz, M. 479, 497
Kemeny, J. 477
Keuth, H. 480, 497
Keynes, J. M. 23—4, 37, 474, 497
Khan, M. K. 494
Kitcher, P. 498, 501
Kleiner, S. A. 478, 497
Klug, V. 24, 474, 497
Knuuttila, S. 477, 497
Koertge, N. 265, 443—4, 470, 482, 497
Kolmogorov, A. N. 8, 494
Krajewski, W. 113, 478, 485—6, 497
Krantz, D. H. 476, 478, 490, 497
von Kries, J. 23
Kripke, S. 38, 93
Kuhn, T. 259, 266, 443, 445, 462, 478, 488, 497
Kuhn, T. S. 470, 497
Kuipers, T. 90, 191, 372—4, 380—2, 445, 477, 481, 489, 497
Kullback, S. 7, 497
Kuratowski, K. 497
Kyburg, H. E. Jr. 477, 491, 497

Lagrange, J. L. 163
Lakatos, I. 259, 443—4, 478, 491, 495, 497—8, 503
Lakoff, G. 498
Lance, G. N. 26
Laplace, P. S. 4, 83, 163, 281, 506
Lasch, R. 494
Laudan, L. xiii, 263, 267, 445—6, 478, 480, 482, 484—6, 498
Laymon, R. 486, 498
Leadbetter, M. R. 478, 492

INDEX OF NAMES

Lebesgue, H. 6, 431
Legendre, A. M. 4
Leibniz, G. W. 35, 166
Lenin, V. I. 164, 166, 168, 174, 498
Leplin, J. 482, 484, 488, 493, 498
Lepore, E. 479, 498
LeSage, G. 166
Levenshtein, V. I. 12, 498
Levi, I. ix, xiv—xvii, 149, 155, 160, 406, 410—1, 413—7, 419—22, 430—3, 436, 442, 470, 477—82, 487—8, 490—1, 498—9
Lévy, P. 8
Lewis, D. 93, 199, 262, 316, 365, 476—7, 484, 499
Licentius 157
Lindenbaum, A. 41, 79, 191, 204, 458
Lindley, D. 409, 428—9, 483, 499
Locke, J. 103, 474
Loux, M. 476, 499
Luce, R. D. 425, 497, 506
Ludwig, G. 106, 472, 499
Lukasiewicz, J. 146

MacCormick, N. 501
Mach, E. 32, 166, 168, 499
Mahalanobis, P. C. 18, 27, 289, 499
Maher, P. L. 482, 487, 489, 499
Manninen, J. 501
Marcus, R. B. 501
Marx, K. 174, 502
Maxwell, G. 491, 495, 503—4
Maxwell, J. C. 23
Mayr, D. 401—2, 486, 499
Mazurkiewicz, S. 11, 190, 244
McMullin, E. 494
Meinong, A. 29
Mellor, D. H. 494
Merrill, G. H. 479, 499
Michalos, A. 451, 499—500
Michener, C. D. 28
Mikenberg, I. 479, 499
Mill, J. S. 23—4, 97, 499
Miller, D. xi, xiii, xvi—xvii, 180, 187—92, 195, 197—8, 202, 205, 233, 235, 243, 258, 268, 276—7, 381, 417, 442—3, 445—6, 449—54, 456—9, 462, 472—3, 480, 483, 487, 489—90, 499—500
Minkowski, H. 4, 5, 33, 107, 288, 293, 302—3, 385, 388, 393, 403, 453
von Mises, R. E. 478
Moisil, G. C. 505
Monk, J. D. 48, 72—3, 476, 500
Montague, R. 42, 476, 500
Montaigne, M. 161
Moore, R. E. 434, 500
Morgenbesser, S. 506
Morgenstern, O. 407, 424, 500
Mortensen, C. 190, 500
Moses, L. E. 409, 487, 492
Mosteller, F. 34
Mostowski, A. 497
Mott, P. 189, 488, 500
Moulines, C. U. 398—9, 486, 500
Musgrave, A. 443, 498, 503

Nagel, E. 491, 500, 503, 506
Nägeli, K. W. 168
von Neumann, J. 407, 424, 500
Neurath, O. 135
Newton, I. 104, 116, 166—7, 264—6, 422, 500
Newton-Smith, W. H. 267, 480, 485, 500
Neyman, J. 428
Nickles, T. 497—8, 501
Niiniluoto, I. xiii, 24, 71, 88, 91, 137, 143, 167, 195—9, 202—5, 208, 223, 226—7, 232, 235, 237—8, 240—1, 247, 249, 253, 257, 263, 267—8, 274, 276—7, 288, 290, 298—9, 302, 310—1, 313, 315, 317, 321—4, 326, 331—3, 342—3, 349—51, 353, 355—7, 363, 365, 368, 370, 373, 377—8, 382, 386—7, 393—4, 396, 401, 413, 419, 421—3, 426, 434—5, 442, 445—7, 449—51, 453—4, 462—3, 466, 469—70, 473—89, 492, 496, 498, 500—2, 504, 506
Nowak, L. 113, 478, 485—6, 502
Nozzolini 163
Nute, D. 477, 502

INDEX OF NAMES

Oddie, G. viii—ix, xii—iii, xvii, 198, 232, 235—9, 241, 245, 249, 310, 318, 323, 332—4, 353—5, 363, 365, 369, 372—8, 445, 469, 476, 480—1, 484—5, 501—2
Olivier, D. C. 14, 347, 353, 474, 491

Panov, S. 501
Parthasarathy, K. R. 502
Parzen, E. 478, 502
Patoluoto, I. 501
Patryas, W. 288, 485, 502
Pearce, D. A. xvii, 458—9, 468, 477, 479, 481, 484, 486, 488—90, 502
Pearce, G. 477, 494
Pearson, E. S. 428
Pearson, K. 15, 26—7, 302, 322, 502
Peirce, C. xi, xv, 20, 24, 28, 34, 136—7, 140, 156, 162, 166—8, 171—2, 178, 421, 433, 478, 480, 490, 498, 503
Perry, C. B. 480, 503
Pfalz, J. L. 482, 504
Pietarinen, J. 413, 479, 496
Pitt, J. C. 104, 501, 503
Plantinga, A. 38, 503
von Plato, J. 478, 500, 503
Platon 103, 136, 157, 160
Poisson, D. 163
Polya, G. 4
Popkin, R. H. 158, 503
Popper, K. R. viii, xi—vi, 147—8, 152, 156, 160—3, 171, 175, 179, 183, 185—98, 201—2, 233, 235—8, 259, 263—5, 267, 277, 406, 415, 417, 420—3, 426, 433, 442—4, 450, 462, 469—70, 477—80, 482—3, 488—9, 494, 496, 498—500, 502—6
Pörn, I. 501
Prade, H. 10, 479, 492
Prezelecki, M. 479, 481, 489, 495—6, 500, 503, 506
Priestley, J. 166, 168
di Prisco, C. A. 492
Ptolemy, C. 387
Pühringer, C. 497
Putnam, H. 136—7, 141—2, 266—7, 445, 478—9, 490, 502—3

Pythagoras 103

Quine, W. V. O. xi, 171, 503

Radnitzky, G. 443, 490, 493—4, 496—7, 503, 507
Raiffa, H. 425, 428, 499, 503
Ramig, P. F. 494
Ramsey, F. P. 134, 494
Rantala, V. 72—5, 206, 401, 476—7, 479, 486, 502, 504
Rao, C. R. 8, 475, 504
Rawls, J. 481, 504
Reichenbach, H. 171, 179—80, 210, 478, 504
Reid, T. 23
Reisinger, L. 481, 504
Rescher, N. 135—6, 182, 445, 504—5
Restle, F. 34
Rice, J. R. 4, 6, 504
Robinson, G. S. 480, 504
Rogers, D. J. 28
van Rootselaar, B. 495
Rorty, R. 136, 504
Rosenfeld, A. 475, 482, 494, 504, 506
Rosenkrantz, R. xiv, 198, 403—4, 482—3, 504
Rudner, R. 487, 504
Ruse, M. 475, 504
Russell, B. 36
van Ryzin, J. 475, 504, 507

Saarinen, E. 496, 500, 505
Salmon, W. 451, 487, 504
Sandkühler, H. J. 501
Sanford, B. 479, 504
Sattath, S. 476, 504
Savage, L. J. xv, 409, 430, 487, 489, 505
Scheibe, E. 505
Schiller, F. C. S. 265—6, 268
Schilpp, P. A. 490, 498, 503, 505
Schlaifer, R. 428, 503
Schmidt, H.-J. 490, 495, 499—500
Schwarz, G. 476, 505
Scott, D. 72, 74, 105, 205—6, 478, 505

INDEX OF NAMES

Scotus, D. 101
Sellars, W. 104, 503, 506
Shannon, C. 151
Shepard, R. N. 32
Shimony, A. 285, 487, 505
Sigwart, C. 23
Simmons, G. F. 402, 474, 505
Simpson, T. 163
Sintonen, M. 478, 484, 488—9, 505
Slomson, A. 41, 477, 490
Smokler, H. 477, 497
Sneath, P. H. A. 25, 475, 505
Sneed, J. D. xiv, 113, 445, 477, 486, 490, 505
Snell, W. 387
Sokal, R. R. 25—7, 475, 505
Spinoza, B. 158, 164, 505
Spohn, W. 477, 505
Sprott, D. A. 494
Staal, J. F. 495
Stalnaker, R. C. 476—7, 494, 505
Stegmüller, W. xiv, 80, 409, 445, 461, 477, 486—8, 505
Stenius, E. 476, 506
Stone, M. 79, 204, 458
Stout, A. K. 476
Strawson, P. 134
Stumpf, C. 29
Suppe, F. 478, 500, 506
Suppes, P. xiv, 105, 286, 477—8, 495—7, 500, 503, 505
Swinburne, R. 482, 485, 506
Szaniawski, K. 503

Tammelo, I. 501
Tanimoto, T. T. 28
Tarski, A. xi, 41, 78, 93, 134—5, 137—41, 143—4, 183, 185, 203, 206, 263, 479—80, 493, 500, 503, 506
Tchebycheff, P. L. 3, 4, 107, 288, 386, 485
Thompson, J. B. 494
Tiao, G. 409, 428—9, 491
Tichý, P. viii—ix, xi—iii, xvii, 187—8, 190, 195, 198, 202, 204—5, 232, 235—41, 250, 303, 310, 314, 318, 323—34, 348—50, 353—6, 373, 423—4, 442, 445—6, 450—2, 469, 481, 483—4, 489, 506
Todhunter, I. 163, 483, 506
Törnebohm, H. 197, 506
Tuomela, R. xii—iii, xvii, 88, 91, 198, 320—2, 343, 357, 422, 423, 445, 447, 451—2, 458, 477, 479, 481—4, 486—9, 498, 501—2, 506
Tversky, A. 32—4, 316, 476, 490, 497, 504, 506

Uchii, S. 94, 96, 98—101, 506
Ullmann, J. R. 30, 506
Urbach, P. 198, 444, 469—70, 507

Vallauri, L. L. 501
Vetter, H. 204, 235, 507
van der Waals, J. D. 114
Wachbroit, R. 488, 507
Wald, A. 407, 428, 431, 507
Watanabe, S. 36, 507
Watkins, J. 425, 444, 488, 496, 507
Weil, A. 19
Weingartner, P. 497, 501
Westfall, R. S. 498
Weston, T. S. 480
Whewell, W. 488
Wilks, S. S. 17, 483, 507
Williams, P. M. 482, 507
Williams, W. T. 26
Wittgenstein, L. 83
Wójcicki, R. 477, 503, 507
Woozley, A. D. 174, 507
von Wright, G. H. 44, 53, 476—7, 507
Wright, W. D. 31
Wundt, W. 23

Xenophanes 160

Zadeh, L. 145, 178, 479, 507
Zadeh, L. A. 475, 507
Ziehen, T. 24, 28, 507
Zinnes, J. L. 506

INDEX OF SUBJECTS

acceptance 160, 415
analogy 23—24
analytic truth 43
answer 124, 129
  complete 129
  partial 129
approximate
  application 398
  explanation 398
  prediction 395
  reduction 401
  truth xi, 176, 200, 279—280, 394—396, 433, 472
  validity 287, 403
attribute space 42, 44
attributive constituent 62
average function 211

background knowledge 126, 259, 284—289
balanced distance 212
Bayes' formula 283
Bayesianism xiv, 161, 283, 309, 406—410, 426—441
Boolean algebra 10, 41, 79
Boolean space 79
breadth
  of generalization 53, 75
  of linkage 323, 354

Cauchy-sequence 19
cell 43
classification 15, 43
Clifford measure 313, 321, 347, 356, 383
closeness (to the truth) 217, see distance, M-function
cladistic analysis 24
cluster analysis 29
cognitive problem 129

coherence 135, 183
colour space 29
combination 127
complement 210
complete
  space 19, 21
  theory 77, 79, 132, 262—364
completion (of a problem) 128
concept 42
conceptual
  enrichment 448, 460
  system 45
concretization 113, 399
confirmation 270, 286, 486
consensus 136
consistency 42, 71
constituent
  infinitely deep 72
  higher-order 374, 485
  monadic 52, 310—345
  nomic 94, 131, 376
  polyadic 66, 131, 346—371
  propositional 204
  reduced 68, 353
constituent-structure 54, 90
content
  empirical 148
  falsity content 184, 193
  information content 148, 444, 448, 488
  logical 148
  truth content 184, 192
content element 148
contingency table 15
continuous 21
convergence 19, 166, 170, 396, 460
correlation coefficient 26
correspondence
  principle 399
  theory of truth 135, 138, 183, 479

INDEX OF SUBJECTS 515

corroboration xiii, 242—43, 264, 483—84
counterfactual 262, 486

deduction 40
dense 210
depth 59, 62, 205—07
diagram 48
disjunctive closure 126
distance 1—17, 209—19, 242—56
  function Δ 209
  between constituents 310—34, 346—61, 374—93
    d_C 312, 321, 347, 356, 383 (Clifford)
    d_D 349
    d_{DJ} 349
    d_J 316, 358, 383 (Jyväskylä)
    d_T 323—34, 355 (Tichý)
    d_W 317, 358 (weighted symmetric difference)
  between laws 382—93, 403—05
  between statements 209—216, 242—255
    Δ_{av} 214 (average)
    Δ_{inf} 213 (infimum)
    Δ_{min} 214 (minimum)
    Δ_{max} 214 (maximum)
    Δ^γ_{mm} 216 (min-max)
    $\Delta^{\gamma\gamma'}_{ms}$ 216, 361 (min-sum)
    Δ_{sup} 213 (supremum)
    δ 243
    δ^γ_{mmm} 253
    $\delta^{\gamma\gamma'}_{nms}$ 248
    δ_O 245 (Oddie)
  between structures (worlds) 365—367
distributive normal form 53, 74, 204
divergence 7, 29, 404
domain 41

elementary
  class 79
  equivalence 72
entourage 20
epistemic utility 406, 410—16, 488—89

equivalence
  logical 40
  of languages 459
error 159, 163, 172, 281—89, 314
essentialism 37—38, 458
estimation 133, 268—78, 309, 426—41, 487
existential sentence 40, 335—41
expansion
  of constituent 76
  of model 143
expectation 268, 407
explication 144, 237
extension 41

fact 135
factual
  sentence 42
  truth 138
fallibilism 156, 267, 421, 482
false 42, 159, 479
family of predicates 44
filter 21
finer than 127
formula 40
function space 4
fuzzy set 10, 145, 258

Gauss' curve 281
generalization 40
  accidental 91, 372
  existential 40
  lawlike 91, 372
  universal 40

Hamming distance 8, 10, 29, 481
Hausdorff
  distance 11
  space 19, 22
Hilbert space 5, 109, 303

idealization 113, 286—89, 399—403, 486
identity 58, 68, 321—23
incommensurability 462—68
inductive logic 80—91, 98—102, 300—02, 341—45, 380, 442, 445, 479, 487

information 148—155
 about the truth 201, 221
 conditional 153, 193
 content 148
 depth 153
 incremental 153, 193, 411
 semantic 151
 substantial 152
 surface 153
 transmitted 154, 193, 197, 413
instrumentalism 104
intended model 80, 112, 397
intension 42
interpretation 41, 479
interval estimation 428
isomorphy
 of orderings 210
 of state descriptions 50
 of structures 23, 72

χ^2 (khi square) 15, 302, 322
knowledge 125, 134, 157

λ-continuum 84
λ-α-continuum 85
language
 modal $L(\Box)$ 92, 374—81
 monadic L_N^k 39—42, 310—45
 monadic with identity $L_N^{k=}$ 58, 321—22
 polyadic 61—76, 346—364
language-shift 447
lattice 10, 78
law
 causal 94
 of coexistence 97, 109—14, 118
 probabilistic 118—21, 133, 403—05
 quantitative 109—21, 382—94
 of succession 97, 114—17, 120
lawlikeness 91
legisimilitude 372—405
likelihood function 202, 409, 427
likeness 24, 27, 160
limit 19
Lindenbaum algebra 41, 79
linguistic invariance 446—59
linkage 323, 354

loss function 428
 posterior 430

M-function 217
 M_{av} 234
 M_{max} 219
 M_{min} 218
 M_{mc}^γ 226
 M_{mm}^γ 222
 M_{mmc}^γ 226
 $M_{ms}^{\gamma\gamma'}$ 228
 M_{sum} 221
Mahalanobis distance 18, 289
matching coefficient 27, 33
mean 16
meaning
 postulate 43, 447, 464
 variance xvi, 464
measurement 106
meet 128
metric
 Boolean valued 10, 190
 Boorman-Olivier 14
 Canberra 26
 city-block 3
 Euclidean 3, 47, 454
 Fu 13
 Levenshtein 12
 L_p 5
 Manhattan 3
 Mazurkiewicz 11, 190
 Minkowski 4—5, 33, 291, 302, 385, 393, 398
 pseudometric 1
 semimetric 2
 trivial 412
metric space 1
metrizable 18
metrization 105
model 42, 78
monotone sequence (of constituents) 77

necessity 91, 374
neighbourhood 18, 21
nomic 91, 113
numerical quantifier 58

INDEX OF SUBJECTS

odd predicate 457
open
  formula 40
  set 18

pattern recognition 29
phenetics 24, 37
plausibility 182
plenitude 101
point
  estimate 428
  hypothesis 213
possibility 92, 118
pragmatics, logical xii, 469—73
pragmatism 92, 136
prediction 395
probability 80, 160—64
  epistemic 81, see inductive logic
  physical 118, see propensity
  regular 82
  surface 82
  problem 129—133
    continuous 209, 254—56, 303—09
    discrete 209
    subproblem 130, 132, 488
problem-solving ability 376, 478, 484
progress 263, 460—65, 482, 488—89
propensity 118, 478
property 35—36, 42, 457, 476
P-set 126, 478

q-dimensional system of inductive logic 89
Q-predicate 43
quality 29, 104
quantity 105, 133
question 122

random variable 7
range 48
realism xii, xiii, 38, 92, 103—04, 136, 141, 156, 267, 445, 458—60, 482, 486
reduction 401, 486
reduction function 198, 217
regression analysis 289
resemblance 25

robustness 336, 470

satisfaction 41
scatter 16
scepticism 156—61
sentence 40
sharp hypothesis 213
similarity xii, 1, 22—38, 199, 213, 291, 469—71, 475, 489
singular sentence 40, 290—309
state
  description 47, 63, 130, 297—302, 328—30
  space 108, 251—56, 303—09
Stone space 79
strength, logical 40
structure 41, 61, 141
structure description 50, 63, 130, 302—03
subordinate (constituent) 75
suspension of judgment 129
symmetric difference 9, 311

target 204—09, 261—62, 448
tautology, see valid
taxonomy 24, 26
theory 72, 77, 80, 368—81
topology 18
Tr 232
trajectory 115
transformation 98, 114
translation 453—59, 468, 484
tree 12, 347—48
true 42, 134—43, 164—171
  almost 176, 204, see approximate truth
  partly 175
  totally 175
truth
  absolute 168, 174
  approximate 176—78, 200, 279, 394
  degree of truth 169—182, 197, 218, 485
  epistemic 482
  factual 138
  indefinite 147, 256—59, 306—09, 369

logical 42
partial 175—81, 220
Peirce's definition 136—37, 167
relative 168, 174
Tarski's definition 42, 137—144
whole 175, 186
truthlikeness, see verisimilitude
  and probability 1, 160—64, 183, 274, 301, 342, 474
  and similarity vii, 1, 155, 196, 198—262
  degrees of 222—34, 257, see M-functions
  epistemic xii, 263
  estimation 263—78, 286—88, 300—02, 341—45, 380, 442
  Hilpinen's definition 202
  Kuipers' definition 373, 380–82
  logical xii, 263
  Miller's definition 190—92
  Popper's definition 183—97, 264—65, 442—45
  probable 278
  properties of Tr 232—34
  Tichý–Oddie definition 234—42, 323—35, 348—56, 374—76

uniformity 20, 398, 402, 473—74
uniformizable space 22
uniform space 20
universal formula 40, 335—41
utility 406—416

vagueness 143—47
valid 42
ver 269
verisimilitude 160—62, 192—97, 278, 374, 474

width
  of constituent 52
  of predicate 44, 46
world 141, 207

SYNTHESE LIBRARY

Studies in Epistemology, Logic, Methodology,
and Philosophy of Science

Managing Editor:
JAAKKO HINTIKKA, Florida State University, Tallahassee

Editors:
DONALD DAVIDSON, University of California, Berkeley
GABRIËL NUCHELMANS, University of Leyden
WESLEY C. SALMON, University of Pittsburgh

1. J. M. Bochénski, *A Precis of Mathematical Logic*. 1959.
2. P. L. Guiraud, *Problèmes et méthodes de la statistique linguistique*. 1960.
3. Hans Freudenthal (ed.), *The Concept and the Role of the Model in Mathematics and Natural and Social Sciences*. 1961.
4. Evert W. Beth, *Formal Methods. An Introduction to Symbolic Logic and the Study of Effective Operations in Arithmetic and Logic*. 1962.
5. B. H. Kazemier and D. Vuysje (eds.), *Logic and Language. Studies Dedicated to Professor Rudolf Carnap on the Occasion of His Seventieth Birthday*.1962.
6. Marx W. Wartofsky (ed.), *Proceedings of the Boston Colloquium for the Philosophy of Science 1961–1962*. Boston Studies in the Philosophy of Science, Volume I. 1963.
7. A. A. Zinov'ev, *Philosophical Problems of Many-Valued Logic*. 1963.
8. Georges Gurvitch, *The Spectrum of Social Time*. 1964.
9. Paul Lorenzen, *Formal Logic*. 1965.
10. Robert S. Cohen and Marx W. Wartofsky (eds.), *In Honor of Philipp Frank*. Boston Studies in the Philosophy of Science, Volume II. 1965.
11. Evert W. Beth, *Mathematical Thought. An Introduction to the Philosophy of Mathematics*. 1965.
12. Evert W. Beth and Jean Piaget, *Mathematical Epistemology and Psychology*. 1966.
13. Guido Küng, *Ontology and the Logistic Analysis of Language. An Enquiry into the Contemporary Views on Universals*. 1967.
14. Robert S. Cohen and Marx W. Wartofsky (eds.), *Proceedings of the Boston Colloquium for the Philosophy of Sciences 1964–1966. In Memory of Norwood Russell Hanson*. Boston Studies in the Philosophy of Science, Volume III. 1967.
15. C. D. Broad, *Induction, Probability, and Causation. Selected Papers*. 1968.
16. Günther Patzig, *Aristotle's Theory of the Syllogism. A Logical-Philosophical Study of Book A of the Prior Analytics*. 1968.
17. Nicholas Rescher, *Topics in Philosophical Logic*. 1968.
18. Robert S. Cohen and Marx W. Wartofsky (eds.), *Proceedings of the Boston Colloquium for the Philosophy of Science 1966–1968*. Boston Studies in the Philosophy of Science, Volume IV. 1969
19. Robert S. Cohen and Marx W. Wartofsky (eds.), *Proceedings of the Boston Colloquium for the Philosophy of Science 1966–1968*. Boston Studies in the Philosophy of Science, Volume V. 1969

20. J. W. Davis, D. J. Hockney, and W. K. Wilson (eds.), *Philosophical Logic*. 1969
21. D. Davidson and J. Hintikka (eds.), *Words and Objections. Essays on the Work of W. V. Quine*. 1969.
22. Patrick Suppes. *Studies in the Methodology and Foundations of Science. Selected Papers from 1911 to 1969*. 1969
23. Jaakko Hintikka, *Models for Modalities. Selected Essays*. 1969
24. Nicholas Rescher et al. (eds.), *Essays in Honor of Carl G. Hempel. A Tribute on the Occasion of His Sixty-Fifth Birthday*. 1969
25. P. V. Tavanec (ed.), *Problems of the Logic of Scientific Knowledge*. 1969
26. Marshall Swain (ed.), *Induction, Acceptance, and Rational Belief*. 1970.
27. Robert S. Cohen and Raymond J. Seeger (eds.), *Ernst Mach: Physicist and Philosopher*. Boston Studies in the Philosophy of Science, Volume VI. 1970.
28. Jaakko Hintikka and Patrick Suppes, *Information and Inference*. 1970.
29. Karel Lambert, *Philosophical Problems in Logic. Some Recent Developments*. 1970.
30. Rolf A. Eberle, *Nominalistic Systems*. 1970.
31. Paul Weingartner and Gerhard Zecha (eds.), *Induction, Physics, and Ethics*. 1970.
32. Evert W. Beth, *Aspects of Modern Logic*. 1970.
33. Risto Hilpinen (ed.), *Deontic Logic: Introductory and Systematic Readings*. 1971.
34. Jean-Louis Krivine, *Introduction to Axiomatic Set Theory*. 1971.
35. Joseph D. Sneed, *The Logical Structure of Mathematical Physics*. 1971.
36. Carl R. Kordig, *The Justification of Scientific Change*. 1971.
37. Milic Capek, *Bergson and Modern Physics*. Boston Studies in the Philosophy of Science, Volume VII. 1971.
38. Norwood Russell Hanson, *What I Do Not Believe, and Other Essays* (ed. by Stephen Toulmin and Harry Woolf). 1971.
39. Roger C. Buck and Robert S. Cohen (eds.), *PSA 1970. In Memory of Rudolf Carnap*. Boston Studies in the Philosophy of Science, Volume VIII. 1971
40. Donald Davidson and Gilbert Harman (eds.), *Semantics of Natural Language*. 1972.
41. Yehoshua Bar-Hillel (ed.), *Pragmatics of Natural Languages*. 1971.
42. Sören Stenlund, *Combinators, λ-Terms and Proof Theory*. 1972.
43. Martin Strauss, *Modern Physics and Its Philosophy. Selected Paper in the Logic, History, and Philosophy of Science*. 1972.
44. Mario Bunge, *Method, Model and Matter*. 1973.
45. Mario Bunge, *Philosophy of Physics*. 1973.
46. A. A. Zinov'ev, *Foundations of the Logical Theory of Scientific Knowledge (Complex Logic)*. (Revised and enlarged English edition with an appendix by G. A. Smirnov, E. A. Sidorenka, A. M. Fedina, and L. A. Bobrova.) Boston Studies in the Philosophy of Science, Volume IX. 1973.
47. Ladislav Tondl, *Scientific Procedures*. Boston Studies in the Philosophy of Science, Volume X. 1973.
48. Norwood Russell Hanson, *Constellations and Conjectures* (ed. by Willard C. Humphreys, Jr.). 1973
49. K. J. J. Hintikka, J. M. E. Moravcsik, and P. Suppes (eds.), *Approaches to Natural Language*. 1973.
50. Mario Bunge (ed.), *Exact Philosophy – Problems, Tools, and Goals*. 1973.
51. Radu J. Bogdan and Ilkka Niiniluoto (eds.), *Logic, Language, and Probability*. 1973.

52. Glenn Pearce and Patrick Maynard (eds.), *Conceptual Change*. 1973.
53. Ilkka Niiniluoto and Raimo Tuomela, *Theoretical Concepts and Hypothetico-Inductive Inference*. 1973.
54. Roland Fraissé, *Course of Mathematical Logic* – Volume 1: *Relation and Logical Formula*. 1973.
55. Adolf Grünbaum, *Philosophical Problems of Space and Time*. (Second, enlarged edition.) Boston Studies in the Philosophy of Science, Volume XII. 1973.
56. Patrick Suppes (ed.), *Space, Time, and Geometry*. 1973.
57. Hans Kelsen, *Essays in Legal and Moral Philosophy* (selected and introduced by Ota Weinberger). 1973.
58. R. J. Seeger and Robert S. Cohen (eds.), *Philosophical Foundations of Science*. Boston Studies in the Philosophy of Science, Volume XI. 1974.
59. Robert S. Cohen and Marx W. Wartofsky (eds.), *Logical and Epistemological Studies in Contemporary Physics*. Boston Studies in the Philosophy of Science, Volume XIII. 1973.
60. Robert S. Cohen and Marx W. Wartofsky (eds.), *Methodological and Historical Essays in the Natural and Social Sciences. Proceedings of the Boston Colloquium for the Philosophy of Science 1969–1972*. Boston Studies in the Philosophy of Science, Volume XIV. 1974.
61. Robert S. Cohen, J. J. Stachel, and Marx W. Wartofsky (eds.), *For Dirk Struik. Scientific, Historical and Political Essays in Honor of Dirk J. Struik*. Boston Studies in the Philosophy of Science, Volume XV. 1974.
62. Kazimierz Ajdukiewicz, *Pragmatic Logic* (transl. from the Polish by Olgierd Wojtasiewicz). 1974.
63. Sören Stenlund (ed.), *Logical Theory and Semantic Analysis. Essays Dedicated to Stig Kanger on His Fiftieth Birthday*. 1974.
64. Kenneth F. Schaffner and Robert S. Cohen (eds.), *Proceedings of the 1972 Biennial Meeting, Philosophy of Science Association*. Boston Studies in the Philosophy of Science, Volume XX. 1974.
65. Henry E. Kyburg, Jr., *The Logical Foundations of Statistical Inference*. 1974.
66. Marjorie Grene, *The Understanding of Nature. Essays in the Philosophy of Biology*. Boston Studies in the Philosophy of Science, Volume XXIII. 1974.
67. Jan M. Broekman, *Structuralism: Moscow, Prague, Paris*. 1974.
68. Norman Geschwind, *Selected Papers on Language and the Brain*, Boston Studies in the Philosophy of Science, Volume XVI. 1974.
69. Roland Fraissé, *Course of Mathematical Logic* – Volume 2: *Model Theory*. 1974.
70. Andrzej Grzegorczyk, *An Outline of Mathematical Logic. Fundamental Results and Notions Explained with All Details*. 1974.
71. Franz von Kutschera, *Philosophy of Language*. 1975.
72. Juha Manninen and Raimo Tuomela (eds.), *Essays on Explanation and Understanding. Studies in the Foundations of Humanities and Social Sciences*. 1976.
73. Jaakko Hintikka (ed.), *Rudolf Carnap, Logical Empiricist. Materials and Perspectives.* 1975.
74. Milic Capek (ed.), *The Concepts of Space and Time. Their Structure and Their Development*. Boston Studies in the Philosophy of Science, Volume XXII. 1976.
75. Jaakko Hintikka and Unto Remes, *The Method of Analysis. Its Geometrical Origin and Its General Significance*. Boston Studies in the Philosophy of Science, Volume XXV. 1974.

76. John Emery Murdoch and Edith Dudley Sylla, *The Cultural Context of Medieval Learning*. Boston Studies in the Philosophy of Science, Volume XXVI. 1975.
77. Stefan Amsterdamski, *Between Experience and Metaphysics. Philosophical Problems of the Evolution of Science*. Boston Studies in the Philosophy of Science, Volume XXXV. 1975.
78. Patrick Suppes (ed.), *Logic and Probability in Quantum Mechanics*. 1976.
79. Hermann von Helmholtz: *Epistemological Writings. The Paul Hertz/Moritz Schlick Centenary Edition of 1921 with Notes and Commentary by the Editors*. (Newly translated by Malcolm F. Lowe. Edited, with an Introduction and Bibliography, by Robert S. Cohen and Yehuda Elkana.) Boston Studies in the Philosophy of Science, Volume XXXVII. 1975.
80. Joseph Agassi, *Science in Flux*. Boston Studies in the Philosophy of Science, Volume XXVIII. 1975.
81. Sandra G. Harding (ed.), *Can Theories Be Refuted? Essays on the Duhem-Quine Thesis*. 1976.
82. Stefan Nowak, *Methodology of Sociological Research. General Problems*. 1977.
83. Jean Piaget, Jean-Blaise Grize, Alina Szeminska, and Vinh Bang, *Epistemology and Psychology of Functions*. 1977.
84. Marjorie Grene and Everett Mendelsohn (eds.), *Topics in the Philosophy of Biology*. Boston Studies in the Philosophy of Science, Volume XXVII. 1976.
85. E. Fischbein, *The Intuitive Sources of Probabilistic Thinking in Children*. 1975.
86. Ernest W. Adams, *The Logic of Conditionals. An Application of Probability to Deductive Logic*. 1975.
87. Marian Przelecki and Ryszard Wójcicki (eds.), *Twenty-Five Years of Logical Methodology in Poland*. 1976.
88. J. Topolski, *The Methodology of History*. 1976.
89. A. Kasher (ed.), *Language in Focus: Foundations, Methods and Systems. Essays Dedicated to Yehoshua Bar-Hillel*. Boston Studies in the Philosophy of Science, Volume XLIII. 1976.
90. Jaakko Hintikka, *The Intentions of Intentionality and Other New Models for Modalities*. 1975.
91. Wolfgang Stegmüller, *Collected Papers on Epistemology, Philosophy of Science and History of Philosophy*. 2 Volumes. 1977.
92. Dov M. Gabbay, *Investigations in Modal and Tense Logics with Applications to Problems in Philosophy and Linguistics*. 1976.
93. Radu J. Bodgan, *Local Induction*. 1976.
94. Stefan Nowak, *Understanding and Prediction. Essays in the Methodology of Social and Behavioral Theories*. 1976.
95. Peter Mittelstaedt, *Philosophical Problems of Modern Physics*. Boston Studies in the Philosophy of Science, Volume XVIII. 1976.
96. Gerald Holton and William Blanpied (eds.), *Science and Its Public: The Changing Relationship*. Boston Studies in the Philosophy of Science, Volume XXXIII. 1976.
97. Myles Brand and Douglas Walton (eds.), *Action Theory*. 1976.
98. Paul Gochet, *Outline of a Nominalist Theory of Proposition. An Essay in the Theory of Meaning*. 1980.
99. R. S. Cohen, P. K. Feyerabend, and M. W. Wartofsky (eds.), *Essays in Memory of Imre Lakatos*. Boston Studies in the Philosophy of Science, Volume XXXIX. 1976.
100. R. S. Cohen and J. J. Stachel (eds.), *Selected Papers of Léon Rosenfield*. Boston Studies in the Philosophy of Science, Volume XXI. 1978.

101. R. S. Cohen, C. A. Hooker, A. C. Michalos, and J. W. van Evra (eds.), *PSA 1974: Proceedings of the 1974 Biennial Meeting of the Philosophy of Science Association.* Boston Studies in the Philosophy of Science, Volume XXXII. 1976.
102. Yehuda Fried and Joseph Agassi, *Paranoia: A Study in Diagnosis.* Boston Studies in the Philosophy of Science, Volume L. 1976.
103. Marian Przelecki, Klemens Szaniawski, and Ryszard Wójcicki (eds.), *Formal Methods in the Methodology of Empirical Sciences.* 1976.
104. John M. Vickers, *Belief and Probability.* 1976.
105. Kurt H. Wolff, *Surrender and Catch: Experience and Inquiry Today.* Boston Studies in the Philosophy of Science, Volume LI. 1976.
106. Karel Kosík, *Dialectics of the Concrete.* Boston Studies in the Philosophy of Science, Volume LII. 1976.
107. Nelson Goodman, *The Structure of Appearance* (Third edition.) Boston Studies in the Philosophy of Science, Volume LIII. 1977.
108. Jerzy Giedymin (ed.), *Kazimierz Ajdukiewicz: The Scientific World-Perspective and Other Essays, 1931-1963.* 1978.
109. Robert L. Causey, *Unity of Science.* 1977.
110. Richard E. Grandy, *Advanced Logic for Applications.* 1977.
111. Robert P. McArthur, *Tense Logic.* 1976.
112. Lars Lindahl, *Position and Change. A Study in Law and Logic.* 1977.
113. Raimo Tuomela, *Dispositions.* 1978.
114. Herbert A. Simon, *Models of Discovery and Other Topics in the Methods of Science.* Boston Studies in the Philosophy of Science, Volume LIV. 1977.
115. Roger D. Rosenkrantz, *Inference, Method and Decision.* 1977.
116. Raimo Tuomela, *Human Action and Its Explanation. A Study on the Philosophical Foundations of Psychology.* 1977.
117. Morris Lazerowitz, *The Language of Philosophy. Freud and Wittgenstein.* Boston Studies in the Philosophy of Science, Volume LV. 1977.
119. Jerzy Pelc, *Semiotics in Poland, 1894-1969.* 1978.
120. Ingmar Pörn, *Action Theory and Social Science. Some Formal Models.* 1977.
121. Joseph Margolis, *Persons and Mind. The Prospects of Nonreductive Materialism.* Boston Studies in the Philosophy of Science, Volume LVII. 1977.
122. Jaakko Hintikka, Ilkka Niiniluoto, and Esa Saarinen (eds.), *Essays on Mathematical and Philosophical Logic.* 1978.
123. Theo A. F. Kuipers, *Studies in Inductive Probability and Rational Expectation.* 1978.
124. Esa Saarinen, Risto Hilpinen, Ilkka Niiniluoto, and Merrill Provence Hintikka (eds.), *Essays in Honour of Jaakko Hintikka on the Occasion of His Fiftieth Birthday.* 1978.
125. Gerard Radnitzky and Gunnar Andersson (eds.), *Progress and Rationality in Science.* Boston Studies in the Philosophy of Science, Volume LVIII. 1978.
126. Peter Mittelstaedt, *Quantum Logic.* 1978.
127. Kenneth A. Bowen, *Model Theory for Modal Logic. Kripke Models for Modal Predicate Calculi.* 1978.
128. Howard Alexander Bursen, *Dismantling the Memory Machine. A Philosophical Investigation of Machine Theories of Memory.* 1978.
129. Marx W. Wartofsky, *Models: Representation and the Scientific Understanding.* Boston Studies in the Philosophy of Science, Volume XLVIII. 1979.
130. Don Ihde, *Technics and Praxis. A Philosophy of Technology.* Boston Studies in the Philosophy of Science, Volume XXIV. 1978.

131. Jerzy J. Wiatr (ed.), *Polish Essays in the Methodology of the Social Sciences.* Boston Studies in the Philosophy of Science, Volume XXIX. 1979.
132. Wesley C. Salmon (ed.), *Hans Reichenbach: Logical Empiricist.* 1979.
133. Peter Bieri, Rolf-P. Horstmann, and Lorenz Krüger (eds.), *Transcendental Arguments in Science Essays in Epistemology.* 1979.
134. Mihailo Marković and Gajo Petrović (eds.), *Praxis, Yugoslav Essays in the Philosophy and Methodology of the Social Sciences.* Boston Studies in the Philosophy of Science, Volume XXXVI. 1979.
135. Ryszard Wójcicki, *Topics in the Formal Methodology of Empirical Sciences.* 1979.
136. Gerard Radnitzky and Gunnar Andersson (eds.), *The Structure and Development of Science.* Boston Studies in the Philosophy of Science, Volume LIX. 1979.
137. Judson Chambers Webb, *Mechanism, Mentalism, and Metamathematics. An Essay on Finitism.* 1980.
138. D. F. Gustafson and B. L. Tapscott (eds.), *Body, Mind, and Method. Essays in Honor of Virgil C. Aldrich.* 1979.
139. Leszek Nowak, *The Structure of Idealization. Towards a Systematic Interpretation of the Marxian Idea of Science.* 1979.
140. Chaim Perelman, *The New Rhetoric and the Humanities. Essays on Rhetoric and Its Applications.* 1979.
141. Wlodzimierz Rabinowicz, *Universalizability. A Study in Morals and Metaphysics.* 1979.
142. Chaim Perelman, *Justice, Law, and Argument. Essays on Moral and Legal Reasoning.* 1980.
143. Stig Kanger and Sven Öhman (eds.), *Philosophy and Grammar. Papers on the Occasion of the Quincentennial of Uppsala University.* 1980.
144. Tadeusz Pawlowski, *Concept Formation in the Humanities and the Social Sciences.* 1980.
145. Jaakko Hintikka, David Gruender, and Evandro Agazzi (eds.), *Theory Change, Ancient Axiomatics, and Galileo's Methodology. Proceedings of the 1978 Pisa Conference on the History and Philosophy of Science,* Volume I. 1981.
146. Jaakko Hintikka, David Gruender, and Evandro Agazzi, *Probabilistic Thinking, Thermodynamics, and the Interaction of the History and Philosophy of Science. Proceedings of the 1978 Pisa Conference on the History and Philosophy of Science,* Volume II. 1981.
147. Uwe Mönnich (ed.), *Aspects of Philosophical Logic. Some Logical Forays into Central Notions of Linguistics and Philosophy.* 1981.
148. Dov M. Gabbay, *Semantical Investigations in Heyting's Intuitionistic Logic.* 1981.
149. Evandro Agazzi (ed.), *Modern Logic – A Survey. Historical, Philosophical, and Mathematical Aspects of Modern Logic and Its Applications.* 1981.
150. A. F. Parker-Rhodes, *The Theory of Indistinguishables. A Search for Explanatory Principles below the Level of Physics.* 1981.
151. J. C. Pitt, *Pictures, Images, and Conceptual Change. An Analysis of Wilfrid Sellars' Philosophy of Science.* 1981.
152. R. Hilpinen (ed.), *New Studies in Deontic Logic. Norms, Actions, and the Foundations of Ethics.* 1981.
153. C. Dilworth, *Scientific Progress. A Study Concerning the Nature of the Relation Between Successive Scientific Theories.* 1981.
154. D. W. Smith and R. McIntyre, *Husserl and Intentionality. A Study of Mind, Meaning, and Language.* 1982.

155. R. J. Nelson, *The Logic of Mind*. 1982.
156. J. F. A. K. van Benthem, *The Logic of Time. A Model-Theoretic Investigation into the Varieties of Temporal Ontology, and Temporal Discourse*. 1982.
157. R. Swinburne (ed.), *Space, Time and Causality*. 1982.
158. R. D. Rozenkrantz, *E. T. Jaynes: Papers on Probability, Statistics and Statistical Physics*. 1983.
159. T. Chapman, *Time: A Philosophical Analysis*. 1982.
160. E. N. Zalta, *Abstract Objects. An Introduction to Axiomatic Metaphysics*. 1983.
161. S. Harding and M. B. Hintikka (eds.), *Discovering Reality. Feminist Perspectives on Epistemology, Metaphysics, Methodology, and Philosophy of Science*. 1983.
162. M. A. Stewart (ed.), *Law, Morality and Rights*. 1983.
163. D. Mayr and G. Süssmann (eds.), *Space, Time, and Mechanics. Basic Structure of a Physical Theory*. 1983.
164. D. Gabbay and F. Guenthner (eds.), *Handbook of Philosophical Logic*. Vol. I. 1983.
165. D. Gabbay and F. Guenthner (eds.), *Handbook of Philosophical Logic*. Vol. II. 1984.
166. D. Gabbay and F. Guenthner (eds.), *Handbook of Philosophical Logic*. Vol. III. 1985.
167. D. Gabbay and F. Guenthner (eds.), *Handbook of Philosophical Logic*. Vol. IV, forthcoming.
168. Andrew, J. I. Jones, *Communication and Meaning*. 1983.
169. Melvin Fitting, *Proof Methods for Modal and Intuitionistic Logics*. 1983.
170. Joseph Margolis, *Culture and Cultural Entities*. 1984.
171. Raimo Tuomela, *A Theory of Social Action*. 1984.
172. Jorge J. E. Gracia, Eduardo Rabossi, Enrique Villanueva, and Marcelo Dascal (eds.), *Philosophical Analysis in Latin America*. 1984.
173. Paul Ziff, *Epistemic Analysis. A Coherence Theory of Knowledge*. 1984.
174. Paul Ziff, *Antiaesthetics. An Appreciation of the Cow with the Subtile Nose*. 1984.
175. Wolfgang Balzer, David A. Pearce, and Heinz-Jürgen Schmidt (eds.), *Reduction in Science. Structure, Examples, Philosophical Problems*. 1984.
176. Aleksander Peczenik, Lars Lindahl, and Bert van Roermund (eds.), *Theory of Legal Science. Proceedings of the Conference on Legal Theory and Philosophy of Science, Lund, Sweden, December 11–14, 1983*. 1984.
177. Ilkka Niiniluoto, *Is Science Progressive?* 1984.
178. Binal Matilal and Jaysankar Lal Shaw (eds.), *Exploratory Essays in Current Theories and Classical Indian Theories of Meaning and Reference*. 1985.
179. Peter Kroes, *Time: Its Structure and Role in Physical Theories*. 1985.
180. James H. Fetzer, *Sociobiology and Epistemology,* 1985.
181. L. Haaparanta and J. Hintikka, *Frege Synthesized. Essays on the Philosophical and Foundational Work of Gottlob Frege*. 1986.
182. Michael Detlefsen, *Hilbert's Program. An Essay on Mathematical Instrumentalism*. 1986.
183. James L. Golden and Joseph J. Pilotta (eds.), *Practical Reasoning in Human Affairs. Studies in Honor of Chaim Perelman*. 1986.
184. Henk Zandvoort, *Models of Scientific Development and the Case of Nuclear Magnetic Resonance*. 1986.